REPUBLIC OF SOUTH AFRICA
REPUBLIEK VAN SUID-AFRIKA

DEPARTMENT OF AGRICULTURE
DEPARTEMENT VAN LANDBOU

FLORA OF SOUTHERN AFRICA

VOLUME 33

PART 7, FASCICLE 2 (FIRST PART)

ISBN 0 621 07943 X

G.P.-S

FLORA OF SOUTHERN AFRICA

which deals with the territories of

SOUTH AFRICA, CISKEI, TRANSKEI, LESOTHO, SWAZILAND, BOPHUTHA-TSWANA, SOUTH WEST AFRICA/NAMIBIA, BOTSWANA AND VENDA

VOLUME 33   ASTERACEAE (Compositae)

PART 7   INULEAE

Fascicle 2 GNAPHALIINAE (First part)

by

O. M. Hilliard

University of Natal, Pietermaritzburg

Edited by

O. A. Leistner

Editorial Committee: B. de Winter, D. J. B. Killick and O. A. Leistner

Botanical Research Institute,
Department of Agriculture

1983

CONTENTS

NEW TAXA AND NEW COMBINATIONS PUBLISHED IN PART 7, FASCICLE 2
  (First part)

Helichrysum albertense *Hilliard*, sp. nov., p. 7,2: 177
Helichrysum asperum *(Thunb.) Hilliard & Burtt* var. **glabrum** *Hilliard,* var. nov.,
    p. 7,2: 128
Helichrysum aureofolium *Hilliard,* sp. nov., p. 7,2: 158
Helichrysum deserticola *Hilliard,* sp. nov., p. 7,2: 168
Helichrysum dunense *Hilliard,* sp. nov., p. 7,2: 143
Helichrysum erubescens *Hilliard,* sp. nov., p. 7,2: 174
Helichrysum fourcadei *Hilliard,* sp. nov., p. 7,2: 131
Helichrysum jubilatum *Hilliard,* sp. nov., p. 7,2: 179
Helichrysum mutabile *Hilliard,* sp. nov., p. 7,2: 302
Helichrysum refractum *Hilliard,* sp. nov., p. 7,2: 133
Helichrysum rosum *(Berg.) Less.* var. **arcuatum** *Hilliard,* var. nov., p. 7,2: 107
Helichrysum saxicola *Hilliard,* sp. nov., p. 7,2: 190
Helichrysum solitarium *Hilliard,* sp. nov., p. 7,2: 149
Troglophyton acocksianum *Hilliard,* sp. nov., p. 7,2: 42
Troglophyton elsiae *Hilliard,* sp. nov., p. 7,2: 40
Troglophyton leptomerum *Hilliard,* sp. nov., p. 7,2: 45
Troglophyton tenellum *Hilliard,* sp. nov., p. 7,2: 44
Vellereophyton felinum *Hilliard,* sp. nov., p. 7,2: 35
Vellereophyton gracillimum *Hilliard,* sp. nov., p. 7,2: 33
Vellereophyton niveum *Hilliard,* sp. nov., p. 7,2: 34
Vellereophyton pulvinatum *Hilliard*, sp. nov. p. 7,2: 35

Edmondia fasciculata *(Andr.) Hilliard,* comb. nov., p. 7,2: 316
Galeomma stenolepis *(S. Moore) Hilliard,* comb. nov., p. 7,2: 13
Helichrysum asperum *(Thunb.) Hilliard & Burtt* var. **appressifolium** *(Moeser) Hilliard,*
    comb. nov., p. 7,2: 128
Helichrysum pumilio *(O. Hoffm.) Hilliard & Burtt* subsp. **fleckii** *(S. Moore) Hilliard,* comb.
    et stat. nov., p. 7,2: 166
Troglophyton capillaceum *(Thunb.) Hilliard & Burtt* subsp. **diffusum** *(DC.) Hilliard,*
    comb. et stat. nov., p. 7,2: 40

Date of publication: December 1983

INTRODUCTION

For a key to the families and the genera not keyed out in this part, the Flora should be used in conjunction with R.A. Dyer's Genera of Southern African Flowering Plants, Vol. 1 (1975) and Vol. 2 (1976), which are arranged on the lines of the Engler system. The genera are numbered, as far as possible, according to the list published by De Dalla Torre and Harms in their Genera Siphonogamarum (1900 − 1907) in order to facilitate reference, though genera in the Flora are not necessarily arranged in this sequence.

The following condensed abbreviations for literature references are used:

C. F. A.	Conspectus Florae Angolensis
R.A. Dyer, Gen.	The Genera of Southern African Flowering Plants by R. A. Dyer. Vol. 1 (1975) and Vol. 2 (1976)
F.C.	Flora Capensis
F.C.B.	Flore du Congo et du Rwanda-Burundi
F.S.W.A.	Prodromus einer Flora von Südwestafrika
F.T.A.	Flora of Tropical Africa
F.T.E.A.	Flora of Tropical East Africa
F.W.T.A.	Flora of West Tropical Africa
F.Z.	Flora Zambesiaca
Burtt Davy, Fl. Transv.	Manual of the Flowering Plants and Ferns of the Transvaal and Swaziland, Vol. 1 (1926) and Vol. 2 (1932).

Localities are sometimes referred to in terms of the degree reference system (Leistner & Morris in Ann. Cape Prov. Mus. 12: 1 − 565; 1976).

This fascicle was compiled in accordance with a Guide to Contributors to the Flora of Southern Africa (Ross, Leistner & De Winter, 1977), which is available from the Librarian, Botanical Research Institute, Private Bag X101, Pretoria, 0001.

Volume 33 of the Flora, of which the present publication is a component, will appear in nine parts of which the seventh is divided into fascicles (see p. ix). The number of the part, which in the present publication is '7', and the number of the fascicle, namely '2', precede the page number on all pages marked with Arabic numerals. This was done with a view to compiling a combined index to the entire volume.

PLAN OF FLORA OF SOUTHERN AFRICA

Cryptogam volumes will in future not be numbered but will be known by the name of the group they·cover. The number assigned to the volume on Characeae therefore becomes redundant.

Alien families are marked with an asterisk.

Published volumes and parts are shown in italics.

Please note that local prices as given below do not include GST unless specifically mentioned.

INTRODUCTORY VOLUMES

The genera of Southern African flowering plants
    Vol. 1: *Dicotyledons* (Published 1975). Price: R11,00. Overseas: R14,00. Post free
    Vol. 2: *Monocotyledons* (Published 1976). Price: R8,00. Overseas: R10,00. Post free
Botanical exploration in Southern Africa (Published 1981). Price R40,00 (Obtainable from booksellers).

CRYPTOGAM VOLUMES

Charophyta (Published as Vol. 9 in 1978). Price: R4,25. Overseas: R5,30. Post free

Bryophyta:

        Part 1: Mosses: Fascicle 1: *Sphagnaceae − Grimmiaceae* (Published 1981). Price: R24,33, GST R0,97,
                  Total R25,30. Overseas: R30,40. Post free
              Fascicle 2: Gigaspermaceae − Bartramiaceae
              Fascicle 3: Erpodiaceae − Hookeriaceae
              Fascicle 4: Fabroniaceae − Polytrichaceae

Pteridophyta

FLOWERING PLANTS VOLUMES

Vol. 1: *Stangeriaceae, Zamiaceae, Podocarpaceae, Pinaceae*, Cupressaceae, Welwitschiaceae, Typhaceae, Zosteraceae, Potamogetonaceae, Ruppiaceae, Zannichelliaceae, Najadaceae, Aponogetonaceae, Juncaginaceae, Alismataceae, Hydrocharitaceae* (Published 1966). Price: R1,75. Overseas: R2,20. Post free

Vol. 2: Poaceae

Vol. 3: Cyperaceae, Arecaceae, Araceae, Lemnaceae, Flagellariaceae

Vol. 4: Restionaceae, Mayacaceae, Xyridaceae, Eriocaulaceae, Commelinaceae, Pontederiaceae, Juncaceae

Vol. 5: Liliaceae, Agavaceae

Vol. 6: Haemodoraceae, Amaryllidaceae, Hypoxidaceae, Tecophilaeaceae, Velloziaceae, Dioscoreaceae

Vol. 7: Iridaceae: Part 1: Nivenioideae, Iridoideae
              Part 2: Ixioideae: Fascicle 1
                            Fascicle 2: *Syringodea, Romulea* (Published 1983).
                                  Price: R3,96, GST R0,24, Total R4,20. Overseas: R5,00. Post free

Vol. 8: Musaceae, Strelitziaceae, Zingiberaceae, Cannaceae*, Burmanniaceae, Orchidaceae

Vol. 9: Casuarinaceae*, Piperaceae, Salicaceae, Myricaceae, Fagaceae*, Ulmaceae, Moraceae, Cannabaceae*, Urticaceae, Proteaceae

Vol. 10: Part 1: *Loranthaceae, Viscaceae* (Published 1979). Price: R2,00. Overseas: R2,50. Post free
Santalaceae, Grubbiaceae, Opiliaceae, Olacaceae, Balanophoraceae, Aristolochiaceae, Rafflesiaceae, Hydnoraceae, Polygonaceae, Chenopodiaceae, Amaranthaceae, Nyctaginaceae

Vol. 11: Phytolaccaceae, Aizoaceae, Mesembryanthemaceae

Vol. 12: Portulacaceae, Basellaceae, Caryophyllaceae, Illecebraceae, Cabombaceae, Nymphaeaceae, Ceratophyllaceae, Ranunculaceae, Menispermaceae, Annonaceae, Trimeniaceae, Lauraceae, Hernandiaceae, Papaveraceae, Fumariaceae

Vol. 13: *Brassicaceae, Capparaceae, Resedaceae, Moringaceae, Droseraceae, Roridulaceae, Podostemaceae, Hydrostachyaceae* (Published 1970). Price: R10,00. Overseas: R12,00. Post free

Vol. 14: Crassulaceae

Vol. 15: Vahliaceae, Montiniaceae, Escalloniaceae, Pittosporaceae, Cunoniaceae, Myrothamnaceae, Bruniaceae, Hamamelidaceae, Rosaceae, Connaraceae

Vol. 16: Fabaceae: Part 1: *Mimosoideae* (Published 1975). Price: R13,50. Overseas: R16,75. Post free
Part 2: *Caesalpinioideae* (Published 1977). Price: R16,00. Overseas: R20,00. Post free

Papilionoideae

Vol. 17: Geraniaceae, Oxalidaceae

Vol. 18: Linaceae, Erythroxylaceae, Zygophyllaceae, Balanitaceae, Rutaceae, Simaroubaceae, Burseraceae, Ptaeroxylaceae, Meliaceae, Aitoniaceae, Malpighiaceae

Vol. 19: Polygalaceae, Dichapetalaceae, Euphorbiaceae, Callitrichaceae, Buxaceae, Anacardiaceae, Aquifoliaceae

Vol. 20: Celastraceae, Icacinaceae, Sapindaceae, Melianthaceae, Greyiaceae, Balsaminaceae, Rhamnaceae, Vitaceae

Vol. 21: Tiliaceae, Malvaceae, Bombacaceae, Sterculiaceae

Vol. 22: *Ochnaceae, Clusiaceae, Elatinaceae, Frankeniaceae, Tamaricaceae, Canellaceae, Violaceae, Flacourtiaceae, Turneraceae, Passifloraceae, Achariaceae, Loasaceae, Begoniaceae, Cactaceae* (Published 1976). Price: R8,60. Overseas: R10,75. Post free

Vol. 23: Geissolomaceae, Penaeaceae, Oliniaceae, Thymelaeaceae, Lythraceae, Lecythidaceae

Vol. 24: Rhizophoraceae, Combretaceae, Myrtaceae, Melastomataceae, Onagraceae, Trapaceae, Haloragaceae, Gunneraceae, Araliaceae, Apiaceae, Cornaceae

Vol. 25: Ericaceae

Vol. 26: *Myrsinaceae, Primulaceae, Plumbaginaceae, Sapotaceae, Ebenaceae, Oleaceae, Salvadoraceae, Loganiaceae, Gentianaceae, Apocynaceae* (Published 1963). Price: R4,60. Overseas: R5,75. Post free

Vol. 27: Part 1: Periplocaceae, Asclepiadaceae (Microloma—Xysmalobium)
Part 2: Asclepiadaceae (Schizoglossum—Woodia)
Part 3: Asclepiadaceae (Asclepias—Anisotoma)
Part 4: *Asclepiadaceae (Brachystelma—Riocreuxia)* (Published 1980). Price: R4,50. Overseas: R6,00. Post free

Asclepiadaceae (remaining genera)

Vol. 28: Cuscutaceae, Convolvulaceae, Hydrophyllaceae, Boraginaceae, Stilbaceae, Verbenaceae, Lamiaceae, Solanaceae, Retziaceae

Vol. 29: Scrophulariaceae

Vol. 30: Bignoniaceae, Pedaliaceae, Martyniaceae, Orobanchaceae, Gesneriaceae, Lentibulariaceae, Acanthaceae, Myoporaceae

Vol. 31: Plantaginaceae, Rubiaceae, Valerianaceae, Dipsacaceae, Cucurbitaceae

Vol. 32: Campanulaceae, Sphenocleaceae, Lobeliaceae, Goodeniaceae

Vol. 33: Asteraceae: Part 1: Lactuceae, Mutisieae, 'Tarchonantheae'
Part 2: Vernonieae, Cardueae
Part 3: Arctotideae
Part 4: Anthemideae
Part 5: Astereae
Part 6: Calenduleae
Part 7: Inuleae: Fascicle 1: Inulinae
Fascicle 2: *Gnaphaliinae (First part)* (Published 1983). Price: R16,70, GST included. Overseas: R19,70. Post free.
Part 8: Heliantheae, Eupatorieae
Part 9: Senecioneae

ASTERACEAE (COMPOSITAE)

Tribe **Inuleae** subtribe **Gnaphaliinae** (first part)

by O. M. HILLIARD*

The tribe Inuleae was recently re-investigated by Merxmüller, Leins & Roessler (in Heywood *et al.* eds, The Biology and Chemistry of the Compositae, London, 1977) who recognized three subtribes, Inulinae *sensu amplo*, Gnaphaliinae *sensu amplo* and Athrixiinae, and postulated a more natural arrangement of generic groups within each subtribe. The genera dealt with here fall into the *Gnaphalium* and *Helichrysum* groups of Gnaphaliinae. *'Helipterum'* (a name thrice illegitimate) also belongs, but is under revision by Professor B. Nordenstam, Stockholm; provision is made for these species in the key below because they may well be sought here.

The sequence of genera given in Dalla Torre & Harms, Genera Siphonogamarum, Leipzig, 1900 − 1907 (and followed by R.A. Dyer, Gen.) is no longer appropriate, and some re-numbering is unavoidable if related genera are to be grouped together.

The structure of the stereome (the thickened region in the lower part of an involucral bract) has been found of considerable value in the classification of the genera (see Hilliard & Burtt in Bot. J. Linn. Soc. 82: 181−232, 1981). Ideally, an inner bract is cleared and examined under a compound microscope, but where the character is used in the following key, the necessary observation can be made with a hand lens, especially if any hairs are scraped away. When 'fenestrated', there is a thin patch in the thick-walled tissue of the stereome and at least the midvein can be seen running through it; when 'undivided', no such thin patch is visible.

Key to genera

1a Leaf margins either flat or revolute: (1b on p. 7,2: 3)

  2a Leaves spathulate and more or less truncate; pappus bristles plumose from the base upwards ....**Facelis** (p. 7,2: 4)

  2b Either leaves not spathulate and truncate or pappus bristles not plumose from the base upwards:

    3a Small annual or weakly perennial herbs with elliptic to ovate leaves on slender petioles and small (up to 6 mm long) white or whitish (rarely purplish) heads on filiform peduncles, shaft of pappus bristles barbellate, tips barbellate to shortly plumose ....**Troglophyton** (p. 7,2: 39)

    3b Plants either perennial or, if annual, leaves and heads not on slender stalks, or shaft of pappus bristles not barbellate:

      4a Loosely branched, tangled subshrubs, leaves small (up to 25 × 7 mm), elliptic to suborbicular on short petioles, heads small (up to 5 mm long), in small terminal clusters, involucral bracts white or whitish, stereome undivided...**Plecostachys** (p. 7,2: 49)

* Department of Botany, University of Natal, Pietermaritzburg.
  Financial assistance from CSIR for field work and overseas study is acknowledged.

4b Characters not as combined above:

5a Heads homogamous:

6a Leaves rigid, tips remarkably mucronate, often hooked or recurved, blade often conduplicate or U-shaped in section .... *'Helipterum'*

6b Leaves soft or otherwise differing from above:

7a Ovaries with relatively large and conspicuous white hairs, pappus bristles fused at the base in a smooth ring, shaft usually plumose or barbellate, rarely smooth .... *'Helipterum'*

7b Ovaries either without large white hairs or pappus bristles not fused at base (though they may cohere strongly by patent cilia):

8a Prostrate woolly annual herb, pappus bristles c. 5, tipped with a tuft of remarkably inflated cells .... **Galeomma** (p. 7,2: 13)

8b Plants either shrubby or herbaceous perennials, or if annual, then pappus bristles many:

9a Prostrate white-woolly herb with tiny (2,5 mm long) heads hidden in woolly glomerules, involucral bracts biseriate, tips white, crisped ....**Vellereophyton** (p. 7,2: 31)

9b Plants either perennial or, if annual, then involucral bracts neither biseriate nor with crisped white tips. **Helichrysum** (p. 7,2: 61)

5b Heads heterogamous:

10a Erect herbs, main stem leaves lanceolate, linear-lanceolate or oblong-lanceolate, broad-based and clasping, briefly or strongly decurrent in stem wings, green and glandular above, white tomentose below, heads more than 12-flowered, involucral bracts white, creamy or straw-coloured, receptacle epaleate ....**Pseudognaphalium** (p. 7,2: 53)

10b Plants not as described above:

11a Involucral bracts with undivided stereome:

12a Leaf margins flat:

13a Receptacle epaleate .... **Gnaphalium** (p. 7,2: 17)

13b Receptacle with paleae resembling the inner involucral bracts... **Tenrhynea** (p. 7,2: 57)

12b Leaf margins strongly revolute.................**Helichrysopsis** (p. 7,2: 47)

11b At least the inner involucral bracts fenestrated (a median vascular strand embedded in thin tissue within the stereome generally visible with a hand lens, particularly if any woolly hairs are scraped off):

14a Female flowers at least five times as many as hermaphrodite:

15a Pappus bristles either plumose from the base upwards or more or less evenly barbellate throughout:

16a Involucral bracts very obtuse or truncate .... **Lasiopogon** (p. 7,2: 5)

16b Involucral bracts long acuminate .... **Galeomma** (p. 7,2: 13)

15b Pappus bristles either scabrid, or subplumose in upper part only:

17a Perennial herbs or shrubs:

18a Flowers 4−6.................................**Achyrocline** (p. 7,2: 59)

18b Flowers at least 9 ....**Helichrysum** (p. 7,2: 61)

17b Annual herbs:

    19a Involucral bracts with opaque white tips; pappus bristles subplumose in upper half .... **Vellereophyton** (p. 7,2: 31)

    19b Involucral bracts translucent; pappus bristles scabrid .... **Pseudognaphalium** (p. 7,2: 53)

14b Female flowers up to 4 times as many as hermaphrodite, or numbers subequal, or hermaphrodite flowers outnumbering female:

    20a Involucral bracts usually biseriate, rarely in 3 series and then pappus bristles abruptly expanded towards base and fused below in a smooth ring .... **Lasiopogon** (p. 7,2: 5)

    20b Involucral bracts usually in at least 3 series, pappus bristles never abruptly expanded and then fused in a ring (though they may be undilated at the base and fused):

        21a Annual or weakly perennial grey-woolly herbs, heads up to 4 × 3 mm, involucral bracts with opaque white tips, pappus bristles subplumose in upper half .... **Vellereophyton** (p. 7,2: 31)

        21b Usually either perennial herbs, suffrutices or shrubs, or coarse biennials; if annual, either bracts without opaque white tips or pappus not subplumose in upper part:

            22a Stereome of the involucral bracts clearly distinguished from the scarious and often coloured lamina:

                23a Flowers 4—6, ♀ outnumbering ☿ .... **Achryrocline** (p. 7,2: 59)

                23b Flowers usually at least 9, if fewer, then ♀ flowers not outnumbering ☿ .... **Helichrysum** (p. 7,2: 61)

            22b Involucral bracts more or less uniform in texture and colour, the limits of the stereome not clearly discernible .... **Genus nov.?** (p. 7,2: 312)

1b Leaf margins involute at least in upper half: (from p. 7,2: 1)

24a Heads very large, 25—30 mm long.................................... **Edmondia** (p. 7,2: 314)

24b Heads less than 15 mm long.. *'Helipterum'*

8986   FACELIS

Facelis *Cass.* in Bull. Soc. philom. 94 (1819); Benth. in Benth. & Hook. f., Gen. Pl. 2,1: 304 (1873); Cabrera, Fl. Prov. Buenos Aires, Comp. 151 (1963). Type species: *F. apiculata* Cass. i.e. *F. retusa* (Lam.) Sch. Bip.

Tufted annual herbs. *Leaves* alternate, spathulate, ± truncate, mucronate, sessile, lower surface white-woolly. *Heads* heterogamous, cylindric, crowded at the ends of the branches. *Involucral bracts* in several series, stereome undivided but vascular strands clearly visible, lamina very thin, pellucid. *Receptacle* smooth. *Flowers* several, ♀ outnumbering ☿, corolla of ♀ flowers filiform, of ☿ narrowly tubular, all with hairs on back of lobes. *Anthers* with an acute apical appendage and tails about equalling the filament collar. *Style branches* narrowly linear, pointed, with sweeping hairs extending to the point of division. *Achenes* turbinate, villous, the hairs unicellular with thickened walls and a basal swelling cushion. *Pappus* bristles plumose, bases connate in a ring.

A South American genus of 4 species, one now naturalized in parts of South Africa.

Facelis retusa *(Lam.) Sch. Bip.* in Linnaea 34: 532 (1866); Cabrera, Fl. Prov. Buenos Aires, Comp. 151 (1963); Henderson & Anderson, Common Weeds in S. Afr. 368, fig. 183 (1966); Hilliard, Compositae in Natal 121 (1977). Type: Argentina and Uruguay, *Commerson* s.n. (P-LAM, holo.!).

Gnaphalium retusum Lam., Encyl. 2: 758 (1788); *Helichrysum retusum* (Lam.) Spreng., Syst. Veg. 3: 484 (1826).

Facelis apiculata Cass. in Bull. Soc. philom. 94 (1819), Dict. Sci. nat. 16: 104 (1820); DC., Prodr. 7: 47 (1838), nom. illegit. Type as above.

Tufted annual herb; stems up to 250 mm, erect or decumbent, closely leafy throughout. *Leaves* sessile, up to 20 x 3 mm, oblong or spathulate-oblong, apex truncate and mucronate, margins entire, somewhat revolute, green above, grey- or white-woolly below. *Heads* oblong, c. 14 x 4 mm, solitary and terminal on very short leafy shoots, many crowded at the ends of the branches in a racemose inflorescence. *Involucral bracts* graded, innermost blotched purple towards the tips. *Flowers* hidden by the copious pappus, nondescript, hermaphrodite ones purple-tipped. *Achenes* 1,5 mm long. *Pappus* bristles fine, white, silky-plumose.

A native of South America, established in Natal, the Transkei and the eastern and southern Cape as a garden weed. Flowering recorded between October and December.

Voucher: *Hilliard & Burtt* 6888 (E; K; NU; PRE; S).

8987 LASIOPOGON

Lasiopogon *Cass.* in Bull. Soc. philom. 75 (1818), in Dict. Sci. nat. 23: 302 (1822); DC., Prodr. 6: 236 (1838) excl. *L. molluginoides* DC.; Harv. in F.C. 3: 264 (1865); Benth. in Benth. & Hook. f., Gen. Pl. 2, 1: 304 (1873); B. Nord. in Jl S. Afr. Bot. 30: 60 (1964); Hilliard & Burtt in Bot. J. Linn. Soc. 82: 212 (1981). Type species: *L. iunatum* Cass., i.e. *L. muscoides* (Desf.) DC.

Comptonanthus B. Nord. in Jl S. Afr. Bot. 30: 54 (1964), p.p., excluding *C. molluginoides (DC.) B. Nord.*, type of the genus.

Small, generally woolly, annual herbs, stems simple or branched, filiform, usually prostrate or decumbent, sometimes erect, laxly leafy but closely so under the heads. *Leaves* very small, oblong to spathulate, obtuse, base narrowed, margins flat, usually loosely enveloped in wool. *Heads* heterogamous, campanulate, very small, closely surrounded by leaves, usually massed in glomerules at the branchlet tips. *Involucral bracts* 2−3-seriate, a few shorter outer bracts occasionally present, otherwise subequal, equalling or slightly exceeding flowers, stereome conspicuously fenestrate, lamina very thin, pellucid, extreme tip sometimes opaque, apex obtuse to truncate, often emarginate, whitish to straw-coloured. *Receptacle* shortly honeycombed or shallowly tuberculate. *Flowers* 10−85, whitish, often tipped reddish, ♀ flowers 5−75, either fewer than, more than, or about equalling the ♀ flowers, sometimes within a single species, narrowly tubular or filiform; ♀ flowers 2−25, tubular below, campanulate above, with 5 deltoid lobes, all lobes glandular hairy on backs. *Anthers* with small obtuse apical appendage and tails more or less equalling or slightly exceeding the filament collar. *Style branches* truncate and penicillate, the sweeping hairs scarcely extending down the backs of the branches. *Achenes* rarely glabrous, usually with minute globular duplex hairs without a swelling cushion, or ellipsoidal unicellular, all myxogenic. *Pappus* bristles either wholly plumose, or with clavate cells at the tip and the shaft wholly or partly plumose, subplumose, barbellate or nude, when not plumose the base with patent cilia, or the bristles expanded in lowest quarter and fused at the base.

A genus of 8 species, 7 of which are confined to the arid areas of Southern Africa, whereas *L. muscoides* shows a remarkable disjunction between its northern range (from central and SE. Spain to North Africa and eastwards through the Saharo-Sindian region to Pakistan and NW. India) and its southern (from South West Africa/Namibia, Botswana and the southern O.F.S. to the Cape as far south as the Great Karoo in the west and Peddie in Ciskei and Komgha in the east).

1a Pappus bristles either free at the base or cohering by patent cilia; achenes either glabrous or with minute globose hairs (sect. *Lasiopogon*):

  2a Leaves thin, loosely woolly:

    3a Pappus bristles delicately plumose from the base and barbs longest there:

      4a Pappus bristles monomorphic, tipped with a pair of pointed cells; ♀ flowers at least six times as many as ♀ .... 1.*L. muscoides*

      4b Pappus bristles dimorphic, those of ♀ flowers tipped with a pair of cells, those of ♀ flowers with a tuft of somewhat thickened cells; ♀ flowers either fewer than or up to three times as many as ♀ .... 2.*L. micropoides*

    3b Pappus bristles either plumose or barbellate only in the upper part, or barbellate throughout:

      5a Pappus plumose in upper part with long delicate barbs degenerating to small thickened cells at the tips .... 3. *L. ponticulus*

      5b Pappus either shortly plumose and without thickened apical cells, or barbellate:

        6a Pappus either plumose or barbellate in upper part only:

          7a Pappus barbellate above; heads c. 4 mm long; flowers c. 40−60.................................4. *L. volkii*

          7b Pappus shortly plumose above; heads up to 2 mm long; flowers c. 10−25 .... 5. *L. brachypterus*

        6b Pappus uniformly barbellate throughout.. 6. *L. glomerulatus*

  2b Leaves somewhat fleshy, more or less glabrous ....7. *L. debilis*

1b Pappus bristles fused at the base in a ring; achenes with relatively large elongate hairs (sect. *Bertilia*) .... 8. *L. minutus*

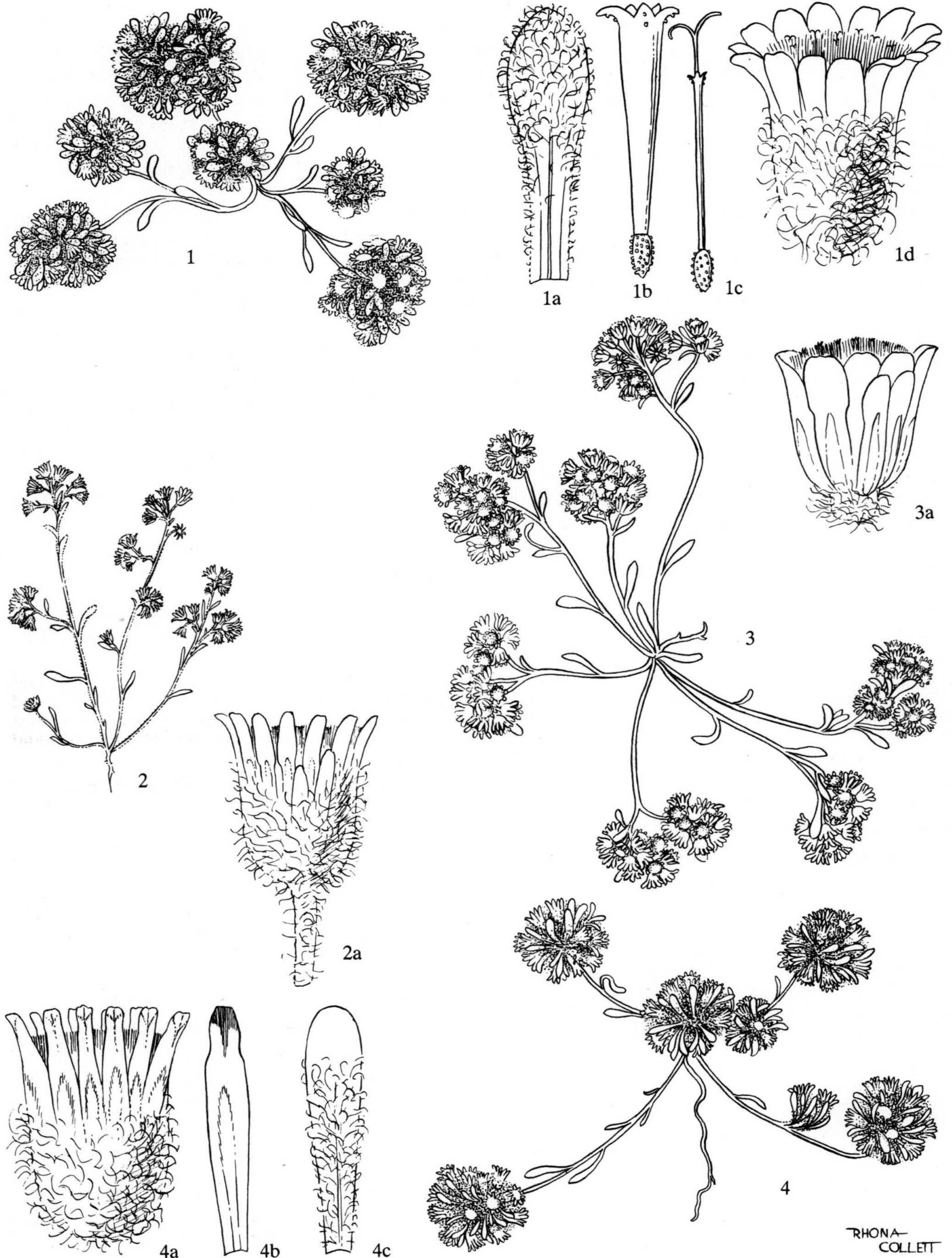
RHONA
COLLETT

Section **Lasiopogon**

1. **Lasiopogon muscoides** (*Desf.*) *DC.*, Prodr. 6:246 (1838); Harv. in F.C. 3: 265 (1865); B. Nord. in Jl S. Afr. Bot. 30: 60 (1964); Merxm., F.S.W.A. 139: 108 (1967); Holub in Tutin et al., Fl. Europ. 4: 131 (1976). Type: Tunisia, herb. Fl. Atlant. (P, holo.!).

Gnaphalium muscoides Desf., Fl. Atlant. 2: 267 t. 231 (1799).

Lasiopogon lanatum Cass. in Bull. Soc. philom. 1818: 75 (1818), nom. illegit.

Annual herb forming small woolly mats up to 10—150 mm across, frequently much smaller, stems few to many from the crown, filiform, simple or branched, prostrate or decumbent, very loosely greyish-white woolly, distantly leafy then closely so under the heads. *Leaves* mostly 3—7 x 1—1,5 mm, smallest under the heads, up to 10 x 1,5 at the root, spathulate or oblong-spathulate, obtuse, base much narrowed, both surfaces loosely greyish-white woolly. *Heads* heterogamous, campanulate, c. 3 × 2 mm, each closely surrounded by leaves then massed in woolly glomerules c. 3—15 mm across at the branchlet tips. *Involucral bracts* biseriate, subequal, slightly exceeding flowers, pellucid, tips very obtuse, pale straw-coloured, sometimes subopaque, whole involucre rotate when achenes shed. *Receptacle* shortly honeycombed. *Flowers* 36—59, 31—53 ♀, 2—6 ☿, whitish, tips sometimes reddish, ♀ 6—25 times as many as ☿. *Achenes* 0,75 mm long, with myxogenic hairs. *Pappus* bristles delicately plumose from the base, terminating in a pair of pointed cilia, a few small patent cilia at base. Fig. 2:4.

Widely distributed from about 23°S in South West Africa/Namibia and Botswana through the semi-desert areas of the Cape as far south as the Ceres-Calvinia Karoo and the Great Karoo and east to the southern O.F.S. and Komgha. Also in SE. and Central Spain, N. Africa and the Saharo-Sindian region to Pakistan and NW. India. Favours bare sandy or gravelly places such as watercourses, the edges of pans, even railway tracks and old fields, probably where there is seasonal water.

Flowering recorded between April and December, most records in September. Map 1.

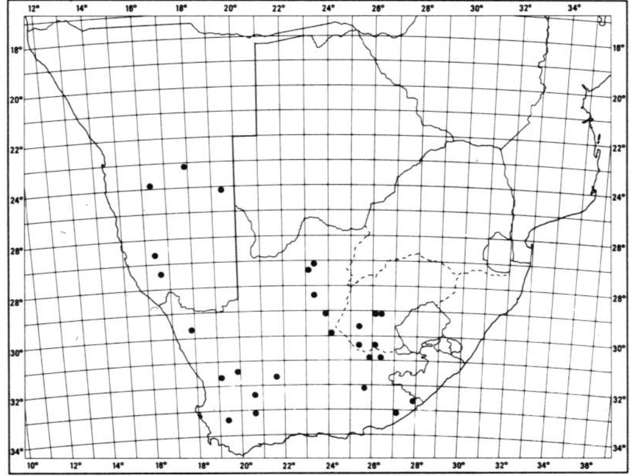

MAP 1.— **Lasiopogon muscoides**

Vouchers: *Acocks* 17202 (PRE); *Flanagan* s.n. (PRE); *Giess* 12816 (WIND).

2. **Lasiopogon micropoides** *DC.*, Prodr. 6: 246 (1838); Harv., Thes. Cap. 2:31 t. 150 (1863), in F.C. 3: 264 (1865), excl. var. β; B. Nord. in Jl S. Afr. Bot. 30: 62 (1964); Merxm., F.S.W.A. 139: 108 (1967). Type : Cape, Namaqualand, between Kouse (Kaus Mts) and Gariep (Orange River), *Drège* 2845 (G-DC, holo.!; E; TCD, iso.!).

Annual herb forming small woolly mats up to 150 mm across, but often much smaller, stems few to many from the crown, filiform, simple or branched, prostrate or decumbent, loosely greyish-white woolly, leafy and densely so under the heads. *Leaves* mostly 2,5—7 × 1—2 mm, smaller under the heads, spathulate, obtuse, base much narrowed, both surfaces loosely greyish-white woolly. *Heads* heterogamous, campanulate, c. 3 × 2 mm, each closely surrounded by leaves, massed in woolly glomerules c. 5—15 mm across at the

---

FIG. 1.—1, **Lasiopogon volkii,** whole plant, × 1; 1a, leaf, × 6,6; 1b, hermaphrodite flower, × 13; 1c, female flower, × 13; 1d, head, × 6,6 (*Merxmüller & Giess* 3166a). 2, **L. brachypterus,** whole plant, × 1,3; 2a, head, × 16 (*Hutchinson* 880). 3, **L. debilis,** whole plant, × 1,3; 3a, head, × 10 (*Acocks* 20610). 4, **L. glomerulatus,** whole plant, × 1; 4a, head, × 6,6; 4b, involucral bract, × 10; 4c, leaf, × 10 (*Acocks* 17128).

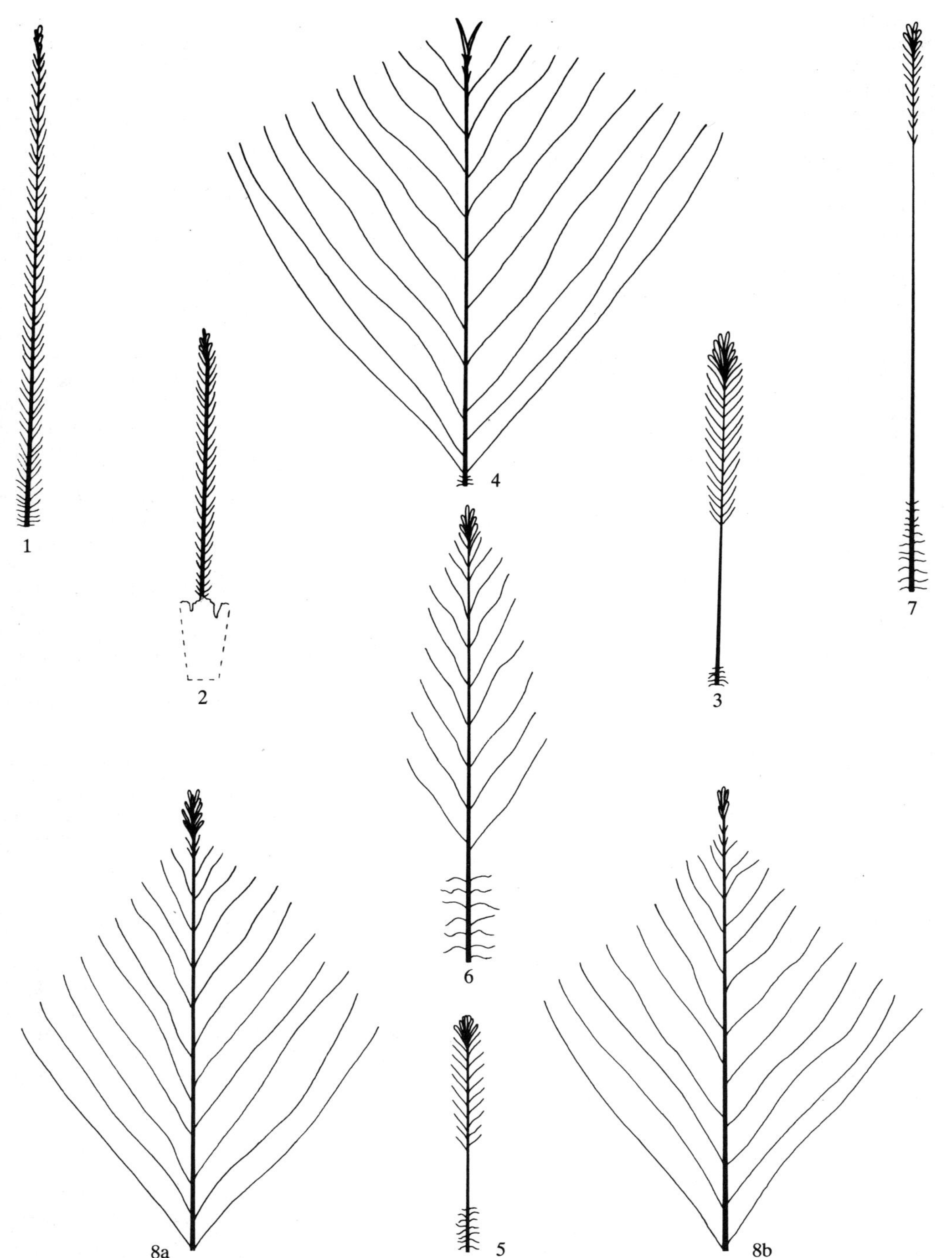

branchlet tips. *Involucral bracts* biseriate, subequal, slightly exceeding flowers, pellucid, tips very obtuse, pale straw-coloured, sometimes subopaque. *Receptacle* shortly honeycombed. *Flowers* 21–35, 9–23 ♀, 7–25 ☿, ♀ either fewer than or up to three times as many as ☿, whitish, sometimes tipped reddish. *Achenes* 0,75 mm, with myxogenic hairs. *Pappus* bristles delicately plumose from the base, the barbs shortening towards the tips, in the ♀ flowers terminating in a pair of cells, in the ☿, in a small tuft of clavate cells, no patent cilia at base. Fig. 2:8.

Recorded mainly from Namaqualand, from the Kareebergen near Vanrhynsdorp north to the Orange River, with one record from central South West Africa/Namibia (2316 Nauchas), but plants are small and easily overlooked. Favours open sandy places such as water courses. Flowering recorded between June and September. Map 2.

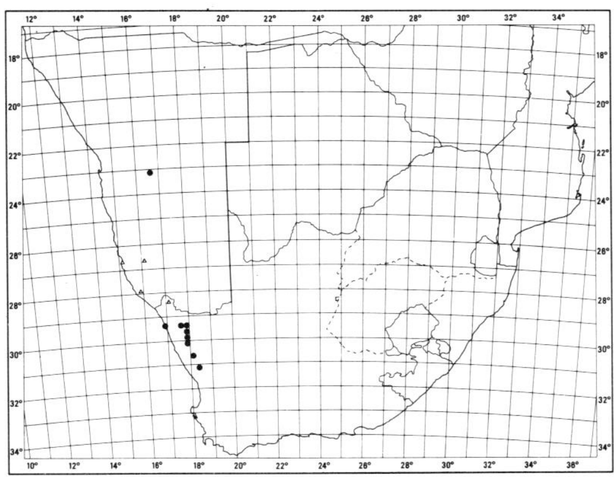

MAP 2.— ● **Lasiopogon micropoides**
△ **Lasiopogon ponticulus**

Vouchers: *Acocks* 19382 (M; PRE), 19551 (PRE); *Bolus* 6561 (E; PRE); *Giess* 13474 (PRE; S; WIND); *Schlechter* 8169 (BOL; E; K; PRE; S), 11321 (E; PRE).

3. **Lasiopogon ponticulus** *Hilliard* in Bot. J. Linn. Soc. 82: 213 (1981). Type:

S.W.A./Namibia, Lüderitz, 1 mile S. of lagoon, 5 viii 1959, *Giess & Van Vuuren* 654 (PRE; holo.!; W; WIND, iso.!).

Annual herb forming small woolly mats, stems few to many from the crown, 10–80 mm long, filiform, simple or branched, prostrate or decumbent, cobwebby-woolly, nude below or distantly leafy but closely leafy under the heads. *Leaves* mostly 3–8 × 1–1,5 mm, smallest under the heads, spathulate, obtuse, base narrowed, both surfaces loosely greyish-white woolly. *Heads* heterogamous, campanulate, c. 3 × 2 mm, each closely surrounded by leaves, then massed in woolly glomerules 5–10 mm across at the branchlet tips. *Involucral bracts* biseriate, subequal, slightly exceeding flowers, tips pellucid, very obtuse, often emarginate, colourless, whitish or palest straw-colour, rotate when achenes shed. *Receptacle* shortly honeycombed. *Flowers* 17–28, 5–15 ♀, 10–16 ☿, ♀ and ☿ equal or subequal, or up to three times as many as ♀ as ☿, or rarely slightly more ♀ than ☿, whitish, sometimes tipped reddish. *Achenes* 0,5 mm long, glabrous. *Pappus* bristles weakly plumose in upper third or half, terminating in a small tuft of clavate cilia, delicate patent cilia at base, cohering lightly.

Known only from southern South West Africa/Namibia, in the environs of Lüderitz, Aus and the Buchuberge (which lie about halfway between Lüderitz and the Orange River), and the Richtersveld, south of the Orange, in Namaqualand, N. Cape. Grows in sand; flowering recorded between July and November. Map 2.

Vouchers: *Dinter* 6433 (BOL; PRE; BM, S and STE mixed with *L. glomerulatus*); *Kinges* 2635 (M; PRE); *Nordenstam* 1689 (S).

4. **Lasiopogon volkii** (*B. Nord.*) *Hilliard* in Bot. J. Linn. Soc. 82: 214 (1981). Type: S.W.A./Namibia, 2 miles W of Schakalskuppe, low black kopje N. of the road, 15 iv 1963, *Nordenstam* 2228 (LD holo.; S, iso.!).

FIG. 2.—**Lasiopogon**, pappus bristles, all × 33. 1, **L. glomerulatus** (*Acocks* 17128). 2, **L. minutus** (*Nordenstam* 1031). 3, **L. debilis** (*Acocks* 20610). 4, **L. muscoides** (*Acocks* 17202). 5, **L. brachypterus** (*Hutchinson* 880). 6, **L. ponticulus** (*Giess & Van Vuuren* 654). 7, **L. volkii** (*Merxmüller & Giess* 3166a). 8, **L. micropoides**, from hermaphrodite flower; 8a, from female flower (*Acocks* 19382).

Gnaphalium volkii B. Nord. in Mitt. bot. StSamml., Münch. 6: 1 (1966); Merxm., F.S.W.A. 139: 84 (1967).

Annual herb forming small woolly mats 30—100 mm across, stems few to many from the crown, prostrate, filiform, well branched, very loosely greyish-white woolly, distantly leafy then closely so near the tips. *Leaves* 3—8 × 1—2 mm, diminishing upwards, spathulate, obtuse, base much narrowed, both surfaces loosely greyish-white woolly. *Heads* heterogamous, campanulate, c. 4 × 3 mm, solitary on short leafy branchlets massed at the branch tips in woolly, leafy glomerules in which the heads are buried. *Involucral bracts* biseriate, subequal, loosely woolly, pellucid, tips exceeding flowers, opaque, white, very obtuse, minutely radiating. *Receptacle* shortly honeycombed. *Flowers* 39—60, 19—41 ♀, 16—21 ☿, ♀ and ☿ subequal, or ♀ up to twice as many as ☿. *Achenes* less than 0,5 mm long, with myxogenic hairs, or hairs sometimes wanting on ☿ flowers. *Pappus* bristles barbellate near tips, shaft smooth, bases cohering strongly by long patent cilia. Fig. 1: 1, 2: 7.

Recorded only from southern South West Africa/Namibia, in 2516, 2616, 2716, 2717 and 2718 degree squares, on flats or hills; flowering between February and September. Map 3.

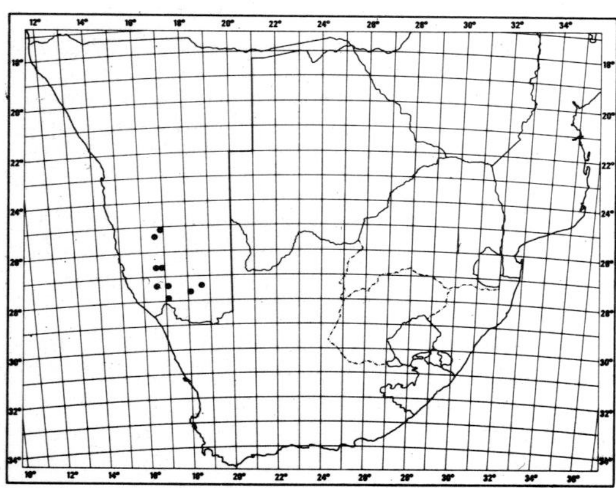

MAP 3.— **Lasiopogon volkii**

Vouchers: *Dinter* 4997 (BOL; G; PRE; SAM); *Giess* 13347 (M; WIND); *Merxmüller & Giess* 3166a (M; PRE; WIND).

5. **Lasiopogon brachypterus** [*O. Hoffm.* ex] *Zahlbr.* in Annln naturh. Mus. Wien 20: 57 (1905); Levyns in Adamson & Salter, Fl. Cape Penins. 777 (1950). Type: Cape, Sir Lowry's Pass, Palmiet River, *Penther* 1145 (W, holo.!).

Comptonanthus brachypterus (Zahlbr.) B. Nord. in Jl S. Afr. Bot. 30: 57 (1964).

Diminutive annual herb 10—50 mm tall, stems filiform, erect or suberect, simple or branched from the base, then simple or sparingly branched above, loosely woolly, leafy. *Leaves* 2—9 × 0,25—1 mm, longest at the root, oblong or oblong-spathulate, apex obtuse or subacute, base slightly narrowed, both surfaces loosely greyish white appressed woolly. *Heads* heterogamous, cylindric-campanulate, c. 1,5 × 1 mm, few to several in loose leafy clusters terminating the branchlets. *Involucral bracts* biseriate, subequal, equalling flowers, tips pellucid, colourless or purplish around the stereome, light golden brown to straw-coloured above, very obtuse, almost truncate, rotate when the achenes are shed. *Receptacle* shortly honeycombed. *Flowers* 9—26, 6—20 ♀, 3—10 ☿, whitish, tipped purplish red. *Achenes* 0,5 mm long, glabrous. *Pappus* bristles shortly plumose in upper half, base with delicate patent cilia cohering lightly, soon caducous. Fig. 1: 2, 2: 5.

On the mountains in the SW. and W. Cape, from the Peninsula N. and NE. to Paarl, Worcester, Ceres, Clanwilliam, Piquetberg and Vanrhynsdorp divisions, with a disjunction (probably real) to the Khamiesberg in Namaqualand, and an isolated record, probably due to lack of collecting in the intervening mountains, from the Klein Swartberg. Grows in damp sandy places, along streams, on the ledges of cliffs or on rock platforms, sometimes in moss cushions on rocks and cliffs; flowering mainly between September and December. Map 4.

Vouchers: *Esterhuysen* 20576 (BOL; K; PRE); *Hutchinson* 880 (BOL; K; PRE); *Schlechter* 5394 (BOL; BM; K; P; W).

6. **Lasiopogon glomerulatus** (*Harv.*) *Hilliard* in Bot. J. Linn. Soc. 82: 213 (1981). Type: Cape, Cradock, auf Bergen und Hügeln zu beiden Seiten des Tarkarivier, 3 000—4 000 Fuss, August, *Ecklon & Zeyher* (S, holo.!).

Gnaphalium glomerulatum [Sond. ex] Harv. in F.C. 3: 262 (1865); Humbert, Fl. Madag. 189: 384, fig. 72: 7—14 (1962); Merxm., F.S.W.A., 139: 83 (1967).

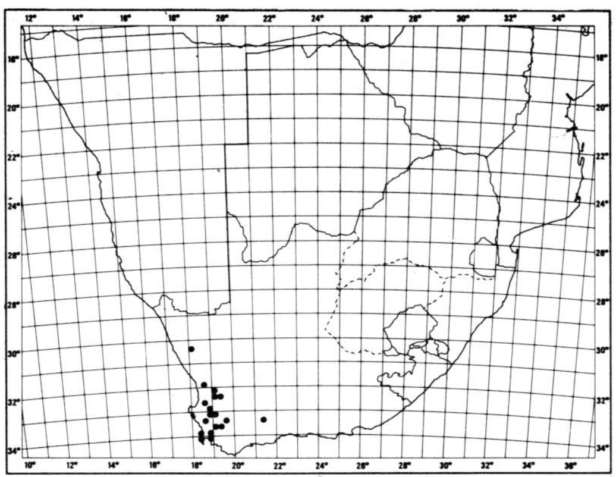

MAP 4.— **Lasiopogon brachypterus**

Gnaphalium araneosum S. Moore in J. Bot., Lond. 51: 209 (1913), non Sch. Bip. (1845). Type: Zambia, Lake Chirengwa, *Rogers* 8406 p.p. (BM, holo.!).

Small annual herb, branches filiform, 10—80 mm long, many from the crown, prostrate, simple or loosely branched, greyish-white woolly, leafy. *Leaves* 3—10 × 0,5—2 mm, oblanceolate to spathulate, subacute to obtuse, loosely greyish-white woolly. *Heads* heterogamous, campanulate, c. 4 × 3 mm, each head surrounded by leaves webbed together with wool, many, or occasionally few, in small woolly glomerules terminating the branchlets. *Involucral bracts* in 2 series, glabrous, outer short, ovate, acute, pellucid, inner lanceolate, exceeding flowers, pellucid sometimes tinged purple, tip obtuse, often minutely notched, opaque, straw-coloured, not radiating. , *Receptacle* domed, tuberculate. *Flowers* 54—84, 49—74 ♀, 3—10 ☿, ♀ flowers purple-tipped, 7—25 times as many as ☿. *Achenes* 0,5 mm long, with myxogenic hairs. *Pappus* bristles equalling corolla, tip scabrid, shaft barbellate, bases cohering strongly by patent cilia. Fig. 1: 4, 2: 1.

Relatively poorly represented in herbaria, but seemingly widely distributed from southern South West Africa/Namibia (south of 27°S) throughout the dry parts of the Cape and the Orange Free State, from the environs of Prieska, Taung, Kimberley and Fauresmith south to Worcester, Montagu and Swellendam divisions in the west, Grahamstown and Redhouse in the east. There is one record from Lake Chirengwa in N.W. Zambia, so the species is probably in Botswana too. Also in Madagascar, on sand banks and coastal dunes fide Humbert (l.c.). Favours stony or gravelly

situations, or hard-packed sand; flowering between May and October. Map 5.

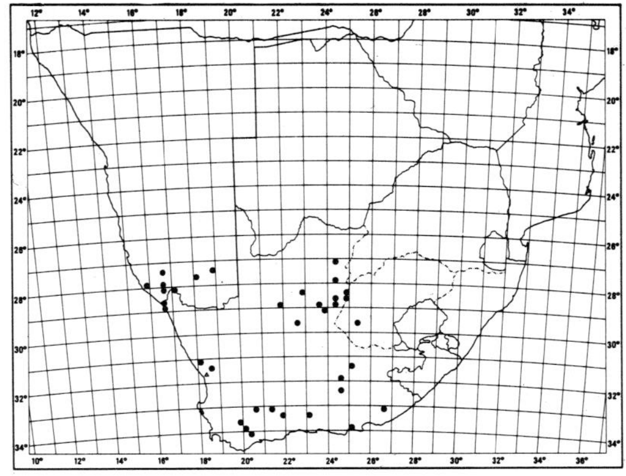

MAP 5.— ● **Lasiopogon glomerulatus**
△ **Lasiopogon minutus**

Vouchers: *Acocks* 17128 (PRE); *Compton* 5631 (NBG; SAM); *Dinter* 6433 (K; M; PRE; BM and E p.p.); *Merxmüller & Giess* 3166b (M; PRE; WIND).

7. **Lasiopogon debilis** (*Thunb.*) *Hilliard* in Bot. J. Linn. Soc. 82: 213 (1981). Type: Cape, in Karoo below Bockland, *Thunberg* (sheet 19201, UPS, holo.!; B, iso.!).

Gnaphalium debile Thunb., Prodr. 149 (1800), Fl. Cap. 651 (1823).

Comptonanthus subcarnosus B. Nord. in Jl. S. Afr. Bot. 30: 59 (1964). Type: Cape, Vanrhynsdorp, Hol River, *Nordenstam* 894 (S!).

Annual herb forming small mats up to c. 150 mm across, stems few to many from the crown, simple or branched, prostrate to suberect, thinly appressed cobwebby, distantly leafy but closely so under the heads. *Leaves* mostly 3—7 × 1—2 mm, smallest under the heads, spathulate or oblong-spathulate, apex rounded, base narrowed, somewhat fleshy, thinly woolly particularly when young. *Heads* heterogamous, campanulate, c. 2 × 2 mm, each surrounded by a few leaves, massed in glomerules c. 5—15 mm across at the branchlet tips. *Involucral bracts* biseriate, subequal, about equalling flowers, tips pellucid, very obtuse, sometimes emarginate, pale straw-coloured, sometimes subopaque at extreme tip.

Receptacle shallowly tuberculate. *Flowers* 19—37, 12—22 ♀, 7—15 ☿, ☿ slightly outnumbering ♀. *Achenes* 0,5 mm long with myxogenic hairs. *Pappus* bristles shortly plumose in upper half, the upper barbs clavate and white, weakly developed patent cilia at base. Fig. 1: 3, 2: 3.

Known from only a few localities between Vanrhynsdorp and Ceres, growing in silty or sandy places that are possibly seasonally damp. Flowering recorded between July and October. Easily recognized by its subfleshy nearly glabrous leaves. Map 6.

Vouchers: *Acocks* 20610 (BOL; K; PRE); *Salter* 3841 (BOL; K); *Schlechter* 8158 (BOL; G; K; P; PRE; S; Z).

Section **Bertilia**

Sect. Bertilia Hilliard & Burtt in Bot. J. Linn. Soc. 82: 214 (1981). Type species: *L. minutus* (B. Nord.) Hilliard & Burtt.

8. **Lasiopogon minutus** (*B. Nord.*) *Hilliard & Burtt* in Bot. J. Linn. Soc. 82: 215 (1981). Type: Cape, Vanrhynsdorp, quartzite area 4 miles NNE of Koekenaap, 25 viii 1962, *Nordenstam* 1031 (LD, holo.; E; S, iso.!).

Gnaphalium minutum B. Nord. in Mitt. bot. St Samml. Münch. 6: 3 (1966).

Small annual herb 15—40 mm tall, stems ± erect, simple or subsimple, tufted, white woolly-cobwebby, leafy. *Leaves* 6—20 × 1—3 mm, oblanceolate, apex subacute, mucronate, base much narrowed, margins flat, both surfaces thinly white woolly-cobwebby. *Heads* heterogamous, campanu-

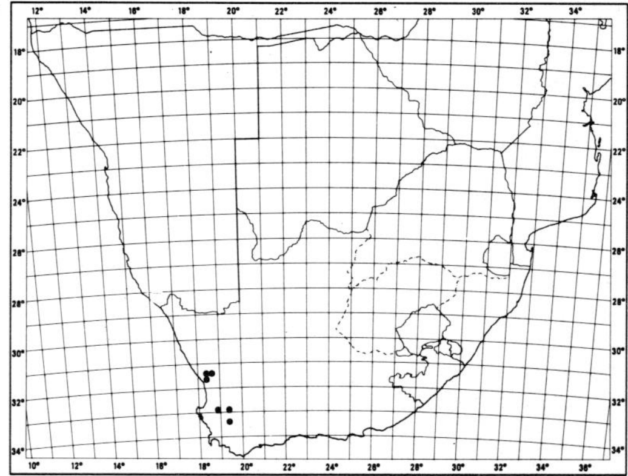

MAP 6.— **Lasiopogon debilis**

late, c. 3 × 2 mm, on short, loosely woolly peduncles clustered at the branchlet tips. *Involucral bracts* in c. 3 series, outer shorter, inner subequal, slightly exceeding flowers, pale straw yellow, tips opaque, pale straw-coloured, obtuse. *Receptacle* shallowly tubercled. *Flowers* c. 25—35, 15—21 ♀, 9—14 ☿ in the proportion 1,5—2: 1, yellow, tipped reddish. *Achenes* not seen, ovaries 0,75 mm long, thickly clad in relatively large elliptical unicellular hairs, highly myxogenic. *Pappus* bristles with barbellate tips and shafts, bases abruptly expanded and scale-like, shortly fused into a smooth ring, tardily deciduous. Fig. 2: 2.

Known only from the type collection. Map 5.

8987a

GALEOMMA

Galeomma *Rauschert* in Taxon 31:557 (1982). Type species: *G. oculus-cati* (L.f.) Rauschert.

Eriosphaera Less., Syn. Comp. 270 (1832); Harv., Thes. Cap. 2: t. 149 (1863), in F.C. 3: 264 (1865); Benth. in Benth. & Hook. f., Gen. Pl. 2,1: 321 (1873); Hilliard & Burtt in Bot. J. Linn. Soc. 82: 211 (1981); non F. G. Dietrich (1817).

Prostrate woolly annual herbs. *Leaves* spathulate or oblong-spathulate, margins flat. *Heads* homogamous or heterogamous, in terminal glomerules embedded in wool. *Involucral bracts* in few series, stereome fenestrated, tip of lamina very acute to acuminate, whitish or pale straw-coloured. *Receptacle* honeycombed. *Flowers* 11−125, female if present outnumbering hermaphrodite, corolla with hairs on backs of lobes. *Style* branches truncate and penicillate. *Achenes* with common duplex hairs, myxogenic or not. *Pappus* bristles either very few, nude below with an apical tuft of much inflated cells, or many, evenly barbed with more or less patent cilia tending to intermingle.

Species 2, in S.W.A/Namibia, Botswana, and the dry parts of the northern and western Cape.

1a Pappus bristles few (c. 5), tipped with a tuft of inflated cells, shaft nude below .... 1. *G. oculus-cati*

1b Pappus bristles many, evenly barbed with long more or less patent cilia tending to intermingle.. 2. *G. stenolepis*

1. Galeomma oculus-cati (L.f.)

Rauschert in Taxon 31:557 (1982). Type: Cape of Good Hope (LINN 989.70 holo.!).

Gnaphalium oculus-cati L.f., Suppl. 364 (1781). *G. oculus* Thunb., Prodr. 151 (1800). Fl. Cap. 657 (1823), nom. illegit.; *Eriosphaera oculus-cati* (L.f.) Less., Syn. Comp. 270 (1832); Harv., Thes. Cap. 2: t. 149 (1863), in F.C. 3:264 (1865).

Prostrate annual herb, stems 10−20 mm long, many radiating from the crown, simple or branched, thinly appressed woolly-cobwebby, glabrescent, sparsely leafy. *Leaves* 7−12 × 3−5 mm, broadly spathulate, very obtuse, mucronate, enveloped in loosely woven white wool. *Heads* homogamous, globose-campanulate c. 4 × 2 mm, several in terminal glomerules surrounded by leaves webbed together with wool, the heads buried in masses of white wool. *Involucral bracts* in 3 series, outermost short, inner subequal, equalling flowers, linear-lanceolate, backs woolly, webbed to subtending leaves, tips glabrous, very acute, sometimes lacerate, straw-coloured, radiating. *Receptacle* honeycombed. *Flowers* c. 11−20. *Achenes* 1 mm long, with minute globose duplex hairs. *Pappus* bristles c. 5, equalling corolla, tips with several much inflated cells, nude below. Fig. 3:1.

Recorded only from Calvinia and Sutherland degree squares (3119, 3220), on rocky slopes, sometimes forming dense patches. Flowers between September and December. Easily recognized by its peculiar pappus. Map 7.

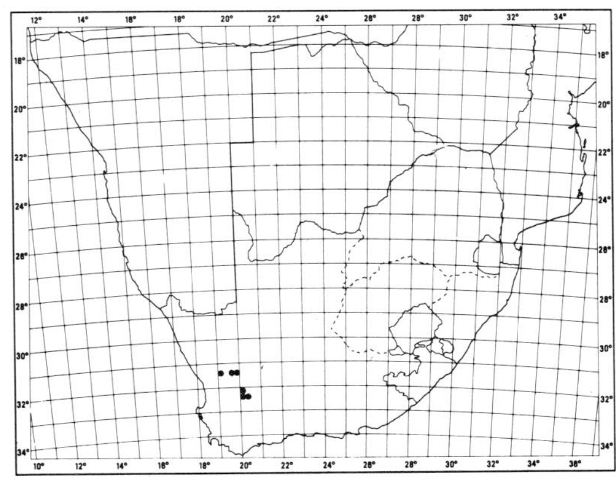

MAP 7.— **Galeomma oculus-cati**

Vouchers: *Acocks* 17756 (PRE); *Galpin* 11149 (K; PRE); *Hutchinson & Pillans* 724 (BOL; K; PRE); *Pearson* 3790 (BOL; K).

2. Galeomma stenolepis (S. Moore)

Hilliard, comb. nov. Type: South West Africa/Namibia, Osire, Sept. 1899, *Dinter* 795 (BM, holo.!; Z, iso.!).

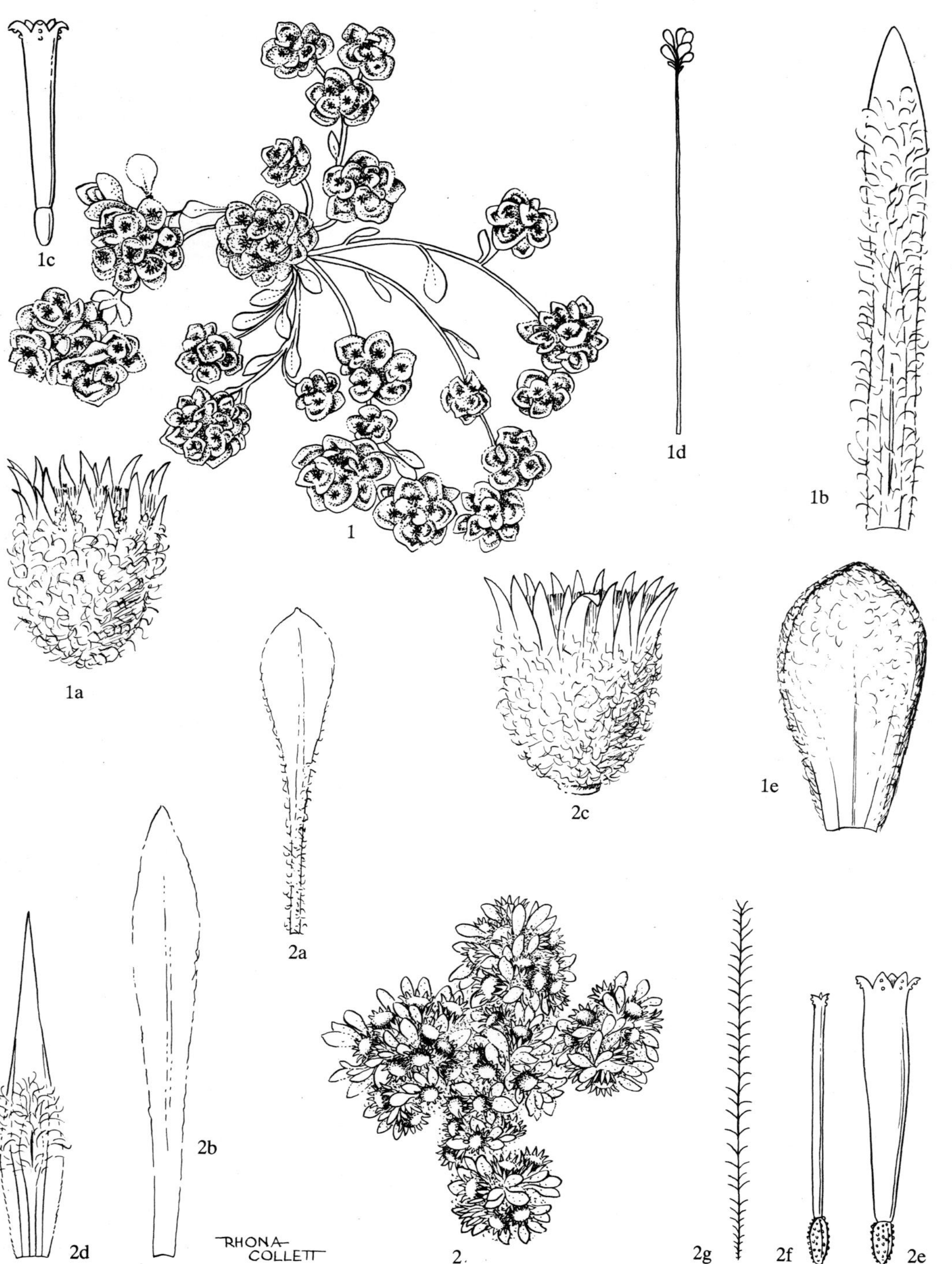

1c
1
1d
1b
1a
2a
2c
1e
2d
2b
RHONA COLLETT
2.
2g
2f
2e

Gnaphalium stenolepis S. Moore in Bull. Herb. Boissier, sér. 2,4: 1015 (1904); Merxm., F.S.W.A. 139: 84 (1967). *Eriosphaera stenolepis* (S Moore) Hilliard & Burtt in Bot. J. Linn. Soc. 82: 212 (1981).

Small woolly annual herb forming dense mats 30—150 mm across, stems many from the crown, filiform, prostrate, simple below, branching above, very loosely white-woolly, leafy at the crown, then nude or nearly so, densely leafy at the branch tips. *Leaves* 4—20 (−30) × 1−5 mm, diminishing upwards, those surrounding the heads always small (c. 4−7 × 1−2 mm), spathulate or oblong-spathulate, apex obtuse in small leaves, obtuse or subacute in larger, mucronate, loosely white woolly, partly glabrescent later. *Heads* heterogamous, campanulate, c. 3 × 3 mm, several in small glomerules terminating the branchlets, each head buried in white wool and closely surrounded by leaves, the tips of which, green or tinged purplish and glabrous or nearly so, emerge from the wool, the glomerules in turn sometimes subtended by long narrow leaves. *Involucral bracts* in 3 series, subequal, backs loosely woolly, tips long-acuminate, glabrous, squarrose, opaque, straw-coloured or whitish. *Receptacle* shortly honeycombed. *Flowers* c. 75−125, 70−120 ♀, 4−11 ☿, yellow, sometimes tipped pink. *Achenes* 0,5 mm long, with myxogenic duplex hairs. *Pappus* bristles

evenly barbed with long more or less patent cilia, tending to intermingle. Fig. 3:2.

Widespread in S.W.A./Namibia from about the 20th parallel S and SW to Botswana and the dry northern Cape, reaching Hay and Kimberley districts, but small and easily overlooked. Grows socially in seasonally damp places such as streambanks, valley bottoms, depressions, and around pans, but possibly thrives when these places are dry, not wet. Flowering recorded in most months; the plants probably appear after rain. Easily recognized by its acuminate involucral bracts and peculiar pappus. Map 8.

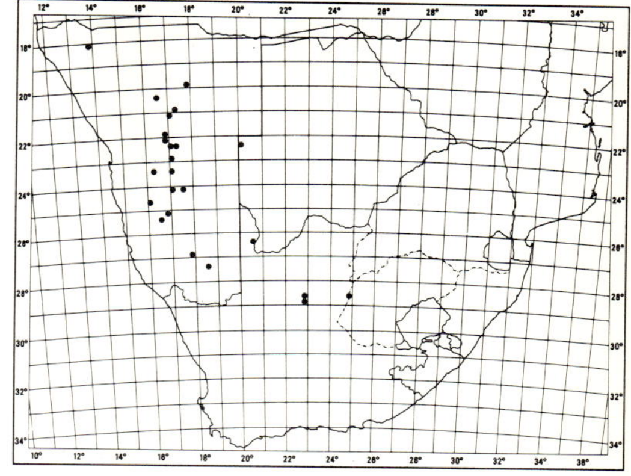

MAP 8.— **Galeomma stenolepis**

Vouchers: *Acocks* 737 (BOL; K; PRE); *Dinter* 610 (K; SAM); *Giess* 13344 (M; PRE; S; WIND).

FIG. 3.—1, **Galeomma oculus-cati,** whole plant, × 0,7; 1a, head, × 6,6; 1b, involucral bract, × 20; 1c, hermaphrodite flower, × 10; 1d, pappus bristle, × 20; 1e, leaf,× 3,3 (*Acocks* 17756). 2, **G. stenolepis**, whole plant, × 1; 2a, leaf, × 6,6; 2b, leaf, × 3,3; 2c, head, × 6,6; 2d, involucral bract, × 17; 2e, hermaphrodite flower, × 20; 2f, female flower, × 20; 2g, pappus bristle, × 27 (*Acocks* 737).

8992   **GNAPHALIUM***

Gnaphalium L., Sp. Pl. 850 (1753), Gen. Pl. ed. 5: 368 (1754); Hilliard & Burtt in Bot. J. Linn. Soc. 82: 190 (1981); Hilliard in Bot. J. Linn. Soc. 82: 267 (1981). Lectotype species: *G. uliginosum* L.

Amphidoxa DC., Prodr. 6: 246 (1838). Type species: *A. gnaphalodes* DC.
Demidium DC., Prodr. 6: 246 (1838). Type species: *D. filagineum* DC.
Gamochaeta Wedd., Chlor. Andina 1: 151 (1856). Type species: *G. americana* (Mill.) Wedd. (=*Gnaphalium americanum* Mill.)

Small annual or perennial herbs, loosely or closely woolly, stems weak. *Leaves* mostly linear, spathulate or oblanceolate, bases narrowed or broad, margins flat. *Heads* heterogamous, usually in clusters, rarely solitary, terminal or racemosely arranged. *Involucral bracts* in few series, stereome undivided or with a few minute thin streaks, flat (not embracing a flower), tip of lamina in inner series opaque or subopaque, or sometimes pellucid, light brown, buff, creamy or white. *Receptacle* nearly smooth or honeycombed. *Flowers* many, ♀ flowers usually outnumbering ☿, corolla of ♀ flowers filiform, of ☿ narrowly tubular, all with hairs on back of lobes. *Style* branches truncate and penicillate. *Achenes* glabrous or with common duplex hairs or with short subglobose duplex hairs. *Pappus* bristles monomorphic or dimorphic, tips scabrid, barbellate, or shortly plumose, bases nude or with small patent cilia not cohering, or lightly fused in groups, or fused into a ring, or pappus wanting in female flowers.

Worldwide; number of species uncertain; 13 species native to our area, 1 pantropical weed possibly native, 5 or 6 American species naturalized.

1a Pappus bristles, if present, either free at the base or cohering lightly by patent cilia:

  2a Heads either solitary or in lax or congested corymbose panicles, bracts in c. 4 series:

    3a Heads at least 5 mm long, base of whole involucre white-felted, only tips of bracts free .... 12. *G. simii*

    3b Heads up to 4 mm long, backs of outer bracts often loosely woolly, but never whole involucre felted with only tips free:

      4a Pappus bristles on ♀ flowers dilated at base; tips of involucral bracts acute to acuminate, or rarely truncate, usually lacerate... 14. *G. pauciflorum*

      4b Pappus bristles either wanting on ♀ flowers or not dilated at the base; tips of involucral bracts acute or obtuse, entire:

        5a Outer involucral bracts of mature heads woolly or cobwebby on backs at least in lower part:

          6a Compound inflorescences carried above leaves on conspicuous cottony peduncles (character not visible in very young material):

            7a Leaves mostly broadest near tip, obtuse or subacute:

              8a Leaves mostly 10−40 mm long, ♀ flowers at least 7 times as many as ☿ ....1. *G. austroafricanum*

              8b Leaves mostly 5−15 mm long, ♀ flowers up to 5 times as many as ☿. High Drakensberg... 3. *G. limicola*

            7b Leaves mostly linear or linear-lanceolate .... 2. *G. griquense*

          6b Compound inflorescence generally closely surrounded by leaves, peduncles, if visible, more or less overlaid by leaves:

            9a Female flowers up to 8 times as many as ☿:

              10a Involucral bracts with snow-white ± obtuse tips 2−4 times as long as broad and often recurved, much exceeding flowers ....8. *G. capense*

              10b Involucral bracts either buff-coloured or dirty white; if pure white then scarcely exceeding flowers:

                11a Leaves mostly 2−6 (−10) mm long, linear to linear-spathulate, obtuse to subacute, mucro often hidden by wool, tips of inner involucral bracts acute or subacute, often ± recurved, much exceeding flowers. S. and E. Cape ....9. *G. vestitum*

---

* Adapted from *Gnaphalium (Compositae) in Africa and Madagascar* (Botanical Journal of the Linnean Society 82(3): 267, 1981) by permission of the Council of the Linnean Society of London.

11b Leaves mostly 6−16 mm long, linear-lanceolate, acute, mucro conspicuous; tips of inner involucral bracts obtuse to subacute, scarcely exceeding the flowers. Transvaal .... 10. *G. nelsonii*

9b Proportion of ♀ flowers to ♀ 17−60: 1 .... 11. *G. confine*

5b Outer involucral bracts of mature heads glabrous on the backs:

12a Leaves clad in silvery 'tissue-paper' indumentum .... 6. *G. englerianum*

12b Leaves woolly or silky-woolly, but not with 'tissue-paper' indumentum:

13a Leaves mostly oblong-spathulate, lingulate or elliptic. S. and E. Cape:

14a Leaves generally oblong-spathulate or oblong-elliptic, tips often folded lengthwise and slightly recurved, indumentum thin and often glabrescent .... 4. *G. gnaphalodes*

14b Leaves generally lingulate or linear-lingulate, sometimes narrowly elliptic, tips not folded lengthwise, both surfaces silky-silvery .... 7. *G. declinatum*

13b Leaves mostly linear or oblong-linear, acute or subacute; interior plateau of Southern Africa .... 5. *G. filagopsis*

2b Heads racemosely arranged, bracts in c. 3 series:

15a Flowers c. 150−200; bases of pappus bristles not dilated .... 13. *G. polycaulon*

15b Flowers c. 20−95; bases of pappus bristles on ♀ flowers slightly dilated .... 14. *G. pauciflorum*

1b Pappus bristles on at least the ♀ flowers fused at the base in a smooth ring:

16a Tips of involucral bracts opaque or subopaque; pappus dimorphic .... 14. *G. pauciflorum*

16b Tips of involucral bracts pellucid; pappus monomorphic:

17a Leaves not markedly discolorous although indumentum on upper surface may be thinner than on lower:

18a Upper leaves folded, tips somewhat recurved; involucral bracts tinged red-purple when young, buff-coloured or palest brown later .... 15. *G. subfalcatum*

18b Upper leaves flat, not recurved; involucral bracts buff-coloured or palest brown:

19a Leaves oblanceolate or narrowly spathulate, less than 10 mm at their broadest:

20a Heads crowded and racemosely arranged .... 16. *G. calviceps*

20b Heads in small glomerules, which may run up into a racemose structure, but the glomerules remain discrete.. 19. *G. sp.*

19b Leaves spathulate, mostly 10−20 mm at their broadest .... 17. *G. pensylvanicum*

17b Leaves markedly discolorous, upper surface glabrescent, lower white-tomentose .... 18. *G. coarctatum*

Section **Gnaphalium**

1. **Gnaphalium austroafricanum** *Hilliard* in Bot. J. Linn. Soc. 82: 277 (1981). Type: Cape, Grahamstown, Douglas Dam, 1820 Settlers Nature Reserve, bare damp ground by stream, 8 xii 1977, *Hilliard & Burtt* 10900 (E, holo.!; NU, iso.!).

Stoloniferous perennial herb, stems many from the crown, loosely branched, particularly above, prostrate or decumbent, branches mostly 150−450 mm long, lower part often nude and rooting, upper thinly greyish-white cottony, leafy. *Leaves* more or less spreading, c. 10−25 (− 35) × 2−4 (−6) mm, narrowly spathulate, apex usually obtuse, sometimes subacute, mucronate, base broad, half-clasping, both surfaces thinly greyish-white appressed tomentose. *Heads* campanulate, c. 3,5 − 4 × 3 mm, in small corymbose clusters generally borne well above the foliage on cottony peduncles,

---

FIG. 4.−1, **Gnaphalium griquense**, whole plant, × 1; 1a, head, × 6,6; 1b, hermaphrodite flower, × 13; 1c, female flower, × 13; 1d, pappus bristle from hermaphrodite flower, × 13 (*Beverly* 141). 2, **G. declinatum**, part of plant, × 1; 2a, head, × 6,6; 2b, pappus bristle from hermaphrodite flower, × 13; 2c, pappus bristle from female flower, × 13 (*Compton* 22176). 3, **G. simii**, part of plant, × 1 (*Sim* 2863). 4, **G. confine**, part of plant, × 1; 4a, head, × 6,6; 4b, hermaphrodite flower, × 13; 4c, female flower, × 13; 4d, pappus bristle from hermaphrodite flower, × 13; 4e, pappus bristle from female flower, × 13 (*Hilliard & Burtt* 11972). 5, **G. gnaphalodes**, part of plant, × 1; 5a, head, × 6,6 (*Holland* 3518). 6, **G. austroafricanum**, part of plant, × 1; 6a, head, × 6,6 (*Hilliard & Burtt* 10145).

RHONA
COLLETT

terminal or subterminal. *Involucral bracts* in c. 4 series, outer short, golden-brown, loosely and lightly woolly, inner bracts exceeding flowers by the length of their snow-white opaque tips, tips obtuse or subacute, 2–3 times as long as broad. *Receptacle* shallowly tuberculate or shortly honeycombed. *Flowers* 83–145, 66–133 ♀, 6–19 ☿, in the proportion 7–20: 1. *Achenes* 0,5 mm long with duplex myxogenic hairs. *Pappus* bristles wanting in ♀ flowers, with shortly plumose tips in ☿ flowers. Fig. 4: 6.

Widely distributed from the Transvaal Lowveld at Barberton and southern Mozambique (Sul do Save) through central and coastal Natal to the E. Cape as far west as the Amatola Mountains and Grahamstown, generally in muddy places. Flowering recorded mainly between October and December. Map 10.

Much confused with *G. declinatum* (no. 7) and *G. capense* (no. 8). *G. austroafricanum* is characterized by its lax habit, spreading and generally obtuse leaves, clusters of heads on conspicuous peduncles that carry the heads well above the foliage, erect involucral bracts, and pappus wanting in the ♀ flowers. In both *G. declinatum* and *G. capense* the compound head is closely surrounded by leaves and the ♀ flowers are pappose. There are differences too in the shapes and set of the leaves and in the involucral bracts.

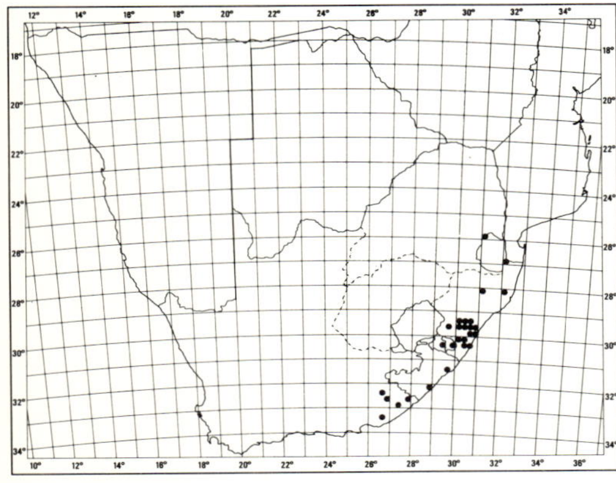

MAP 9.— **Gnaphalium austroafricanum**

Vouchers: *Hilliard & Burtt* 9082 (E; K; MO; NU; PRE; S); *Rudatis* 762 (BM; E; K; P; S); *Story* 5173 (PRE); *Tyson* 1303 (BOL; SAM).

2. **Gnaphalium griquense** *Hilliard & Burtt* in Bot. J. Linn. Soc. 82: 193, 278 (1981). Type: Natal, East Griqualand, near Franklin, Newmarket, *Krook* Pl. Penther. 1026 (W, lecto.!).

Amphidoxa adscendens [O. Hoffm. ex] Zahlbr. in Annln naturh. Mus. Wien 20: 56 (1905), non *Gnaphalium adscendens* Thunb.; Hilliard, Compositae in Natal 129 (1977). Type as for *G. griquense*.

Stoloniferous perennial herb, stems many from each crown, loosely branched, prostrate, decumbent or suberect, mostly c. 150–450 mm long, white-tomentose, leafy. *Leaves* ascending, mostly 15–30 × 1–2 mm, linear or linear-lanceolate, acute to acuminate, base broad, half-clasping, both surfaces appressed pale greyish-white tomentose. *Heads* campanulate, c. 4–5 × 3–4 mm, in small corymbose clusters borne well above the leaves on cottony peduncles, terminal or subterminal. *Involucral bracts* in c. 4 series, outer short, pale brown, copiously white-woolly, inner bracts exceeding flowers by the length of their snow-white opaque tips, tips subacute, 2–3 times as long as broad. *Receptacle* shallowly tuberculate or shortly honeycombed. *Flowers* 70–144, 50–112 ♀, 20–32 ☿, in the proportion 2,5–4:1. *Achenes* 0,5 mm long with duplex myxogenic hairs. *Pappus* bristles generally wanting in the ♀ flowers, or sometimes a few present, with shortly plumose tips in ☿ flowers. Fig. 4:1.

Recorded only from southern Natal and nearby Sehlabathebe in Lesotho, in damp places; flowering between November and February. Map 10.

G. griquense is characterized by its linear or linear-lanceolate acute to acuminate leaves, heads borne well above the foliage leaves, outer involucral bracts white-woolly on the backs and inner bracts with long snow-white tips overtopping the flowers. The ♀ flowers are generally epappose. *G. griquense* is sometimes sympatric with *G. austroafricanum*, which is easily distinguished by its spreading, narrowly spathulate and generally obtuse leaves.

Vouchers: *Bayliss* 5438 (NBG; PRE); *Beverly* 141 (NU); *Hilliard* 2076 (NU); *Hoener* 1821 (E; NU).

3. **Gnaphalium limicola** *Hilliard* in Notes R. bot. Gdn. Edinb. 32: 336 (1973), Compositae in Natal 131 (1977), Bot. J. Linn. Soc. 82: 279 (1981). Type: Cape, Barkly East distr., Naude's Nek, c. 2 500 m, 19 ii 1971, *Hilliard & Burtt* 6639 (E, holo.!; K; M; MO; NBG; NU; PRE; S, iso.!).

Creeping, mat-forming perennial herb, stems thinly grey-woolly, rooting, leafy. *Leaves* 5–15 × 1–3 mm, oblong-spathulate

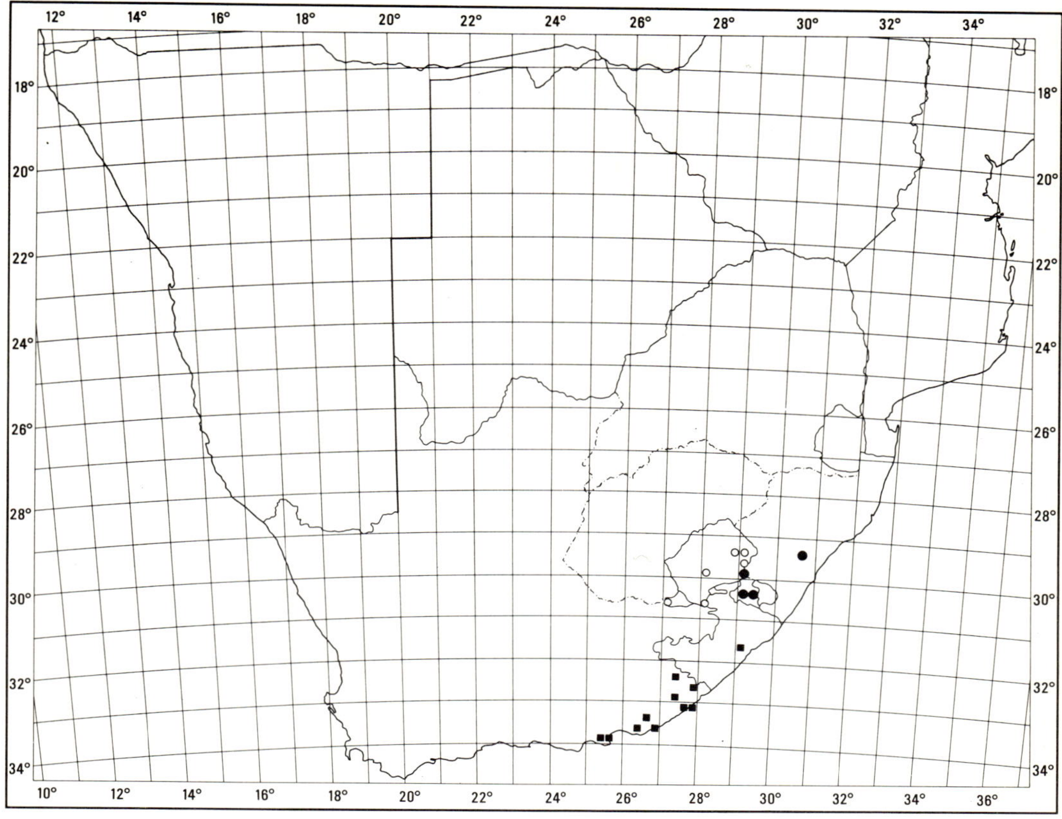

MAP 10.—
- **Gnaphalium griquense**
- **Gnaphalium limicola** (open circle)
- **Gnaphalium gnaphalodes** (filled square)

or lingulate, apex obtuse, mucronate, base half-clasping, both surfaces silky grey-tomentose. *Heads* c. 4 × 3−4 mm, up to 4 on conspicuous peduncles borne well above the leaves. *Involucral bracts* in c. 4 series, outer short, brown, woolly at base, tips of inner bracts oblong, c. 2−3 times as long as broad, ± obtuse, opaque, snow-white, exceeding the flowers. *Receptacle* shallowly tuberculate or honeycombed. *Flowers* 57−87, 38−66 ♀, 13−22 ☿, in the proportion 2−4,5: 1. *Achenes* c. 0,75 mm long with duplex hairs. *Pappus* present on both ♀ and ☿ flowers.

Recorded chiefly from the mountains of Lesotho, and the high Drakensberg and the Witteberg immediately south of Lesotho. Forms mats in muddy streambeds; flowering between October and March. Map 10.

G. limicola is characterized by its creeping rooting stem, lingulate or narrowly spathulate leaves, heads on conspicuous peduncles, and involucre woolly near the base.

Vouchers: *Compton* 21513 (NBG); *Galpin* 6667 (BOL; K; NH; PRE).

4. **Gnaphalium gnaphalodes** *(DC.) Hilliard & Burtt* in Bot. J. Linn. Soc. 82: 193 (1981); Hilliard in Bot. J. Linn. Soc. 82: 280 (1981). Type: Cape, Uitenhage, Zwartkops River, 'Thal und angrenzende Hügel von Villa Paul Maré bis Uitenhaag, 50−500', Dec.', *Ecklon* 1192 (G−DC, holo.!; S mixed with *G. micranthum* var. *spretum*; W, iso.!).

Amphidoxa gnaphalodes DC., Prodr. 6:246 (1838); Harv. in F.C. 3: 263 (1865).

?Gnaphalium micranthum Thunb. var. *spretum* DC., Prodr. 6: 229 (1838). Type: Cape, between the Bushman's and Gouritz Rivers, *Burchell* 4222 (G-DC, holo.!; K, iso.!).

Stoloniferous perennial herb, stems mostly 100−300 mm long, simple or branching, prostrate or decumbent, thinly greyish cottony-tomentose, leafy. *Leaves* ascending, appressed, mostly c. 8−20 × 2,5−4 (−5) mm, oblong-spathulate or rarely narrowly oblong-elliptic, apex usually obtuse, conspicuously mucronate, often folded lengthwise and somewhat recurved, margins more or less undulate, both surfaces loosely greyish-white appressed tomentose, often somewhat glabrescent. *Heads* c. 3−3,5 × 3 mm, in small corymbose clusters at or near the branch tips, eventually carried clear of the leaves on cottony peduncles. *Involucral bracts* in c. 4 series, outer short, golden-brown, glabrous or nearly so, inner with opaque, snow-white, oblong, obtuse tips exceeding the flowers. *Receptacle* shallowly tuberculate. *Flowers* 112−188, 90−159 ♀, 20−50 ☿, in the proportion 2,5−7:1. *Achenes* c. 0,5 mm long with duplex myxogenic hairs. *Pappus* bristles either wanting in ♀ flowers or scabrid, bristles on ☿ flowers shortly plumose at tips. Fig. 4:5.

Grows in marshy places in the E. Cape and Transkei, from about Libode (near Umtata, Transkei) south and west to Grahamstown, Port Elizabeth and Uitenhage. Flowering recorded between September and May. Map 10.

G. gnaphalodes is characterized by its oblong-spathulate leaves often folded lengthwise in the upper part and there slightly recurved, margins often somewhat undulate, and heads with glabrous or nearly glabrous involucres and snow-white bract tips scarcely exceeding the flowers. Pappus may be present or absent on the ♀ flowers.

Vouchers: *Comins* 1849 (K; PRE); *Dahlstrand* 3144 (PRE); *Hutchinson* 1730 (BOL; K); *Rogers* 28044 (PRE); *Schlechter* 2745 (B; BM; K; PRE; S; Z).

The status of *G. micranthum* var. *spretum*, which is included in the synonymy with reservation, will remain uncertain until the plant is better known; see Hilliard (l.c.) for details.

5. Gnaphalium filagopsis *Hilliard & Burtt* in Bot. J. Linn. Soc. 82: 193 (1981); Hilliard in Bot. J. Linn. Soc. 82: 281 (1981). Type: Angola, Upper Zambezi, Ninda

River (14°46′S 20°56′E), *Serpa Pinto* 14 (LISU, holo.!).

Amphidoxa filaginea Fical. & Hiern in Trans. Linn. Soc. Lond., Bot. ser. 2, 2: 21, t. 4 (1881) non *Gnaphalium filagineum* DC.; Merxm., F.S.W.A. 139: 22 (1967). Type as for *G. filagopsis*.

A. filaginea var. *transiens* Merxm. in Mitt. bot. St Samml., Münch. 2: 32 (1954). Type: S.W.A./Namibia, Okavango Valley, Runtu, *Volk* 1925a (M, holo.!; PRE, iso.!).

Silvery grey perennial herb, stems many from the crown, simple to well-branched, erect or decumbent, leafy. *Leaves* usually appressed-ascending, mostly 4−10 (−15) × 1−2 mm, linear or oblong-linear, very rarely lingulate, apex acute or subacute, mucronate, both surfaces silvery grey tomentose or silky-tomentose. *Heads* 2,5−3,5 × 2−2,5 mm, few to several subcorymbosely arranged, terminal, usually ± surrounded by leaves. *Involucral bracts* in c. 4 series, outer short, golden-brown, glabrous or nearly so, tips of inner bracts oblong, obtuse, or rarely subacute, opaque, snow-white, exceeding the flowers. *Receptacle* shallowly tuberculate. *Flowers* 67−140, 54−107 ♀, 12−35 ☿ in the proportions 3−5 (−7):1. *Achenes* 0,5 mm long, with myxogenic duplex hairs. *Pappus* usually wanting in the ♀ flowers, or occasionally a few bristles present.

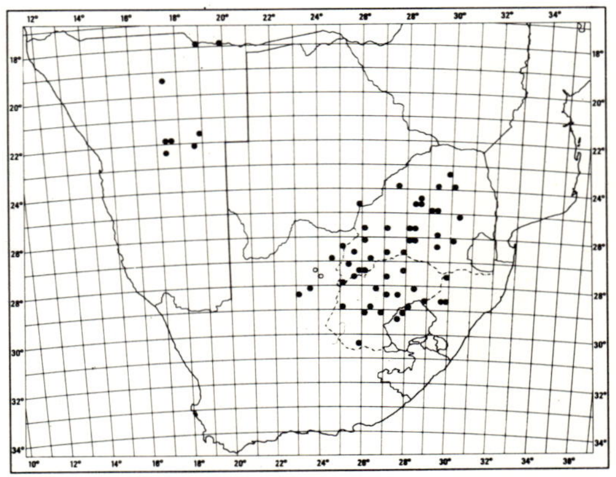

MAP 11.— ● **Gnaphalium filagopsis**
○ **Gnaphalium englerianum**

Widely distributed on the high interior plateau of Southern Africa from southern Angola and

S.W.A./Namibia in the west through Botswana to Zimbabwe in the east and south to the Transvaal, Orange Free State, the Tugela basin in Natal, western Lesotho and nothern Cape, always in wet or seasonally wet places; flowering between September and May. Map 11.

G. filagopsis is characterized by its narrow, acute, appressed-ascending leaves, heads with glabrous involucres and generally obtuse snow-white bract tips exceeding the flowers, and ♀ flowers usually epappose; it is mostly in Zimbabwe and S.W.A./Namibia that some heads may have a few bristles on the ♀ flowers.

Vouchers: *Dieterlen* 703 (PRE; SAM); *Flanagan* 1460 (BOL); *Giess* 9493 (M; WIND); *Hafström & Acocks* 1622 (PRE; S); *Wilms* 699 (BM; E; K; NU; PRE).

6. **Gnaphalium englerianum** (*O. Hoffm.*) *Hilliard & Burtt* in Bot. J. Linn. Soc. 82: 193 (1981); Hilliard in Bot. J. Linn. Soc. 82: 282 (1981). Type: Cape, Kachun, near Kuruman, 1 200 m, Feb. 1886, *Marloth* 1004 (E; PRE; SAM, iso.!.).

Amphidoxa engleriana O. Hoffm. in Bot. Jb. 10: 274 (1889).

G. englerianum is distinguished from *G. filagopsis* (no. 5) by its markedly spathulate leaves with silvery 'tissue-paper' indumentum, and by having fewer flowers in the heads, 45—50, 29—32 ♀, 16—19 ♀, in the proportion 2: 1. However, it is known from only the type collection and another from nearby (Hol River at Blesmanspost near Kuruman, *Acocks* 2485, PRE), and may prove to be no more than an extreme variant of *G. filagopsis*. Map 11.

7. **Gnaphalium declinatum** *L. f.*, Suppl. 365 (1781); Hilliard & Burtt in Bot. J. Linn. Soc. 82: 191 (1981); Hilliard in Bot. J. Linn. Soc. 82: 283 (1981). Type: Cape of Good Hope, *Thunberg* (LINN 989.80 holo.!).

Spiralepsis declinata (L. f.) D. Don in Loudon, Hort. Brit. 340 (1830). *Helichrysum declinatum* (L. f.) Less., Syn. Comp. 278 (1832); Harv. in F.C. 3: 218 (1865), p.p.

G. achilleoides Lam., Encycl. 2: 753 (1788). Type: Cape of Good Hope (P—LAM, holo.!).

G. pygmaeum Thunb., Prodr. 150 (1800), Fl. Cap. 652 (1823), non (L.) Lam. (1788). Type: Cape of Good Hope, *Thunberg* (sheet 19239, UPS holo.!).

Gnaphalium pentheri Gand. in Bull. Soc. Bot. Fr. 55: 42 (1918); Type: Cape, Stromvlei [Stormsvlei, W of Swellendam?], 20 x 1894, *Penther* 1137 (W: M: iso.!).

Stoloniferous mat-forming perennial herb, stems prostrate or decumbent, flowering branchlets mostly 30—100 mm long, simple or subsimple, leafy. *Leaves* imbricate ascending, mostly 4—10 × 1—3 mm, linear, linear-lingulate or narrowly elliptic, tips more or less acute, both surfaces silky-silvery. *Heads* c. 3—4 × 2,5 mm, sometimes solitary, usually several corymbosely arranged at the branch tips, individual peduncles often conspicuous but compound inflorescence generally closely surrounded by leaves. *Involucral bracts* in c. 4 series, outer short, golden-brown, glabrous, tips of inner lanceolate or sometimes oblong, acute or subacute, snow-white, exceeding the flowers by their length. *Receptacle* shallowly tuberculate. *Flowers* 57—100 (—190), 45—75 (—160) ♀, 12—34 ♀, in the proportion 2—3 (—5): 1. *Achenes* 0,5 mm long, with myxogenic duplex hairs. *Pappus* present on all flowers, scabrid on ♀, tips shortly plumose on ♀. Fig. 4:2.

Recorded from the southern Cape, between the coast and a line drawn from Grabouw (19°E) through Swellendam to Riversdale. Forms mats in damp places; flowering between October and February. Map 12.

G. declinatum is characterized by its flat silky-silvery imbricate ascending leaves, individual heads usually on conspicuous peduncles, outer involucral bracts golden-brown and glabrous, tips of inner bracts snow-white, more or less acute and at least twice as long as broad, and all the flowers with pappus. The name is frequently misapplied, especially to *G. austroafricanum*, *G. capense*, *G. gnaphalodes* and *G. filagopsis*.

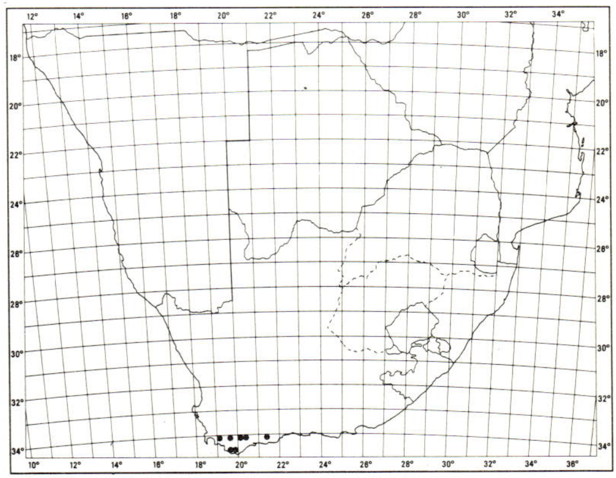

MAP 12.— **Gnaphalium declinatum**

Vouchers: *Bolus* 4152 (BM; BOL; K); *Compton* 22176 (BOL; NBG); *Rogers* 28943 (SAM); *Salter* 4091 (BM; BOL; K).

8. **Gnaphalium capense** *Hilliard* in Bot. J. Linn. Soc. 82: 284 (1981). Type: Cape, Klein Drakensteen und Dal Josaphat, unter 1000 Fuss, Nov. Dec. Jan., *Drège* (E, holo.!; BM; BOL; G—DC; P; PRE; S; TCD; iso.!). Distributed as *Helichrysum declinatum* Less. a.

Perennial herb, stems many from the crown, often c. 100—150 mm long, but up to 450 mm, simple or branched, suberect, decumbent or prostrate, greyish-white appressed woolly, densely leafy. *Leaves* ascending, mostly 6—20 × 1,5—3 mm, linear to narrowly oblanceolate, apex acute to obtuse, mucronate, base broad, half-clasping, both surfaces greyish-white appressed-woolly. *Heads* turbinate-campanulate, c. 3,5—4,5 × 2,5—3 mm, in small corymbose clusters at the branch tips, usually closely surrounded by leaves. *Involucral bracts* in c. 4 series, outer short, pale golden-brown or straw-coloured, lightly and loosely woolly on backs, inner bracts exceeding flowers by the length of their snow-white opaque tips, tips oblong, obtuse or rarely subacute, 2—3 (—4) times as long as broad, often recurved. *Receptacle* shallowly tuberculate. *Flowers* 68—143, 57—124 ♀, 9—20 ☿, in the proportion 3—7,5:1, yellow. *Achenes* 0,5 mm long with duplex myxogenic hairs. *Pappus* on ♀ flowers scabrid, on ☿ flowers shortly plumose at tips.

Recorded only from the Cape, where it ranges from about the Berg River (Piquetberg distr.) south through Ceres, Worcester, Caledon and Swellendam districts then east to Albany district, Somerset East district, the Amatola Mountains and the mountains near Queenstown, the coastal districts as far as King William's Town and inland on the mountains about Graaff-Reinet and Nelspoort. Grows in damp or seasonally damp places; flowering mainly between September and January. Map 13.

Characterized by having the corymbs closely surrounded by leaves, heads with pale outer bracts loosely woolly on the backs, inner bracts with snow-white, oblong obtuse tips 2—4 times as long as broad and often recurved, and pappus present on all flowers. Can be confused with *G. vestitum*, but that species has the inner bracts buff or dirty white and generally acute or subacute. The leaves in *G. vestitum* are usually more closely imbricate than in *G. capense*, with the mucro generally hidden in wool.

Vouchers: *Dyer* 2214 (K; PRE); *Flanagan* 2170 (NBG; PRE; SAM); *Galpin* 2614 (PRE); *Oliver* 5665 (PRE); *Rogers* 17847 (BM).

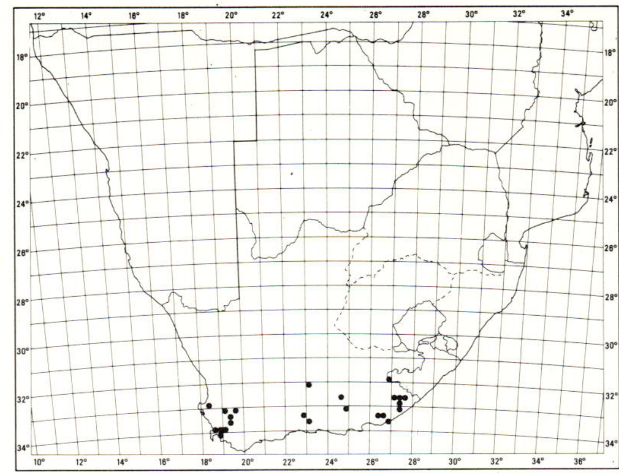

MAP 13.— **Gnaphalium capense**

9. **Gnaphalium vestitum** *Thunb.*, Prodr. 148 (1800), Fl. Cap. 647 (1823); Hilliard & Burtt in Bot. J. Linn. Soc. 82: 261 (1981); Hilliard in Bot. J. Linn. Soc. 82: 285 (1981). Type: Cape of Good Hope, *Thunberg* (sheet 19286, UPS, holo.!).

G. repens L. var. *vestitum* (Thunb.) DC., Prodr. 6: 230 (1838).

G. acilepis DC., Prodr. 6: 229 (1838). Lectotype: Albany, around Grahamstown, *Ecklon* 1191 (G-DC!).

G. repens sensu Thunb., Fl. Cap. 647 (1823); Less., Syn. Comp. 330 (1832); DC., Prodr. 6: 229 (1838); Harv. in F.C. 3: 261 (1965); non L.

Greyish-white woolly herb with tufts of erect or suberect stems c. 50—150 mm long and trailing stems up to 400 mm, all simple or branched, leafy. *Leaves* ascending, more or less appressed, 2—6 (—10) × 0,5—2 (—3) mm, linear or linear-spathulate, obtuse to subacute, mucronate but mucro generally hidden in wool, both surfaces greyish-white tomentose. *Heads* 2,5—3,5 × 2 mm, few to several crowded at the branch tips closely surrounded by leaves. *Involucral bracts* in c. 4 series, outermost short, woolly, tips of inner bracts lanceolate, about twice as long as broad, acute to subacute, sometimes somewhat squarrose, exceeding the flowers by their whole length, opaque dirty white or pale buff. *Receptacle* shallowly tuberculate. *Flowers* 35—81, 25—67 ♀, 5—14 ☿ in the proportion 2—6 (—7) :1. *Achenes* 0,5 mm long with myxogenic duplex hairs. *Pappus* present on all the flowers.

Confined to the S. and E. Cape, from Komgha, south of the Kei River, to the Tsitsikama Mountains near the coast, and inland to Grahamstown and Bruintjeshoogte near Somerset East. Favours damp places in grass or grass and shrub communities, and will behave as a weed; flowering recorded in most months. Map 14.

The small grey appressed leaves with the mucro often hidden by wool and small heads with dirty white or buff-coloured, mostly acute or subacute, almost squarrose involucral bracts, are diagnostic.

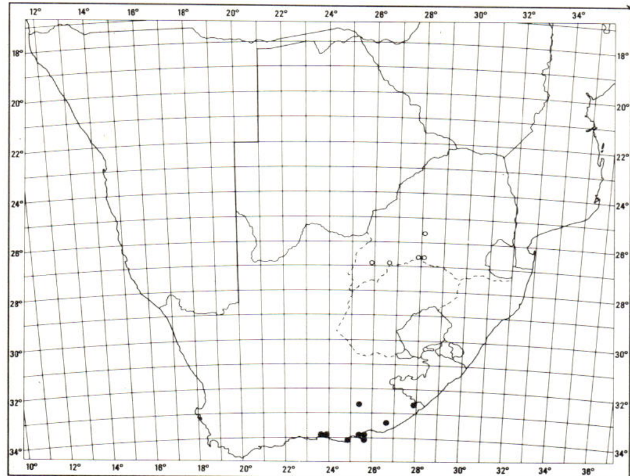

MAP 14.— ● **Gnaphalium vestitum**
        ○ **Gnaphalium nelsonii**

Vouchers: *Acocks* 20112 (M; PRE); *Esterhuysen* 10677 (BOL; NBG; PRE); *Hilliard & Burtt* 10903 (E; K; M; NU; PRE; S); *MacOwan* 1817 (P); *Story* 2270 (BM).

10. **Gnaphalium nelsonii** *Burtt Davy* in Jl S. Afr. Bot. 1: 107 (1935); Hilliard in Bot. J. Linn. Soc. 82: 286 (1981). Type: Transvaal, Pretoria distr., Van der Walt, in native mealie gardens, August, *Nelson* 289 (K, holo.!; PRE, iso.!).

White-woolly perennial herb, branches 40−200 mm long, erect or decumbent, simple or subsimple, many tufted from the crown, leafy. *Leaves* 6−16 × 1,5−2,5 mm, ascending, more or less appressed, linear-lanceolate, apex acute, mucronate, both surfaces white-tomentose. *Heads* c. 3 × 3 mm, several in congested corymbose clusters at the leafy branch tips. *Involucral bracts* in c. 4 series, outer straw-coloured, lightly cobwebby, tips of inner obtuse or subacute, opaque dirty white or white, scarcely exceeding the flowers. *Receptacle*

shallowly tuberculate. *Flowers* 56−118, 39−89 ♀, 17−29 ☿ in the proportion 2−3,5:1. *Achenes* not seen, ovaries with myxogenic duplex hairs. *Pappus* present on all flowers.

Seldom collected and known only from the SW. Transvaal, from Pretoria and Heidelberg west to Klerksdorp and Wolmaransstad. Favours seasonally wet places; flowering between October and December. The linear-lanceolate white-woolly leaves, the rather inconspicuous (less than 1 mm long) white or dirty white tips to the involucral bracts, and all the flowers with pappus are characteristic. The size and shape of the leaves and short, more or less obtuse, tips to the involucral bracts, which scarcely overtop the flowers, easily distinguish it from *G. vestitum*, the species it most closely resembles. Map 14.

Vouchers: *Botha* 711 (PRE); *Hanekom* 1628 (PRE; K); *Jacobsen* 1040 (PRE).

11. **Gnaphalium confine** *Harv.* in F.C. 3: 263 (1865); Hilliard in Bot. J. Linn. Soc. 82: 288 (1981). Type: Cape of Good Hope, Linde, *Zeyher* (S, holo.!).

G. drakensbergense Markötter in Ann. Univ. Stellenbosch 8A, 1: 45 (1930); Hilliard, Compositae in Natal 130 (1977). Lectotype: Orange Free State, Koolhoek, c. 6 000 ft, Oct. ? 1907, *Thode* in STE 3009!

G. declinatum sensu Merxm., F.S.W.A. 139: 83 (1967) quoad spec. *Merxmüller & Giess* 2275b, non L. f.

Tufted grey-woolly annual herb, stems decumbent or erect, up to c. 200 mm long, simple or sparingly branched, leafy. *Leaves* erect or spreading, mostly 8−30 × 2−4 mm, generally oblanceolate, the smaller ones linear-oblong, apex obtuse to subacute, mucronate. *Heads* c. 3 × 3−4 mm, few to several clustered at the branch tips surrounded by leaves, sometimes several clusters on short leafy branches corymbosely arranged. *Involucral bracts* in 4 series, outer pale brown, woolly outside, tips of inner acute or subacute, exceeding the flowers, opaque, dirty white or palest buff, sometimes tinged purple above the stereome. *Receptacle* shallowly tuberculate. *Flowers* 143−244, 139−240 ♀, 4−11 ☿, in the proportion 17−40 (−60) : 1. *Achenes* 0,5 mm long with myxogenic duplex hairs. *Pappus* on ♀ flowers scabrid, on ☿ flowers with subplumose tips. Fig. 4:4.

Widely distributed from S.W.A./Namibia through the N.Cape to the S. and SE. Transvaal, Orange Free State, Lesotho, Natal, Transkei, E. and central Cape. Grows in damp places such as river- or streambanks, ditches, seepage lines, even the drip line of Cave Sandstone overhangs. Can behave as a weed. Flowering recorded in every month, but mainly September to November. Map 15.

Readily confused with *G. polycaulon* (no. 13) but distinguished by the involucral bracts with opaque, not translucent, tips, and dimorphic pappus. The name *G. micranthum* Thunb. has frequently been misapplied to this species, but it belongs to *Vellereophyton dealbatum* (Thunb.) Hilliard & Burtt.

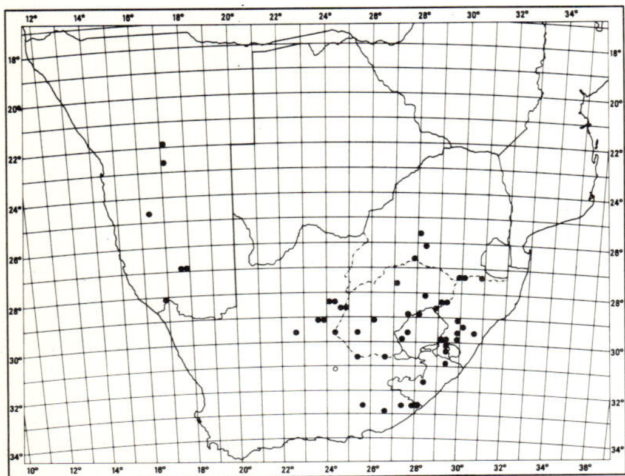

MAP 15.— ● **Gnaphalium confine**
       ○ **Gnaphalium simii**

Vouchers: *Acocks* 9050 (K; PRE); *Flanagan* 311 (NBG; PRE; SAM); *Hilliard & Burtt* 10743 (E; K; M; NU; PRE; S); *Merxmüller & Giess* 28872 (M; PRE; WIND); *Schmitz* 6211 (PRE).

12. **Gnaphalium simii** (*H. Bol.*) Hilliard & Burtt in Bot. J. Linn. Soc. 82: 193 (1981); Hilliard in Bot. J. Linn. Soc. 82: 289 (1981). Type: Cape, near Hanover, *Sim* 2863 (BOL, holo.!; E; K; NU; PRE; iso.!).

Helichrysum simii H. Bol. in Trans. S. Afr. phil. Soc. 18,3: 380 (1907).

Profusely branched mat-forming perennial herb, main branches prostrate, rooting, ultimate branchlets very short, up to 30 mm long, white-felted, leafy. *Leaves* mostly 10−25 × 3−4 mm, spathulate, subacute, much narrowed below to a petiole-like base, then expanded, half-clasping, both surfaces enveloped in thick, silky-woolly white tomentum. *Heads* c. 5 × 3 mm, solitary on nude white-woolly peduncles up to 20 mm long. *Involucral bracts* in 4 series, subequal, base of whole

involucre white-felted, only tips free, oblong, obtuse, opaque white. *Receptacle* shortly honeycombed. *Flowers* c. 25−31, 9−12 ♀, 16−19 ☿, in the proportion 1 : 1,5−2. *Achenes* 0,75 mm long with duplex hairs. *Pappus* bristles with scabrid tips, monomorphic. Fig. 4:3.

Known from only two collections made near Hanover in the E. central Cape. It was last collected in March 1952, when it was mostly in fruit, in a 'calcareous vlei, locally abundant, silvery mats up to 3′ diam.' (*Acocks* 16349, PRE). It is a most distinctive plant. Map 15.

13. **Gnaphalium polycaulon** *Pers.*, Syn. 2: 421 (1807); Grierson in Notes R. bot. Gdn Edinb. 31: 137 (1971); Hilliard, Compositae in Natal 126 (1977), Bot. J. Linn. Soc. 82: 289 (1981); Merxm. & Roessl. in Mitt. bot. St Samml., Münch. 15: 363 (1979). Type: India, *Roxburgh* in Herb. Willdenow (sheet 15500, B!).

G. multicaule Willd., Sp. Pl. 3,3: 1888 (1803), non Lam. (1789). Type as for *G. polycaulon*.

G. indicum sensu Humbert, Fl. Madag. 189: 382 (1962); Merxm., F.S.W.A. 139: 83 (1976), non L.

Annual herb, stems branched from the base, decumbent or erect, up to c. 250 mm long, thinly white-tomentose, leafy. *Leaves* up to 70 × 8 mm, but often much less than half that size, spathulate or oblanceolate, apex acute to obtuse, mucronate, both surfaces thinly grey-tomentose sometimes glabrescent. *Heads* c. 3 × 2,5 mm, in small stalked or sessile clusters racemosely arranged. *Involucral bracts* in 3 series, pellucid, pale buff, ageing palest brown, sometimes purplish around the stereome. *Receptacle* tuberculate. *Flowers* c. 150−225, 140−220 ♀, 4−6 ☿, in the proportion of at least 25 : 1. *Achenes* 0,5 mm long with duplex hairs. *Pappus* bristles of ♀ flowers scabridulous, of ☿ flowers with more or less barbellate tips.

A palaeotropical weed, ranging south through Africa to S.W.A./Namibia and the environs of the Orange River, the Bushveld of the northern and eastern Transvaal, and the Mozambique coastal plain, thence reaching Tongaland in NE. Natal. Also in Madagascar. Favours muddy or sandy streamsides and similar damp places; flowering recorded in most months. Map 16.

Can be confused with *G. confine* (no. 11).

Vouchers: *Codd* 7331 (P; PRE); *Giess & Wiss* 3255 (M); *Vahrmeijer* 823 (PRE); *Wilms* 724 (BM; K).

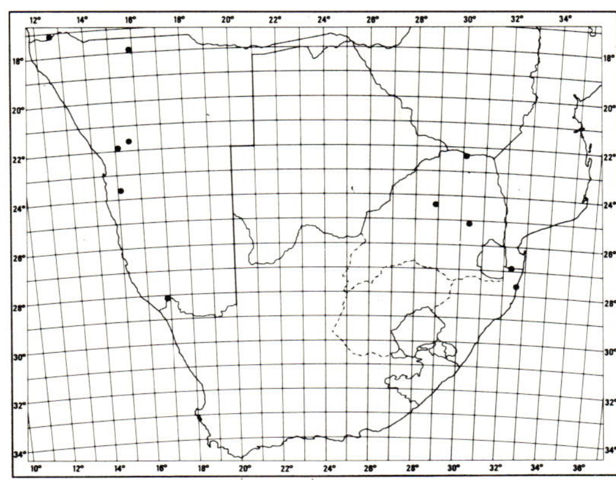

MAP 16.— **Gnaphalium polycaulon**

and then persistent, or dilations ill-developed and then bristles free and caducous; of ☿ flowers with subplumose tips, bases scarcely dilated, not cohering, soon caducous.

Endemic to the SW. Cape, from the Peninsula to Piquetberg, Ceres, Worcester, Caledon, Bredasdorp and Riversdale districts, on both mountains and flats; flowering in September and October. Map 17.

Without close allies and easily recognized by its generally narrow leaves and peculiar pappus.

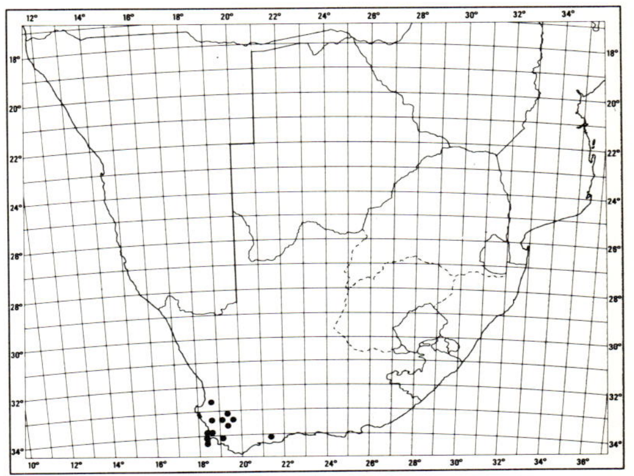

MAP 17.— **Gnaphalium pauciflorum**

Vouchers: *Esterhuysen* 17473 (BOL); *Hafström & Acocks* 1815 (PRE, S); *Hutchinson* 1059 (BOL; K); *Schlechter* 5207 (BOL; G; Z); *Wolley-Dod* 2825 (BM; BOL; K; M).

Section **Gamochaeta** (Wedd.) O. Hoffm.

Gamochaeta Wedd. Type species: *G. americanum* (Wedd.) Mill.

14. **Gnaphalium pauciflorum** *DC.*, Prodr. 6: 229 (1838); Harv. in F.C. 3: 262 (1965); Levyns in Adamson & Salter, Fl. Cape Penins. 777 (1950); Hilliard in Bot. J. Linn. Soc. 82: 290 (1981). Type: Cape, Bergplätze bei der Kapstadt, bis 2000', Sept., *Ecklon* 1036 (G−DC, holo.!; S; SAM, iso.!).

Small annual herb, stems 10−130 mm long, several from the crown, erect or decumbent, simple or branched, white-woolly, leafy. *Leaves* 5−15 (−20) × 1−1,5 (−3) mm, linear-spathulate, or rarely oblong-spathulate, acute, mucronate, both surfaces white-tomentose. *Heads* c. 3 × 1,5−2,5 mm, solitary or few opposite the upper leaves, generally much congested, forming small clusters overtopped by leaves, or, in well-grown specimens, racemosely arranged. *Involucral bracts* in 3 series, outermost elliptic, inner lanceolate, suba-cute to acuminate or occasionally truncate, lamina pellucid, straw-coloured or golden-brown tinged purple, tips exceeding flowers, buff-coloured, opaque or subopaque, often somewhat recurved when acuminate, usual-ly ± lacerate. *Receptacle* tuberculate. *Flowers* 20−94, 16−84 ♀, 4−10 ☿, in the proportion 3−10 : 1, all, or ☿ only, purple-tipped. *Achenes* 0,75 mm long, with duplex hairs. *Pappus* bristles on ♀ flowers scabridulous, bases either dilated and fused

15. **Gnaphalium subfalcatum** *Cabrera* in Revta Mus. La Plata, Bot. 4: 174 (1941); Hilliard, Compositae in Natal 125 (1977). Type: Argentina, Maciel, *Cabrera* 944 (LP).

Gamochaeta subfalcata (Cabrera) Cabrera in Boln Soc. argent. Bot. 9: 383 (1961), in Fl. Prov. Buenos Aires, Comp. 177 (1963); Holub in Fl. Europ. 4: 127 (1976).

Annual herb, branched from the base, stems decumbent, then erect to 350 mm, thinly grey-green silky-woolly, leafy. *Lower leaves* 15−50 × 3−12 mm, spathulate, upper smaller, linear-lanceolate, often folded, tips recurved, apiculate, both surfaces grey-green silky-woolly. *Heads* c. 3 × 3 mm, in small woolly glomerules racemosely arranged at the branch tips. *Involucral bracts* in c. 3 series, pellucid, tips tinged red-purple at first, palest brown later. *Flowers* c. 100 ♀, 4 ☿, white, tipped red-purple. *Achenes* 0,5 mm long, minutely hairy, myxogenic. *Pappus* bristles many, scabrid, bases fused in a smooth ring.

A native of Argentina, naturalized in various parts of the world. In South Africa, recorded from the Cape, including the Peninsula, Natal and Transvaal, in cultivated places and other disturbed ground. First recorded in 1896 (Claremont, Cape).

Vouchers: *Acocks* 23486 (NBG; PRE); *Dyer* 4149 (NU); *Esterhuysen* 10687 (BOL; NBG; PRE); *Hilliard* 5156 (E; K; NU; PRE; S).

16. **Gnaphalium calviceps** (errore *calvescens*) *Fernald* in Rhodora 37: 449 t. 405 (1935); Hilliard, Compositae in Natal 126 (1977). Type described from Virginia.

Gamochaeta calviceps (Fernald) Cabrera in Boln Soc. argent. Bot. 9: 368 (1961), Fl. Prov. Buenos Aires, Comp. 178 (1963).

Annual herb, stems simple or sub-simple, erect, up to c. 300 mm tall, thinly grey silky-woolly, leafy throughout. *Leaves* up to 47 × 7 mm, becoming smaller upwards, oblanceolate, uppermost oblong-lanceolate, passing into long, narrow inflorescence bracts, both surfaces closely grey silky-woolly. *Heads* c. 3 × 2 mm, in small glomerules or short spikes racemosely arranged. *Involucral bracts* in c. 3 series, pellucid, tips tinged light brown, woolly at the base. *Flowers* c. 120 ♀, 3 ☿, whitish. *Achenes* 0,5 mm long, minutely hairy, myxogenic. *Pappus* bristles many, scabridulous, bases fused in a smooth ring.

A native of the United States; in South Africa, recorded only from Ndumu on the Natal-Mozambique border, growing on the side of a track through woodland (1973).

Voucher: *Pooley* 6865 (E; NU; PRE).

17. **Gnaphalium pensylvanicum** *Willd.*, Enum. Hort. Berol. 867 (1809); Grierson in Notes R. bot. Gdn Edinb. 31: 137 (1971); Hilliard, Compositae in Natal 124 (1977). Type: Cult. Jard. Bot. Paris (P-LAM, holo.!).

G. spathulatum Lam., Encycl. 2: 758 (1788), nom. illegit., non *G. spathulatum* Burm. f. (1768). *Gamochaeta pensylvanica* (Willd.) Cabrera in Boln Soc. argent., Bot. 9: 375 (1961), Fl. Prov. Buenos Aires, Comp. 175, fig. 48 (1963).

Gnaphalium peregrinum Fernald in Rhodora 45: 479, t. 795 (1943). Type: Louisiana, *D.S. & H. B. Correll* 9937 (GH).

G. purpureum sensu Hook. f., Fl. Br. Ind. 3: 289 (1881); non L.

Soft herb, branched from the base, probably annual, all parts loosely white-woolly, stems to about 400 mm, often the central one erect, the laterals decumbent then erect, rooting where they touch the ground. *Leaves* up to c. 80 × 20 mm, narrowly spathulate, decreasing in size upwards and passing into inflorescence bracts. *Heads* c. 3 × 2 mm, in small axillary glomerules racemosely arranged. *Involucral bracts* in c. 3 series, innermost about equalling the flowers, pellucid, palest brown or buff. *Flowers* c. 80−100 ♀, 2−3 ☿, whitish, tipped red-purple. *Achenes* 0,5 mm long, minutely hairy, myxogenic. *Pappus* bristles many, scabrid, bases fused in a smooth ring.

A widespread weed in the warmer parts of the world; in Southern Africa, recorded from the Transvaal, Natal, Transkei and E. Cape, a weed in gardens and other damp places. Recorded before 1865 by Gerrard in Zululand.

Vouchers: *Dyer* 4152 (NU); *Hilliard & Burtt* 6947 (E; K; NU; S); *Pegler* 2070 p.p. (PRE); *Wood* 95 (SAM).

18. **Gnaphalium coarctatum** *Willd.*, Sp. Pl. 3: 1886 (1804). Type: Uruguay, Montevideo, *Commerson* (P−LAM, holo.!).

G. spicatum Lam., Encycl. 2: 757 (1788), non Mill. (1768); Grierson in Notes R. bot. Gdn Edinb. 31: 138 (1971); Hilliard, Compositae in Natal 124 (1977). *Gamochaeta spicata* Cabrera in Boln. Soc. argent. Bot. 9: 380 (1961), Fl. Prov. Buenos Aires Comp. 174 (1963). Type as for *G. coarctatum*.

Annual or biennial herb with a basal leaf rosette and several radiating usually strongly decumbent lateral leafy flowering branches up to 500 mm tall. *Leaves* up to 120 × 20 mm, narrowly spathulate, becoming progressively smaller upwards and passing into inflorescence bracts, markedly discolorous, green and glabrescent above (often drying brown), white-felted below. *Heads* c. 4 × 2 mm, oblong, in small glomerules or spikes arranged in a long racemose inflorescence. *Involucral bracts* in c. 3 series, innermost about equalling the flowers, pellucid, tips reddish purple at first, later golden-brown. *Flowers* c. 80−100 ♀, 2−3 ☿, whitish, tipped reddish purple. *Achenes* c. 0,5 mm long, minutely hairy, myxogenic. *Pappus* bristles many, scabrid, bases fused in a smooth ring.

A native of South America, now naturalized in several parts of the world; in Southern Africa recorded from the Cape, Transvaal, Natal and Swaziland. Favours damp places and sometimes a troublesome weed in lawns. Easily recognized by its strongly discolorous leaves. Originally recorded in the Cape Peninsula in 1897.

Vouchers: *Bolus* 10751 (BOL); *Codd* 1366 (NU; PRE); *Compton* 25694 (NBG); *Hilliard & Burtt* 10858 (E; K; MO; NU; S).

19. **Gnaphalium sp.**

Small annual herb, stems 50−120 mm long, erect or decumbent, simple or shortly branched above, solitary or tufted from the base and then sparingly branched below, loosely greyish-white woolly, distantly or more closely leafy. *Leaves* mostly 10−25 × 1−5 mm, smallest under the heads, oblanceolate or narrowly spathulate, apex obtuse to acute, mucronate, much narrowed below to a flat petiolar part, both surfaces thinly greyish-white woolly. *Heads* 2,5−3 × 1,5−2 mm, c. 2−6 in small glomerules terminating the branchlets. *Involucral bracts* in c. 3 series, pellucid, palest buff. *Flowers* c. 69−86, 66−84 ♀, 2−3 ☿, whitish. *Achenes* 0,5 mm long, minutely hairy, myxogenic. *Pappus* bristles many, scabridulous, bases fused in a smooth ring.

Known from 3 collections: *Compton* 25348 (NBG; PRE) from a hill NE. of Mbabane, Swaziland; *Gibbs Russell* 3126 A (PRE), Fort Hare, Ciskei; *Hilliard & Burtt* 10524 A (NU), Pillar Cave, Garden Castle Forest Reserve, Underberg district, Natal.

The plants are introductions from America, and I failed to match them at Kew. Two different species may be involved: the Compton and Gibbs Russell specimens have erect, distantly leafy, solitary stems, unbranched except near the tip in the Gibbs Russell specimen, which was collected on a 'dry hillside facing east'. The Hilliard & Burtt specimen has decumbent, sparingly branched and closely leafy stems tufted from the base, and came from bare wet earth along the dripline of a Cave Sandstone overhang.

Dr Cabrera, Instituto de botanica Darwinion, Argentina, kindly examined the specimens and wrote 'number 10524A resembles the dwarf forms of *G.* [*amochaeta*] *spiciformis* (Sch. Bip.) Cabrera from southern Patagonia and Tierra del Fuego, but it is impossible for me to establish its identity without doubt'.

Excluded species

Amphidoxa glandulosa Klatt = Denekia capensis Thunb.

A. lasiocephala O. Hoffm. = Artemisiopsis villosa (*O. Hoffm.*) *Schweick.*

A villosa O. Hoffm. = Artemisiopsis villosa (*O. Hoffm.*) *Schweick.*

Gnaphalium acuminatum Link, sp. dubia, no specimen traced.

G. cicatrisatum Vahl, sp. dubia, no specimen traced.

G. denudatum L. f., sp. dubia, no specimen traced.

G. lasiocaulon Link, sp. dubia, no specimen traced.

G. tephrodes Link, sp. dubia, no specimen traced.

8992a   **VELLEREOPHYTON**

Vellereophyton *Hilliard & Burtt* in Bot. J. Linn. Soc. 82: 210 (1981). Type species: *V. dealbatum* (Thunb.) Hilliard & Burtt.

Grey- or white-woolly annual or perennial herbs, stems few to many from the crown, prostrate, ascending or erect, simple or branched, leafy. *Leaves* small to medium sized, oblanceolate or spathulate, apex subacute or obtuse, base much narrowed, petiole-like, margins flat, both surfaces thinly or thickly grey- or white-woolly. *Heads* heterogamous, or homogamous in one species, very small, few or many in woolly glomerules subcorymbosely arranged at the branchlet tips. *Involucral bracts* in (2−) 3−5 series, backs woolly, inner bracts subequal, about equalling or exceeding the flowers, stereome fenestrate in upper half, lower part concave, embracing adjacent flower, lamina thin, pellucid, sometimes purplish above stereome, tip opaque white, obtuse or rounded, radiating. *Receptacle* nearly smooth. *Flowers* 10−40, yellow, often tipped red, ♀ flowers (0−) 1−30, either fewer than, about equalling, or up to 10 times as many as ♂, narrowly tubular; ♂ flowers tubular, scarcely widened above with 5 deltoid lobes markedly mamillate on outer surface except in one species, glandular hairs with biseriate stalk and small 2-celled globular head on backs of all lobes. *Anthers* with small obtuse apical appendage and tails about equalling or exceeding the filament collar. *Style branches* truncate and penicillate. *Achene* wall smooth or nearly so or the cells imbricate, oblong duplex hairs with swelling cushion frequently present, often myxogenic. *Pappus* bristles subplumose in upper part with clavate hairs, delicate patent cilia at base, cohering or not.

A genus of 7 species endemic to the Cape; one species naturalized in Australia and New Zealand.

1a Heads heterogamous:

  2a Heads up to 3 × 2,5 mm:

    3a Female flowers outnumbering the hermaphrodite .... 1. *V. dealbatum*

    3b Female flowers fewer than hermaphrodite:

      4a Female flowers at least 4, all the ovaries hairy:

        5a Flowers less than 20 in a head; cilia at tips of pappus bristles white ....2. *V. gracillimum*

        5b Flowers at least 20 in a head; cilia at tips of pappus bristles colourless:

          6a Plants more or less erect, loosely branched; hermaphrodite flowers 2−3 times as many as female...3. *V. niveum*

          6b Plants prostrate, much-branched and cushion-forming; up to half as many more hermaphrodite flowers than female .... 5. *V. pulvinatum*

      4b Female flowers 1 or 2, ovaries glabrous; ovaries of hermaphrodite flowers hairy .... 4. *V. felinum*

  2b Heads c. 4 × 3 mm ....6. *V. vellereum*

1b Heads homogamous ....7. *V. lasianthum*

1. **Vellereophyton dealbatum** *(Thunb.)* *Hilliard & Burtt* in Bot. J. Linn. Soc. 82: 210 (1981). Type: Cape, in damp depressions around Cape Town, April, *Thunberg* (sheet 19125, UPS, holo.!).

Gnaphalium dealbatum Thunb., Prodr. 149 (1800), Fl. Cap. 652 (1823).

G. candidissimum Lam., Encycl. 2: 754 (1788); Less., Syn. Comp. 328 (1832); DC., Prodr. 6: 229 (1838); Harv. in F.C. 3: 261 (1865); Levyns in Adamson & Salter, Fl. Cape Penins. 776 (1950), nom. illegit.

G. maculatum Thunb., Prodr. 149 (1800), Fl. Cap. 651 (1823). Type: Cape of Good Hope, *Thunberg* (sheet 19191, UPS, holo.!).

G. micranthum Thunb., Prodr. 149 (1800), Fl. Cap. 651 (1823); Harv. in F.C. 3: 261 (1865). Lectotype: Cape of Good Hope, *Thunberg* (sheet 19197, UPS!).

G. carroënse Schrank in Denkschr. K. Akad. Wiss. Münch. 8: 159 (1824). Type: Cape of Good Hope, *Brehm* s.n. (M, holo.! mixed with *Pseudognaphalium luteo-album;* S, iso.!).

White-woolly herb, root possibly perennial, stems tufted from the crown, very variable in length, mostly 30−450 mm,

2a
2
3
3a
3b
3c
4
4a
4b
4d
4c
1
1a
1b
1c
RHONA
COLLETT
1d
3d

prostrate to weakly erect, subsimple to well-branched, dense or lax, woolly, leafy. *Leaves* mostly 10−45 × 2−6 mm decreasing in size upwards, oblanceolate or narrowly spathulate, subacute to obtuse, base much attenuated, petiole-like in lower leaves, both surfaces white silky-woolly, wool sometimes thin. *Heads* heterogamous, narrowly campanulate, 2,5−2,75 (−3) × 1,75−2 mm, many in dense very woolly glomerules 5−10 mm across, subcorymbosely arranged at the branch tips. *Involucral bracts* in 3 series, backs woolly, inner bracts subequal, equalling flowers, stereome purplish red in upper part, tips oblong, obtuse, opaque white, radiating, eventually whole bract spreading horizontally. *Receptacle* nearly smooth. *Flowers* 16−41, 10−34 ♀, 2−10 ☿, yellow, tipped red-purple. *Achenes* 0,5 mm long, with myxogenic duplex hairs. *Pappus* bristles many, equalling corolla, tips subplumose, bases with patent cilia, cohering lightly. Fig. 5: 4.

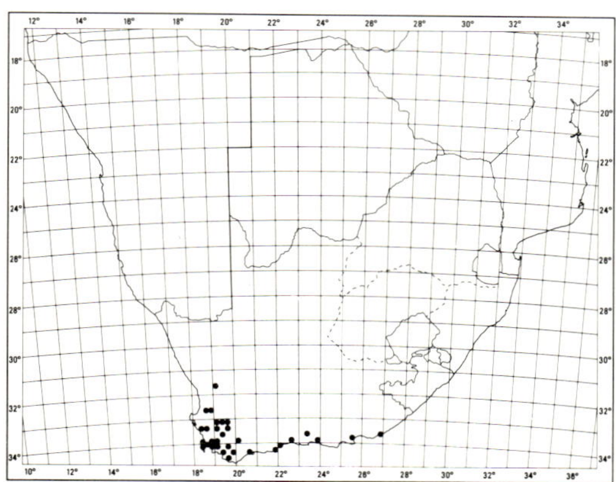

MAP 18.— **Vellereophyton dealbatum**

From about Kamieskroon in Namaqualand and Lokenburg, Calvinia, S. through the SW. districts and E. through the coastal districts to Alexandria, on the flats and in the mountains from sea level to c. 750 m. Grows in damp, often sandy, places such as streambeds, slacks in sand dunes, marshes, seepages, even margins of salt pans. Flowers between July and March. Adventive in New Zealand, Tasmania, S. Australia and Victoria. Map 18.

Very variable in stature but easily recognized.

Vouchers: *Acocks* 17559 (PRE); *Compton* 15038 (NBG); *Esterhuysen* 441 (NBG; PRE); *MacOwan* 963 (BOL; SAM); *Tyson* 818 (PRE; SAM).

2. **Vellereophyton gracillimum** *Hilliard, sp. nov.* V. dealbato (*Thunb.*) Hilliard & Burtt affinis sed involucri bracteis supra stereomate haud rubro-notatis, floribus femineis paucioribus (5−6, nec 10−34) in proportione 1 ♀ : 2 ☿ (nec 2−10 ♀ : 1 ☿).

Herba annua albo-lanata; caules 40−50 mm longi, filiformes, verosimiliter primum decumbentes demum erecti, solitarii vel plures, simplices vel subsimplices, albo-lanati, remote foliati. Folia plerumque 5−10 × 2−4 mm, spatulata, apice obtusa vel subacuta apiculata, basi multo angustata, utrinque albo-pannoso-lanata. Capitula heterogama, campanulata, 2,5 × 1,5 mm, sessilia vel breviter pedunculata, pluria subcorymbose aggregata ad apices ramulorum superiorum. Involucri bracteae 3-seriatae, exteriores breves, interiores subaequales dorso lanatae, apicibus ovatis obtusissimis opacis albis flores superantibus radiantibus. Receptaculum fere laeve. Flores 15−17, 4−6 feminei, 10−12 hermaphroditi, flavi rubro-purpureo-apiculati. Achaenia 0,5 mm longa, pubescentia, pilis verosimiliter non myxogenis. Pappi setae plures, corollam aequantes, superne subplumosae, basi pilis patentibus non cohaerentibus.

Type: Cape, Calvinia distr., Lokenburg, c. 2 100 ft., 26 ix 1953, arid fynbos of sandy slopes, abundant in small patches, under bushes in wet places, *Acocks* 17255 (PRE, holo.!).

White-woolly annual herb, stems 10−50 mm long, filiform, probably decumbent then erect, solitary, or several from the root, simple or subsimple, white-woolly,

---

FIG. 5.−1, **Vellereophyton vellereum,** part of plant, × 1,3; 1a, hermaphrodite flower, × 13; 1b, female flower, × 13; 1c, pappus bristle, × 20; 1d, head, × 6,6 (*Hilliard & Burtt* 11109). 2, **V. gracillimum,** whole plant, × 1,3; 2a, head, × 10 (*Acocks* 17255). 3, **V. felinum,** part of plant, × 1; 3a, pappus bristle, × 20; 3b, hermaphrodite flower, × 16; 3c, female flower, × 16; 3d, head, × 13 (*Taylor* 6113). 4, **V. dealbatum,** part of plant, × 1,3; 4a, head, × 13; 4b, hermaphrodite flower, × 17; 4c, female flower, × 17; 4d, pappus bristle, × 20 (*Hilliard & Burtt* 13077).

distantly leafy. *Leaves* mostly 5−10 × 2−4 mm, spathulate, apex obtuse to subacute, apiculate, base much narrowed, both surfaces white woolly-felted. *Heads* heterogamous, campanulate, 2,5−3 × 1,5 mm, sessile or shortly pedunculate, several together subcorymbosely arranged at the tips of the apical branchlets. *Involucral bracts* in 3 series, outer short, inner subequal, backs woolly, tips ovate, very obtuse, opaque white, exceeding flowers, radiating. *Receptacle* nearly smooth. *Flowers* 15−17, 4−6 ♀, 10−12 ☿, yellow, tipped red-purple. *Achenes* 0,5 mm long, hairy, probably not myxogenic. *Pappus* bristles several, equalling corolla, subplumose above, the cilia white, bases with patent cilia, not cohering. Fig. 5: 2.

Known only from the type collection and one other, from Suurvlakte in the Cold Bokkeveld, in damp and partly shaded places; flowering in September. Map 19.

Can be confused with *V. niveum* (below), which is a coarser plant with longer, narrower leaves and translucent, not opaque white, cilia at the tips of the pappus bristles.

Voucher: *Esterhuysen 33950* (BOL).

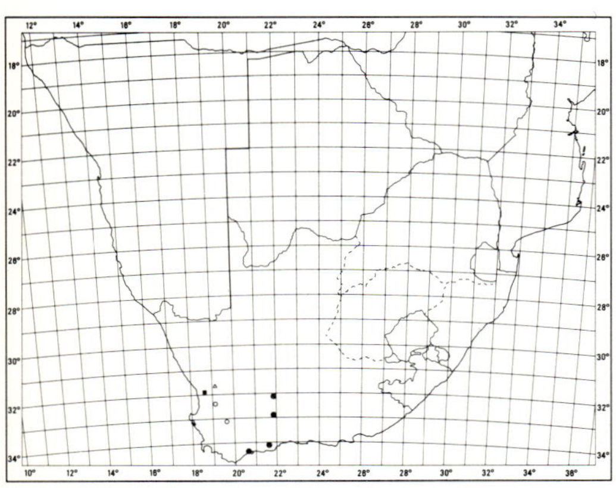

MAP 19.— △ **Vellereophyton gracillimum**
● **Vellereophyton niveum**
○ **Vellereophyton felinum**
■ **Vellereophyton pulvinatum**

3. **Vellereophyton niveum** *Hilliard, sp. nov. a* V. *dealbato (Thunb.) Hilliard & Burtt capitulis late campanulatis 2,5 mm latis* (nec anguste campanulatis 2 mm latis), bracteis involucri apicibus late ovatis vel suborbicularibus (nec oblongis), floribus femineis hermaphroditos multo superantibus (nec vice versa).

Herba albo-lanata, radice fortasse perenni; caules e caudice caespitosi, longitudine variabiles plerumque 30−450 mm, prostrati vel infirme erecti, subsimplices vel laxe ramosi, lanati, foliati. Folia plerumque 6−35 (−50) × 2−5 mm, sursum decrescentia, oblanceolata vel anguste spatulata, apice obtuso vel subacuto, ad basin multo attenuata praecipue in foliis inferioribus petiolaria, utrinque tenuiter albo-lanata. Capitula heterogama, campanulata, 2,5−3 × 2−2,5 mm, in glomerulos parvos 5−10 mm diametro ad ramulorum apices subcorymbose disposita. Involucri bracteae c. 5-seriatae, dorso lanatae, exteriores breviores, interiores subaequales flores paulo superantes, interdum supra stereomate rubro-purpureae, demum horizontaliter patentia. Receptaculum fere laeve. Flores 20−37, 5−13 feminei, 15−24 hermaphroditi, flavi, roseo-apiculati. Achaenia 0,5 mm longa, pilis duplicibus myxogenis. Pappi setae numerosae, corollam aequantes, apicibus subplumosis, basi pilis patentibus vix inter se cohaerentes.

Type: Cape, Prince Albert distr., 2 miles S. of Kruidfontein station, c. 1 800 ft. bed of Gamka River, occasional, white *not* silvery, 11 xi 1963, *Acocks* 23437 (PRE, holo.!; K, iso.!).

Fraserburg district, Layton (Vergenoegd) c. 3 200 ft, 8 xii 1964, Karroid Broken Veld, Koekemoer River, rare *not* silvery, *Acocks* 23536 (PRE); ibid., sand dam in Rietvlei, 3 000 ft, 25 ix 1967, *Shearing* 140 (PRE). Riversdale district, near Albertinia, 150 ft., growing in sand, Jan. 1915, *Muir* 1931 (PRE); ibid., 30 i 1951, *Compton* 22592 (NBG). Bredasdorp district, Potberg, SE. of main peak, 91 m, 2 i 1971, sandy areas around pan, *Thompson* 1141 (PRE).

White-woolly herb, root possibly perennial, stems tufted from the crown, very variable in length, mostly 30−450 mm, prostrate to weakly erect, subsimple to laxly branched, woolly, leafy. *Leaves* mostly 6−35 (−50) × 2−5 mm, smaller upwards, oblanceolate or narrowly spathulate, apex obtuse to subacute, base much attenuated, petiole-like especially in lower leaves, both surfaces thinly white woolly. *Heads* heterogamous, campanulate, 2,5−3 × 2−2,5 mm, in small glomerules 5−10 mm across, subcorymbosely arranged at the branch tips. *Involucral bracts* in c. 5 series, backs woolly, outer shorter, inner subequal, slightly exceeding flowers, sometimes purplish red above stereome, tips broadly ovate to suborbicular, very obtuse, opaque white, radiating, eventually whole bract rotate. *Receptacle* nearly smooth. *Flowers* 20−37, 5−13 ♀, 15−24 ⚥, yellow, tipped pink. *Achenes* 0,5 mm long, with duplex hairs. *Pappus* bristles many, equalling corolla, tips subplumose, bases with patent cilia, scarcely cohering.

Known from the environs of Vergenoegd and Kruidfontein (3221 Merweville), Potteberg near Bredasdorp and Albertinia near Riversdale, from near sea level to c. 975 m, growing in damp sand in river beds or around pans. Flowering recorded between September and January. Map 19.

Can be confused with *V. dealbatum* (no. 1) and *Helichrysum indicum* (p. 7, 2: 180); distinguished from *V. dealbatum* by its broader heads in which ⚥ flowers outnumber ♀, and by the involucral bracts, not or scarcely coloured red about the stereome and with white tips distinctly broader than the shaft; distinguished from *H. indicum* by its heterogamous heads (in *H. indicum* the heads are generally homogamous; only very rarely are 1 or 2 ♀ flowers present) and smooth, not crisped, involucral bracts.

4. **Vellereophyton felinum** *Hilliard, sp. nov. a* V. *dealbato (Thunb.) Hilliard & Burtt floribus femineis paucioribus 1 −2 (nec 10 −34), hermaphroditis femineis numerosioribus (nec paucioribus), apicibus bractearum involucralium late ovatis vel subrotundis (nec oblongis), achaeniis dimorphis distinguenda.*

Herba griseo-lanata, verosimiliter perennis; rami 50 −150 mm longi, decumbentes, inferne radicantes, superne lanati, crebre foliati. Folia plerumque 10 −30 × 3 −10 mm, sursum decrescentia, spatulata, apice obtuso vel subacuto apiculato, basi multo angustata, utrinque griseo-lanato-pannosa. Capitula heterogama, campanulata, c. 3 × 2 mm, pluria in glomerulos parvos 5 −10 mm diam. solitarios vel subcorymbose ad ramulorum apices disposita, foliis circumcincta. Involucri bracteae 3-seriatae, dorso lanatae, exteriores breves, interiores subaequales, inferne rubro-purpureae, apicibus late ovatis vel subrotundis obtusissimis opacis albis flores superantibus radiantibus. Receptaculum fere laeve. Flores 11 −13, 1 −2 feminei,

9 −11 hermaphroditi, flavi rubro-apiculati. Achaenia dimorpha, florum femineorum 0,75 mm longa glabra, hermaphroditorum 0,5 mm longa pilis duplicibus verosimiliter haud myxogenis induta. Pappi setae numerosae, corollam aequantes, apicibus subplumosis, basi pilis patentibus leviter inter se cohaerentibus.

Type: Cape, Ceres distr., Swartruggens, 4 000 ft., 2 xii 1964, arid fynbos on the highest parts of the level sandy plateau and broken, bouldery slope 5,4 miles along the farm road going N. from the summit of Katbakkies Pass, on deep orange sand, *Taylor* 6113 (PRE, holo.!; STE, iso.!).

Grey-woolly herb, probably perennial, branches 5−150 mm long, decumbent, rooting in lowermost part, woolly above, closely leafy. *Leaves* mostly 10−30 × 3−10 mm, diminishing upwards, spathulate, apex obtuse to subacute, apiculate, base much narrowed, both surfaces grey woolly-felted. *Heads* heterogamous, campanulate, c. 3 × 2 mm, several in small glomerules 5−10 mm across, solitary or subcorymbosely arranged at the branch tips, surrounded by reduced leaves. *Involucral bracts* in 3 series, backs woolly, outermost short, inner subequal, shaft purple-red, tips broadly ovate to subrotund, very obtuse, opaque white, exceeding flowers, radiating. *Receptacle* nearly smooth. *Flowers* c. 11−13, 1−2 ♀, 9−11 ⚥, yellow, tipped red. *Achenes* dimorphic, of ♀ flowers 0,75 mm long, glabrous, of ⚥ flowers 0,5 mm long, with duplex hairs, probably not myxogenic. *Pappus* bristles many, equalling corolla, tips subplumose, bases with patent cilia, cohering lightly. Fig. 5: 3.

Known only from the type collection. Closely allied to *V. niveum* (no. 3), but distinguished by its grey, not white, indumentum, heads containing c. 11−13 flowers, not 20−27, with only 1−2 ♀ flowers, not 5−13, and the achenes of *V. felinum* are dimorphic. Further collecting is needed to test the validity of these apparent distinctions. Map 19.

5. **Vellereophyton pulvinatum** *Hilliard, sp. nov. a* V. *dealbato (Thunb.) Hilliard & Burtt floribus hermaphroditis femineos paulo excedentibus (nec femineis hermaphroditos in proportione 2:1 excedentibus).*

Herba griseo-albo-lanata, verosimiliter perennis; caules e caudice multi, multiramo-

si, prostrati, tegetes densos c. 100 mm diam. formantes, lanati, foliati. Folia plerumque 5—10 × 1,5—2,5 mm, circum dimidio longitudinis petiolari, oblanceolata vel spatulata, ad apicem et ad basin attenuata, utrinque griseo-albo-lanata. Capitula heterogama, campanulata, 2,5 × 2 mm, pluria in glomerulos subcorymbosos ad ramulorum apices disposita. Involucri bracteae 3-seriatae, dorso lanatae, exteriores breviores, interiores subaequales flores superantes, interdum supra stereomate pallide purpureo-tincta, apicibus oblongis obtusissimis opacis albis radiantibus, demum bracteae totae horizontaliter patentes. Receptaculum fere laeve. Flores 26 — 30, 11 — 15 feminei, 17 — 21 hermaphroditi. Achaenia 0,5, pilis duplicibus myxogenis induta. Pappi setae multae, corollam circum aequantes, apicibus subplumosis, basi pilis patentibus sed vix inter se cohaerentes.

Type: Cape, Clanwilliam distr., along railway line near station at Klaver, c. 150 ft, not too frequent, stems prostrate from a perennial rootstock, forming cushion growths, 13 iii 1926, *Smith* 2602 A (PRE, holo.!).

Greyish-white woolly herb, probably perennial, stems many from the crown, much branched, prostrate, forming dense mats c. 100 mm in diam., woolly, leafy. *Leaves* mostly 5—10 × 1,5—2,5 mm, roughly half the length petiolar, oblanceolate or spathulate, tapering at both ends, both surfaces greyish-white woolly. *Heads* heterogamous, campanulate, 2,5 × 2 mm, several in small subcorymbose clusters at the branch tips. *Involucral bracts* in 3 series, backs woolly, outer shorter, inner subequal, exceeding flowers, sometimes with faint purplish red tinge above stereome, tips oblong, very obtuse, opaque white, radiating, eventually entire bract spreading horizontally. *Receptacle* nearly smooth. *Flowers* c. 26—30, 11—15 ♀, 17—21 ☿. *Achenes* 0,5 mm long with myxogenic duplex hairs. *Pappus* bristles many, about equalling corolla, tips subplumose, bases with patent cilia, scarcely cohering.

Known only from the type collection. Easily recognized by its cushion growth form and heads in which the ☿ flowers only slightly outnumber the ♀. Map 19.

6. **Vellereophyton vellereum** *(R.A. Dyer) Hilliard* in Bot. J. Linn. Soc. 82: 211 (1981). Type: Cape, Bathurst div., sand dunes near mouth of Great Fish River, *MacOwan* 1441 (K, holo.!; BM; BOL; PRE; SAM; iso.!).

Helichrysum vellereum R.A. Dyer in Kew Bull. 1934: 266 (1934).

Soft-wooded perennial herb, stems up to 600 mm long, decumbent or ascending, often rooting near the base, branched, branches silky white-felted, leafy. *Leaves* mostly 25—40 × 5—10 mm, diminishing upwards, oblong-spathulate or oblanceolate, apex obtuse, base much narrowed, sessile, both surfaces white silky-woolly-felted. *Heads* heterogamous, campanulate, 4 × 3 mm, many in small woolly-felted cymose clusters at the branch tips, surrounded by a few small leaves. *Involucral bracts* in 4 series, slightly graded, outer webbed together with white wool, inner slightly exceeding the flowers, tips rounded, smooth, opaque milk-white, minutely radiating. *Receptacle* nearly smooth. *Flowers* 25—40, 8—14 ☿, 16—26 ♀. *Achenes* 0,75 mm long, glabrous. *Pappus* bristles many, equalling corolla, subplumose in upper half, bases cohering lightly by patent cilia. Fig. 5: 1.

Found along the southern Cape coast from Humansdorp district, west of Port Elizabeth, to the environs of East London, on the beach, in dune slacks; flowering between November and January. Map 20.

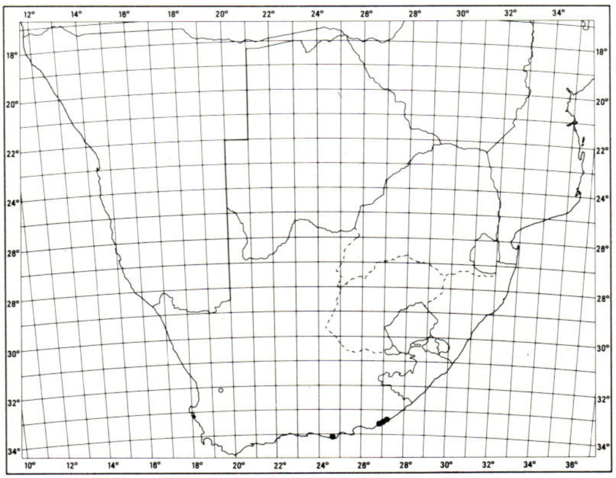

MAP 20.— ● **Vellereophyton vellereum**
○ **Vellereophyton lasianthum**

V. vellereum has larger heads than any of the other species, and this character, together with the distinctive pappus and glabrous ovaries, make recognition easy.

Vouchers: *Acocks* 21818 (PRE); *Dyer* 3351 (PRE); *Galpin* 2931 (PRE); *Hilliard & Burtt* 11109 (E; K; M; MO; NU; PRE; S).

7. **Vellereophyton lasianthum** *(Schltr. & Moeser) Hilliard* in Bot. J. Linn. Soc. 82: 211 (1981). Type: Cape, Klein-Namaland, Koude Bokkeveld, Klyn Vley [Klein Vlei, the last farm before the turning to the central Cedarberg], 4 000 ft, 20 i 1897, *Schlechter* 10061 (BM; BOL; E; G; K; M; P; PRE; S; Z, iso.!).

Helichrysum lasianthum Schltr. & Moeser in Bot. Jb. 44: 296 (1910).

Annual (?) herb, whole plant densely white-woolly, stems several from the crown, up to 150 mm long, simple or subsimple, prostrate, leafy. *Radical leaves* up to 20 × 5 mm, subspathulate, obtuse; *stem leaves* mostly 5−15 × 3−5 mm, spathulate, obtuse. *Heads* homogamous, campanulate, c. 2,5 × 2 mm, shortly pedunculate, in few-headed, very woolly glomerules at the branch tips and opposite the upper leaves, the axillary branchlets sometimes elongating and over-topping the primary glomerule. *Involucral bracts* in 2 series, outer shorter than inner, inner about equalling flowers, minutely radiating, oblong, backs woolly, only extreme tips glabrous, spreading, crisped, sub-opaque, white, reddish above the stereome. *Receptacle* smooth. *Flowers* c. 11−12. *Achenes* elliptic in outline, 0,75 mm long, glabrous or hairy. *Pappus* bristles many, about equalling corolla, tips subplumose, bases with patent cilia, not cohering.

V. lasianthum is known only from the type collection. It is very distinctive by virtue of its homogamous heads, and biseriate, very woolly, involucral bracts of which only the extreme tips are glabrous. Map 20.

8992b

TROGLOPHYTON

Troglophyton *Hilliard & Burtt* in Bot. J. Linn. Soc. 82: 208 (1981). Type species: *T. capillaceum* (Thunb.) Hilliard & Burtt.

Delicate annual or short-lived perennial herbs well-branched from the base, stems filiform, prostrate, diffuse or weakly erect, thinly white-woolly, distantly leafy. *Leaves* small, on filiform petioles, blade elliptic or ovate, margins flat, both surfaces cobwebby or thinly woolly, upper sometimes glabrescent. *Heads* homogamous or heterogamous, small, on filiform peduncles racemosely or subcorymbosely arranged, or rarely nearly sessile and crowded. *Involucral bracts* in 3(−5) series, subequal, slightly exceeding flowers, stereome undivided or with 2 or 3 thin linear patches in upper part, lower part canaliculate, embracing adjacent flower, lamina delicate, pellucid, sometimes purplish above stereome, tip subopaque or opaque white, obtuse to subacute, either radiating or whole involucre rotate at maturity. *Receptacle* shallowly tubercled. *Flowers* 8−30, yellow, sometimes tipped pink, ♀ flowers (0−)2−23, fewer than, or about equalling, or 2−3(−7) times as many as ☿, narrowly tubular; ☿ flowers narrowly infundibuliform or tubular below, narrowly campanulate above with 5 deltoid lobes, glandular hairs with biseriate stalk and small globular 2-celled head on backs of all lobes. *Anthers* with small obtuse apical appendage and tails shorter than, about equalling or slightly exceeding filament collar. *Style* branches truncate and penicillate. *Achene* wall with slightly raised cells, moderately imbricate or not, oblong duplex hairs with swelling cushion frequently present, apparently not always emitting mucilage (neither imbricate cells nor duplex hairs a specific character). *Pappus* bristles barbellate or subplumose above, these cilia ± clavate and with delicate wall thickenings, shaft markedly barbellate, bases cohering by patent cilia.

A genus of 6 species endemic to Southern Africa.

1a Heads homogamous:

  2a Heads up to 3 mm long .... 1. *T. capillaceum*

  2b Heads c. 6 mm long .... 3. *T. acocksianum*

1b Heads heterogamous:

  3a Heads c. 3−4(−5) mm long .... 2. *T. elsiae*

  3b Heads up to 3 mm long:

    4a Female flowers either fewer than hermaphrodite or slightly outnumbering them, cilia at tips of pappus bristles often noticeably thickened or elongated:

      5a Tips of involucral bracts very obtuse .... 4. *T. tenellum*

      5b Tips of involucral bracts subacute .... 5. *T. leptomerum*

    4b Female flowers mostly at least twice as many as the hermaphrodite, cilia at tips of pappus bristles scarcely thickened .... 6. *T. parvulum*

1. **Troglophyton capillaceum** *(Thunb.) Hilliard & Burtt* in Bot. J. Linn. Soc. 82: 209 (1981). Type: Cape of Good Hope, *Thunberg* (sheet 19108, UPS, holo.!).

Gnaphalium capillaceum Thunb., Prodr. 152 (1800), Fl. Cap. 660 (1823). *Helichrysum capillaceum* (Thunb.) Less., Syn. Comp. 275 (1832); DC., Prodr. 6: 170 (1838) excl. var. *erectum;* Harv. in F.C. 3: 215 (1865); Levyns in Adamson & Salter, Fl. Cape Penins. 780 (1950); Merxm., F.S.W.A. 139: 93 (1967); Hilliard, Compositae in Natal 208 (1977).

H. oreophilum Dinter in Fedde, Repert. (Beih.) 53: 13, 69 (1928), nom. nud., non Klatt (1896). Type: S.W.A./Namibia, Klein Karas, *Dinter* 4749 (BOL; G; K; SAM; PRE; iso.!).

H. capillaceum var. *majus* DC., Prodr. 6: 170 (1838). Lectotype: Cape, Graaff-Reinet distr., along Sundays River near Monkey Ford, *Burchell* 2873 (G-DC!; K, isolecto.!).

Two subspecies are recognized:

(a) subsp. **capillaceum.**

Delicate annual herb 50−300 mm tall, stems loosely branched, diffuse or weakly erect. *Leaves* mostly 4−30(−50) × 2−15(−20) mm, $\frac{1}{3}-\frac{1}{2}$ the length petiolar, blade elliptic to elliptic-ovate, more or less tapering at both ends, mucronate. *Heads* homogamous, campanulate, 2,5−3 × 1,5−2 mm, solitary on long (up to 10—15 mm) peduncles subracemosely arranged. *Involucral bracts* in 3 series, tips obtuse, generally emarginate or erose, pellucid, only extreme tip sometimes opaque white, occasionally purplish above the stereome, rarely whole bract purplish. *Flowers* 14—30. *Achenes* 0,75 mm long, hairy. *Pappus bristles* with barbellate shaft and tip. Fig. 7:3.

Widely distributed from southernmost S.W.A./Namibia E. to Bloemfontein in the O.F.S. and S. to the Cape Peninsula in the W. and Komgha, near the Great Kei River, in the E. Grows in damp shade, either under rocks or shrubs; flowering between April and, November, but at its peak in August and September. Map 21.

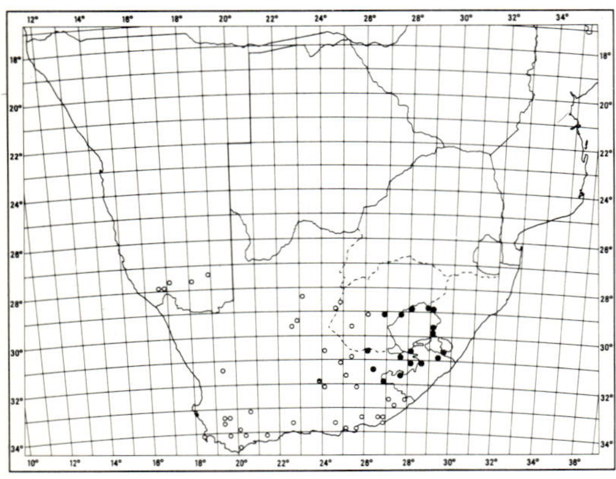

MAP 21.— ○ **Troglophyton capillaceum**
        subsp. **capillaceum**
      ● **Troglophyton capillaceum**
        subsp. **diffusum**
      ◐ sympatric

Introduced into Britain with wool shoddy (recorded from N. Hants., Blackmoor).

Easily recognized by its small homogamous heads subracemosely arranged, which characters at once distinguish it from *T. parvulum* (no. 6) with which it is frequently confused.

Vouchers: *Flanagan* 1340 (BOL; SAM); *Fourcade* 3075 (BOL; NBG; PRE); *Giess* 13058 (M; PRE; S;

WIND); *Hutchinson* 533 (BM; BOL; K); *Rehmann* 3829 (BM; K).

(b) subsp. **diffusum** *(DC.) Hilliard*, comb. et stat. nov. Lectotype: Cape, Bruintjeshoogte, *Burchell* 3047 (G-DC!; K).

Helichrysum capillaceum (Thunb.) Less. var. *diffusum* DC., Prodr. 6: 170 (1838).

Distinguished from the typical plant by the involucral bracts being markedly opaque white at the tips and smoothly rounded, and by the pappus bristles tipped with several clavate cilia. There is also a strong tendency for the leaves to be ovate with base truncate to broadly rounded; however, in the lectotype the base tapers to the petiole. Fig. 6: 1, 7: 5.

Subsp. *diffusum* is more easterly in its distribution than subsp. *capillaceum,* and occurs mainly on the Drakensberg and neighbouring mountains from Bergville district in Natal, northern Lesotho and Thaba Nchu Mountain in the eastern Orange Free State S. and W. to the high mountains of Transkei and the Cape Drakensberg, Sneeuwberg, Koudeveldberg and Oudeberg near Graaff-Reinet, and the Amatola Mountains, between 900 and 2 100 m. In the central Cape mountains in particular there seems to be a gradual transition to subsp. *capillaceum,* and plants with nearly pellucid involucral bracts and barbellate pappus tips may grow with others having opaque white bracts and pappus with clavate cilia at the tips, or specimens may be intermediate in these characters. Similarly, there is some gradation in leaf form. Map 21.

Subsp. *diffusum* grows in damp partially shaded places under overhanging rocks and is particularly common in the Cave Sandstone overhangs of the Drakensberg. Flowers mainly between November and April, but possibly flowers can be found in any month.

Vouchers: *Acocks* 13820 (PRE); *Esterhuysen* 10201 (BOL; NBG; NU; PRE); *Hilliard* 5511 (E; K; MO; NU; S); *Hilliard & Burtt* 6537 (E; K; M; NBG; NU).

2. **Troglophyton elsiae** *Hilliard, sp. nov. T. capillaceo (Thunb.) Hilliard & Burtt affinis sed capitulis c. 3—4 mm longis heterogamis (non 2,5 mm, homogamis), et apicibus lacteis bractearum involucri magis conspicuis.*

FIG. 6.—1, **Troglophyton capillaceum** subsp. **diffusum,** whole plant, × 1; 1a, head, × 6,6; 1b, hermaphrodite flower, × 16 (*Davis & Davis* 171). 2, **T. acocksianum,** part of a plant, × 1; 2a, head, × 5,3 (*Goldblatt* 5832). 3, **T. parvulum,** whole plant, × 1,5; 3a, head, × 6,6; 3b, hermaphrodite flower, × 16; 3c, female flower, × 16 (*Goldblatt* 5881).

RHONA
COLLETT

Herba delicatula tegetes intricatas formans; rami majores ad 120 mm longi, filiformes, repentes, radicantes, tenuiter griseo-albo-lanati, distanter foliati. Folia plerumque 5 −25 × 1,5 − 8 mm, dimidio longitudinis petiolare, apice obtuso vel subacuto mucronato, basi in petiolo attenuata, utrinque tenuiter griseo-lanata, supra glabrescentia vel fere glabrescentia. Capitula heterogama, campanulata, c. 3 −4(−5) × 1,5 −2 mm, solitaria vel pauca in pedunculis filiformibus laxe lanatis subcorymbose disposita. Involucri bracteae 3-seriatae, dorso laxe lanatae, seriebus duabus interioribus subaequalibus flores superantibus, apicibus ellipticis obtusis opacis lacteis radiantibus. Receptaculum breviter tuberculatum. Flores 13 −28, 2 −7 feminei, 10 −22 hermaphroditi. Achaenia c. 0,75 mm longa, fere dimorpha, ea florum femineorum pubescentia, ea florum hermaphroditorum glabra, raro omnia pubescentia. Pappi setae numerosae, corollam circum aequantes, apicibus subplumosis pilis clavatis, scapo barbellato, basi inter se pilis patentibus cohaerentes.

Type: Cape, Clanwilliam div., overhanging rock shelter on summit of Krakadouw Peak, Cedarbergen, 1 700 m, 27 xii 1947, *Esterhuysen* 14314 (NBG, holo.!; BOL; K; PRE, iso.!).

Cedarberg, plateau S. of Tafelberg, 5 000 ft, 28 xii 1962, *Esterhuysen* 29994 (BOL; K; PRE; S); S. Cedarberg, Apollo Peak, 5 500 ft, 13 xii 1950, *Esterhuysen* 18085 (BOL); ibid., 13 xii 1975, *Esterhuysen* 34155 (BOL). Ceres div., Bokkeveld Tafelberg, 5 500 ft, 8 xii 1940, *Esterhuysen* 3943 (BOL). Worcester div., Hex River Mountains, Mt Brodie, 8 iii 1943, *Esterhuysen* 8752 (BOL); ibid., Milner Peak, 6 600 ft, 2 i 1959, *Esterhuysen* 28082a (BOL; K; PRE; S).

A delicate herb forming tangled mats, main branches up to 120 mm long, filiform, creeping, rooting, thinly greyish-white woolly, distantly leafy. *Leaves* mostly 5 −25 × 1,5 −8 mm, up to half the length petiolar, blade elliptic, apex obtuse to subacute, mucronate, base tapering into petiole, both surfaces thinly grey-woolly, upper glabrescent or nearly so. *Heads* heterogamous, campanulate, c. 3 −4 (−5) × 1,5 −2 mm, solitary or few on filiform, loosely woolly peduncles subcorymbosely arranged. *Involucral bracts* in 3 series, backs loosely woolly, inner two series subequal, exceeding flowers, tips elliptic, obtuse, opaque milk-white, radiating. *Receptacle* shallowly tubercled. *Flowers* 13 −28, 2 −7 ♀, 10 −22 ☿. *Achenes* c. 0,75 mm long, usually dimorphic, those of ♀ flowers hairy, of ☿ flowers glabrous, or rarely all hairy. *Pappus* bristles many, about equalling corolla, tips subplumose with clavate hairs, shaft barbellate, bases cohering by patent cilia. Fig. 7: 1.

Recorded from the Cedarberg, Bokkeveld Tafelberg in Ceres division, and the Hex River Mountains, between 1 500 and 2 000 m above sea level, growing in damp shade under rock overhangs. Flowers in December and January. Map 22.

The specific epithet honours Miss Elsie Esterhuysen who has gathered many a good plant off the Cape Mountains.

3. **Troglophyton acocksianum** *Hilliard, sp. nov.* T. elsiae *Hilliard affinis, sed capitulis majoribus 6 × 3,5 mm (non (3 −4) −5 × 1,5 −2 mm) homogamis.*

Herba tegetes formans, verosimiliter perennis; caules filiformes, prostrati, radicantes, distanter foliati, partibus juvenilibus laxe albo-lanati. Folia plerumque 8 −16 × 3 −6 mm, dimidio longitudinis petiolari, ovata vel ovato-elliptica, apice acuto mucronato, basi abrupte in partem petiolarem angustata, utrinque tenuiter albo-lanata. Capitula homogama, campanulata, c. 6 × 3,5 mm, solitaria, vel ad 6 in pedunculis brevibus ad ramulorum apices corymbose disposita. Involucri bracteae c. 5-seriatae

---

FIG. 7.—**Troglophyton**, involucral bracts and pappus bristles of all the species. 1, **T. elsiae**, involucral bract, × 13; 1a, pappus bristle, × 27 (*Esterhuysen* 14314). 2, **T. parvulum**, involucral bract, × 13 (*Acocks* 17774); 2a, involucral bract, × 13 (*Esterhuysen* 13028); 2b, pappus bristle, × 27 (*Acocks* 17774); 2c, involucral bract, × 17 (*Hardy* 762 : see text); 2d, pappus bristle, × 27 (*Hardy* 762). 3, **T. capillaceum** subsp. **capillaceum**, involucral bracts, × 13; 3a, pappus bristle, × 27 (*Fourcade* 3075). 4, **T. tenellum**, involucral bract, × 17; 4a, pappus bristle, × 27 (*Maguire* 1893). 5, **T. capillaceum** subsp. **diffusum**, involucral bract, × 13; 5a, pappus bristle, × 27 (*Esterhuysen* 10201). 6, **T. leptomerum**, involucral bract, × 13; 6a, pappus bristle from hermaphrodite flower, × 27; 6b, pappus bristle from female flower, × 27 (*Barker* 6631). 7, **T. acocksianum**, involucral bract, × 13; 7a, pappus bristle, × 27 (*Acocks* 18625).

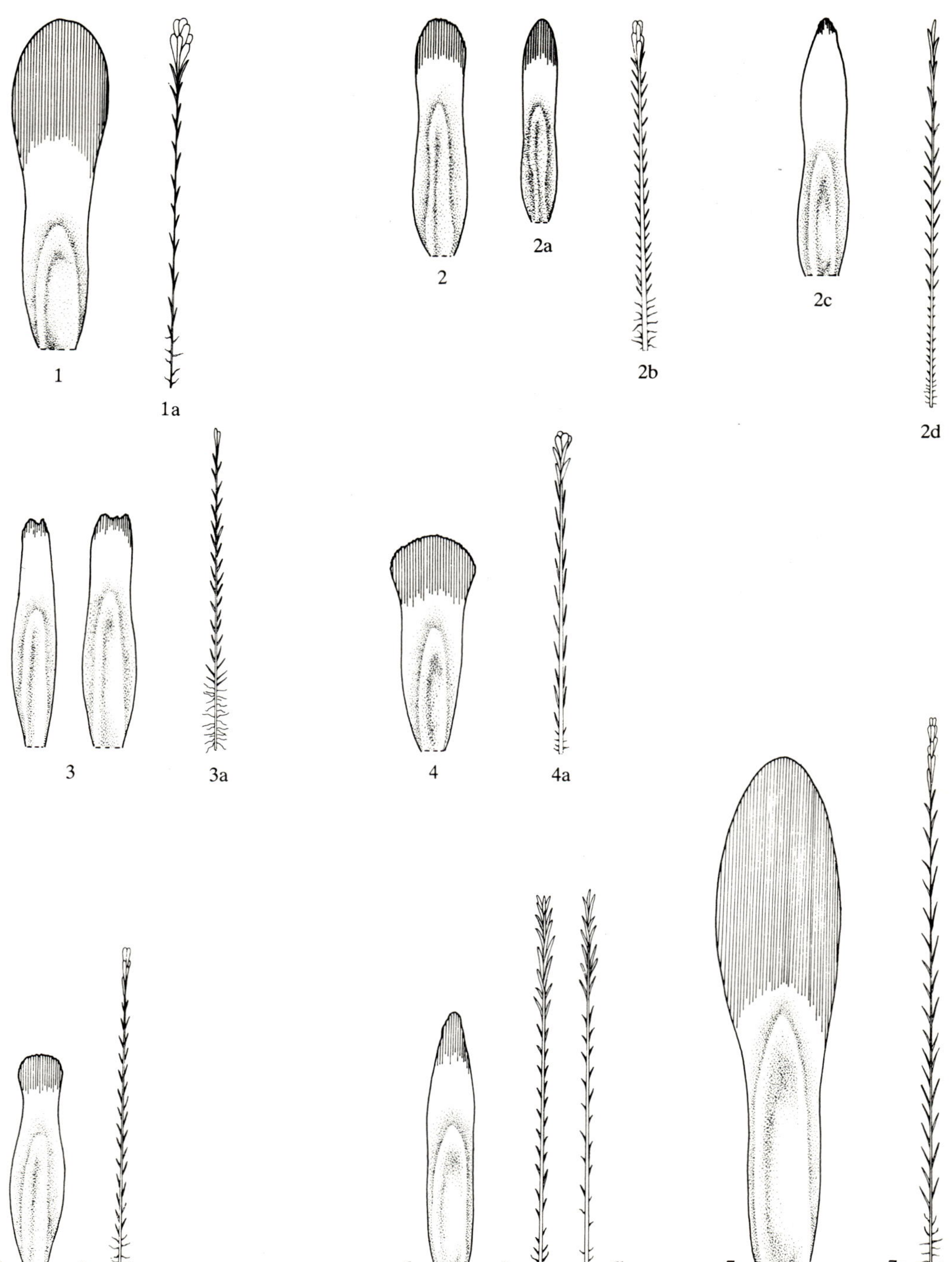

extremae breves, interiores subaequales flores superantes, dorso lanatae, apicibus lanatis opacis lacteis radiantibus. Receptaculum *breviter tuberculatum.* Flores *18 −21.* Ovarium *pubescens (achaeniis non visis).* Pappi setae *numerosae, corollam circum aequantes, apicibus subplumosis pilis clavatis, scapo barbellato, basi inter se pilis patentibus cohaerentes.*

Type: Cape, Calvinia distr., Akkerendam, c. 4 700 ft., 14 xi 1955, foot of shale krantzes on S. aspect, rare, dense mats, *Acocks 18625* (PRE, holo.!).

Mat-forming herb, probably perennial, stems filiform, prostrate, rooting, young parts loosely white woolly, distantly leafy. *Leaves* mostly 8−16 × 3−6 mm, about half the length petiolar, ovate or ovate-elliptic, apex acute, mucronate, base abruptly narrowed to the petiolar portion, both surfaces thinly white-woolly. *Heads* homogamous, campanulate, c. 6 × 3,5 mm, solitary or up to 6 on short peduncles corymbosely arranged at the branch tips. *Involucral bracts* in c. 5 series, outermost short, inner subequal, exceeding flowers, backs woolly, tips elliptic, obtuse, opaque milk-white, radiating. *Receptacle* shallowly tubercled. *Flowers* 18−21. *Achenes* not seen, ovaries hairy. *Pappus* bristles many, about equalling corolla, tips subplumose with clavate hairs, shaft barbellate, bases cohering by patent cilia. Fig. 6: 2, 7: 7.

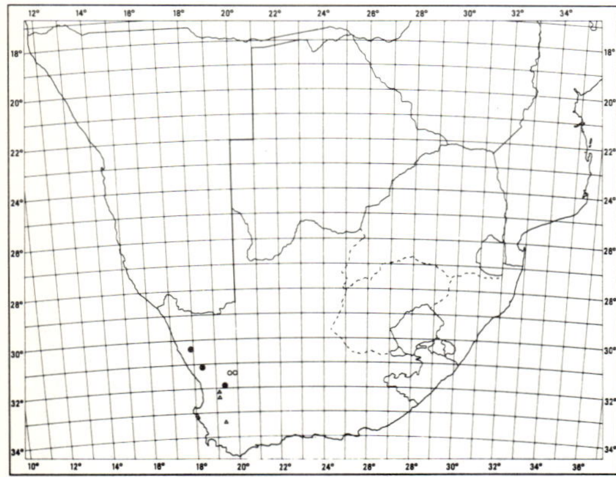

MAP 22.— △ **Troglophyton elsiae**
    ○ **Troglophyton acocksianum**
    ● **Troglophyton tenellum**

Known only from the type collection and one from the Hantamsberg, also near Calvinia. Closely allied to *T. elsiae* (no. 2), but a coarser plant with larger, homogamous, heads. Map 22.

Voucher: *Goldblatt 5832* (MO; NU).

The specific name commemorates the late Mr John Acocks, a notable student of the South African flora.

4. **Troglophyton tenellum** *Hilliard, sp. nov.* T. capillaceo *(Thunb.) Hilliard & Burtt affinis sed capitulis heterogamis, nec homogamis, bracteis involucri distincte alboapiculatis nec pellucidis nec fere pellucidis.*

Herba annua, delicatula, erecta (?), 30 −100 mm alta, e basi ramosa, ramis filiformibus arachnoideis distanter foliatis. Folia radicalia, si praesentia, 7 −25 × 2,5 −13 mm, dimidio longitudinis petiolari; caulina 6 −17 × 1,5 −7 mm, sessilia vel subsessilia; lamina elliptica, utrinque attenuata, apice acuto vel subacuto mucronato, supra et subtus arachnoidea. Capitula heterogama, campanulata, 2,5 × 2 mm, solitaria; pedunculi ad 15 mm filiformes foliis oppositi, ad ramulorum apices decrescescentes capitulis corymbosis. Involucri bracteae 3-seriatae, dorso laxe lanatae, apicibus flores paulo superantibus obtusissimis leviter crispatis opace lacteis minute radiantibus. Flores 10 −26, 6 −14 feminei, 6 −16 hermaphroditi. Achaenia 0,75 mm longa, pubescentia. Pappi setae numerosae, corollam plus minusve aequantes, apicibus subplumosae pilis clavatis, scapo barbellato, basi pilis patentibus inter se cohaerentes.

Type: Cape, Calvinia distr., top of Botterkloof Pass, 24 ix 1952, *Maguire 1893* (NBG, holo.!; BOL, iso.!).

Namaqualand, Klipfontein Station, 1 000 m, 8 ix 1925, *Marloth* 12528 (PRE); Khamiesberg div., Namaroup, 12 ix 1911, *Pearson* 6627 (BOL). Vanrhynsdorp distr., Bitterfontein, 1 ix 1897, *Schlechter* 11021 (BM; BOL; K; PRE). Clanwilliam distr., Cedarberg, Wolfberg Plateau between The Arch and The Cracks, 10 x 1976, *Esterhuysen* 34398 (BOL). Piquetberg div., east slopes of Avontuur Hill, 3 500 ft. ix 1934, *Pillans* 7389 (BOL).

A delicate erect (?) annual herb 30−100 mm tall, branching from the base, branches filiform, cobwebby, distantly leafy. *Radical leaves,* if present, 7−25 × 2,5−13 mm, half the length petiolar, *cauline leaves* subtending each dichotomy, 6−17 ×

1,5—7 mm, sessile or nearly so, blade elliptic, tapering at both ends, apex acute or subacute, mucronate, both surfaces cobwebby. *Heads* heterogamous, campanulate, 2,5 × 2 mm, solitary on filiform leaf-opposed peduncles up to 15 mm long, peduncles becoming much shorter and corymbosely arranged towards the branchlet tips. *Involucral bracts* in 3 series, backs loosely woolly, tips slightly exceeding flowers, obtuse, slightly crisped, opaque milk-white, minutely radiating. *Receptacle* shallowly tubercled. *Flowers* 10—26, 6—14 ♀, 6—16 ☿. *Achenes* 0,75 mm long, with duplex hairs. *Pappus* bristles many, about equalling corolla, tips subplumose with clavate cilia, shaft barbellate, bases cohering by patent cilia. Fig. 7: 4.

Recorded from Khamiesberg, Calvinia, Vanrhynsdorp, Clanwilliam and Piquetberg districts, in shady damp places on the hills and mountains, probably always under rocks, flowering in September and October. Map 22.

Can be confused with *T. capillaceum* (no. 1) but easily distinguished by its heterogamous heads subcorymbosely rather than racemosely arranged.

5. **Troglophyton leptomerum** *Hilliard, sp. nov. ab peraffini* T. tenello *Hilliard bracteis involucri apicibus laevibus lanceolatis subacutis (haud obtusissimis) et ab* T. capillaceo *(Thunb.) Hilliard & Burtt capitulis heterogamis et apicibus bractearum differt.*

Herba annua delicatula e basi multiramosa; rami ad 120 mm longi, verosimiliter infirme erecti vel prostrati, filiformes, tenuiter arachnoidei, distanter foliati. Folia radicalia ad 25 × 6 mm, dimidio longitudinis petiolari; caulina radicalibus similia sed sursum decrescentia (ad 5 × 2 mm redacta) et petiolo quartum vel trientem folii tantum aequante; lamina elliptica, utrinque attenuata, apice obtuso vel subacuto mucronato, supra et subtus tenuissime arachnoidea. Capitula heterogama, campanulata, c. 3 × 2 mm, ad ramulorum apices corymbose disposita in pedunculis filiformibus (ad 15 mm longis sed sursum brevioribus) tenuissime arachnoideis. Involucri bracteae · 3-seriatae, dorso laxe arachnoideae, apicibus flores superantibus lanceolatis subacutis albis opacis vel subopacis. Receptaculum breviter

tuberculatum. Flores 8—15, 3—8 feminei, 4—11 hermaphroditi. Ovarium pubescens (achaenio haud viso). Pappi setae numerosae, corollam aequantes, apice subplumoso, scapo barbellato, basi inter se pilis patentibus cohaerentes.

Type: Cape, Namaqualand, 20 miles NE. of Springbok, 8 ix 1950, *Barker* 6631 (NBG, holo.!).

Granitkuppe 17 Meilen von Springbok nach Pofadder, 13 ix 1963, *Merxmüller & Giess* 3782 (M; WIND); Ookiep, September, *Dümmer* s.n. (K); Khamiesberg near Garies, 16 x 1954, *Esterhuysen* 23701 (BOL). Clanwilliam distr., S. Cedarberg, Apollo Peak, 4 500 ft, 13 xii 1950, *Esterhuysen* 18080 (BOL). Top of Piquetberg Mountain, 9 xi 1934, *Pillans* 7389 (K).

A delicate annual herb much-branched from the base, branches up to 120 mm long, probably weakly erect or prostrate, filiform, thinly cobwebby, distantly leafy. *Radical leaves* up to 25 × 6 mm, half the length petiolar, *cauline leaves* subtending each dichotomy, similar to radical but diminishing in size upwards (as small as 5 × 2 mm), and petiole reduced to ⅓ to ¼ total length, blade elliptic, tapering at both ends, apex obtuse to subacute, mucronate, both surfaces very thinly cobwebby. *Heads* heterogamous, campanulate, 3 × 2 mm, solitary at the branchlet tips on long (up to 15 mm), filiform, very thinly cobwebby peduncles, much shorter upwards and corymbosely arranged. *Involucral bracts* in 3 series, backs loosely cobwebby, tips exceeding flowers, lanceolate, subacute, more or less opaque white. *Receptacle* shallowly tubercled. *Flowers* 8—17, 3—8 ♀, 4—11 ☿. *Achenes* 0,75 mm long with duplex hairs. *Pappus* bristles many, equalling corolla, tips subplumose, shaft barbellate, bases cohering by patent cilia. Fig. 7: 6.

Recorded from the environs of Springbok, the Khamiesberg, S. Cedarberg and Piquetberg Mountain. Grows in moist shady places under rocks; flowering between September and December. Closely allied to *T. tenellum* (no. 4) but distinguished by its subacute, not very obtuse, involucral bracts. Map 23.

6. **Troglophyton parvulum** *(Harv.) Hilliard & Burtt* in Bot. J. Linn. Soc. 82: 209 (1981). Lectotype: Cape, Tulbagh, Steendahl, *Pappe* s.n. (TCD!; K; S, isolecto.!).

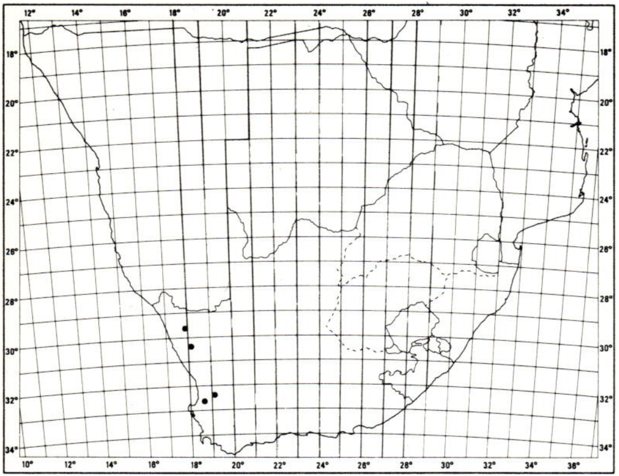

MAP 23.— **Troglophyton leptomerum**

Gnaphalium parvulum Harv. in F.C. 3: 262 (1865); Levyns in Adamson & Salter, Fl. Cape Penins. 777 (1950); Merxm., F.S.W.A. 39: 84 (1967).

Helichrysum capillaceum (Thunb.) Less. var. *erectum* DC., Prodr. 6: 170 (1838). Lectotype: Cape, Paarlberg, *Drège* 5813 (G—DC!; BM, isolecto.!).

Small annual herb c. 10—150 mm tall, stems simple or branched, more or less erect. *Leaves* mostly 3—15(—30) × 1—6(—12) mm, $\frac{1}{3}$—$\frac{1}{2}$ the length petiolar, uppermost subsessile, blade oblanceolate to elliptic tapering at both ends, subacute to obtuse, mucronate. *Heads* heterogamous, campanulate, 2,5—3 × 1—2 mm, pedunculate to nearly sessile, few to many in corymbose clusters terminating the branchlets or occasionally subsolitary on diminutive plants. *Involucral bracts* in 3 series, pellucid, occasionally reddish above the stereome, tips very obtuse to subacute, opaque or subopaque, white, minutely radiating. *Flowers* 13—27, 9—23 ♀, 3—8 ⚥. *Achenes* 0,5 mm long, hairy. *Pappus* bristles with tip and shaft barbellate. Fig. 6: 3, 7: 2.

T. parvulum ranges from the district of Lüderitz Süd in southernmost S.W.A./Namibia to the W. Cape

as far south as the Cape Peninsula, on mountains and hills in damp shady places, often under rocks and cliffs, or sometimes under shrubs and trees, but occasionally in the open where the ground is suitably moist; flowering between August and November, mainly in October. Map 24.

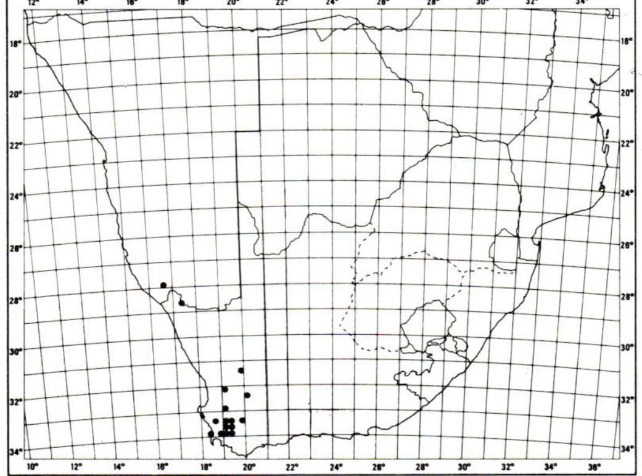

MAP 24.— **Troglophyton parvulum**

Vouchers: *Acocks* 17774 (PRE); *Esterhuysen* 13028 (BOL; NBG; PRE), 14618 (BOL; E); *Schlechter* 1589 (BOL); *Wilms* 3261 (BM; G; Z).

The type of *T. parvulum,* and nearly all other specimens, have heads crowded in glomerules, the very short peduncles hidden by wool and the surrounding leaves, the involucral bracts very obtuse with opaque or subopaque white tips. Specimens with heads mostly on distinct peduncles and involucral bracts with subacute semipellucid tips may represent another species (e.g. *Hardy* 762, K; PRE; *Merxmüller & Giess* 32269 (M) mixed with *T. capillaceum; Taylor* 2815, M; NBG). However, field studies to assess the variability of *T. parvulum* are needed.

T. parvulum is frequently confused with *T. capillaceum* (no. 1), but it is readily distinguished by its heterogamous heads corymbosely, not racemosely, arranged. The species it most resembles are *T. tenellum* (no. 4) and *T. leptomerum* (no. 5), which can be distinguished by different proportions of female and hermaphrodite flowers in the heads and by details of bract tips and pappus form.

8992c

HELICHRYSOPSIS

Helichrysopsis *Kirp.* in Trudy bot. Inst. Akad. Nauk SSSR ser. 1, 9: 32 (1950); Hilliard & Burtt in Bot. J. Linn. Soc. 82: 207 (1981). Type species: *H. septentrionalis* (Vatke) Hilliard.

Perennial herb, stems much-branched, spreading, ashy-grey, closely leafy. *Leaves* linear, up to 15 × 1 mm, margins strongly revolute, white-cottony, contrasting strongly with the silky-grey upper leaf surface and the midrib below, giving a striped effect to the leaves. *Heads* heterogamous, narrowly campanulate, c. 5 × 3 mm, few clustered at the branchlet tips. *Involucral bracts* in c. 4 series, outer short, pellucid, palest brown, inner with stereome undivided, tips opaque milk-white, exceeding the flowers. *Receptacle* honeycombed. *Flowers* c. 45–65, 30–50 ♀, corolla filiform, 12–13 ☿, corolla cylindric, all lobes glandular-hairy on backs. *Anthers* with small obtuse apical appendage and tails exceeding the filament collar. *Style branches* truncate and penicillate. *Achenes* less than 0,5 mm long, with minute 2-celled globular hairs without a swelling cushion, myxogenic. *Pappus* bristles many, tips plumose, shaft nude, 2 or 3 bristles fused together near the base and then, a little lower, these compound bristles fused into a smooth ring.

Monotypic. Endemic to the coastal plain of southern Mozambique and its southward extension into Zululand, as far south as the shores of Lake St Lucia. An isolated genus, without obvious allies.

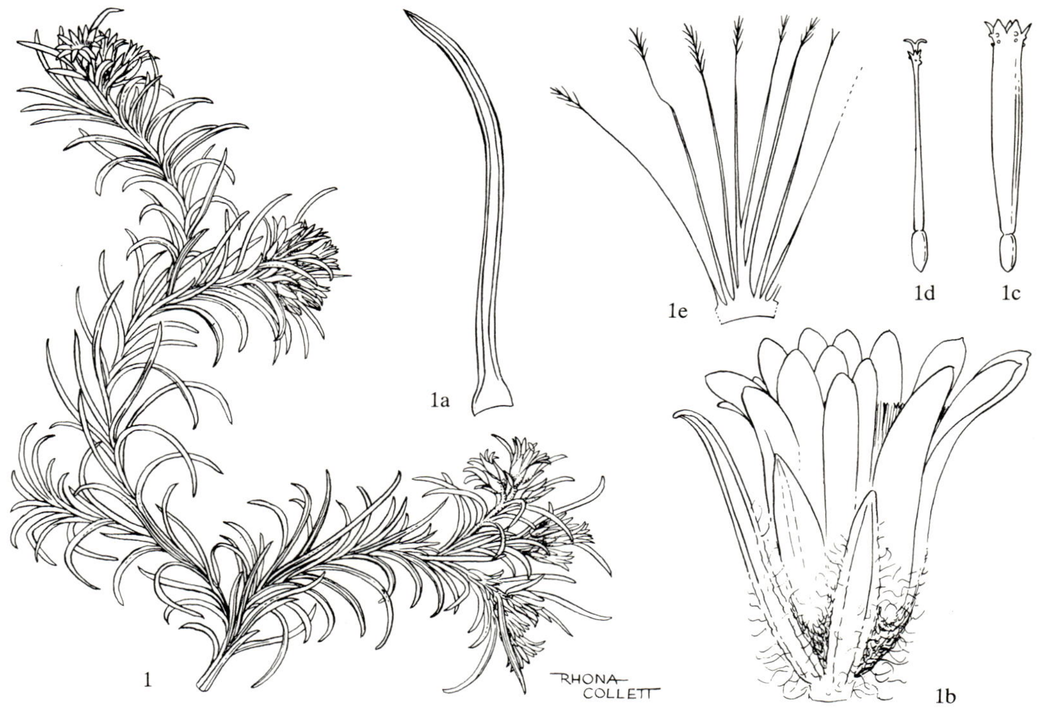

FIG. 8.—1, **Helichrysopsis septentrionale,** part of a plant, × 1; 1a, leaf, margins strongly revolute, × 2,6; 1b, head, × 6,6; 1c, hermaphrodite flower, × 8; 1d, female flower, × 8; 1e, part of ring of pappus bristles, × 12 (*Taylor* 66).

Helichrysopsis septentrionale *(Vatke) Hilliard* in Bot. J. Linn. Soc. 82: 207 (1981). Type: Mozambique, Inhambane, in dry sandy fields, *Peters* s.n. (B†).

Anaxeton septentrionalis Vatke in Öst. Bot. Zeitschrift 27: 194 (1877). *Gnaphalium septentrionale* (Vatke) Hilliard in Notes R. bot. Gdn Edinb. 31: 8 (1971), Compositae in Natal 127 (1977). *Gnaphalium stenophyllum* Oliv. & Hiern in F.T.A. 3: 344 (1877), nom. illegit. *Helichrysopsis stenophylla* Kirp. in Trudy bot. Inst. Akad. Nauk SSSR. ser. 1, 9: 32 (1950), nom. illegit.

The only species, and a most distinctive plant in both general aspect and peculiar pappus. Found in sandy places; flowering between December and April. Fig. 8: 1. Map 25.

Vouchers: *Pooley* 269 (E; K; NH; NU; S; COI); *Schlechter* 11589 (BM; BOL; E; LEN).

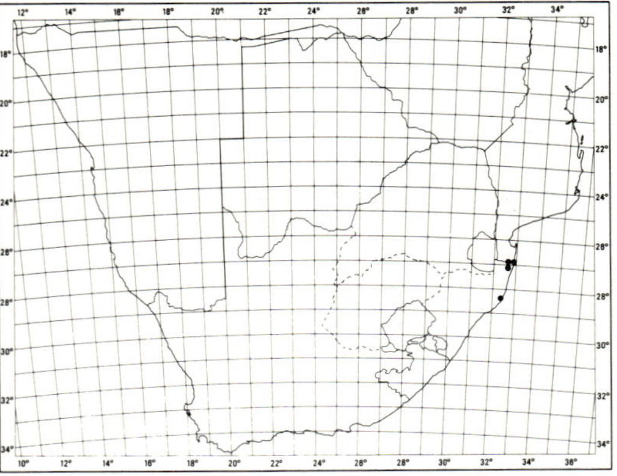

MAP 25.— **Helichrysopsis septentrionale**

8992d

PLECOSTACHYS

Plecostachys *Hilliard & Burtt* in Bot. J. Linn. Soc. 82: 206 (1981). Type species: *P. serpyllifolia* (Berg.) Hilliard & Burtt.

Subshrubs, stems long, slender, tangled, thinly white-tomentose, leafy. *Leaves* small, elliptic to suborbicular, apex obtuse or rounded, sessile or shortly petioled, margins flat or undulate, upper surface generally cobwebby, glabrescent, lower white-tomentose. *Heads* heterogamous, small, many in small compact corymbose clusters terminating the branchlets. *Involucral bracts* in 3—4 series, inner about equalling or slightly exceeding flowers, stereome undivided, tips opaque white, obtuse, radiating, *Receptacle* shortly honeycombed. *Flowers* c. 10—40, ♀ flowers 3—30, fewer than, about equalling, or more than ♂ flowers, narrowly tubular; ♂ flowers tubular below, slightly expanded above with 5 deltoid lobes; glandular hairs on backs of all lobes. *Anthers* with small obtuse or subacute deltoid apical appendage, and tails either shorter or slightly longer than filament collar. *Style* branches truncate and penicillate, the short sweeping hairs extending briefly down the backs of the branches. *Achene* wall flat, apparently without crystals, oblong duplex hairs with swelling cushion always present, myxogenic. *Pappus* bristles scabrid above, the cells with delicate wall thickenings, bases cohering by patent cilia, lightly fused as well in one species.

Endemic, species 2, ranging from Swaziland and Natal to the Cape Peninsula.

1a Leaves suborbicular or broadly elliptic, margins crisped-undulate; inner involucral bracts with conspicuous milk-white tips exceeding the flowers; ♀ flowers 3—8..................................... 1. *P. serpyllifolia*

1b Leaves elliptic, margins flat; inner involucral bracts with small dirty white tips about equalling the flowers; ♀ flowers 10—28.. 2. *P. polifolia*

1. **Plecostachys serpyllifolia** *(Berg.) Hilliard & Burtt* in Bot. J. Linn. Soc. 82: 207 (1981). Type: ex hort. per Kallström (STB holo.!).

Gnaphalium serpyllifolium Berg., Descr. Pl. Cap. 250 (1767); Lam., Encycl. 2: 743 (1788), excl. syn. Volck.; Schrank in Denkschr. K. Akad. Wiss. Münch. 8: 158 (1824). *Helichrysum serpyllifolium* (Berg.) Pers., Syn. 2: 416 (1807); Less., Syn. Comp. 277 (1832); DC., Prodr. 6: 172 (1838); Harv. in F.C. 3: 218 (1865); Moeser in Bot. Jb. 44: 307 (1910); Hilliard & Burtt in Notes R. Bot. Gdn Edinb. 32: 359 (1973); Hilliard, Compositae in Natal 215 (1977).

G. orbiculare Thunb., Prodr. 152 (1800), Fl. Cap. 659 (1823). *H. orbiculare* (Thunb.) Druce in Rep. bot. Soc. Exch. Club Br. Isl. 1916: 626 (1917). *H. serpyllifolium* var. *orbiculare* (Thunb.) DC., Prodr. 6: 173 (1838). Type: Cape, hills below Table Mtn, *Thunberg* (sheet 19222, UPS, holo.!).

Much-branched straggling subshrub, branches slender, young parts thinly white-tomentose, leafy. *Leaves* up to 10 × 7 mm, broadly elliptic to suborbicular, apex mucronate, slightly recurved, petiole minute, margins crisped-undulate, upper surface thinly white-tomentose or glabrous, lower white-tomentose. *Heads* heterogamous, turbinate, c. 5 mm long, base woolly, many in small congested rounded clusters terminating the branchlets. *Involucral bracts* in c. 3 series, loosely imbricate, inner shortly exceeding flowers, tips obtuse, milk-white, radiating. *Receptacle* shortly honeycombed. *Flowers* 8—16, 3—8 ♀, 4—9 ♂, usually fewer ♀ flowers than ♂, yellow, sometimes tipped pink. *Achenes* c. 0.75 mm long, with myxogenic duplex hairs. *Pappus* bristles many, equalling or slightly exceeding corolla, scabrid, bases cohering by patent cilia, lightly fused as well. Fig. 9: 2.

Ranges from the Cape Peninsula through the coastal districts to Grahamstown and East London, then surely along the Transkei coast (but no records) to its northernmost stations on the Natal coast between Port Edward and Umdoni Park, Umzinto district. Grows in damp sandy places often near the sea, but in the southern and western Cape also on the lower slopes of the mountains. Flowering recorded between November and August, but at its peak in April. Map 26.

Vouchers: *Acocks* 10985 (PRE); *Compton* 14560 (NBG; PRE); *Esterhuysen* 14388 (BOL; PRE); *Hilliard* 1682 (E, NU); *MacOwan* 737 (BOL; PRE; SAM).

2. **Plecostachys polifolia** *(Thunb.) Hilliard & Burtt* in Bot. J. Linn. Soc. 82: 207 (1981). Type: Cape of Good Hope, *Thunberg* (sheet 19231, UPS, holo.!).

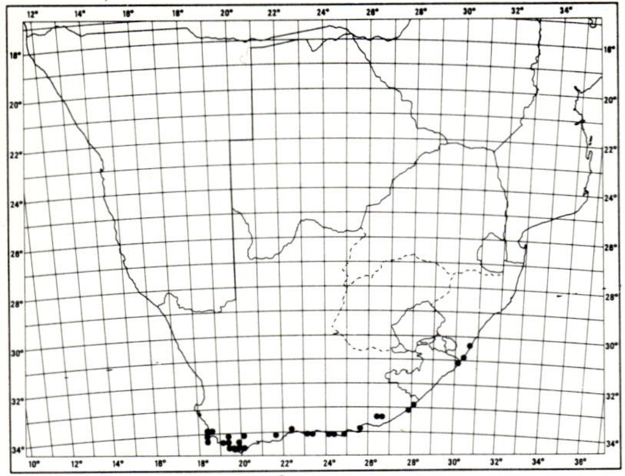

MAP 26.— **Plecostachys serpyllifolia**

minutely radiating. *Receptacle* shortly honeycombed. *Flowers* 17—41, 10—28 ♀, 4—20 ☿, ♀ flowers usually outnumbering ☿, yellow, sometimes tinged purple. *Achenes* 0,5 mm long with myxogenic duplex hairs. *Pappus* bristles many, a little shorter than corolla, scabrid, bases cohering lightly by patent cilia. Fig. 9: 1.

Ranges from the mountainous parts of the Cape Peninsula NE. to the Matroosberg and Hex River Mountains and E. over the Cape fold mountains to the mountains of the E. Cape and Transkei and the Noodsberg, Ndwedwe and Table Mountain near Durban, in Natal, with one record from Havelock Mine in Swaziland. Grows in damp and often partially shaded places such as streamsides, forest margins and damp cliff faces, usually near forest, often forming very large tangles. Flowers between September and December, mainly November and December. Map 27.

Gnaphalium polifolium Thunb., Prodr. 151 (1800), Fl. Cap. 656 (1823); Hilliard & Burtt in Notes R. Bot. Gdn Edinb. 32,3: 337 (1973); Hilliard, Compositae in Natal 128 (1977). *Helichrysum serpyllifolium* var. *polifolium* (Thunb.) DC., Prodr. 6: 173 (1838); Moeser in Bot. Jb. 44: 307 (1910).

Gnaphalium dodii Levyns in Jl S. Afr. Bot. 7: 84 (1941), in Adamson & Salter, Fl. Cape Penins. 776 (1950). Type: Cape, Orange Kloof, *Levyns* 6369 (BOL, holo.!).

Much-branched, straggling subshrub, branches slender, young parts thinly white-tomentose, leafy, often with axillary leaf tufts. *Leaves* up to 25 × 7 mm, elliptic tapering to a short petiole-like base, apex obtuse, apiculate, somewhat recurved, margins flat or nearly so, upper surface cobwebby and soon glabrous, or persistently white-tomentose, lower white-tomentose. *Heads* heterogamous, campanulate, c. 2,5—3 × 2 mm, base woolly, many in congested rounded clusters terminating the branchlets. *Involucral bracts* in c. 4 series, closely imbricate, inner about equalling flowers, tips obtuse, commonly dirty-white,

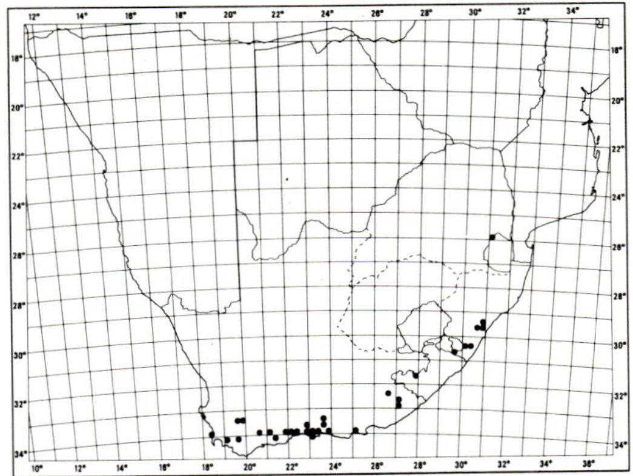

MAP 27.— **Plecostachys polifolia**

Vouchers: *Esterhuysen* 10891 (BOL; K; PRE); *Flanagan* 2147 (PRE); *Hilliard & Burtt* 7287 (E; K; MO; NU; PRE; S); *Hutchinson* 1200 (BM; BOL; K; PRE).

FIG. 9.—1, **Plecostachys polifolia,** flowering branch, × 1; 1a, leaf, × 6,6; 1b, head, × 10; 1c, involucral bract, × 6,6; 1d, hermaphrodite flower, × 13; 1e, female flower, × 13; 1f, pappus bristle, × 20 (*Hilliard & Burtt* 10989). 2, **P. serpyllifolia,** flowering branch, × 1; 2a, leaf, × 6,6; 2b, head, × 6,6; 2c, involucral bract, × 6,6 (*Breen* 183).

1
1a
1b
1c
1d
1e
1f
2
2a
2b
2c
RHONA COLLETT

8992e

PSEUDOGNAPHALIUM

Pseudognaphalium *Kirp.* in Trudy bot. Inst. Akad. Nauk SSSR ser. 1, 9: 33 (1950); Hilliard & Burtt in Bot. J. Linn. Soc. 82: 204 (1981). Type species: *P. oxyphyllum* (DC.) Kirp., from Mexico.

Annual, biennial or perennial herbs. *Leaves* lanceolate, linear-lanceolate or oblanceolate, sometimes decurrent, margins flat or undulate, both surfaces woolly, or glandular above, woolly below. *Heads* heterogamous, in small clusters corymbosely or cymosely arranged. *Involucral bracts* in 3—4 series, inner about equalling or slightly exceeding the flowers, white to yellow, often buff or cream, stereome usually fenestrated, rarely undivided *(P. oligandrum)*. *Receptacle* smooth or honeycombed. *Flowers* few to many, ♀ flowers more than the ☿, corolla filiform or narrowly tubular, ☿ tubular, scarcely broadened above, 5-lobed; all flowers glandular hairy on backs of lobes. *Anthers* with a small obtuse or subacute apical appendage, and tails slightly longer or shorter than the filament collar. *Style branches* truncate and penicillate. *Achenes* either glabrous or with duplex myxogenic hairs, epidermal cells tabular, often imbricate. *Pappus* bristles scabrid, apical cells sometimes inflated, bases cohering by patent cilia.

Species 50—60, in Africa, SE. Asia, Europe and the Americas.

1a Leaves with acute to acuminate tips, discolorous, heads in clusters arranged in corymbose panicles, achenes glabrous, but often papillate by protruding tips of imbricate epidermal cells (subgen. *Pseudognaphalium*)

  2a At least the main stem leaves decurrent, wings often reaching to the node below, heads containing up to 100 flowers .... 1. *P. undulatum*

  2b Leaves not decurrent, at most cordate-clasping, heads containing at least 145 flowers .... 2. *P. oligandrum*

1b Lower leaves oblanceolate, obtuse, concolorous, heads in clusters cymosely arranged (best seen when plant fully developed), achenes hairy (subgen. *Laphangium*) ....3. *P. luteo-album*

Subgenus **Pseudognaphalium**

1. **Pseudognaphalium undulatum** *(L.) Hilliard & Burtt* in Bot. J. Linn. Soc. 82: 205 (1981). Type: Cape of Good Hope, *Gnaphalium* No. 14 in herb. Cliff. (BM, holo.!).

Gnaphalium undulatum L., Sp. Pl. 852 (1753); DC., Prodr. 6: 226 (1838); Harv. in F.C. 3: 261 (1865); Holub in Tutin et al., Fl. Europ. 4: 128 (1976); Hilliard, Compositae in Natal 131 (1977).

Helichrysum decurrens Moench, Meth. 576 (1794), nom. illegit.

H. montosicolum Gand. in Bull. Soc. bot. Fr. 65: 44 (1918). Type: Natal, Camperdown, *Schlechter* 3282 (BM; E; G; P; PRE; S; WU; Z, iso.!).

A bushy spreading annual herb up to c. 600 mm tall, stems thinly greyish-white woolly, glabrescent, leafy. *Leaves* up to 80 × 12 mm, oblong-lanceolate or lanceolate, the upper ones linear-lanceolate, apex very acute to acuminate, base cordate or subcordate in the upper leaves, broad in the main stem leaves and decurrent, the wings often reaching to the node below or sometimes lower, margins entire, somewhat undulate, upper surface glandular, green drying brown, lower white-felted. *Heads* c. 3 mm long, campanulate, several together in tight clusters, these arranged in a large, wide-spreading corymbose panicle. *Involucral bracts* in c. 4 series, loosely imbricate, innermost about equalling the flowers, whitish, woolly at the base. *Receptacle* shortly honeycombed. *Flowers* c. 36—100, c. 33—90 ♀, 3—15 ☿, yellowish. *Achenes* c. 0,5 mm long, glabrous. *Pappus* bristles many, scabrid, bases cohering lightly by patent cilia. Fig. 10: 2.

Ranges from the SW. Cape, including the Peninsula, across the mountains (including the Khamiesberg) and along the coast to the mountainous parts of Lesotho, the Transkei, the Natal Drakensberg and Midlands as far north as Laing's Nek, and the mountainous NE. corner of the O.F.S.; there is an isolated record from Bloemfontein, and another from the Magaliesberg (*Mogg* 15150, PRE); and a few from S.W.A./Namibia in Windhoek and Grootfontein districts. Also recorded from southernmost Madagascar, and naturalized in parts of Europe (Channel Islands, Finisterre peninsula, Cherbourg and S. Italy) and N. Africa (Tunis). Occurs as a wool alien in Britain. Map 28.

Favours damp places, particularly around rock outcrops or on scree, stream- and river-banks, or near forest margins; flowering mainly between December and April.

Vouchers: *Bolus* 613 (BOL); *Compton* 22647 (NBG); *Galpin* 6688 (BOL; PRE); *Giess* 11764 (M); *Hilliard & Burtt* 8742 (E; K; NU; PRE; S).

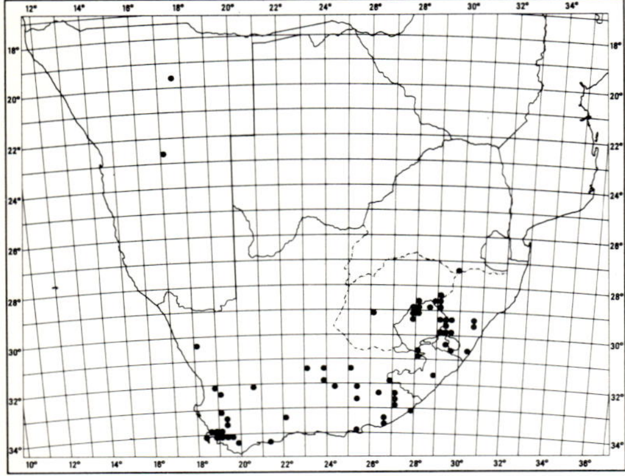

MAP 28.— **Pseudognaphalium undulatum**

2. **Pseudognaphalium oligandrum** (*DC.*) *Hilliard & Burtt* in Bot. J. Linn. Soc. 82: 204 (1981). Type: Madagascar, Emirne Province, in cultivated places, *Bojer* 13 (G−DC, holo.!).

Anaphalis oligandra DC., Prodr. 6: 275 (1838). *Gnaphalium oligandrum* (DC.) Hilliard & Burtt in Notes R. bot. Gdn Edinb. 34: 256 (1976); Hilliard, Compositae in Natal 132 (1977); Merxm. & Roessl. in Mitt. bot. StSamml., Münch. 15: 364 (1979), p.p.

Helichrysum steudelii [Sch. Bip. ex] A. Rich., Tent. Fl. Abyss. 1: 421 (1848). *Gnaphalium steudelii* (A. Rich.) Oliv. & Hiern in F.T.A. 3: 343 (1877). Type: Ethiopia, Adoua, *Schimper* 1: 231 (BM; G; K; S, iso.!).

Gnaphalium undulatum sensu Humbert, Fl. Madag. 189: 385, fig. 72, 25−35 (1962); Henderson & Anderson, Common Weeds in S. Afr. 376, fig. 187 (1966); Merxm., F.S.W.A. 139: 84 (1967), p.p., non L.

Annual herb, stems simple or branching from the base, stiffly erect to c. 1 m, branching above into the compound inflorescence, loosely white-woolly, closely leafy. *Leaves* up to 80 × 4 (−8) mm, oblong, oblong-lanceolate or lanceolate, apex sub-acute to acute, base broad, more or less cordate-clasping, scarcely decurrent, margins entire, often somewhat undulate, upper surface glandular, green drying brown, lower white-felted. *Heads* c. 4 mm long, broadly campanulate, several together in tight clusters, these arranged in a large corymbose panicle. *Involucral bracts* in c. 4 series, imbricate, innermost slightly exceeding the flowers, stereome undivided, lamina whitish, the outer 2 series white-woolly on the backs. *Receptacle* shortly honeycombed. *Flowers* c. 150−440 in our area, 135−415 ♀, 10−30 ☿. *Achenes* c. 0,5 mm long, glabrous. *Pappus* bristles many, scabrid, bases cohering lightly by patent cilia. Fig. 10: 1.

Widespread in Africa, including Madagascar, from Ethiopia and Eritrea south through Kenya, Uganda, Rwanda, Katanga, Tanzania, Zambia, Malawi and Zimbabwe to Botswana, the Transvaal, Swaziland, W. Lesotho, NE. Orange Free State, Natal, Transkei, E. Cape at Cathcart, Komgha and Coombs Valley near Grahamstown, N. Cape in Vryburg, Barkly West and Hay divisions. Also recorded from S.W.A./Namibia (Waterberg, Auros and Okahandja) and Huambo district in Angola. Map 29.

Favours dry sandy or stony places and is often found on roadsides; flowering mainly between December and May, but as early as September.

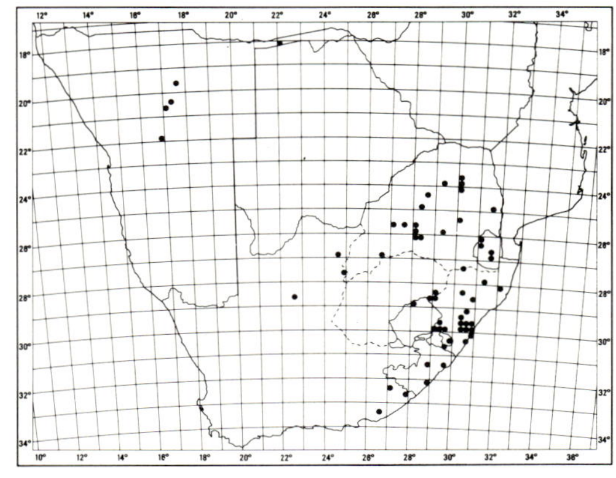

MAP 29.— **Pseudognaphalium oligandrum**

FIG. 10.−1, **Pseudognaphalium oligandrum,** part of plant showing clusters of heads arranged in a corymbose panicle, × 1; 1a, head, × 6,6; 1b, female flower, × 16; 1c, hermaphrodite flower, × 16; pappus bristle, × 16 (*Hilliard & Burtt* 7717). 2, **P. undulatum,** part of stem to show decurrent leaf, × 1,3 (*Strey* 9513). 3, **P. luteo-album,** part of plant showing clusters of heads cymosely arranged; 3a, head, × 6,6 (*Hilliard & Burtt* 7495).

1
1b
1c
1d
1a
2
3a
3
RHONA COLLETT

Much confused with *P. undulatum,* but distinguished by the leaves, not or only shortly decurrent on the stems, the heads with generally more flowers, and the stereomes of the inner bracts undivided (not fenestrated as in *P. undulatum).*

Where the two species are sympatric, they usually occupy different habitats, *P. oligandrum* in the drier situations, *P. undulatum* in the wetter.

Vouchers: *Acocks* 2180 (BOL; PRE); *Hilliard & Burtt* 7946 (E; K; MO; NU; PRE; S); *Merxmüller & Giess* 30086 (M; WIND).

Subgenus **Laphangium**

Subgen. Laphangium *Hilliard & Burtt* in Bot. J. Linn. Soc. 82: 205 (1981). Type species: *P. luteo-album* (L.) Hilliard & Burtt.

Gnaphalium section *Calolepis* Kirp. in Bot. Mater. Gerb. bot. Inst. V.A. Komarova 20: 309 (1960).

3. **Pseudognaphalium luteo-album** *(L.) Hilliard & Burtt* in Bot. J. Linn. Soc. 82: 206 (1981). Type: Herb. Van Royen (L, sheet 900. 286−294!).

Gnaphalium luteo-album L., Sp. Pl. 851 (1753); DC., Prodr. 6: 230 (1838); Harv. in F.C. 3: 262 (1865); Oliv. & Hiern in F.T.A. 3: 343 (1877); Humbert, Fl. Madag. 189: 385 fig. 72, 21−24 (1962); Adams in F.W.T.A. edn 2, 2: 266 (1963); Henderson & Anderson, Common Weeds in S. Afr. 374, fig. 186 (1966); Merxm., F.S.W.A. 139: 84 (1967); Holub in Tutin et al., Fl. Europ. 4: 128 (1976); Hilliard, Compositae in Natal 127 (1977). Type as above.

G. trifidum Thunb. Fl. Cap. 652 (1823). Type: Cape of Good Hope, *Thunberg* (sheet 19278, UPS, holo.!).

Annual herb, all parts thinly greyish-white woolly, stems to 500 mm, several from the base, usually decumbent and occasionally rooting at the base, then erect, simple below, often cymosely branched above, leafy. *Leaves* up to 80 × 10 mm, oblanceolate, becoming progressively smaller and lanceolate to linear upwards, obtuse or subacute, uppermost acute to acuminate, base clasping. *Heads* ovoid, c. 4 mm long, in dense glomerules, one or several at the branch tips. *Involucral bracts* in c. 3 series, about equalling the flowers, stereome divided, lamina translucent, palest golden-brown, buff or whitish in our area. *Receptacle* tuberculate. *Flowers* c. 70−175 ♀, 5−15 ☿, whitish, tipped reddish. *Achenes* c. 0,5 mm long, with myxogenic hairs. *Pappus* bristles many, scabrid, bases cohering lightly by patent cilia. Fig. 10: 3.

Described from southern Europe, but a widespread weed, found throughout southern Africa, nearly always on sandy or clayey soils near streams and marshes, or in gardens. Flowers more or less throughout the year.

Vouchers: *Dyer* 4153 (NU; PRE); *Esterhuysen* 11760 (NBG); *Giess & Wiss* 3260 (WIND); *Hilliard & Burtt* 7177 (E; K; MO; NU; PRE; S); *Wright* 247 (E; K; NU; NH).

8994

TENRHYNEA

Tenrhynea *Hilliard & Burtt* in Bot. J. Linn. Soc. 82: 232 (1981). Type species: *T. phylicifolia* (DC.) Hilliard & Burtt.

Rhynea DC., Prodr. 6: 154 (1838); Harv. in F.C. 3: 204 (1865); Hilliard & Burtt in Notes R. bot. Gdn Edinb. 32, 3: 374 (1973), in Bot. J. Linn. Soc. 82: 196 (1981); Hilliard, Compositae in Natal 134 (1977), non Scop. (1777).

A shrubby perennial up to 1 m tall, branches closely leafy. *Leaves* alternate, up to 30 × 10 mm, often much smaller, lanceolate, margins entire, bases decurrent in long narrow wings, green above drying grey, with thin skin-like indumentum, white-tomentose below. *Heads* heterogamous, cylindric, c. 5 × 2 mm, in small terminal clusters, often further aggregated into a spreading corymbose panicle. *Involucral bracts* in several series, imbricate, graded, inner ones with stereome undivided, tips shortly exceeding the flowers, radiating, opaque milk-white or occasionally pink. *Receptacle* with paleae resembling the inner involucral bracts. *Flowers* 17−26, 3−11 ♀, corolla narrowly tubular, 10−17 ☿, outnumbering the ♀, corolla tubular below, campanulate above, 5-lobed, all lobes with glandular hairs on backs. *Anthers* with a lanceolate apical appendage and tails shorter or longer than the filament collar. *Style* branches with slightly rounded tips, sweeping hairs short, extending briefly down the backs. *Achenes* with elongate duplex hairs with a basal swelling cushion, myxogenic. *Pappus* bristles scabrid above, the cilia more or less fused, shaft scabridulous, base nude, not cohering, caducous.

A monotypic genus endemic to eastern South Africa. Allied to *Cassinia* R. Br., endemic to Australia and New Zealand.

Tenrhynea phylicifolia *(DC.) Hilliard & Burtt* in Bot. J. Linn. Soc. 82: 232 (1981). Type: Transkei, between the Umsikaba and Umzimvubu Rivers, *Drège* 5013 (G-DC, holo.!; E; K; iso.!).

Rhynea phylicifolia DC., Prodr. 6: 154 (1838), in Del., Ic. Sel. 4, t. 52 (1840); Harv. in F.C. 3: 205 (1865); Hilliard, Compositae in Natal 134 (1977). *Cassinia phylicifolia* (DC.) Wood, Natal Plants 4,3: t. 355 (1906).

The only species, ranging from the Zoutpansberg to the highlands of the E. Transvaal and Swaziland, through Natal from near sea level to c. 1 220 m, to about Port St Johns in the Transkei. Grows in grassland, often in coarse herbage near forest margins; flowering between February and May. Fig. 11:1. Map 30.

Sometimes mistaken for *Helichrysum* but easily recognized by its narrow decurrent leaves and heads with paleae resembling the involucral bracts.

Vouchers: *Acocks* 12712 (NU; PRE); *Codd* 4196 (NU; PRE); *Hilliard & Burtt* 9958 (E; K; MO; NU; S).

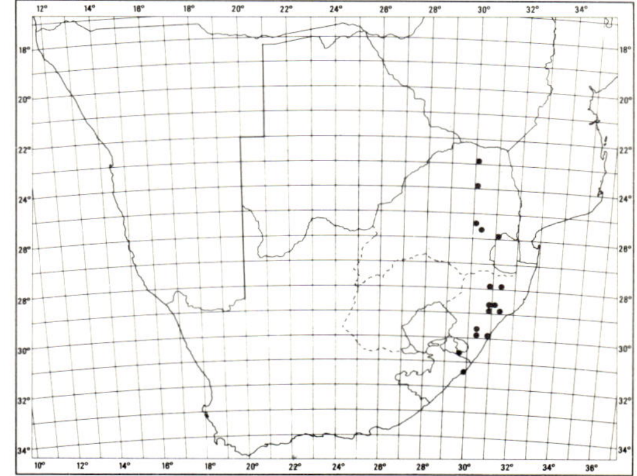

MAP 30.— **Tenrhynea phylicifolia**

2c
2d
2e
2b
2a
1d
1e
1f
1a
1b
1c
2
1
RHONA
COLLETT

8994a

ACHYROCLINE

Achyrocline *(Less.) DC.*, Prodr. 6: 219 (1838); Benth. in Benth. & Hook. f., Gen. Pl. 2,1: 305 (1873); Oliv. & Hiern in F.T.A. 3: 339 (1877); Hutch. & Dalz., F.W.T.A. 2,1: 158 (1931); Phill., Gen. 2: 792 (1951); Humbert, Fl. Madag. 189: 386, fig. 73, 1−7 (1962); Cabrera, Fl. Prov. Buenos Aires 4,6: 153, fig. 39 (1963); Hilliard & Burtt in Bot. J. Linn. Soc. 82: 201 (1981). Type species: *A. satureioides* (Lam.) DC.

Gnaphalium subgenus *Achyrocline* Less., Syn. Comp. 332 (1832).

Half-shrubs or perennial herbs, erect or straggling. *Leaves* alternate, linear, lanceolate or elliptic, more rarely ovate, acute to acuminate, petiolate or sessile, bases sometimes decurrent in long stem wings, woolly or glandular. *Heads* heterogamous, small, in crowded clusters, usually further arranged in corymbose panicles, or rarely solitary. *Involucral bracts* in several series, graded, loosely imbricate, stereomes of inner fenestrated, lamina white, yellow or rufous, about equalling the flowers. *Receptacle* smooth, honeycombed or shortly fimbrilliferous. *Flowers* few (c. 4−10), ♀ outnumbering the ☿, ☿ corolla narrowly tubular, ♀ flowers tubular scarcely widened above, 5-lobed, all flowers glandular-hairy on backs of lobes. *Anthers* with a small lanceolate apical appendage, tails about equalling the filament collar. *Style* branches truncate and penicillate. *Achenes* usually glabrous or rarely hairy. *Pappus* bristles scabridulous, apical cells slightly inflated, bases cohering by patent cilia.

About 25 species, mostly in tropical and subtropical America, a few in tropical Africa, one reaching South Africa, one endemic to Madagascar and the Comoros.

Achyrocline stenoptera *(DC.) Hilliard & Burtt* in Bot. J. Linn. Soc. 82: 202 (1981). Type: Natal, between the Umkomaas and Umzimkulu Rivers, *Drège* 5011 (G-DC, holo.!; BM; K, iso.!).

Helichrysum stenopterum DC., Prodr. 6: 201 (1838); Harv. in F.C. 3: 244 (1865); Moeser in Bot. Jb 44: 241 (1910); Compton, Fl. Swaziland 635 (1976); Hilliard, Compositae in Natal 149 (1977). *Gnaphalium stenopterum* (DC.) Sch. Bip. in Bot. Ztg 3: 172 (1845).

Achyrocline hochstetteri [Sch. Bip. ex] A. Rich., Tent. Fl. Abyss. 1: 429 (1848); Oliv. & Hiern in F.T.A. 3: 339 (1877); Hutch. & Dalz., F.W.T.A. 2,1: 159 (1931). *Helichrysum hochstetteri* (A. Rich.) Hook. f. in J. Linn. Soc. Bot. 6: 13 (1861). Syntypes: Ethiopia, Mt Kubbi, *Schimper* I: 237 (K!): ibid. *Schimper* II: 1058 (K!; S!).

H. gerrardii Harv. in F.C. 3: 244 (1865); Moeser in Bot. Jb. 44: 242 (1910). Type: Natal, Nototi (sic: Non-oti?), *Gerrard* 1001 (TCD, holo.! BM; K, iso.!).

A straggling, strongly aromatic perennial herb, stems up to 2m long when supported by other vegetation, simple below, branching above, branches long, terminating in corymbose panicles, gland-dotted, commonly greyish-white woolly, sometimes glabrous, leafy. *Leaves* up to 120 × 15 mm, decreasing in size upwards and passing into distant inflorescence bracts, lanceolate or linear-lanceolate, apex very acute to acuminate, base narrowed, decurrent on stem in long narrow wings, margins revolute, gland-dotted below, both surfaces glabrous or, more commonly, scabrous above and lightly cobwebby at least initially, white-felted below. *Heads* heterogamous, c.

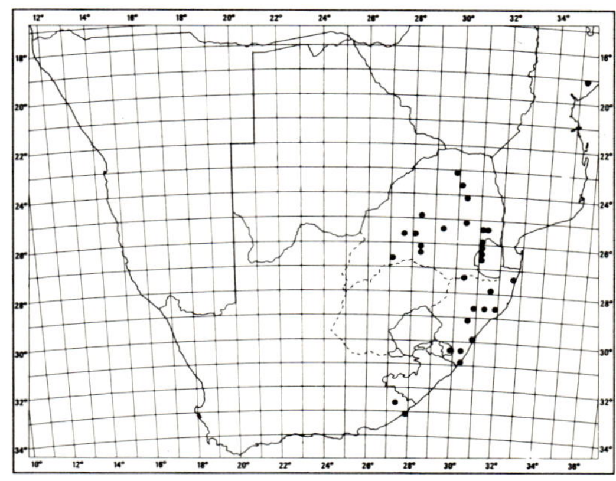

MAP 31.— **Achyrocline stenoptera**

FIG. 11.−1, **Tenrhynea phylicifolia,** part of plant, × 1; 1a, leaf showing decurrent base, × 2; 1b, head, × 6,6; 1c, palea, × 10; 1d, hermaphrodite flower, × 10; 1e, female flower, × 10; 1f, pappus bristle, × 16 (*Hilliard* 2812). 2, **Achyrocline stenoptera,** part of plant, × 1; 2a, leaf showing decurrent base, × 1,3; 2b, head, × 13; 2c, pappus bristle, × 16; 2d, hermaphrodite flower, × 10; 2e, female flower, × 10 (*Hilliard* 5020).

3 × 1 mm, very many in congested cymose clusters corymbose-paniculately arranged. *Involucral bracts* in 2−3 series, loosely imbricate, about equalling flowers, not radiating, translucent, inner tipped bright canary-yellow. *Receptacle* with fimbrils at least equalling ovaries. *Flowers* 4−6, 3−5 ♀, 1−2 ☿. yellow. *Achenes* 0,75 mm, glabrous. *Pappus* bristles many, delicate, bases cohering by patent cilia. Fig. 11:2.

Widespread in the E. half of S. Africa from East London and Komgha in the eastern Cape, through the more coastal parts of Natal to the Transvaal, Swaziland, Mozambique, Zimbabwe, Zambia, East Africa, Ethiopia, Yemen and West Africa. Grows in streamside or marshside scrub, straggling up through rank vegetation. Flowers in April and May. Much confused with *Helichrysum odoratissimum* (p. 7,2: 74) and *H. gymnocomum* (p. 7,2: 75) but easily distinguished by its very acute to acuminate leaves and heterogamous heads in which the female flowers outnumber the hermaphrodite. Map 31.

Vouchers: *Compton* 27767 (PRE); *Gerstner* 8309 (PRE); *Hilliard* 5020 (E; K; S; NBG; NH; NU; PRE; SRGH); *Hilliard & Burtt* 10305 (E; K; NU; MO; S; PRE); *Wilms* 740 (E; NU).

9006   **HELICHRYSUM**

Helichrysum *Mill.*, Gard. Dict. abr. edn 4 (1754) (*'Elichrysum'*), corr. Pers., Syn. Pl. 2: 414 (1807) nom. cons.; Harv. in F.C. 3: 207 (1865), p.p.; Benth. & Hook. f., Gen. Pl. 2, 1: 309 (1873), p.p.; Hoffmann in Natürl. Pfl Fam. 4,5: 190 (1897), p.p.; Moeser in Bot. Jb. 44: 239 (1910), p.p.; Hilliard & Burtt in Bot. J. Linn. Soc. 82: 193 (1981). Lectotype species: *H. orientale* (L.) Gaertn.

Lepiscline Cass. in Bull. Soc. philom. 1818: 31 (1818). Type species: *L. cymosum* (L.) Cass.

Leontonyx Cass., Dict. sci. Nat. 25: 466 (1822). Lectotype species: *L. tomentosus* Cass., nom illegit. = *L. squarrosus* DC., nom. illegit. = *Helichrysum spiralepis* Hilliard & Burtt.

Euchloris D. Don in Mem. Wern. nat. Hist. Soc. 5: 548 (1826). Type species: *E. nudifolia* (L.) D. Don.

Pentataxis D. Don in Mem. Wern. nat. Hist. Soc. 5: 550 (1826). Type species: *P. micrantha* D. Don.

Spiralepis D. Don in Mem. Wern. nat. Hist. Soc. 5: 551 (1826). Lectotype species: *S. squarrosa* D. Don.

Manopappus Sch. Bip. in Flora, Regensburg 27: 677 (1844). Type species: *M. anomalus* (Less.) Sch. Bip.

Annual, biennial or perennial herbs, suffrutices or shrubs, generally erect, sometimes prostrate, occasionally more or less scandent, usually woolly or cobwebby, often glandular as well, rarely glabrous. *Leaves* variously shaped, sessile or petiolate, margins entire, flat or revolute, or rarely weakly involute. *Heads* cylindric, turbinate, campanulate, subglobose or depressed-globose, homogamous or heterogamous with ♀ flowers generally fewer than, sometimes more than, or about equalling ☿, often in lax or congested corymbose panicles, or sometimes in glomerules or sometimes solitary. *Involucral bracts* in few to many series, stereome usually fenestrated, rarely undivided, lamina equalling or exceeding flowers, often radiant, white, yellow, straw-coloured, brown, pink or red, sometimes a combination of colours, pellucid to opaque, glabrous or hairy. *Receptacle* smooth, honeycombed or fimbrilliferous, rarely paleate. *Flowers* 2—2 500, corolla of ♀ flowers often narrowly tubular, sometimes with a well-developed limb, or rarely filiform; of ☿ flowers funnel-shaped or tubular and then often more or less campanulate above, lobes 5; hairs always present on backs of corolla lobes. *Anthers* generally with apical appendages lanceolate, rarely with mucronate tip; thickenings in endothecial tissue polarized. *Style* branches truncate and penicillate. *Achenes* usually glabrous or with common duplex hairs, these often myxogenic, or rarely duplex hairs without a swelling cushion; unique multicellular hairs known only in *H. sphaeroideum;* outer walls of epidermal cells smooth or imbricate. *Pappus* generally present, rarely wanting, generally uniseriate, rarely in more than one series, setae few to many, shaft smooth, scabridulous, scabrid or barbellate below, tips or upper part of shaft scabrid, barbellate, sub-plumose or plumose, the uppermost cells often clavate, bases often free or cohering by patent cilia, more rarely partly or completely fused in a ring; soon caducous.

About 500 species, mainly African, including Madagascar, also southern Europe, SW. Asia, S. India, Ceylon, Australia; c. 245 species in S. Africa, widely distributed.

The retention of the species outside Africa and the Mediterranean in the genus as defined here needs confirmation.

The Southern African species display great morphological diversity and it has been found convenient to divide them into groups. A key to the groups is given below; keys to the species follow the description of each group. These groups cannot be given formal recognition until the genus has been examined over its whole range.

1a Heads cylindric, turbinate, campanulate or subglobose, never depressed-globose: (1b on p. 7,2: 65)

2a Heads 4—5 mm long containing 5 flowers, involucral bracts opaque white or sometimes rosy, ovaries glabrous. Small shrubs, leaf margins flat or undulate ....Group 1 (p. 7,2: 65)

2b Combination of characters not as above:

3a Heads either homogamous containing less than 30 flowers or heterogamous with less than 20 flowers, involucral bracts about as long as the flowers, not radiating (though occasionally squarrose), bright yellow, receptacle smooth or honeycombed (not fimbrilliferous):

4a Ovaries glabrous:

  5a Heads generally heterogamous ((0−) 1−3 female flowers)............................ Group 2 (p. 7,2: 68)

  5b Heads homogamous:

    6a Pappus either wanting or bristles few, nude below, shortly plumose above:

      7a Pappus wanting or if present leaf bases decurrent in long stem wings .... Group 3 (p. 7,2: 71)

      7b Pappus present, leaf bases not decurrent... Group 6 (p. 7,2: 82)

    6b Pappus bristles barbellate or shortly subplumose in upper part, bases cohering by patent cilia .... Group 7 (p. 7,2: 90)

4b Ovaries hairy:

  8a Heads very many, felted together to form one or a few flat umbrella-like discs... Group 5 (p. 7,2: 79)

  8b Heads either free from one another or, if felted together, not forming a flat umbrella-like disc:

    9a Pappus either wanting or if present leaf bases decurrent in long narrow stem wings .... Group 4 (p. 7,2: 74)

    9b Pappus present, leaf bases not decurrent in long wings:

      10a Pappus bristles nude below, not cohering .... Group 6 (p. 7,2: 82)

      10b Pappus bristles with patent cilia at base, cohering or not:

        11a Heads containing 5 flowers; summit of the Drakensberg .... Group 7 (p. 7,2: 90)

        11b Heads containing 6−30 flowers:

          12a Stems more or less uniformly leafy throughout, leaves mostly 4−20 × 1−8 mm... Group 12 (p. 7,2: 122)

          12b Plants with tufts of radical leaves up to c. 80−250 × 15−50 mm .... Group 21 (p. 7,2: 212)

3b Involucral bracts either not bright yellow, or heads with more than 30 flowers, or bracts radiating, or receptacle fimbrilliferous, or characters otherwise not as combined in 3a:

  13a Heads homogamous, receptacle fimbrilliferous, ovaries glabrous:

  14a Shrubs or subshrubs, hard- or soft-wooded:

    15a Leaves (ignore reduced upper ones) linear, lanceolate, oblanceolate or narrowly elliptic:

      16a Leaves sessile, sometimes subdecurrent; heads 3 (−4, 5) × 1−2 (−3) mm. Group 8 (p. 7,2: 92)

      16b Leaves minutely petiolate; heads c. 4 × 3 mm................................... Group 11 (p. 7,2: 119)

    15b Leaves (ignore upper reduced ones) broadly elliptic, ovate or rhomboid-elliptic to orbicular, or panduriform:

      17a Soft-wooded shrubs or subshrubs, branches often tangled or more or less scandent; in shrub communities and on forest margins .... Group 18 (p. 7,2: 198)

      17b Erect twiggy shrubs; among rock outcrops..................................... Group 19 (p. 7,2: 203)

  14b Perennial herbs, frequently with tufts of radical leaves:

    18a Stems tufted, wiry, decumbent... Group 22 (p. 7,2: 215)

    18b Stems solitary or subsolitary, stout, erect:

      19a Involucral bracts about equalling the flowers, not radiating, though squarrose in one species .... Group 23 (p. 7,2: 226)

      19b Involucral bracts exceeding the flowers, radiating, sometimes squarrose as well in one species .... Group 24 (p. 7,2: 244)

  13b Heads heterogamous or if homogamous, either receptacle smooth or honeycombed but not fimbrilliferous, or ovaries hairy:

  20a Receptacle paleate:

    21a Heads c. 5 mm long, many, either clustered or in a large corymbose panicle, involucral bracts white or brownish yellow...Group 20 (p. 7,2: 209)

    21b Heads c. 10 mm long, solitary or few in lax corymbs, involucral bracts lemon-yellow .... Group 29 (p. 7,2: 279)

  20b Receptacle smooth, honeycombed or fimbrilliferous, but not paleate:

22a Leaves on petioles up to 70 mm long, blade ovate to subrotund, up to c. 130 × 110 mm, base cordate; heads 2−3 mm long, many in small clusters arranged in spreading corymbose panicles .... Group 16 (p. 7,2: 181)

22b Either leaves or heads not as above:

  23a Heads 7−20 mm long, involucral bracts exceeding flowers, radiating, usually white, pink, rose or combinations of these colours, often with brown as well, sometimes yellow or light brown:

    24a Annuals; arid areas .... Group 15 (p. 7,2: 153)

    24b Herbaceous perennials, subshrubs or shrubs:

      25a Ovaries glabrous, pappus bristles without patent cilia at base, not cohering:

        26a Involucral bracts mostly white with some brown or crimson, if yellow, then plants either mat-forming or herbaceous perennials.......................... Group 29 (p. 7,2: 279)

        26b Involucral bracts yellow or light brown; shrubs or coarse woody perennials .... Group 30 (p. 7,2: 287)

      25b Ovaries hairy or rarely glabrous and then pappus bristles either with minute patent cilia at base, cohering or not, or bristles fused together at base:

        27a Pappus bristles either lightly fused at base in a smooth ring, or bases cohering by patent cilia with some light fusion as well; involucral bracts combinations of white and golden-brown, rose and golden-brown, or white and rose, receptacle smooth or shortly honeycombed .... Group 26 (p. 7,2: 259)

        27b Pappus bristles either free at the base or cohering by patent cilia, if fused or with some fusion in bundles as well, involucral bracts either white or rose but not combinations of colours, or receptacle fimbrilliferous:

          28a Herbaceous perennials with one or more tufts of radical leaves, flowering stems simple, erect, usually l-headed, rarely (in 2−3 species) bearing several heads in a lax corymbose panicle, involucral bracts usually self-coloured, sometimes parti-coloured in *H. adenocarpum* (no. 209) and *H. petraeum* (no. 210) .... Group 28 (p. 7,2: 270)

          28b Habit, arrangement of heads or bracts not as described above:

            29a Plants mat-forming or tufted, leaves linear to lanceolate-elliptic, sessile; involucral bracts white or sometimes palest yellowish buff outside, bases of pappus bristles cohering strongly by patent cilia .... Group 25 (p. 7,2: 256)

            29b Plants differing from above in habit, or leaf shape (ignore uppermost reduced leaves), or bracts variously coloured, or bases of pappus bristles not cohering strongly by patent cilia:

              30a Heads 10−15 mm long, involucral bracts white, rose and brown, or if only c. 7 mm long, then a hard cushion-forming shrub with pale (sometimes very pale) lemon-yellow involucral bracts .... Group 27 (p. 7,2: 265)

              30b Heads 7−11 mm long, involucral bracts white or white tipped brown .... Group 17 (p. 7,2: 184)

  23b Heads up to 6 mm long, involucral bracts either equalling or longer than the flowers, erect or radiating, variously coloured, or if heads 7 mm long, then bracts either equalling the flowers and not radiating, or bright canary-yellow:

    31a Involucral bracts in 4 series, 2 outer short, webbed together with wool, often more or less persistent, 2 inner soon caducous, tip of lamina with an opaque median patch often drawn out into an acute or acuminate and then recurved tip, or tip truncate and often more or less emarginate; woolly herbs, often prostrate or mat-forming, leaves elliptic-oblong to spathulate, obovate or subrotund....................... Group 14 (p. 7,2: 142)

    31b Involucral bracts not as described above:

      32a Involucral bracts silvery, white, white and rose, or wholly rose, or white with brown tips or pale brown outside; receptacle smooth or honeycombed, or rarely fimbrilliferous and then heads homogamous:

        33a Twiggy shrublet with stiff branches and hard spathulate leaves; heads c. 4 × 1 mm, tightly clustered... Group 2 (p. 7,2: 68)

        33b Habit, foliage or heads not as above:

34a Outer involucral bracts white, spathulate, larger than inner; prostrate perennial herb with slightly falcate silvery leaves and heads in congested terminal clusters. Natal Drakensberg.. Group 27 (p. 7,2: 265)

34b Involucral bracts neither spathulate nor outer larger than inner:

35a Tips of inner involucral bracts either about equalling the flowers and not radiating or minutely radiating; leaves generally linear with revolute margins, if broader, receptacle either fimbrilliferous or with flattened tubercles .... Group 12 (p. 7,2: 122)

35b Tips of involucral bracts mostly exceeding the flowers, radiating or squarrose; leaves mostly broad with flat margins, or if linear with revolute margins, receptacle honeycombed; if bracts about equalling flowers and either not radiating or minutely radiating, leaves not linear with revolute margins:

36a Annual herbs or if perennial or shrubby, involucral bracts either acute to acuminate or squarrose, or both; or leaves linear with revolute margins. Most species confined to S.W.A./Namibia and the arid parts of the Cape; a few ranging eastwards but always in dry sandy places...... Group 15 (p. 7,2: 153)

36b Either shrubs or perennial herbs with obtuse involucral bracts. Cape, Natal and Lesotho, most species on the mountains, one (*H. crispum*, no. 122) a strand plant .... Group 17 (p. 7,2: 184)

32b Involucral bracts straw-coloured, tawny or various shades of yellow or brown, red tints sometimes associated with these colours; if white or whitish, or with some rose or wholly rosy, then heads heterogamous and receptacle fimbrilliferous:

37a Shrubs or bushy perennials, leaf blade woolly, elliptic to ovate or obovate contracted to a petiole-like base, heads homogamous.............. Group 19 (p. 7,2: 203)

37b Either habit or foliage not as described above, or heads heterogamous:

38a Involucral bracts white or sometimes partly or wholly rosy, receptacle fimbrilliferous:

39a Ovaries glabrous... Group 8 (p. 7,2: 92)

39b Ovaries hairy, or sometimes hairy or glabrous in the same species (*H. teretifolium* no. 35) and then leaves linear with revolute margins Group 9 (p. 7,2: 101)

38b Involucral bracts straw-coloured, tawny, or various shades of yellow or brown, red tints sometimes associated with these colours:

40a Heads heterogamous, few-flowered (up to 8 flowers), involucral bracts not radiating, receptacle smooth, ovaries glabrous..................... Group 2 (p. 7,2: 68)

40b Characters not combined as above:

41a Receptacle fimbrilliferous, ovaries glabrous, corolla of ♀ flowers funnel-shaped .... Group 8 (p. 7,2: 92)

41b Receptacle smooth or honeycombed, or if fimbrilliferous, either ovaries hairy or corolla of ♀ flowers campanulate above:

42a Heads homogamous, containing less than 30 flowers, corolla funnel-shaped, ovaries usually hairy, occasionally hairs few or wanting on the same plant.. Group 12 (p. 7,2: 122)

(and occasional specimens from Group 15 will run down here)

42b Heads either heterogamous (♀ flowers may be very few or only one) or if homogamous, either corollas campanulate above or ovaries glabrous, or both

43a At least the inner involucral bracts clear yellow, minutely radiating but not squarrose, receptacle honeycombed.................. Group 22 (p. 7,2: 215)

43b Involucral bracts straw-coloured, tawny, or shades of golden brown, sometimes with red or purple admixtures, rarely clear yellow and then either involucral bracts more or less squarrose or receptacle fimbrilliferous:

44a Corollas of ♀ flowers funnel-shaped.................... Group 15 (p. 7,2: 153)

44b Corollas of ♀ flowers campanulate above:

45a Ovaries glabrous, receptacle smooth or honeycombed, involucral bracts with conspicuous glabrous tips............... Group 13 (p. 7,2: 136)

45b Ovaries hairy, or if glabrous, then receptacle fimbrilliferous or involucral bracts hairy all over the backs with only the extreme tips occasionally glabrous:

46a Heads either heterogamous with involucral bracts shortly radiating or not, leaf margins flat or revolute; or rarely homogamous and then tips of bracts minutely radiating or leaf margins strongly revolute:

47a Leaves mostly linear to lanceolate or narrowly oblong with revolute or subrevolute margins; corollas of ♀ flowers narrowly campanulate above .... Group 9 (p. 7,2: 101)

47b Leaves mostly oblong, elliptic, ovate or obovate (ignore reduced uppermost leaves) with flat margins; or linear to oblong or linear-lanceolate with revolute or subrevolute margins in 2 species, but corollas of ♀ flowers (as in all species in this group) broadly campanulate, almost hemi-spherical, above .... Group 10 (p. 7,2: 111)

46b Heads homogamous, involucral bracts not radiating but leaf margins flat .... Group 11 (p. 7,2: 119)

1b Heads depressed-globose, large (containing at least 300 flowers), achenes glabrous though epidermis often rough-looking from imbricate cells. Coarse annuals, biennials or perennials or shrubs; involucral bracts frequently bright yellow, sometimes brown outside or wholly brown, very rarely white, whitish, or pink .... Group 30 (p. 7,2: 287)

Group 1

Small shrubs, *leaves* small, oblong or obovate, often minutely petiolate; *heads* homogamous, c. 4−5 × 2 mm, in small congested or lax corymbose panicles; *involucral bracts* not radiating, white or sometimes rosy; *receptacle* fimbrilliferous; *flowers* usually 5, corolla cylindric; *achenes* glabrous; *pappus* bristles scabrid, bases cohering strongly by patent cilia.

Species 1−2, in Angola, S.W.A./Namibia, Botswana, and the arid parts of the SW. Transvaal, Cape, O.F.S. and Lesotho.

1a Leaves mostly 6−23 × 2−9 mm, linear-oblong, oblong or obovate, bases usually broad and often subcordate-clasping, petiole-like only in the larger obovate leaves, margins often undulate or crisped .... 1. *H. zeyheri*

1b Leaves mostly 20−80 × 6−15 mm, lanceolate or oblanceolate with a small petiole distinct in all but the uppermost reduced leaves, margins flat or slightly undulate near the base .... 2. *H. tomentosulum*

1. **Helichrysum zeyheri** *Less.*, Syn. Comp. 309 (1832); DC., Prodr. 6: 196 (1838); Harv. in F.C. 3: 253 (1865); Moeser in Bot. Jb. 44: 254 (1910); Merxm., F.S.W.A. 139: 99 (1967). Lectotype: S. Africa, Vaal River, Rhenosterkop, *Zeyher* 885 (K!; BM; PRE; S; SAM, isolecto.!).

Gnaphalium zeyheri (Less.) Sch. Bip. in Bot. Ztg 3: 171 (1845).

Pentataxis micrantha D. Don in Mem. Wern. nat. Hist. Soc. 5: 550 (1826), non *Helichrysum micranthum* DC. Type: Cape of Good Hope, *Labillardière* s.n.

Helichrysum burchellii DC., Prodr. 6: 196 (1838) var. *α complicatum* DC. (i.e. var. *burchellii*). *Gnaphalium burchellii* (DC.) Sch. Bip. in Bot. Ztg 3: 171 (1845). *Helichrysum zeyheri* var. *burchellii* (DC.) Harv. in F.C. 3: 254 (1865). Lectotype: Trans Gariep, near source of Kuruman River at Little Klibbolikhonni, *Burchell* 2506 (G-DC!).

H. burchellii var. *β intermedium* DC., l.c. *Helichrysum zeyheri* var. *intermedium* (DC.) Harv. in F.C. 3: 254 (1865). Lectotype: Zwartberge, *Drège*, 773 (G-DC!; K, isolecto.!).

Dense or open shrublet c. 150−1 000 mm tall, branches virgate, greyish-white woolly, leafy. *Leaves* mostly 6−23 × 2−9 mm, diminishing upwards, linear-oblong, oblong or obovate, apex obtuse to acute, mucronate, base sometimes petiole-like in the larger obovate leaves, usually broad and often subcordate-clasping, margins often undulate or crisped, both surfaces closely greyish-white woolly, wool sometimes thin, only the midvein visible. *Heads* homogam-ous, cylindric, c. 4 × 2 mm, many in terminal compact corymbose panicles. *In-volucral bracts* in 4−5 series, graded, inner

1b
1a
2a
2b
1c 1d
1
2
H. Wouda-du Toit

about equalling the flowers, not radiating, imbricate, oblong-lanceolate, more or less distinctly keeled, blunt or subacute, often more or less truncate and emarginate, tips opaque white or inner sometimes rosy. *Receptacle* with fimbrils nearly equalling ovaries. *Flowers* nearly always 5. *Achenes* 0,75 mm, glabrous. *Pappus* bristles many, about equalling corolla, scabrid, bases cohering strongly by patent cilia. Fig. 12: 1.

Widespread, and common in the drier parts of the Cape from Tulbagh, Paarl and Stellenbosch east through Worcester, Laingsburg, Ladismith, Oudtshoorn and Prince Albert districts to Humansdorp and Uitenhage, in the west northwards through Clanwilliam, Springbok and Kenhardt districts to the 21st parallel in South West Africa/Namibia, in the east, northwards through Graaff-Reinet, Cradock, Victoria West, Middelburg, Colesburg, Postmasburg, Kuruman and Vryburg to Messina, Zeerust and Potgietersrus districts in the Transvaal, Bethulie, Fauresmith, Bloemfontein and Ficksburg in the O.F.S. Also in Botswana and Lesotho. On sandy and stony soils; flowering mainly between November and May. Recorded as palatable to stock. Map 32.

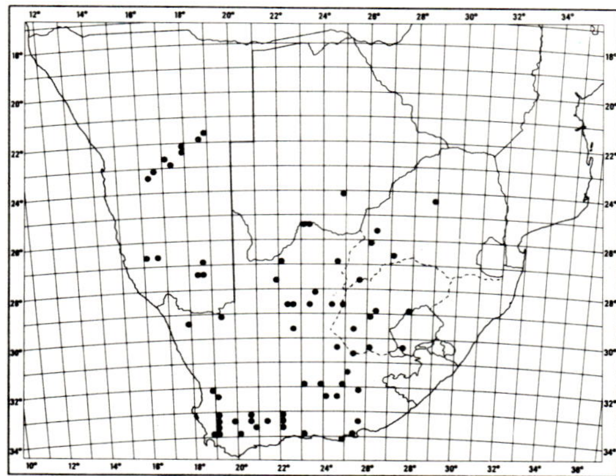

MAP 32.— **Helichrysum zeyheri**

Very variable particularly in stature and leaf shape and Harvey (l. c.) recognized three varieties based primarily on leaf shape and the degree of undulation of the leaf margins, but these are scarcely tenable now a wide range of material is available. Not easily confused with any species in our area other than its close ally *H. tomentosulum* (no. 2), but that species has generally much larger leaves (mostly 20−80 × 6−15 mm), differently shaped and with generally flat, not undulate nor crisped, margins, and all but the uppermost reduced leaves distinctly though often minutely petiolate, not sessile. Also, the involucral bracts in *H. zeyheri* are generally blunter than in *H. tomentosulum*.

Vouchers: *Acocks* 12420 (PRE); *Codd* 3537 (PRE); *Esterhuysen* 15419 (BOL; NBG); *Giess* 12453A (PRE; WIND).

2. **Helichrysum tomentosulum** *(Klatt) Merxm.* in Mitt. bot. StSamml., Münch. 1: 410 (1954), F.S.W.A. 139: 98 (1967); Merxm. & Roessl. in Mitt. bot. StSamml., Münch. 15: 367 (1979). Type: [S.W.A./Namibia] Hereroland, Nosob, *Fleck* 261 (Z, holo.!).

Stenocline tomentosula Klatt in Bull. Herb. Boissier 3: 434 (1895).

Helichrysum benguellense Hiern var. *latifolium* [S. Moore ex] Moeser in Bot. Jb. 44: 254 (1910). Lectotype: S.W.A./Namibia, Okahandja, *Dinter* 565 (Z!; E; K; SAM, isolecto.!;).

Two subspecies are recognized:

(a) subsp. **tomentosulum**.

Shrublet c. 450−1 000 mm tall, branches virgate, greyish-white woolly, often yellowish in dry state, leafy. *Leaves* mostly 20−80 × 6−15 mm, diminishing upwards, lanceolate or oblanceolate, apex acute to subobtuse, mucronate, petiole often small, but distinct in all but the uppermost reduced leaves, margins flat or somewhat undulate near the base, both surfaces closely greyish-white woolly, wool sometimes thin, 3 main veins of the larger leaves often visible on one or both surfaces, sometimes secondary nerves as well. *Heads* homogamous, cylindric, c. 4−5 × 2 mm, many in terminal congested or spreading corymbose panicles. *Involucral bracts* in 4 series, graded, inner about equalling flowers, not radiating, imbricate, oblong-lanceolate, sometimes keeled, subacute to acuminate, opaque white or inner sometimes rosy. *Receptacle* with short fimbrils. *Flowers* nearly always 5. *Achenes* 0,75 mm, glabrous. *Pappus* brist-

---

FIG. 12.−1, **Helichrysum zeyheri**, part of plant, × 1; 1a, part of twig to show sessile leaves with undulate margins, × 1,3; 1b, head, × 10; 1c, flower, × 13; 1d, pappus bristle, × 13 (*Brueckner* 1085). 2, **H. tomentosulum** subsp. **tomentosulum**, flowering branch, × 1; 2a, part of twig to show petiolate leaves with flat margins, × 1,3; 2b, head, × 10 (*Oliver* s.n.).

les many, about equalling corolla, scabrid, bases cohering by patent cilia. Fig. 12: 2.

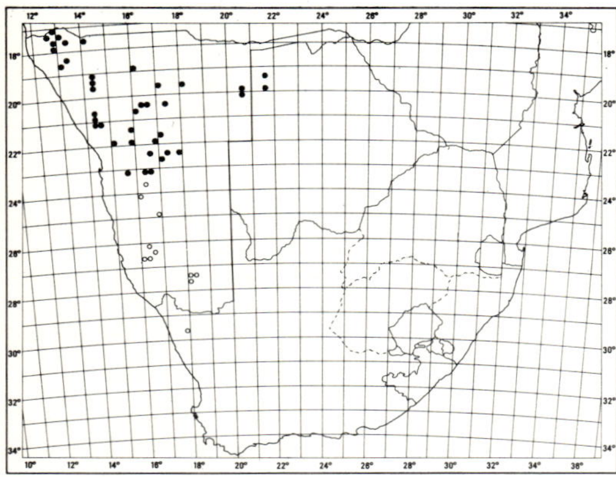

MAP 33.— ● **Helichrysum tomentosulum** subsp. **tomentosulum**
○ **Helichrysum tomentosulum** subsp. **aromaticum**

Recorded from about the 23rd parallel in South West Africa/Namibia northwards to the Kaokoveld in the north-west and nearby parts of Angola, and north-east to the Waterberg, Grootfontein, Nama Pan and the Aha Hills in Botswana (19° 45'S' 21° 8'E). Grows on sandy and stony soils, flowering between March and August. Map 33.

Closely allied to *H. zeyheri* (no. 1): see under that species.

Vouchers: *Giess* 12378 (M; PRE; WIND); *Seydel* 1153 (M; PRE; WIND); *De Winter & Leistner* 5448 (K; M; PRE; WIND).

(b) subsp. **aromaticum** *(Dinter)* *Merxm.* in Mitt. bot. StSamml., Münch. 2: 330 (1957), in F.S.W.A. 139: 98 (1967). Type: S.W.A./Namibia, Aus, *Dinter* 1070.

H. aromaticum Dinter in Fedde, Repert. 18: 248 (1922).

Differs from the typical plant in its twiggy habit, with short, stiff branches, each terminating in a small cluster of heads. The heads can be as tightly clustered in subsp. *tomentosulum* as in subsp. *aromaticum,* but the open, loosely branched habit of the former is diagnostic.

The two subspecies appear to be completely allopatric, subsp. *aromaticum* having been recorded only south of the 23rd parallel in South West Africa/Namibia and extending into Namaqualand. Map 33.

Vouchers: *Dinter* 5014 (BOL; G; K; PRE); *Kinges* 2469 (M; PRE; WIND); *Merxmüller & Giess* 28189 (M; PRE; WIND).

Group 2

Small shrubs or bushy perennial herbs; *leaves* small, oblong to spathulate; *heads* heterogamous (or rarely homogamous in the same species), 3−4 × 1−2 mm, in congested corymbose panicles; *involucral bracts* not radiating, silvery or lemon-yellow; *receptacle* smooth; *flowers* 3−8, (0−) 1−2 ♀, corolla of ♀ flowers narrowly campanulate above, *achenes* glabrous; *pappus* bristles scabrid, bases with small patent cilia, not cohering.

Species 3−5, mainly S. Africa, in the Cape, O.F.S., Lesotho, Transvaal Highveld and E. Highlands and Natal near the Drakensberg, *H. callicomum* reaching E. highlands of Zimbabwe. On poor stony or sandy soils.

1a Leaves closely woolly .... 3. *H. callicomum*

1b Leaves enveloped in 'tissue-paper' indumentum, becoming woolly only with age:

　2a Involucral bracts pale lemon-yellow, not webbed together .... 4. *H. rutilans*

　2b Involucral bracts silvery, webbed together at the base with delicate 'tissue-paper' indumentum
　.. 5. *H. dasycephalum*

3. Helichrysum callicomum *Harv.* in F.C. 3: 247 (1865); Moeser in Bot. Jb. 44: 251 (1910); Hilliard, Compositae in Natal 165 (1977). Lectotype: [Lesotho], Basutoland, *Cooper* 730 (TCD!; BM; E; G; K; PRE; W; Z, isolecto.!).

Tufted perennial herb up to 400 mm high, stems woody at base, mostly branching there, then above into the compound inflorescence, closely greyish-white felted, densely leafy. *Leaves* up to 25 × 6 mm, smaller and more distant upwards, passing into inflorescence bracts, spathulate, oblong-spathulate or oblong upwards, apex obtuse, upper leaves sometimes acute, base half-clasping, both surfaces closely greyish-felted. *Heads* heterogamous, cylindric, c. 4,5 × 1 mm, very many closely aggregated in corymbose panicles up to 60—80 mm across. *Involucral bracts* in c. 4 series, graded, closely imbricate, inner about equalling flowers, not radiating, tips acuminate, glossy, pellucid, pale straw-coloured. *Receptacle* smooth. *Flowers* 4—6, 1—3 ♀, 2—3 ☿, yellow. *Achenes* 0,75 mm, glabrous. *Pappus* bristles several, delicate, scabrid, bases with patent cilia, not cohering. Fig. 13:1.

Ranges from the E. highlands of Zimbabwe to the Transvaal Highveld and eastern highlands, E. Orange Free State and neighbouring Lesotho, Midlands and Uplands of Natal, Transkei in the foothills of the Drakensberg, the Cape Drakensberg, and from Queenstown and Dohne south to Grahamstown. Often forms large stands in overgrazed grassland. Flowers from March to May. Map 34.

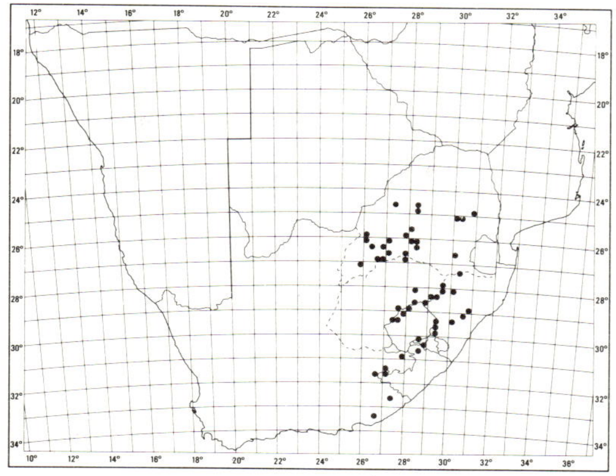

MAP 34.— **Helichrysum callicomum**

Vouchers: *Acocks* 24188 (PRE); *Hilliard* 4973 (E; K; NH; NU); *Wilms* 741 (E; NU); *Wright* 476 (E; K; M; NH; NU).

4. **Helichrysum rutilans** (*L.*) *D. Don* in Sweet, Hort. Brit. 223 (1826); Hilliard & Burtt in Bot. J. Linn. Soc. 82: 262 (1981). Type: *Gnaphalium* sheet no. 6 in herb. Cliff. (BM!).

Gnaphalium rutilans L., Sp. Pl. 854 (1753).

G. parviflorum Lam., Encycl. 2: 748 (1788). *Helichrysum parviflorum* (Lam.) DC., Prodr. 6: 203 (1838); Harv. in F.C. 3: 247 (1865). Type: Cape of Good Hope (P-LAM!).

G. citrinum Schrank in Denkschr. K. Akad. Wiss. Münch. 8: 152 (1824), non Lam. (1778). Type: Cape of Good Hope, *Brehm* s.n. (M, holo.!).

Helichrysum parviflorum var. *longifolium* DC., l.c. Type: Cape, *Drège* 5755 (G-DC, holo.!).

H. parviflorum var. *latifolium* DC., l.c. Type: Cape, Sneeuwberg, 3 000—4 000 ft, *Drège* 5786 (G-DC, holo.!).

H. manopappum O. Hoffm., Pl. Penther. in Annln naturh. Mus. Wien 24: 299 (1910). Type: Cape, Zonder Einde Rivier, Loskraal, *Penther* 1179 (BM; BOL; M; S, iso.!).

H. niveum sensu Moeser in Bot. Jb. 44: 252 (1910); Levyns in Adamson & Salter, Fl. Cape Penins. 783 (1950), non (L.) Less.

Dense twiggy shrublet up to 600 mm tall and as much across, or sometimes lax, branches very leafy, both they and the leaves enveloped in closely woven silvery grey indumentum with 'tissue-paper' surface, becoming woolly when old, somewhat viscid. *Leaves* very variable in size and shape, mostly 6—30 × 1—3 (—6) mm, spathulate, narrowly obovate, oblong or linear, often folded lengthwise, apex subacute, mucronate, often recurved, base very shortly decurrent. *Heads* heterogamous, or rarely homogamous on the same plant, cylindric, c. 3—4 × 1,5 (—2,5) mm, many in clusters crowded in terminal corymbose panicles. *Involucral bracts* in 4 series, graded, inner equalling flowers, imbricate, ovate-lanceolate, acute, pellucid, glossy, pale lemon-yellow, not radiating. *Receptacle* smooth. *Flowers* 3—5 (—8), (0—) 1 (—2) ♀, 2—4 (—7) ☿, bright yellow. *Achenes* less than 1 mm long, glabrous. *Pappus* bristles 2—4, scabrid, nearly equalling corolla, or occasionally wanting. Fig. 13:3.

Ranges from the Cape Peninsula through Piquetberg, Malmesbury, Ceres, Worcester, Caledon, Robertson, Bredasdorp, Swellendam and Riversdale districts to Humansdorp and Uitenhage, thence north to Cathcart, Cradock, Tarka, Middelburg, Queenstown, Aliwal North, and Hay districts and the southern O.F.S. near Fauresmith, Philippolis and Bethulie. A constituent of shrub communities on sandy, shaly or stony soils, flowering in any month. Map 35.

Readily confused with *H. dasycephalum* (no. 5) which, however, has mainly silvery, not lemon-yellow, involucral bracts, and these are webbed together with delicate 'tissue-paper-like' indumentum composed of woolly hairs. The area of *H. dasycephalum* lies east of that of *H. rutilans,* overlapping only in the Queenstown area (c. 27°E).

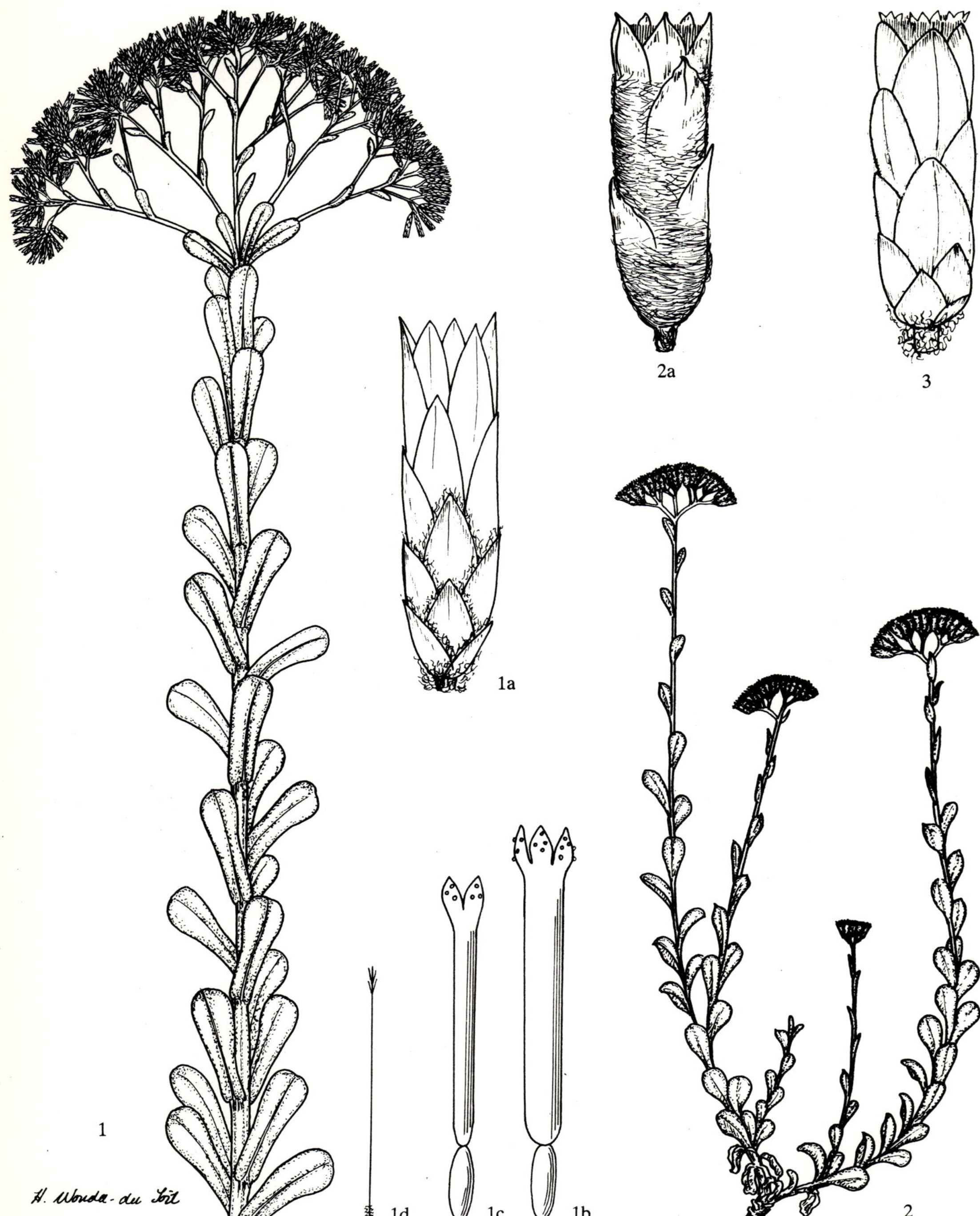

Vouchers: *Acocks* 17639 (PRE); *Compton* 18508 (NBG); *Esterhuysen* 23806 (BOL; PRE); *Leistner* 1110 (PRE); *Nordenstam* 3228 (NU; S).

5. Helichrysum dasycephalum *O.Hoffm.*

in Annln naturh. Mus. Wien 24: 299 (1910); Moeser in Bot. Jb. 44: 251 (1910); Hilliard, Compositae in Natal 165 (1977). Type: Natal, Van Reenen's Pass, *Krook*, no. 1437 in herb. Penther (W, holo.!; BM; BOL; M; NH; S, iso.!).

Twiggy dwarf shrub up to 300 mm tall, old stems woody, up to 8 mm diam., main branches prostrate or decumbent, ultimate branches stiffly erect, densely leafy, they and the leaves enveloped in peculiar silvery grey felt with a 'tissue-paper' surface. *Leaves* rigid, up to 20 × 5 mm, narrowly to broadly spathulate, becoming smaller and oblong-spathulate or oblong upwards, apex obtuse to subacute, mucronate, hooked, base half-clasping. *Heads* heterogamous, cylindric, c. 4 × 1 mm, many in tight clusters usually webbed together with woolly hairs, arranged in compact corymbose panicles up to 20 mm across at the branch tips. *Involucral bracts* in c. 3 series, closely imbricate, about equalling flowers, not radiating, pellucid, silvery, the tips sometimes washed palest brown or palest yellow, woolly hairs webbing the bracts in a thin skin. *Receptacle* smooth. *Flowers* 4−5 (−6), 1−3 ♀, 3−4 ☿, bright yellow. *Achenes* not seen, ovaries glabrous. *Pappus* bristles few (up to 8), scabridulous, shorter than corolla, bases with patent cilia, not cohering. Fig. 13:2.

Recorded from the mountainous E. part of the Orange Free State, the Witteberg near Lady Grey, Lesotho, Transkei and E. Cape and south to the mountains around Queenstown, Cathcart and Keiskammahoek. Never re-collected on Van Reenen's Pass, Natal (the type locality), but occurs on the edge of the escarpment below The Sentinel, SW. of Van Reenen. Forms stiff twiggy mats and cushions on rock sheets and hard bare earth, flowering between February and April. Map 35.

Closely allied to *H. rutilans* (above); see comments under that species. Can also be confused with *H. callicomum* (no. 3), but distinguished by its different indumentum and different involucre.

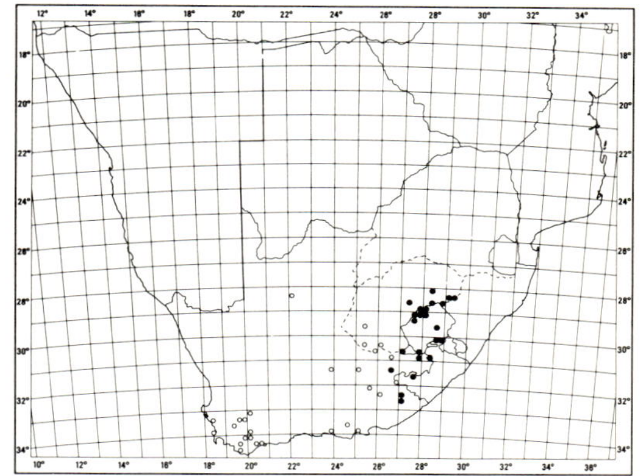

MAP 35.— ○ **Helichrysum rutilans**
      ● **Helichrysum dasycephalum**

Vouchers: *Hilliard & Burtt* 6667 (E; K; MO; NU; PRE; S); *Pegler* 1619 (BM; BOL; PRE); *Story* 3339 (PRE).

Group 3

Stoloniferous perennial herbs; *leaves* small, lanceolate, bases decurrent or not; *heads* homogamous, c. 3 × 1−1,5 mm, many crowded and felted together in dense glomerules, which may be further arranged in corymbose panicles; *involucral bracts* not radiating, creamy or pale to bright yellow; *receptacle* honeycombed; *flowers* 3−11, corolla campanulate above; *achenes* glabrous; *pappus* bristles either wanting, or few, tips subplumose, bases nude.

Species 6−7, confined to eastern Southern Africa from the E. Transvaal through Natal and the Transkei to the E. Cape, in seepage lines or along streamlets, often associated with rank grass and other herbage.

FIG. 13.−1, **Helichrysum callicomum,** part of plant, × 1; 1a, head, × 13; 1b, hermaphrodite flower, × 20; 1c, female flower, × 20; 1d, pappus bristle, × 20 (*Wright* 527). 2, **H. dasycephalum,** flowering branch, × 1; 2a, head, × 13 (*Hilliard & Burtt* 6667). 3, **H. rutilans,** head, × 13 (*Hilliard & Burtt* 13058).

1d
1c
1b
2b
2a
1
1a
2
H. Wouda. du Toit

1a Leaf bases decurrent on the stems in long narrow wings, involucral bracts creamy or very pale straw-coloured, pappus bristles 2–several ....6. *H. natalitium*
1b Leaf bases not decurrent, inner involucral bracts bright canary-yellow, pappus wanting ....7. *H. epapposum*

6. **Helichrysum natalitium** *DC.*, Prodr. 6: 201 (1838); Harv. in F.C. 3: 243 (1865); Moeser in Bot. Jb. 44: 240 (1910); Hilliard, Compositae in Natal 148 (1977). Type: Natal, Port Natal [Durban], *Drège* 5009 (G-DC, holo.!; P, iso.!).

Gnaphalium natalitium (DC.) Sch. Bip. in Bot. Ztg 3: 172 (1845).

A stoloniferous perennial herb, stem erect to c. 1 m, simple below, branching above into a corymbose panicle, thinly greyish-white woolly or cobwebby, very leafy. *Leaves* up to 200 × 20 mm, decreasing rapidly in size upwards and passing into distant inflorescence bracts, lanceolate or linear-lanceolate, tapering at both ends, decurrent on stem in long, narrow wings, glandular above, sometimes thinly greyish-white woolly as well, thinly greyish-white woolly below. *Heads* homogamous, c. 3 × 1 mm, very many matted together at base with whitish wool to form dense cymose-corymbose clusters up to 20 m across terminating the inflorescence branches. *Involucral bracts* in 2–3 series, subequal, about equalling flowers, loosely imbricate, not radiating, creamy white or very pale straw-colour, pellucid becoming opaque. *Receptacle* deeply honeycombed. *Flowers* c. 8–11, dull yellow. *Achenes* 0,75 mm, glabrous. *Pappus* bristles 2 to several, tips subplumose, bases nude, not cohering. Fig. 14:1.

Nearly confined to Natal, and recorded from Babanango in the north to Alfred and Port Shepstone districts in the south and Umzimkulu district in the Transkei, on Natal's southern border, from near sea level to 2 300 m. Favours damp and partly shaded situations particularly along streams and at forest margins, flowering between January and April. Map 36.

Vouchers: *Acocks* 10177 (PRE); *Hilliard* 4818 (E; K; M; NH; NU; PRE); *McClean* 700 (PRE); *Tyson* 1164 (BOL; SAM); *Wright* 434 (E; K; M; NH; NU).

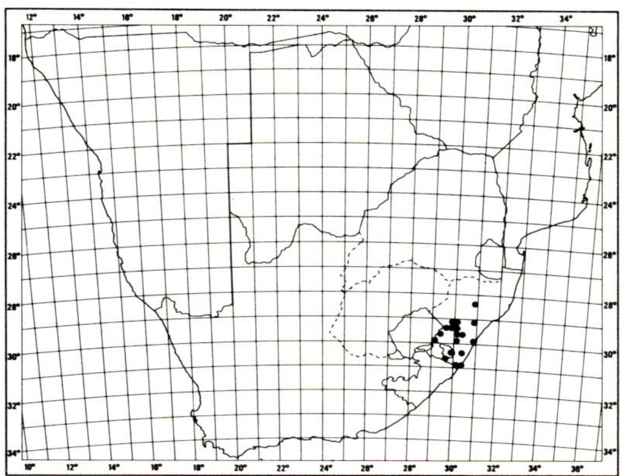

MAP 36.— **Helichrysum natalitium**

7. **Helichrysum epapposum** *H. Bol.* in Trans. R. Soc. S. Afr. 1: 155 (1909), excl. var. *robustum* H. Bol.; Compton, Fl. Swaziland 630 (1976); Hilliard, Compositae in Natal 160 (1977). Lectotype: Natal, Great Noodsberg, *Wood* 4131 (BOL!; E; K, isolecto.!).

Humea epapposa (H. Bol.) S. Moore in J. Linn. Soc. Bot. 40: 112 (1911). *Calomeria epapposa* (H. Bol.) Heine in Adansonia 7: 138 (1967).

H. inerme Moeser in Bot. Jb. 44: 246 (1910) excl. var. *brachycladum* Moeser. Lectotype: Natal, Great Noodsberg, *Wood* 4131 (Z!; E; K, isolecto.!).

Helichrysum flavum Burtt Davy in Jl. S. Afr. Bot. 1: 108 (1935). Type: Transvaal, Barberton distr., White River, *Rogers* 20144 (K, holo.!; BOL, iso.!).

Perennial herb, stems very slender (c. 1–2 mm diam.), elongate, branching mostly near the base, decumbent and rooting there, then erect to c. 600 mm, thinly greyish-white woolly, the indumentum 'stringy' and stripping from the older parts, closely leafy becoming pedunculoid upwards. *Leaves* rather rigid, more or less appressed, up to

FIG. 14.—1, **Helichrysum natalitium,** part of whole plant, × 1; 1a, detail to show decurrent leaf bases, × 1; 1b, head, × 10; 1c, flower, × 13; 1d, pappus bristle, × 13 (*Hilliard* 4818). 2, **H. epapposum,** part of whole plant, × 1; 2a, detail to show leaves, not decurrent, × 1; 2b, head, × 13 (*Hilliard & Burtt* 14513).

22 × 4 (−5) mm, becoming smaller upwards, lanceolate, apex acuminate, base broad, half-clasping, margins revolute, upper surface cobwebby, glabrescent, lower greyish-white woolly, 'stringy'. *Heads* homogamous, very narrowly campanulate, c. 3 × 1,5 mm, very many in dense cymose clusters felted together at the base in compact, flattish corymbs up to 20 mm across terminating the branches. *Involucral bracts* in 2−3 series, loosely imbricate, translucent, outer tipped pale brown, inner bright canary-yellow, about equalling flowers, not radiating. *Receptacle* honeycombed. *Flowers* (3−) 4−5, yellow. *Achenes* 0,75 mm, glabrous. *Pappus* wanting. Fig. 14:2.

Ranges from the Waterberg, Magaliesberg (Rustenburg) and eastern highlands of the Transvaal through Natal and the Transkei to the Amatola Mountains in the E. Cape. Absent from coastal Natal. Grows along muddy streamsides and in marshes, often on tussocks, forming tangled clumps or the slender

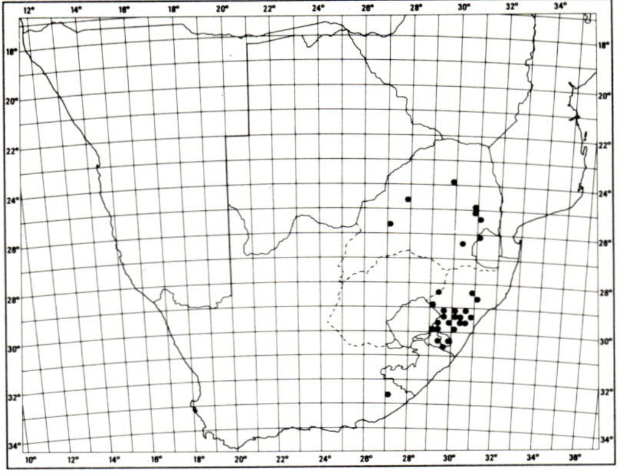

MAP 37.— **Helichrysum epapposum**

stems may be supported by other vegetation. Flowers between February and April. Map 37.

Vouchers: *Hilliard* 5478 (E; K; NU; S); *Moll* 655 (NU; PRE).

Group 4

Shrublets or bushy or straggling perennial herbs; *leaves* small, linear to spathulate; *heads* homogamous or heterogamous (not always a species character), 2,5−3,5 × 1−2 mm, in dense clusters further aggregated into compact or lax corymbose panicles; *involucral bracts* not radiating, bright yellow; *receptacle* smooth or shortly fimbrilliferous; *flowers* 3−15, corollas campanulate above; *achenes* hairy; *pappus* bristles either wanting or, if present, scabrid, bases with patent cilia, cohering or not.

Species 8−11, mainly in the eastern and southern parts of Southern Africa; *H. odoratissimum* (no. 8) is also further north, in tropical Africa. All favour rocky grassy mountain slopes, well watered.

1a Leaves generally broadest above the middle, bases decurrent in narrow or broad stem wings (ignore uppermost reduced leaves):

  2a Heads 2,5−3 mm long, heterogamous, flowers usually c. 7−15 ....8. *H. odoratissimum*

  2b Heads c. 3−4 mm long, usually homogamous, flowers 4−7, occasionally 1 or 2 of them ♀
    .... 9. *H. gymnocomum*

1b Leaves either linear or broadest in lower half, bases not decurrent:

  3a Leaves linear..10. *H. infaustum*

  3b Leaves oblong, elliptic or lanceolate-elliptic ....11. *H. griseolanatum*

8. **Helichrysum odoratissimum** *(L.) Sweet*, Hort. Brit. 223 (1826); Less., Syn. Comp. 301 (1832); DC., Prodr. 6: 202 (1838); Harv. in F.C. 3: 245 (1865) incl. vars; Moeser in Bot. Jb. 44: 242 (1910); Hilliard & Burtt in Notes R. Bot. Gdn Edinb. 32: 353 (1973); Hilliard, Compositae in Natal 150 (1977). Lectotype: Cape of Good Hope (LINN 989.48!).

Gnaphalium odoratissimum L., Sp. Pl. 855 (1753), edn 2: 1196 (1763). *G. odorum* Salisb., Prodr. 192 (1796), nom. illegit. *G. odoratum* Thunb., Prodr. 150 (1800), Fl. Cap. 654 (1823), nom. illegit.

G. aureofulvum Berg., Descr. Pl. Cap. 257 (1767). Type: Cape of Good Hope, *Grubb* (STB, holo.!).

G. pedunculare L., Mant. alt. 284 (1771). *Helichrysum pedunculare* (L.) DC., Prodr. 6: 198 (1838) quoad syn., excl. descript. Type: Cape of Good Hope (LINN 989.47!).

G. strigosum Thunb., Prodr. 150 (1800), Fl. Cap. 654 (1823). Type: hills near Cape Town, *Thunberg* (sheet 19266, UPS, holo.!).

Helichrysum odoratissimum var. *acuminatum* DC., Prodr. 6: 202 (1838). Type: specimen received from Berlin (G-DC, holo.!).

H. odoratissimum var. *undulaefolium* DC., l.c. Lectotype: Cape, Table Mountain, *Drège* 5759 (G-DC!; BM, isolecto.!).

H. odoratissimum var. *lanatum* Sond. in F.C. 3: 245 (1865). Type: Cape, Uitenhage, Van Stadensberg, Oct., *Zeyher* 2874 (S, holo.!; PRE; SAM, iso.!).

A much branched bushy or straggling aromatic perennial herb, stems often decumbent and rooting at the base and becoming woody there, erect or sprawling, glandular, generally thinly greyish-white woolly, wool sometimes wanting, very leafy, the branchlets becoming pedunculoid upwards with distant reduced leaves or bracts. *Leaves* very variable, 5−60 × 1,5−15 mm, linear-oblong, lanceolate, lingulate or spathulate, apex generally obtuse, sometimes acute, mucronate, base narrowed or broad, clasping, generally decurrent in long narrow or broad stem wings, wings sometimes wanting, glandular and setose-scabrid above, generally thinly or thickly greyish-white woolly on both surfaces, sometimes without wool or glabrescent above. *Heads* heterogamous, c. 2,5−3 mm long, very many matted together with wool at the base in dense or loose, flattened or rounded, terminal cymose clusters. *Involucral bracts* closely imbricate, innermost about equalling the flowers, not radiating, pellucid, outer pale brown, inner bright or pale canary-yellow. *Receptacle* with fimbrils about equalling the ovaries. *Flowers* (3−) 7−15, (1−) 2−4 ♀, (2−) 5−11 ☿, yellow. *Achenes* 0,75 mm, barrel-shaped, with duplex hairs. *Pappus* bristles many, scabrid, bases cohering by patent cilia. Fig. 15: 1.

Ranges from the Soutpansberg through the highlands of the E. Transvaal and W. Swaziland to the Midlands and Uplands of Natal, the NE. Orange Free State, Lesotho, the Cape Drakensberg, the mountains and coastal areas of the Transkei and E. Cape thence across the Cape fold mountains as far as the Peninsula, the Cedarberg and the Giftberg in Vanrhynsdorp district. Also on the mountains in Mozambique, Zimbabwe, Malawi and further north. Map 38.

Forms large clumps on grassy or rocky slopes and will also colonize bare areas such as roadside and pathside banks. Flowering recorded in all months, mainly from August to December in the SW. Cape, January to June elsewhere.

H. odoratissimum shows great variation in leaf size and indumentum; particularly robust plants with large leaves are found in the Transvaal and Swaziland, but these can be matched elsewhere in the geographical range of the species. The distinguishing features of *H. odoratissimum* are its leaves, generally broadest in the upper part, and decurrent on the stems in long wings, and the heterogamous heads with hairy achenes and pappus bristles cohering strongly by patent cilia. *H. gymnocomum* (below) is very closely allied.

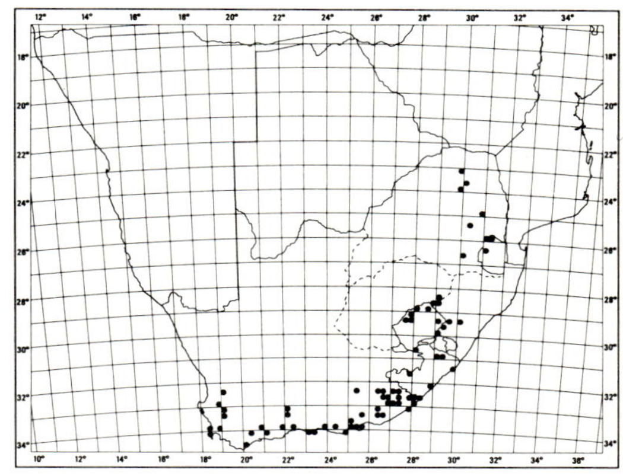

MAP 38.— **Helichrysum odoratissimum**

Vouchers: *Compton* 19154 (NBG); *Esterhuysen* 22538 (BOL; PRE); *Flanagan* 366 (SAM); *Galpin* 594 (BOL; PRE); *Hilliard & Burtt* 10917 (E; K; M; MO; NU; PRE; S).

9. **Helichrysum gymnocomum** *DC.,* Prodr. 6: 202 (1838); Harv. in F.C. 3: 244 (1865); Hilliard, Compositae in Natal 151 (1977). Lectotype: Cape, between Bushman's and Gouritz Rivers, *Burchell* 4334 (G-DC!; K, isolecto.!).

Gnaphalium gymnocomum (DC.) Sch. Bip. in Bot. Ztg 3: 172 (1845).

H. gymnocomum var. *acuminatum* DC., Prodr. 6: 202 (1838). Type: Cape, between Van Staadensberg and Bethelsdorp, *Drège* 59496 (G-DC, holo.!; BM; E; PRE, iso.!).

Very closely allied to *H. odoratissimum* (above), but distinguished by its *heads,* often homogamous, c. 3−4 mm long, *involucral bracts* loosely imbricate; *flowers* 4−7, 1 or 2 of these occasionally ♀, *pappus* often wanting or consisting of a few bristles

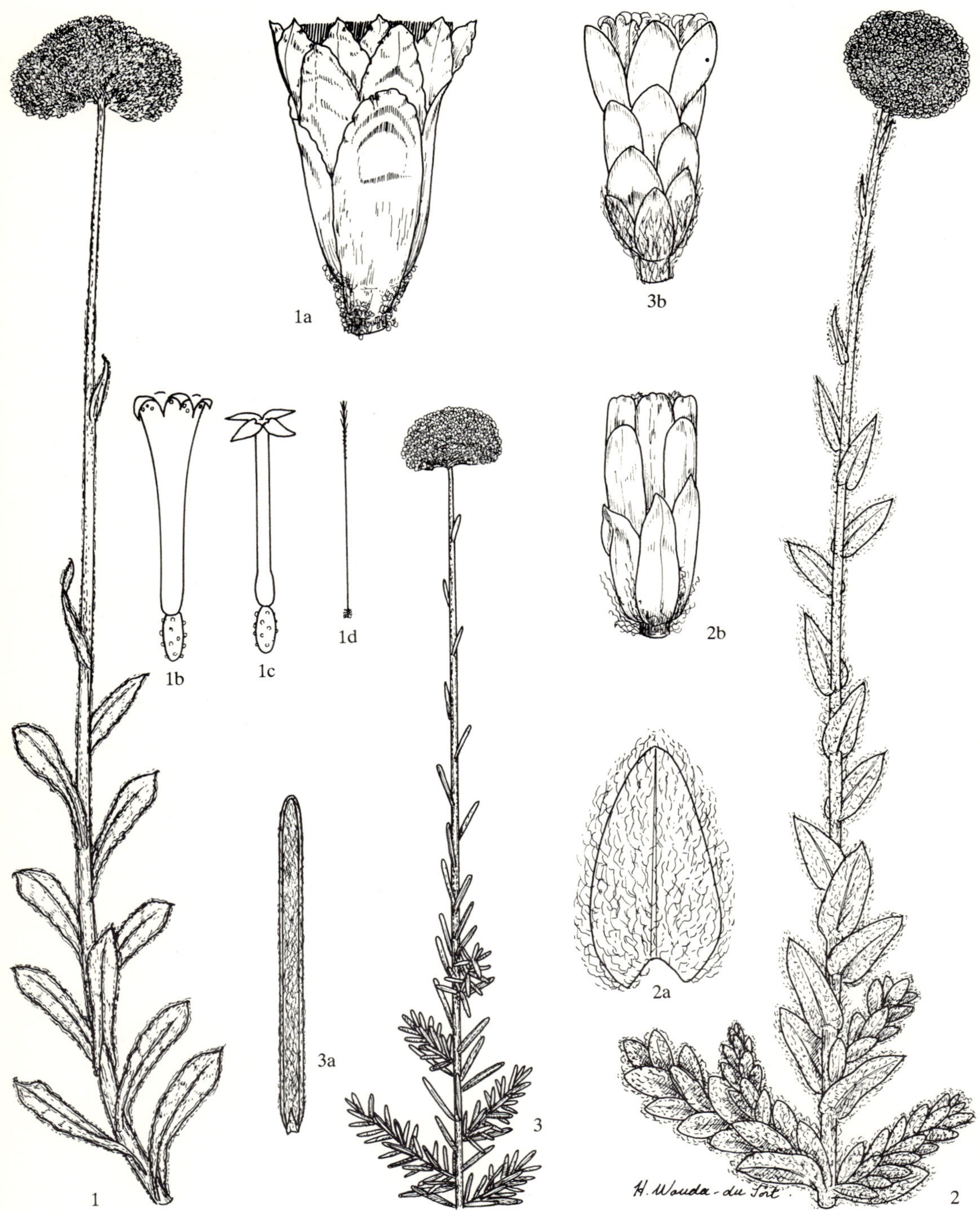

1a

1b 1c 1d

3b

2b

2a

3a

3

H. Wouda-du Toit.

1

2

free at the base, but may be copious and cohering at the base in Cape material.

Found in Natal between c. 1 500 and 3 000 m above sea level and ranging southwards to the Cape Coast as far west as the mouth of the Gouritz River. In Natal, partly sympatric with *H. odoratissimum* but reaching higher altitudes than that species, although this is not so in Lesotho. Occupies the same sort of habitats as *H. odoratissimum* and flowers between February and July. Map 39.

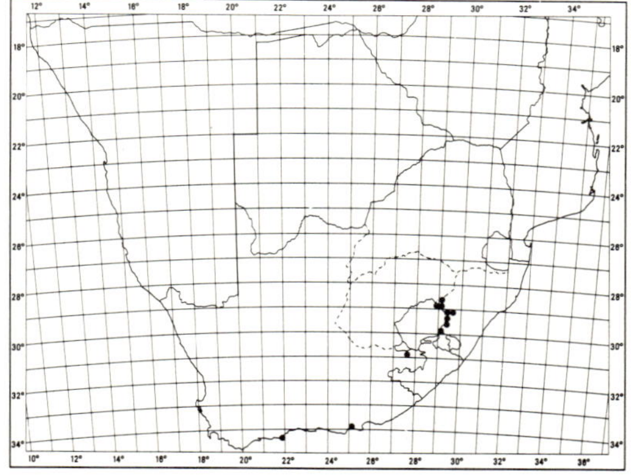

MAP 39.— **Helichrysum gymnocomum**

Vouchers: *Codd* 2785 (NU; PRE); *Esterhuysen* 21695 (BOL; PRE); *Hilliard* 4802 (E; K; MO; NH; NU; S); *West* 812 (BM; PRE); *Wright* 550 (E; NH; NU; S).

10. **Helichrysum infaustum** *Wood & Evans* in J. Bot., Lond. 35: 351 (1897); Moeser in Bot. Jb. 44: 245 (1910), excl. var. *discolor;* Hilliard, Compositae in Natal 152 (1977). Type: Natal, near Van Reenen's Pass, Drakensberg, 5 000—6 000 ft., *Wood* 6973 (NH, holo.!; BOL; E, iso.!).

Humea infausta (Wood & Evans) S. Moore in J. Linn. Soc., Bot. 40: 112 (1911). *Calomeria infausta* (Wood & Evans) Heine in Adansonia 7: 138 (1967).

A dwarf shrublet forming mats or cushions up to c. 200 mm high, old branches woody, decumbent, nude, rooting, branchlets closely grey-woolly, densely leafy,

becoming nude and remotely leafy upwards. *Leaves* up to 12 × 1,5 mm, smaller and more distant upwards, linear, apex obtuse, apiculate, base broad, half-clasping, margins strongly revolute, upper surface thinly grey-woolly, lower densely white-woolly. *Heads* homogamous, c. 3 × 1 mm, cylindric, many in compact cymose clusters 10—15 mm across at the branch tips. *Involucral bracts* in 2—4 series, graded, closely imbricate, outer tipped pale brown, inner bright canary-yellow, about equalling flowers, not radiating. *Receptacle* with pit margins slightly produced. *Flowers* 2—5, yellow. *Achenes* 0,75 mm, with duplex hairs. *Pappus* wanting. Fig. 15: 3.

Recorded along the low Drakensberg on the Transvaal-Natal border and the nearby Skurweberg in Vryheid district, thence southwards along the escarpment and nearby mountains in both Natal and the Orange Free State to Lesotho as far south as Roma in the west and Sehlabathebe in the east, and Underberg district in Natal. A specimen from the sandstone outcrops above the Umtamvuma in southernmost Natal has spent heads and flowering specimens should be sought to confirm the determination. Forms low, spreading cushions or mats over bare rock sheets and bare, eroded or sparsely-grassed areas, flowering from January to April. Map 40.

Distinguished from *H. griseolanatum* (below) by its differently shaped leaves, and from *H. cymosum* subsp. *calvum* (no. 26), by its woolly obtuse leaves with strongly revolute margins and homogamous heads.

Vouchers: *Devenish* 1697 (E; K; MO; NU; PRE; S); *Hilliard & Burtt* 8916 (E; K; MO; NU; S); *Scheepers* 1846 (PRE).

11. **Helichrysum griseolanatum** *Hilliard* in Notes R. bot. Gdn. Edinb. 32: 349 (1973), Compositae in Natal 153 (1977). Type: Cape, summit Drakensberg near Luhana Pass, alt. 8 500 ft., 25 May 1897, *Galpin* 2325 (BOL, holo.!; PRE, iso.!).

H. epapposum var *robustum* H. Bol. in Trans. R. Soc. S. Afr. 1: 155 (1909). Type: lecto as above.

H. inerme var. *brachycladum* Moeser in Bot. Jb. 44: 246 (1910). Lectotype: Natal, East Griqualand, Mt Currie, *Tyson* 1255 (Z!; K!).

Well-branched subshrub forming mats up to 250 mm tall and 1 m across, old stems

---

FIG. 15.—1, **Helichrysum odoratissimum,** flowering branch, × 1; 1a, head, × 13; 1b, hermaphrodite flower, × 13; 1c, female flower, × 13; 1d, pappus bristle, × 13 (*Hilliard & Burtt* 10917). 2, **H. griseolanatum,** flowering branch, × 1; 2a, leaf, × 3,3; 2b, head, × 10 (*Hilliard & Burtt* 9751). 3, **H. infaustum,** flowering branch, × 1; 3a, leaf, × 3,3; 3b, head, × 10 (*Hilliard* 2859).

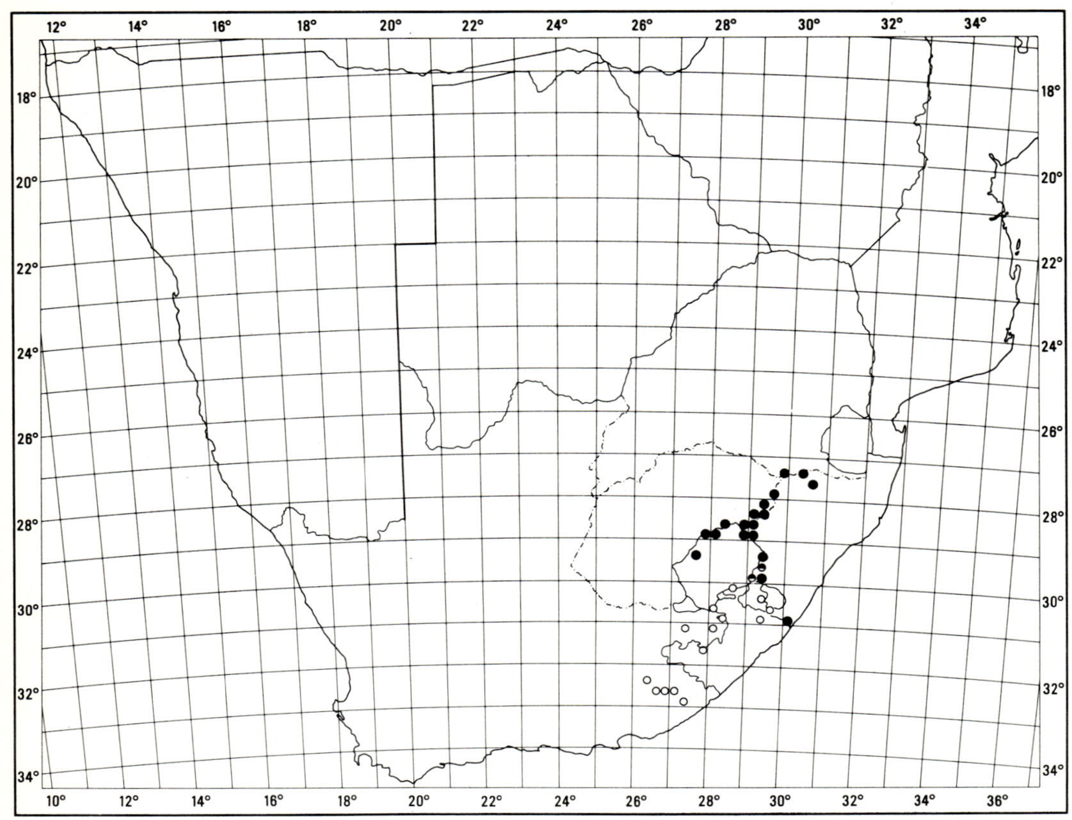

MAP 40.— ● **Helichrysum infaustum**
○ **Helichrysum griseolanatum**
◐ sympatric

up to 8 mm diam., main branches becoming woody, bare, decumbent and rooting, ultimate branchlets greyish-white woolly, closely leafy at first, more distantly so at flowering. *Leaves* more or less imbricate, up to 15 × 6 mm, oblong, elliptic, or lanceolate-elliptic, apex obtuse or acute, base broad, half-clasping, margins revolute, both surfaces greyish-white woolly. *Heads* homogamous, cylindric, c. 3,5 × 1 mm, very many felted together at the base in dense cymose clusters disposed in compact, flattish corymbs terminating the branches. *Involucral bracts* in c. 4 series, graded, closely imbricate, about equalling flowers, pellucid, not radiating, canary-yellow. *Receptacle* smooth. *Flowers* 3—5, yellow.

Achenes 0,75 mm, with duplex hairs. *Pappus* wanting. Fig. 15: 2.

On the mountains, from southern Natal (Drakensberg, Zuurberg, Ngeli and Mt Currie), through the western Transkei to the Cape Drakensberg about Maclear and Barkly East and the mountains of the E. Cape (King William's Town, Keiskammahoek, Katberg, Hogsback and Great Winterberg). Favours rock sheets and other bare stony areas on mountain tops and slopes: flowering between December and April. Map 40.

Sometimes confused with *H. odoratissimum,* but that species has leaves generally broadest in the upper half and decurrent on the stem in broad or narrow wings.

Vouchers: *Baur* s.n. (SAM); *Hilliard* 3901 (BOL; E; K; NU); *Hilliard & Burtt* 6622 (E; K; NBG; NU; MO; PRE; S).

Group 5

Tufted perennial herbs; *leaves* linear to elliptic or rhomboid, medium-sized; *heads* homogamous, c. 3−4 ×1−3 mm, felted together in umbrella-like clusters; *involucral bracts* not radiating, bright yellow; *receptacle* smooth, nude or with some paleae; *flowers* 3−18, corollas narrowly campanulate above; *achenes* hairy; *pappus* bristles either wanting or few, scabrid, bases nude.

Species 12−13; *H. umbraculigerum* ranges from Zimbabwe to the E. Cape, *H. krookii* is endemic to the Drakensberg Centre; both favour well-watered grass slopes.

1a Stems leafy throughout, becoming bracteate then nude only near the heads .... 12. *H. umbraculigerum*

1b Stems leafy only in the lower half, leaves frequently rosetted ....13. *H. krookii*

12. **Helichrysum umbraculigerum**

Less., Syn. Comp. 284 (1832); DC., Prodr. 6: 186 (1838); Harv. in F.C. 3: 236 (1865); Moeser in Bot. Jb. 44: 248 (1910); Batten & Bokelmann, Wild Fl. E. Cape Prov. 160, plate 127, 6 (1966); Hilliard, Compositae in Natal 153 (1977). Type: Cape, *Krebs* 147.

Gnaphalium umbraculigerum (Less.) Sch. Bip. in Bot. Ztg 3: 171 (1845). *G. umbraculigerum* var. α *heterophyllum* O. Kuntze, Rev. Gen. Pl. 3,2: 155 (1898).

G. umbraculigerum var. β *aequilatum* [O. Hoffm. ex] O. Kuntze, Rev. Gen. Pl. 3,2: 155 (1898). *Helichrysum umbraculigerum* var. *aequilatum* (O. Kuntze) Moeser in Bot. Jb. 44: 248 (1910). Type: Natal, Van Reenen's Pass, *O. Kuntze* s.n. (K, iso.!).

H. coactum M. D. Henderson in Bothalia 6: 421 (1954). Type: Natal, Bergville distr., Cathedral Peak Forest Research Station, *Killick* 1354 (PRE, holo.! BM; K; NH; NU; S, iso.!).

Tufted perennial herb, stems often decumbent and rooting, then erect to c. 1 m, young parts thinly grey-woolly, leafy. *Leaves* mostly 20−80 × 3−25 mm, becoming smaller upwards and passing into bracts, very variable in shape, ranging from linear-lanceolate to elliptic and tapering at both ends; or lanceolate- or oblong-spathulate to spathulate, acute or subacute, base slightly to much narrowed and ear-clasping; or ovate, elliptic-ovate to rhomboid-ovate and rather abruptly narrowed to a petiole-like ear-clasping base; upper surface with stout or delicate glandular hairs, erect or appressed, mostly cobwebby or thinly to thickly greyish white-woolly as well, lower surface often thickly greyish-white woolly, or wool sometimes thin or wanting. *Heads* homogamous, cylindric, c. 3 × 1 mm, very many crowded and webbed together with wool forming a flattened, umbrella-like disc, the flattened branches of the cymose inflorescence visible

through the wool. *Involucral bracts* biseriate, subequal, loosely imbricate, about equalling the flowers, not radiating, pellucid, canary-yellow, outer often golden-brown. *Receptacle* nearly smooth. *Flowers* 3−4 (−6), yellow. *Achenes* 0,5−0,75 mm long, barrel-shaped, with duplex hairs. *Pappus* bristles 0−7, shorter than corolla, scabrid, bases not cohering. Fig. 16: 1.

Ranges from the E. highlands of Zimbabwe to the Soutpansberg and the E. highlands of the Transvaal to Natal, the Transkei and the Cape as far west as the Zuurberg near Port Elizabeth and the mountains about Graaff-Reinet. Grows in rough grassland or scrub, often on forest margins or in damp gullies and along streambanks, flowering between January and April. Map 41.

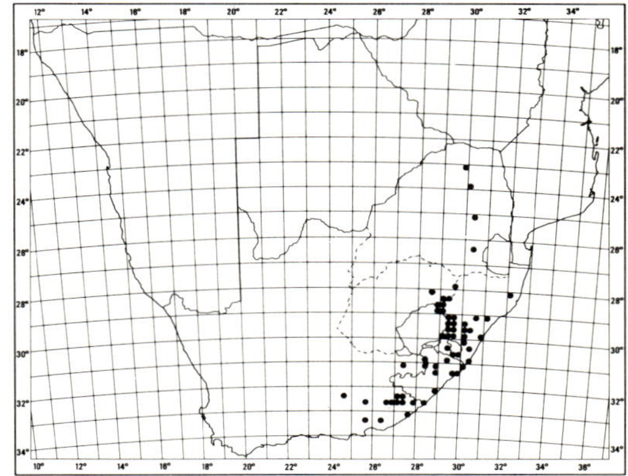

MAP 41.— **Helichrysum umbraculigerum**

H. umbraculigerum displays remarkable variation in leaf shape, and there is some geographical partitioning of this variation. For instance, plants with ovate or rhomboid-ovate leaves much narrowed to a distinctly petiolar part are very nearly confined to

1
1e
1a
1c
1b
2b
2a
1d
2
H. Wanda du Toit

Natal, where they have a wide altitudinal range from near sea level to c. 2 000 m. But in the Natal Drakensberg between c. 1 500 and 2 300 m, leaves may range from linear-lanceolate through all intermediates to ovate or rhomboid-ovate; narrow-leaved plants may occur in the same area as broad-leaved ones, but the former are generally found in wetter situations than the latter.

In the Transvaal and Zimbabwe, leaves are mostly more or less elliptic, tapering at the base, but I have seen one very narrow-leaved specimen from near Lake Chrissie (*Rogers* 11536, BOL).

In the Cape, within an area roughly circumscribed by a line encircling Grahamstown, Cathcart, Barkly Pass, Qumbu, Port St Johns (Transkei) and East London, leaves tend to be oblong-spathulate, slightly narrowed towards the base then expanded and ear-clasping, and the flowers frequently lack a pappus. Elsewhere in the Cape, as well as in the area just mentioned, leaves may taper to a more markedly petiolar base but are never ovate nor rhomboid as in Natal; the flowers generally have a pappus. We have not been able to trace the type specimen, but Lessing described the leaves as spathulate-oblong-obovate, slightly acute, much narrowed below, and the flowers as having a pappus. Krebs lived in Grahamstown for a while, and his specimen was possibly collected thereabouts.

The recognition of infraspecific categories is scarcely practicable, but the varietal name *aequilatum* (= *H. coactum*) applies to specimens with linear to narrowly elliptic leaves.

Vouchers: *Codd* 2772 (NU; PRE); *Hilliard* 4970 (E; K; MO; NU; PRE; S); *Hilliard & Burtt* 7992 (E; K; M; NU; PRE; S); *Pegler* 675 (BM; K; NU; PRE); *Schlechter* 4733 (BM).

13. **Helichrysum krookii** *Moeser* in Bot. Jb. 44: 248 (1910) and in Annln naturh. Mus. Wien 24: 317 (1910); Hilliard, Compositae in Natal 156 (1977). Lectotype: Natal, Mt Currie distr., Newmarket, *Krook* 1029 (W!).

A tuberous-rooted rhizomatous perennial herb, flowering stem solitary, simple, erect, glandular, thinly cobwebby. *Leaves* mostly radical, more or less rosetted, up to 100 × 50 mm, elliptic, apex subacute, abruptly narrowed to a broad, clasping, petiole-like base up to 30 × 6 mm, both surfaces glandular-pubescent, stem leaves smaller, narrower, sessile, passing rapidly upwards into distant bracts. *Heads* homogamous, c. 3−4 × 2−3 mm, very many felted together to form very compact corymbose clusters, these further felted together into a flattish disc 20−70 mm across. *Involucral bracts* in 2−3 series, loosely imbricate, about equalling flowers, not radiating, bright canary-yellow. *Receptacle* with paleae, equalling the corolla, subtending up to half the flowers. *Flowers* 7−18, yellow, sometimes tinged red. *Achenes* 1,25 mm, with myxogenic duplex hairs. *Pappus* bristles few, slightly shorter than corolla, tips barbellate, bases free. Fig. 16: 2.

Along the low Drakensberg on the Natal-Transvaal border, thence south along the Drakensberg and its outliers (Bamboo and Mawahqua Mountains) to Ngeli Mountain (Natal-Transkei border), the mountains near Kokstad and neighbouring Transkei, and the Drakensberg on the Lesotho-Transkei-Cape border, thence south-west as far as the Barkly Pass area. Forms small colonies in grassland, particularly near streams or on steep moist slopes; flowering in January and February. Map 42.

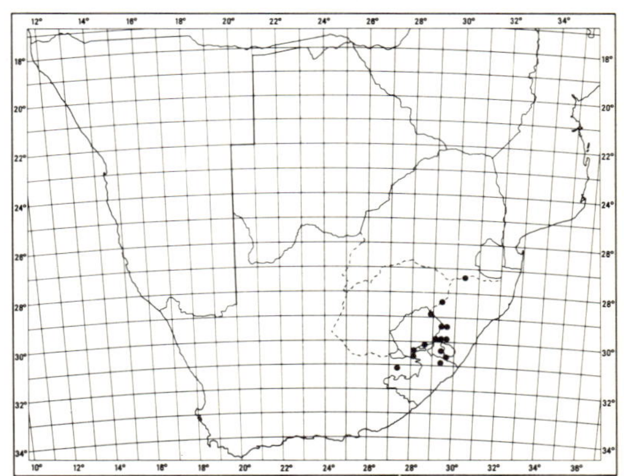

MAP 42.— **Helichrysum krookii**

Vouchers: *Galpin* 6687 (PRE); *Hilliard* 4974 (E; K; NH; NU); *Hilliard & Burtt* 6552 (E; K; MO, NBG; NU; PRE); *Wright* 403 (E; K; NH; NU).

FIG. 16.—1, **Helichrysum umbraculigerum,** part of whole plant, × 1 (*Rennie* 352); 1a, leaf, form common in Natal from sea level to c. 2000 m, × 1 (*Hilliard* 4937); 1b, leaf, form common in the Natal Drakensberg, × 1 (*Hilliard* 4983); 1c, leaf, form occasional in the Drakensberg, × 1 (*Hilliard & Burtt* 7992); 1d, leaf, form prevailing in E. Cape, × 1 (*Hilliard & Burtt* 3750); 1e, head, × 17 (*Rennie* 352). 2, **H. krookii,** whole plant, × 1; 2a, flower, × 13; 2b, pappus bristle, × 13 (*Wright* 403).

Group 6

Subshrubs or perennial herbs, frequently mat-forming; *leaves* linear to elliptic or spathulate, small or medium-sized; *heads* usually homogamous, one species heterogamous, c. 2,5–5 × 1–3 mm, many felted together in dense corymbose clusters; involucral bracts either erect or minutely radiating, bright yellow; *receptacle* shortly honeycombed; *flowers* 4–26, corolla narrowly funnel-shaped; *achenes* often glabrous, or hairy in a few species; *pappus* bristles few, tips shortly plumose, bases nude.

Species 14–22; numbers 15–20 are closely allied. *H. subglomeratum* ranges from Angola to the Okavango and Zimbabwe thence south to the Transvaal, Orange Free State, Lesotho, Natal and the Cape; the others are confined to S. Africa, Lesotho and Swaziland, several being endemic to the Drakensberg Centre. Mostly in short grassland or on rock sheets.

1a Ovaries glabrous:

  2a Cauline leaves more or less spreading, or if appressed then reduced in size:

    3a Either leaf margins flat or outer involucral bracts obtuse, or both:

      4a Leaves concolorous, heads 2,5–5 mm long:

        5a Tufted perennial herbs, heads 2,5–3,5 mm long:

          6a Radical leaves lingulate, oblong or elliptic... 14. *H. subglomeratum*

          6b Radical leaves linear (2–3 mm broad)... 15. *H. oligopappum*

        5b Mat-forming perennial herb or subshrub; heads 4–5 mm long .... 16. *H. spodiophyllum*

      4b Leaves discolorous (thin papery indumentum above, drying dark, silky white below); heads 5 mm long ....20. *H. ephelos*

    3b Leaf margins strongly revolute and outer (brown) involucral bracts very acute.................. 19. *H. nanum*

  2b Cauline leaves erect, closely imbricate, all more or less the same size:

    7a Leaves elliptic, apex obtuse or subacute, thin skin-like indumentum above, silky-tomentose below; flowers 8–12 in each head...17. *H. albanense*

    7b Leaves lanceolate, apex acute to very acute, both surfaces enveloped in silky often 'stringy' indumentum, rarely woolly-felted; flowers 5–6 in each head................................... 18. *H. glomeratum*

1b Ovaries hairy:

  8a Leaves narrowly lanceolate or linear, broader leaves often conduplicate and slightly falcate, discolorous..21. *H. subfalcatum*

  8b Leaves spathulate, concolorous .... 22. *H. albirosulatum*

14. Helichrysum subglomeratum *Less.,* Syn. Comp. 283 (1832); DC., Prodr. 6: 186 (1838), excl. var. *imbricatum* DC.; Harv. in F.C. 3: 235 (1865); Moeser in Bot. Jb. 44: 249 (1910); Merxm., F.S.W.A. 139: 249 (1967); Hilliard, Compositae in Natal 156 (1977). Lectotype: Cape, *Krebs* 146 (G-DC!; S, isolecto.!).

Gnaphalium subglomeratum (Less.) Sch. Bip. in Bot. Ztg 3: 171 (1845); *Helichrysum subglomeratum* var. *lingulatum* Harv. in F.C. 3: 236 (1865).

Perennial herb, stock stout, crowned with one or several leaf rosettes, flowering stems several, lateral to each rosette, 80–600 mm long, often decumbent at the base or ascending, commonly simple, sometimes forking once or twice near the base, more rarely near the apex, densely leafy. *Radical leaves* up to 120 × 15 mm, but often less than half that, lingulate, oblong or elliptic, apex obtuse or subacute, base broad, clasping, both surfaces clothed in

FIG. 17.–1, **Helichrysum spodiophyllum,** part of whole plant, × 1; 1a, leaf, margins flat, but blade often U-shaped in section, × 1,3; 1b, head, × 10; 1c, pappus bristle, × 10; 1d, flower, × 10 (*Hilliard & Burtt* 8390). 2, **H. nanum,** leaf, margins strongly revolute, × 1,3; 2a, head, × 10 (*Solomon* 35).

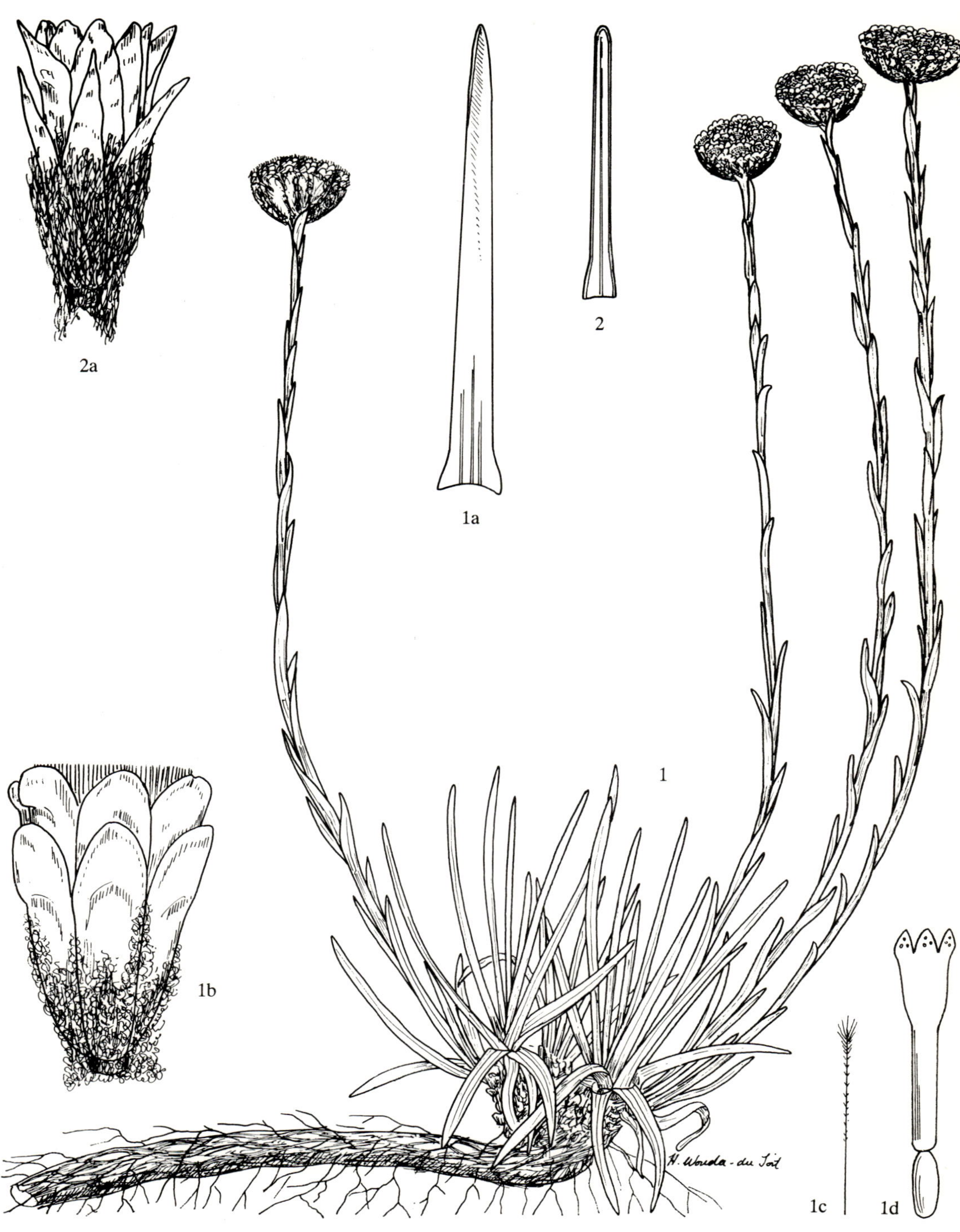

2a
1a
2
1
1b
1c
1d
H. Wonda-du Toit

very close, silvery silky-felted indumentum, occasionally more woolly than silky; *cauline leaves* similar but smaller and decreasing rapidly in size upwards, spreading or more or less imbricate, enveloped in the same silvery, skin-like indumentum as the radical leaves and the stem. *Heads* homogamous, narrowly cylindric, 2,5—3,5 × 1,5—2 mm, very many in small corymbose clusters congested in a flattened spreading cymose corymb c. 20—40 mm across, the whole matted together below with closely interwoven wool. *Involucral bracts* in 3—4 series, subequal, loosely imbricate, outer brown or straw-coloured, inner tipped bright canary yellow, about equalling the flowers, not radiating, translucent. *Receptacle* shortly honeycombed. *Flowers* 4—13, yellow. *Achenes* 0,75 mm long, narrow, glabrous. *Pappus* bristles c. 5—7, delicate, tips shortly plumose, bases nude, not cohering.

Ranges from Zimbabwe to the Transvaal, where it is recorded from the E. highlands, the Waterberg, and the Highveld, to Natal, where it is commoner along the Drakensberg than in the coastal districts, the mountainous NE. corner of the Orange Free State, Lesotho, and the E. Cape, both in the mountains and along the coast, as far west as Uitenhage and Graaff-Reinet. Also recorded from Huila in Angola and the Okavango Valley in NE. S.W.A./Namibia. Grows scattered in stony grassland or, in the Cape, in fynbos, or may form small mats over rock sheets. Flowers from March to June. Map 43.

Vouchers: *Codd* 2903 (NU; PRE); *De Winter* 4368 (PRE; WIND); *Hilliard & Burtt* 10031 (E; K; M; MO; NU; S): *Wright* 486 A (E; K; M; NH; NU; SRGH).

15. **Helichrysum oligopappum** H. Bòl.

in Trans. S. Afr. phil. Soc. 18: 388 (1907); Moeser in Bot. Jb. 44: 249 (1910); Hilliard, Compositae in Natal 157 (1977). Lectotype: Natal, Great Noodsberg, *Wood* 4142 (BOL!; E; K; NH; PRE, isolecto.!).

Distinguished from *H. subglomeratum* (above) by its linear radical leaves up to c. 100 × 2 (—3) mm. In Natal, there are consistently fewer flowers in the heads of *H. subglomeratum* than in those of *H. oligopappum* (4—8, not 10—13), but this distinction does not hold over the whole geographical range of *H. subglomeratum*. It may well be that *H. oligopappum* is no more than a minor variant of *H. subglomeratum* with narrow radical leaves. At present, *H. oligopappum* is known from a few collections in New Hanover, Lion's River, Mpendhle and Camperdown districts in Natal, where it grows in poor stony grassland; flowering between February and April. Map 43.

Vouchers: *Hilliard* 4856 (E; K; NH; NU; PRE; SRGH); *Wood* 5280 (E; Z).

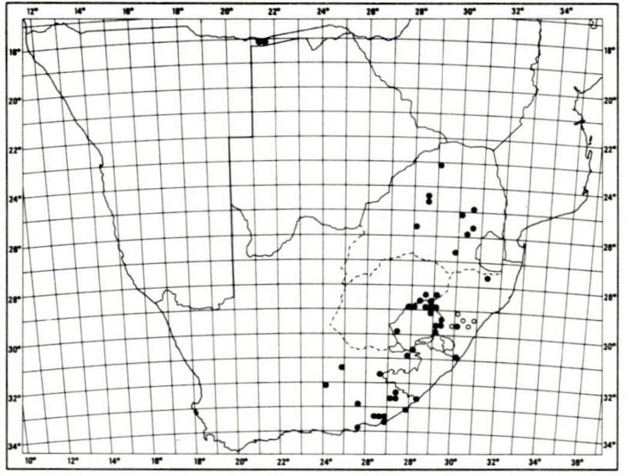

MAP 43.— ● **Helichrysum subglomeratum**
○ **Helichrysum oligopappum**

16. **Helichrysum spodiophyllum** *Hilliard & Burtt*

in Notes R. bot. Gdn Edinb. 34: 83 (1975); Compton, Fl. Swaziland 635 (1976); Hilliard, Compositae in Natal 158 (1977). Type: Natal, Ngotshe distr., Ngome Forest Reserve, c. 1 200 m, *Hilliard & Burtt* 5933 (NU, holo.!; E; K; NH; S, iso.!).

Mat-forming closely tufted perennial herb or subshrub, main stem woody, up to 10 mm diam., with numerous prostrate, ultimately bare and rooting, branches becoming thickened at the nodes and there producing tufts of leaves, flowering stems mostly decumbent at the base, then erect to c. 100—250 mm, simple, silky-silvery, leafy becoming bracteate upwards. *Radical leaves* 10—100 × 1—2 mm, linear, flat or channelled, apex subacute, base expanded, clasping, both surfaces closely grey silkywoolly, *cauline leaves* up to 20 mm long, smaller upwards, linear-lanceolate to lanceolate, broad-based, clasping, spreading or appressed, grey silky-woolly. *Heads* homogamous, cylindric, 4—5 × 2 mm, many in a tightly congested corymbose cluster 10—20 mm across, heads webbed together at the base with tissue-paper-like indumentum. *Involucral bracts* in 3—4 series, closely imbricate, about equalling the flowers, translucent, outer tipped brown, inner bright yellow becoming opaque, minutely radiating. *Receptacle* honeycombed. *Flowers* 8—12. *Achenes* 0,75 mm, obscurely ribbed, glabrous. *Pappus* bristles c. 5, equalling corolla, tips shortly plumose, bases nude, not cohering. Fig. 17: 1.

Recorded from Lydenburg in the E. Transvaal, the mountains in Mbabane district, Swaziland, the low Drakensberg on the Natal-Transvaal border, Itala, Ngome and Ceza on the Natal-Zululand border, Mt Gilboa, Lion's River district, Natal, then a disjunction to Dohne and Mt Kemp east of the Amatola Mts in the E. Cape. Forms large woody mats on rock outcrops and rock sheets in grassland, rooting in the crevices and sprawling over the adjacent rock surfaces. Flowers between December and February in the northern part of its range, but as late as May in the Cape. Can be confused with *H. nanum* (no. 19), which is readily distinguished by its leaves with strongly revolute margins and very acute outer involucral bracts. Map 44.

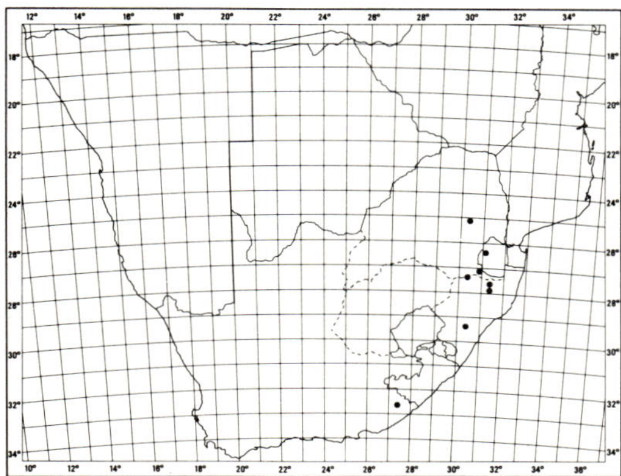

MAP 44.— **Helichrysum spodiophyllum**

Vouchers: *Devenish* 538 (M; PRE); *Hilliard & Burtt* 10007 (E; K; MO; NU; S); *Hilliard & Burtt* 8390 (E; K; NU; PRE; S); *Sim* 19746 (NU).

17. **Helichrysum albanense** *Hilliard* in Notes R. bot. Gdn Edinb. 40: 249 (1982). Type: Cape, Albany distr., between Riebeek East and Grahamstown, *Burchell* 3534 (G-DC, holo.!; K; PRE, iso.!).

Helichrysum subglomeratum Less. var. *imbricatum* DC., Prodr. 6: 186 (1838); Harv. in F.C. 3: 236 (1865); Moeser in Bot. Jb. 44: 249 (1910); Hilliard & Burtt in Notes R. bot. Gdn Edinb. 32: 362 (1973). Type as above.

Perennial herb spreading by means of woody underground runners, flowering stems up to 400 mm long, erect or decumbent, simple or virgately branched,

silky, closely leafy. *Leaves* erect, imbricate, mostly 8−20 × 2−6 mm, elliptic, apex obtuse or subacute, mucronate, somewhat hooked, base half-clasping, upper surface with thin skin-like indumentum, green drying grey, lower surface white silky-tomentose. *Heads* homogamous, turbinate, 3 × 2 mm, very many in dense, roundish, terminal corymbose clusters matted together with wool at the base, these clusters congested in a flattish corymbose cyme mostly 15−20 mm across. *Involucral bracts* in c. 4 series, graded, outer brown, inner bright canary-yellow, subopaque, about equalling flowers, not radiating. *Receptacle* very shortly honeycombed. *Flowers* 8−12. *Achenes* not seen, ovaries glabrous. *Pappus* bristles few, tips shortly plumose, bases nude, not cohering. Fig. 18: 3.

Confined to the E. Cape, ranging from Queenstown south and west to Albany, Uitenhage and Humansdorp districts, in grassland or shrub communities, flowering between July and October. Can be confused with *H. glomeratum* (below) but easily distinguished by its differently shaped leaves with different indumentum and heads containing more flowers; it also flowers earlier. Map 45.

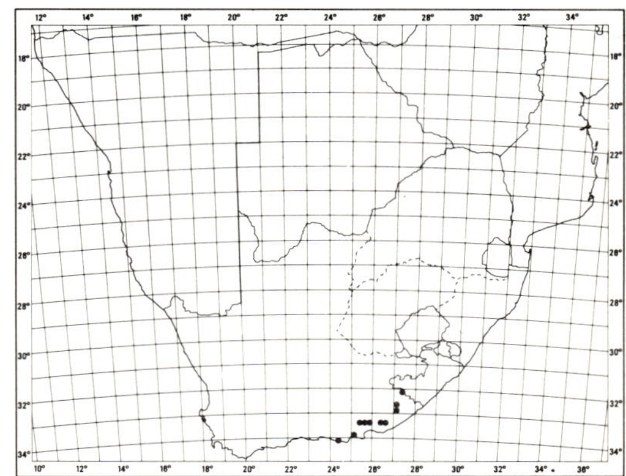

MAP 45.— **Helichrysum albanense**

Vouchers: *Acocks* 21253 (K; PRE); *Hilliard & Burtt* 10824 (E; K; MO; NU; S); *Long* 708 (K; PRE); *Taylor* 9540 (PRE); *Thode* A878 (K; PRE).

18. **Helichrysum glomeratum** *Klatt* in Bull. Herb. Boissier 4: 460 (1896); Moeser

1a
4a
1
4
2
2d
2c
2b
2a
3
4b
4c
4d
H. Wouda-de Jong

in Bot. Jb. 44: 249 (1910); Hilliard, Compositae in Natal 159 (1977). Lectotype: Natal, East Griqualand, grassy, stony places on Malowe Mt, 4 500 ft, *Tyson* 740 (Z!; BM; BOL; K; SAM; UPS, isolecto.!).

A rhizomatous perennial herb, stems solitary or 2 to 3 together, erect to c. 450 mm, commonly simple, rarely forking, loosely grey-cottony, densely leafy. *Radical leaves* (seldom represented in herbaria) rosetted, spreading, up to 30 × 10 mm, lanceolate, apex acute to very acute, base broad, half-clasping, both surfaces enveloped in silvery silky, rather 'stringy' indumentum, rarely woolly-felted; *cauline leaves* similar but smaller (up to 20 × 8 mm), not decreasing in size upwards, erect, closely imbricate. *Heads* homogamous, cylindric, 3−4 × 1−1,5 mm, very many in congested, roundish, corymbose clusters matted together with wool at the base, these in turn congested (but not matted) in a flattish corymbose cyme 20−50 mm across. *Involucral bracts* in c. 4 series, graded, imbricate, outer brownish, inner tipped bright canary-yellow, about equalling flowers, not radiating, translucent or sub-opaque. *Receptacle* with pit margins slightly produced. *Flowers* 5−6, yellow. *Achenes*

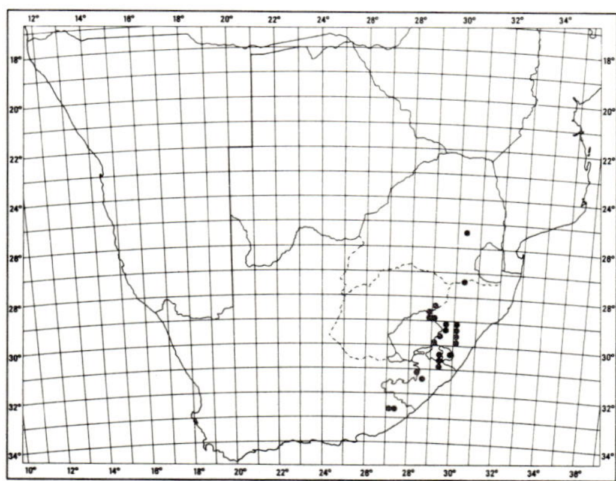

MAP 46.— **Helichrysum glomeratum**

not seen, ovaries glabrous. *Pappus* bristles c. 6, tips shortly plumose, bases nude, not cohering. Fig. 18:2.

From Lydenburg in the E. highlands of the Transvaal south to the Orange Free State around Harrismith and Witsieshoek, the Midlands and Uplands of Natal, East Griqualand and the Transkei to the Amatola Mountains in the E. Cape. Often in large colonies in open grassland; flowering between February and May, but mainly in March and April. Map 46.

Vouchers: *Hilliard & Burtt* 6549 (E; K; MO; NU; PRE); *Wright* 137 (E; M; NU).

19. Helichrysum nanum *Klatt* in Bull. Herb. Boissier 4: 461 (1896); Moeser in Bot. Jb. 44: 249 (1910); Hilliard, Compositae in Natal 159 (1977). Type: Natal [Bulwer], Amawahqua [Mtn], *Wood* 4593 (Z, holo.!; BM; K; NH, iso.!).

A mat-forming perennial herb, main branches stoloniferous, slender (c. 1,5 mm diam.), nude, rooting, producing numerous erect dwarf shoots with closely rosetted leaves, flowering stems terminal to each rosette, solitary, erect, 50−120 mm high, simple or very rarely forked, closely leafy. *Leaves* stiff, erect, up to 35 mm long, c. 1 mm broad, linear, apex obtuse, base broadened, clasping, margins strongly revolute, both surfaces enveloped in silvery, sericeous, closely felted indumentum. *Heads* homogamous, cylindric, 4−5 × 1,5 mm, many felted together in small corymbose clusters, these congested, but not felted, in a terminal cymose-corymbose cluster 15−20 mm across. *Involucral bracts* in c. 2−3 series, subequal, loosely imbricate, outer brownish, inner tipped bright canary-yellow, about equalling flowers, not radiating, translucent. *Receptacle* smooth. *Flowers* 4−6, yellow. *Achenes* c. 1 mm long, obscurely 5-ribbed, glabrous. *Pappus* bristles c. 5, slender, tips shortly plumose, bases nude, not cohering. Fig. 17: 2.

Ranges from Harrismith district in the Orange Free State through the foothills of the Natal and

---

FIG. 18.−1, **Helichrysum albirosulatum,** part of whole plant, × 1; 1a, head, × 8 (*Hoener* 2186). 2, **H. glomeratum,** whole plant, × 0,5; 2a, leaf, × 1; 2b, head, × 13; 2c, flower, × 13; 2d, pappus bristle, × 13 (*Hilliard & Burtt* 6549). 3, **H. albanense,** leaf, × 1 (*Hilliard & Burtt* 10824). 4, **H. subfalcatum,** part of whole plant, × 1, 4a, head, × 8; 4b, hermaphrodite flower, × 10; 4c, female flower, × 10; 4d, pappus bristle, × 10 (*Hilliard* 4809).

Transskei Drakensberg to the Cape Drakensberg at Naudés Nek, Maclear and Ugie. Forms extensive mats in poor stony grassland or over rock sheets, and will colonize bare eroded areas. Flowers in March and April, sometimes as late as July. Map 47.

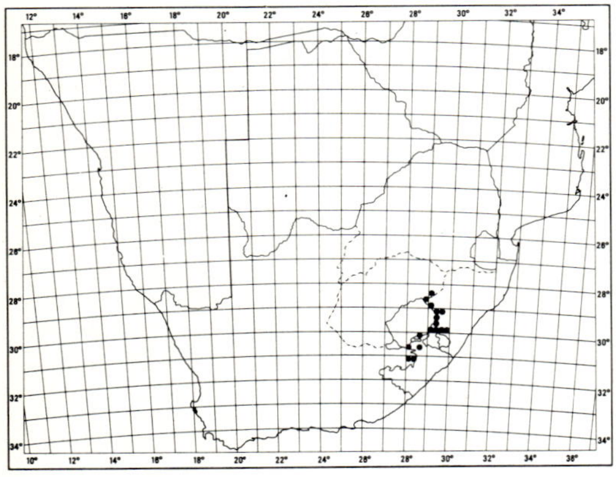

MAP 47.— **Helichrysum nanum**

Vouchers: *Edwards* 3091 (NU; PRE); *Hilliard* 5520 (E; K; NU; S); *Wright* 146 (E; NH; NU).

20. **Helichrysum ephelos** *Hilliard* in Notes R. bot. Gdn Edinb. 40: 254 (1982). Type: Natal, Lion's River distr., Fort Nottingham Commonage, c. 1675 m, 4 v 1977, *Hilliard & Burtt* 10329 (NU, holo.!; E; K; MO; PRE; S, iso.!).

Mat-forming perennial herb, main stems stoloniferous, rooting and branching freely, producing numerous leaf rosettes, flowering stems decumbent then erect, 300−400 mm long, simple, silvery sericeous, appressed-leafy. *Radical leaves* rosetted, mostly 40−100 × 3−10 mm, linear-lanceolate, apex obtuse or subacute, base broad, clasping, upper surface with thin 'tissue-paper' indumentum, lower white silky-felted; *cauline leaves* much reduced, c. 10−30 × 1,5−3 mm, degenerating upwards into bracts, linear or linear-lanceolate, acute to acuminate, enveloped in silky indumentum webbed to the stem, only leaf tips free. *Heads* homogamous, c. 5 × 2 mm, many lightly webbed together in a very congested, terminal, cymose cluster c. 15−25 mm across. *Involucral bracts* in c. 3 series, loosely imbricate, outer shorter,

pellucid, tips light brown, inner subequal, equalling flowers, tips very obtuse, canary-yellow, not radiating. *Receptacle* shortly honeycombed. *Flowers* 7−10. *Achenes* 1 mm long, glabrous. *Pappus* bristles many, shortly plumose in upper half, bases nude, not cohering.

Recorded only from Fort Nottingham Commonage in Natal and the top of Mt Insizwa, Mt Ayliff district, Transkei. Forms large mats on damp earth banks and tussocks at the marshy sources of streams; flowering in May. Map 48.

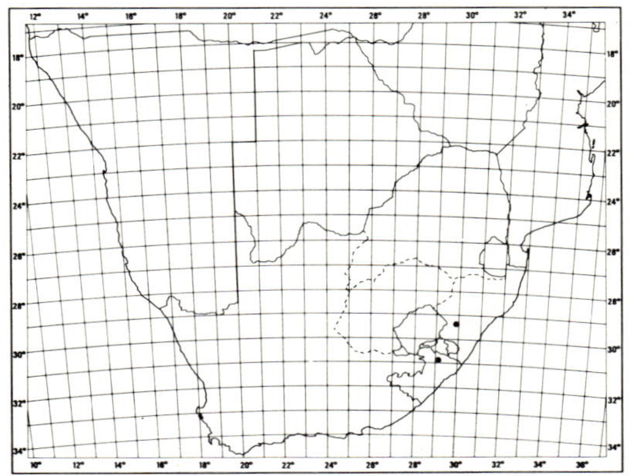

MAP 48.— **Helichrysum ephelos**

Voucher: *Hilliard & Burtt* 7307 (E; NU).

21. **Helichrysum subfalcatum** *Hilliard* in Notes R. bot. Gdn Edinb. 32: 361 (1973), Compositae in Natal 163 (1977). Type: Natal, Estcourt distr., Highmoor Forest Reserve, spur running E. from Giant's Castle, c. 8 000 ft., *Hilliard* 4809 (NU, holo.!; E, iso.!).

Mat-forming perennial herb, main stems branching, prostrate, rooting, producing numerous congested leaf rosettes, flowering stems terminal, 10−200 mm long, simple, distantly leafy upwards. *Leaves* 30−60 × 2−7 mm, generally narrowly lanceolate, rarely linear, often conduplicate and slightly falcate, apex acute, base broad, clasping, indumentum smooth, sericeous, stripping like tissue-paper, upper surface green drying grey, lower silvery white. *Heads* heterogamous, cylindric, c. 4 × 3 mm, many webbed together in a congested terminal cluster 15—20 mm across. *Involucral bracts* in c. 3 series, subequal, loosely imbricate, equalling the flowers, translucent, outer brownish, all tipped bright canary-yellow, obtuse, not radiating. *Receptacle* shortly honeycombed. *Flowers* c. 10—17, 2—4 ♀, 7—14 ☿, yellow. *Achenes* 0,75mm long, with myxogenic duplex hairs. *Pappus* bristles 3—5, tips shortly plumose, bright yellow, shaft nude, bases not cohering. Fig. 18:4.

Recorded along the Natal Drakensberg from the headwaters of the Umzimkulu (Mashai Pass and Wilson's Cave area) and Sani Pass, all in Underberg district, to Mont aux Sources, Bergville district, c. 2 100 to 3 200 m. Grows on steep stony mountain slopes and tops, often in large colonies; flowering between January and March. Map 49.

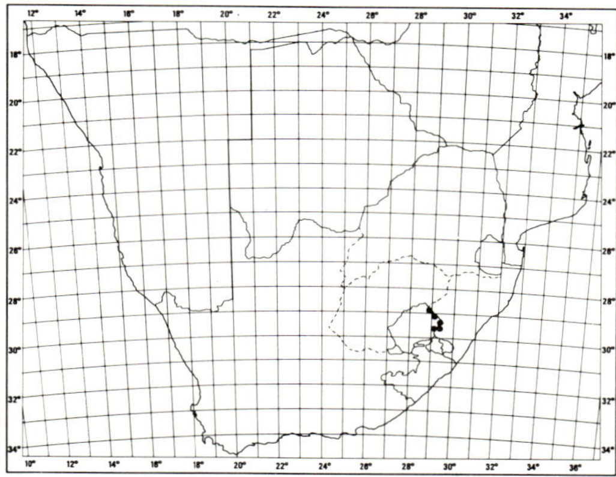

MAP 49.— **Helichrysum subfalcatum**

Vouchers: *Hilliard* 4815 (E; K; NU; PRE); *Schelpe 1387 (NU); Wright* 383 (E; K; M; NU; S).

22. **Helichrysum albirosulatum** *Killick*

in Bothalia 7: 23 (1958); Hilliard, Compositae in Natal 162 (1977). Type: Natal, Bergville distr., Cathedral Peak Forest Research Station, *Killick* 1919 (PRE, holo.!; K, iso.!).

Prostrate well-branched shrublet, main branches bare, up to 10 mm diam., with numerous erect leafy dwarf lateral branches, flowering branches up to 120 mm long, often decumbent then erect, silvery-white felted, closely leafy throughout. *Leaves* at first rosetted at the branch tips, up to 25 × 6 mm, spathulate, apex rounded or subacute, slightly recurved, base half-clasping, both surfaces enveloped in silvery white, slightly glossy, very closely felted indumentum, stem leaves similar but often smaller and narrower. *Heads* homogamous, subcylindric, c. 4 × 2—3 mm, many felted together at the base in a congested, rounded, corymbose cluster 10—20 mm across. *Involucral bracts* in c. 3 series, loosely imbricate, about equalling flowers, obtuse, not radiating, translucent, pale yellow washed golden-brown outside. *Receptacle* shortly honeycombed. *Flowers* c. 8—26, yellow. *Achenes* 1 mm, with myxogenic duplex hairs. *Pappus* bristles c. 6—10, tips plumose, shaft nude, bases not cohering. Fig. 18: 1.

Recorded along the Natal Drakensberg from Cathedral Peak (Bergville district) to Underberg district, with one collection from Bazeia Mountain near Engcobo, Transkei. Appears to be confined to Cave Sandstone, sprawling over platforms of this rock. Flowers between January and March. Map 50.

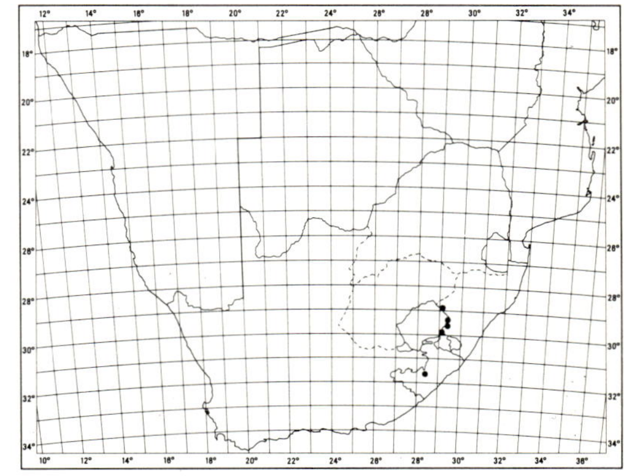

MAP 50.— **Helichrysum albirosulatum**

Vouchers: *Hilliard & Burtt* 9719 (E; NU); *Hoener* 2186 (E; NU); *Wright* 455 (E; K; M; NH; NU).

Group 7

Similar to Group 6, differing in pappus of many bristles, with subplumose tips and bases cohering by patent cilia.

Species 23—24, on the mountains from Natal and Lesotho to the E. Cape on rocks or in short stony turf.

1a Leaves broadly spathulate, ovaries glabrous .... 23. *H. alticolum*

1b Leaves elliptic-lanceolate, often conduplicate and falcate, ovaries hairy .... 24. *H. nimbicola*

23. **Helichrysum alticolum** *H. Bol.* in Trans. S. Afr. phil. Soc. 18: 386 (1907); Moeser in Bot. Jb. 44: 250 (1910). Type: Cape, Stockenstrom div., old Katberg Pass, c. 1650 m, *Galpin* 2397 (BOL, holo.!; K; PRE, iso.!).

H. alticolum var. *montanum* H. Bol. in Trans. S. Afr. phil. Soc. 18: 386 (1907); Moeser in Bot. Jb. 44: 250 (1910); Hilliard, Compositae in Natal 161 (1977). Lectotype: Natal, Drakensberg, summit Mont aux Sources, 11 000 ft, March 1898, *Evans* 742 (BOL!; K; NH; PRE, isolecto.!).

A mat-forming perennial herb, main stems up to 4 mm diam., older parts nude, younger densely clothed in greyish-white silky-woolly indumentum, stoloniferous, giving rise to numerous dwarf branches with closely rosetted leaves, flowering stems lateral, c. 20—50 (—100) mm long, leafy. *Leaves* spathulate, up to 45 × 15 mm, apex rounded, base broad, half-clasping, both surfaces closely greyish-white woolly-felted, the upper surface sometimes merely cob-webby (but glandular) with age. *Heads* homogamous, c. 5—6 mm long, 3 mm across, subcylindric, many in a subglobose terminal cluster 10—20 mm across, their bases felted together with grey wool. *Involucral bracts* in c. 4 series, subequal, very loosely imbricate, about equalling the flowers, spreading at the top but not radiating, glossy, bright canary-yellow, the outermost sometimes brownish. *Receptacle* with pit margins slightly produced. *Flowers* 4—6, yellow. *Achenes* c. 1,5 mm, ellipsoid, obscurely ribbed, glabrous. *Pappus* bristles many, slightly flattened, tips subplumose, bases cohering lightly by patent cilia. Fig. 19: 2.

Recorded from Khotjoane Pass in northern Lesotho and the nearby Mont aux Sources area, thence along the Natal Drakensberg as far south as Loteni, then a disjunction to old Katberg Pass between Seymour and Cathcart. The plants generally form big mats on cliff faces, rock platforms or stony ground between c. 1 900 and 3 200 m above sea level; the type specimen, from old Katberg Pass, is laxer, with longer flowering stems than usual, but it is surely no more than a particularly well-grown plant from a lower altitude (c. 1650 m). May flower in September and October, but more usually between February and April. Map 51.

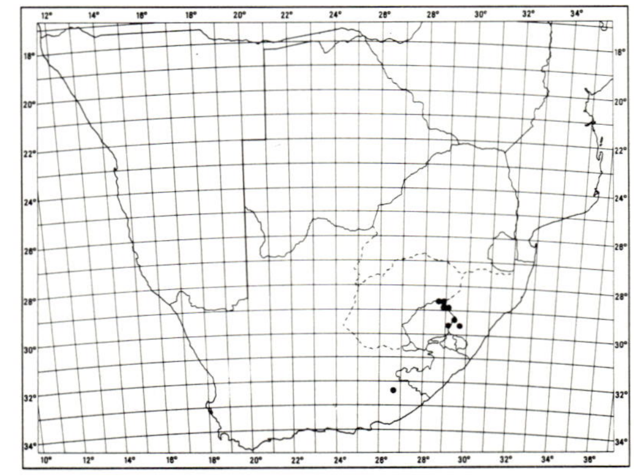

MAP 51.— **Helichrysum alticolum**

Vouchers: *Esterhuysen* 18680 (BOL); *Hilliard* 5001 (E; K; NH; NU); *Schelpe* 1341 (E; NU); *Trauseld* 773 (NU); *Wright* 506 (E; NU).

24. **Helichrysum nimbicola** *Hilliard* in Notes R. bot. Gdn Edinb. 40: 262 (1982). Type: Natal-Lesotho border, Sani Pass, 2850 m, on steep stony grass slopes, 18 ii 1973, *Hilliard* 5334 (NU, holo.!; E; K; MO; PRE; S, iso.!).

Tufted perennial herb forming small mats, main stems decumbent, rooting,

FIG. 19.—1, **Helichrysum nimbicola,** part of whole plant, × 0,7; 1a, head, × 10; 1b, flower, × 8; 1c, pappus bristle, × 8 (*Hilliard* 5334). 2, **H. alticolum,** part of whole plant, × 1; 2a, head, × 6,6; 2b, flower, × 6,6; 2c, pappus bristle, × 6,6 (*Trauseld* 773).

1a
1b 1c
2
2c 2b
2a
1
H. Wolfe du Toit

flowering stems erect, 100−150 mm long, simple, densely leafy near base, becoming distantly so then bracteate near tips. *Leaves* mostly 25−40 × 4−7 mm, half the length petiolar, blade elliptic-lanceolate, often conduplicate and falcate, tip acute, base narrowed to a broad flat clasping petiole, both surfaces grey woolly-felted. *Heads* homogamous, cylindric, 3−4 × 1,5−2 mm, many closely congested and felted together into a flat-topped glomerule 15−20 mm across at the stem tip. *Involucral bracts* in 3−4 series, graded, loosely imbricate, about equalling flowers, pellucid, pale golden brown, tips very obtuse, concave, not radiating. *Receptacle* shortly honeycombed. *Flowers* 4−6, yellow. *Achenes* 0,75 mm long, with myxogenic duplex hairs. *Pappus* bristles many, equalling corolla, upper part subplumose, yellow, bases cohering strongly by patent cilia. Fig. 19: 1.

Known from only four collections on the high Lesotho plateau: the summit of Mount aux Sources, at the top of Mashai Pass and the top of Sani Pass at c. 2 800 m, and from Tselanyane, Butha Buthe district, c. 2 750 m, in stony turf, flowering in February. Similar in aspect to *H. subfalcatum* (no. 21), with which it may be found growing, but easily distinguished by its different indumentum, homogamous heads, and different pappus. Map 52.

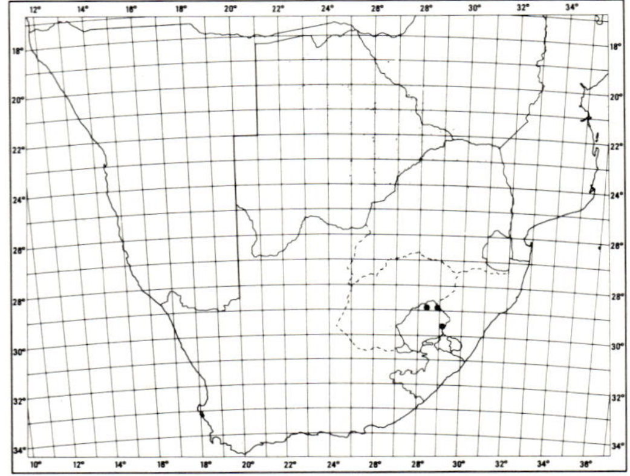

MAP 52.— **Helichrysum nimbicola**

Vouchers: *Hilliard & Burtt* 10479 (NU); *Jacot Guillarmod* 4223 (RUH).

Group 8

Bushy shrubs, subshrubs, or rarely perennial herbs; *leaves* linear to lanceolate, small; *heads* homogamous or heterogamous (sometimes in the same species), 2,5−4 × 1−3 mm, in cymose clusters arranged in corymbose panicles; *involucral bracts* either not radiating or minutely radiating, white, yellow, or tawny; *receptacle* fimbrilliferous; *flowers* 6−45, 0−18 ♀, ♀ sometimes outnumbering ⚥, corolla of ⚥ flowers funnel-shaped, of ♀ flowers narrowly tubular with a conspicuous limb; *achenes* glabrous; *pappus* bristles scabrid, bases cohering by patent cilia, or rarely wanting.

Species 25−33, mainly in Southern Africa, but *H. kraussii* is mainly tropical (Angola, Zambia, Zimbabwe and Mozambique) reaching the Transvaal and coastal Natal, *H. polycladum* occurs in Zimbabwe as well as the Transvaal, Natal and Swaziland, *H. aureonitens* reaches Angola; while *H. melanacme* is closely allied to species in tropical Africa. Found mostly in rough grassland and shrub communities.

1a Involucral bracts brownish, tawny, straw-coloured or yellow:

  2a Leaves linear, margins strongly revolute; stiffly branched shrublet .... 25. *H. kraussii*

  2b Leaves (ignore uppermost reduced ones) linear-lanceolate, linear, oblong or more or less elliptic, margins flat or weakly revolute; soft-wooded tufted perennial herbs or subshrubs, branches often tangled:

    3a Leaves oblong or oblong-spathulate, obtuse or subacute, both surfaces appressed-woolly; heads campanulate ....27. *H. aureonitens*

    3b Leaves either linear-lanceolate or lanceolate, very acute to acuminate, heads cylindric or campanulate; or linear-oblong to elliptic-oblong, acute or acuminate and then upper surface with thin papery indumentum and heads cylindric:

      4a Tips of at least the inner involucral bracts yellow:

5a Leaves (ignore uppermost reduced ones) linear-oblong to elliptic-oblong, acute or rarely acuminate, upper surface clad in paper-like indumentum .... 26. *H. cymosum*

5b Leaves linear-lanceolate, lanceolate or ovate-lanceolate, very acute to acuminate, upper surface glabrous, cobwebby or woolly:

  6a Leaves glabrous or cobwebby above, tips of involucral bracts smooth or nearly so .... 28. *H. tenuiculum*

  6b Leaves woolly-felted on both surfaces, tips of involucral bracts well crisped, more or less squarrose .... 32. *H. simillimum*

4b Tips of at least the inner involucral bracts straw-coloured or tawny:

  7a Female flowers either wanting or fewer than the hermaphrodite:

    8a Leaves glabrous or thinly cobwebby above, tips of involucral bracts smooth or nearly so, flowers 8−14 .... 28. *H. tenuiculum*

    8b Leaves woolly or at least cobwebby above, tips of involucral bracts crisped, flowers (15−) 20−45 .... 29. *H. melanacme*

  7b Female flowers 2−4 times as many as hermaphrodite.................................... 30. *H. interjacens*

1b Involucral bracts white or dirty white:

  9a Leaves up to 2 mm broad; female flowers outnumbering hermaphrodite .... 31. *H. polycladum*

  9b Leaves mostly 3−5 mm broad (ignore uppermost reduced leaves); female flowers fewer than hermaphrodite:

    10a Leaves acuminate; heads c. 3−4 mm long, involucral bracts more or less squarrose .... 32. *H. simillimum*

    10b Leaves acute; heads c. 2,5 mm long, involucral bracts not squarrose .... 33. *H. helianthemifolium*

25. **Helichrysum kraussii** *Sch. Bip.* in Flora, Regensburg 27, 2: 679 (1844); Harv. in F.C. 3: 249 (1865); Wood, Natal Plants 3, 3: 21, t. 269 (1902); Moeser in Bot. Jb. 44: 252 (1910); Hilliard, Compositae in Natal 166 (1977). Type: Natal Bay [Durban], *Krauss* 459 (P, holo.!; G; K; M; TCD; fragment PRE, iso.!).

Gnaphalium kraussii (Sch. Bip.) Sch. Bip. in Bot. Ztg 3: 173 (1845).

Achyrocline steetzii Vatke in Öst. bot. Zeitschr. 27: 194 (1877). *Helichrysum steetzii* (Vatke) O. Hoffm. in Bolm Soc. Broteriana 13: 25 (1896). Type: Mozambique, Inhambane, *Peters* s.n. (B†).

A. batocana Oliv. & Hiern in F.T.A. 3: 339 (1877). Type: Zambia, Batoka country, *Kirk* (K, holo.!).

Bushy aromatic shrublet up to 1 m tall, branches stiff, thinly greyish-white felted, glabrescent, closely leafy. *Leaves* spreading or reflexed, up to 20 × 2 mm, sessile, linear, apex acute, mucronate, margins strongly revolute, sometimes nearly obscuring the lower surface, upper surface cobwebby, glabrescent, lower white-felted. *Heads* heterogamous or homogamous, sometimes on one plant, cylindric, c. 3,5 × 1 mm, very many in dense cymose clusters corymbose-paniculately arranged at the branch tips. *Involucral bracts* in c. 5 series, graded, closely imbricate, about equalling flowers, not radiating, glossy, pale lemon-yellow or straw-coloured. *Receptacle* with fimbrils equalling ovary. *Flowers* 5−6, 0−2 ♀, 3−5 ☿, pale yellow. *Achenes* not seen, ovaries glabrous. *Pappus* bristles many, scabrid, a little shorter than corolla, bases cohering strongly by patent cilia. Fig. 20:1.

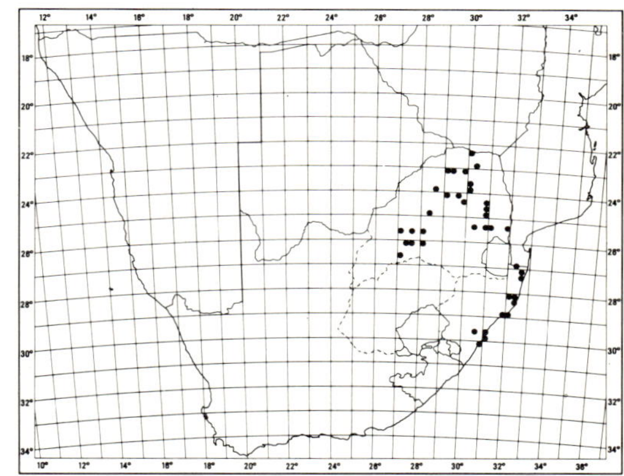

MAP 53.— **Helichrysum kraussii**

Widespread in south tropical Africa, from southern Angola through Zambia and Zimbabwe to

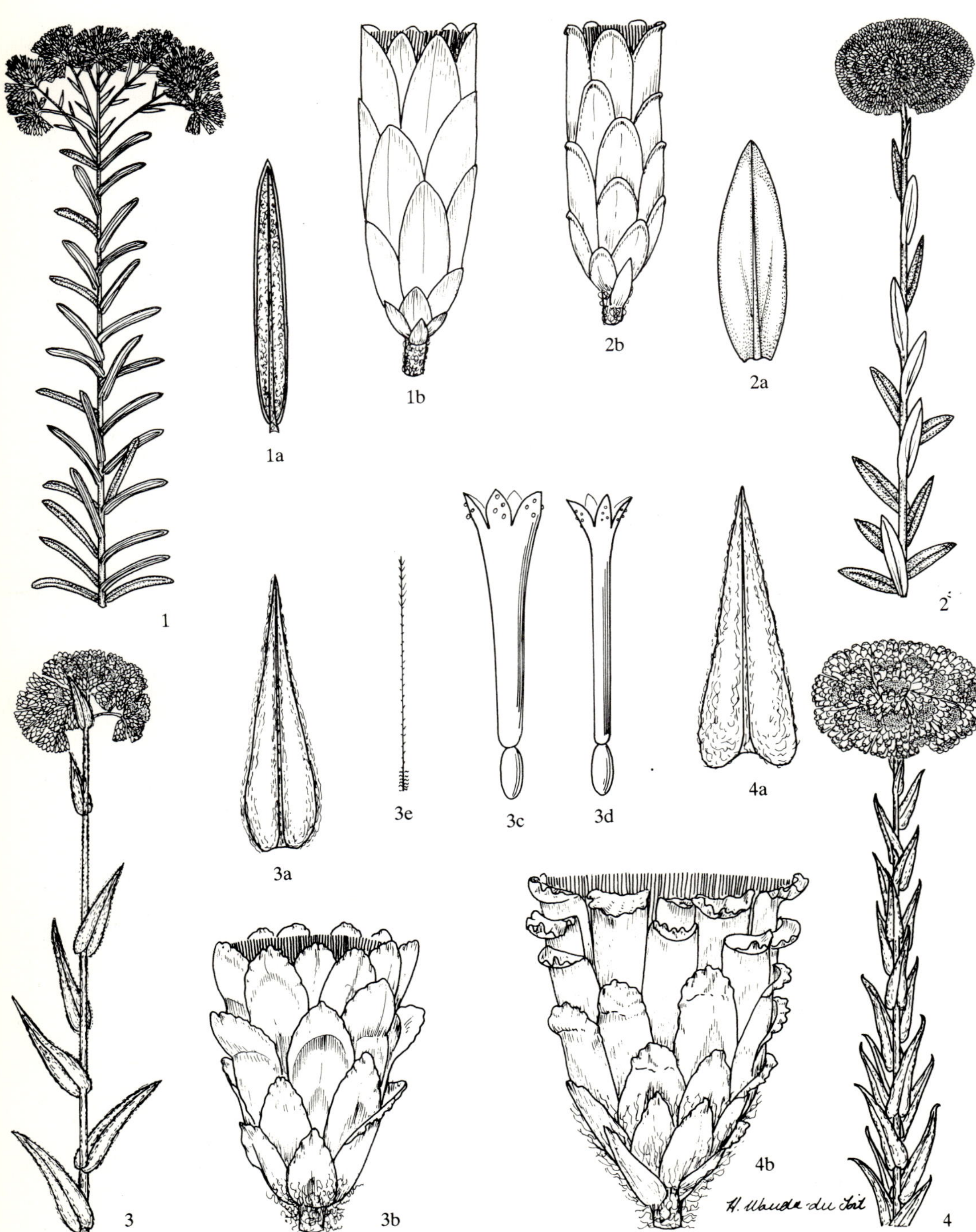
1
1a
1b
2a
2b
2
3
3a
3b
3c
3d
3e
4a
4b
4
H. Maude du Toit

Mozambique and the Transvaal, reaching the southern limit of its range in coastal Natal just south of Durban. Usually in dense stands, in sandy grassland or open woodland; flowering from June to September. Often has fluffy white galls at the branch tips. Map 53.

Sometimes confused with *H. cymosum* (below), but easily distinguished by its very different leaves.

Vouchers: *Acocks* 10878 (PRE); *Hilliard* 4761 (E; K; NU; PRE); *Strey* 6790 (NU).

26. Helichrysum cymosum *(L.) D. Don* in Sweet, Hort. Brit. 223 (1826), in Loudon, Hort. Brit. 342 (1830); Less., Syn. Comp. 302 (1832); Harv. in F.C. 3: 245 (1865); Moeser in Bot. Jb. 44: 256 (1910); Batten & Bokelmann, Wild Flow. E. Cape Prov. 152, plate 122, 2 (1966); Hilliard, Compositae in Natal 168 (1977). Type: Cape of Good Hope, *Gnaphalium* sheet no. 7 in herb. Hort. Cliff. (BM!).

Gnaphalium cymosum L., Sp. Pl. 855 (1753). *Lepiscline cymosum* (L.) Cass., Dict. Sci. nat. 26: 49 (1823).

G. spadiceum Lam., Encycl. 2: 753 (1788). Type: Cape of Good Hope (P-LAM!).

G. cernuum Thunb., Prodr. 150 (1800), Fl. Cap. 654 (1823). Type: Table Mountain, *Thunberg* (sheet 19113, UPS!).

Helichrysum subdecurrens DC., Prodr. 6: 202 (1838). *Gnaphalium subdecurrens* (DC.) Sch. Bip. in Bot. Ztg 3: 172 (1845), non DC. (1834). Type: Cape, near George, on Cradock Berg, 12 ix 1814, *Burchell* 5970 (G-DC, holo.!; K, iso.!).

H. parviflorum var. *aureum* DC., Prodr. 6: 203 (1838). *H. cymosum* var. *pauciflorum* Harv. in F.C. 3: 246 (1865). Lectotype: Cape, Uitenhage, *Burchell* 4226 (G-DC! misquoted by DC. as 4216).

H. cymosum var. *minus* [Sond. ex] Harv. in F.C. 3: 246 (1865). Type: Cape, *Zeyher* 883 (S, holo.!; E; K, iso.!).

Two subspecies are recognized:

1a Heads with 6–20 flowers, fimbrils more than twice as long as ovary, pappus copious .... (a) subsp. *cymosum*
1b Heads with 4–7 flowers, fimbrils about as long as ovary, pappus wanting...(b) subsp. *calvum*

(a) subsp. **cymosum.**

A well-branched spreading subshrub up to 1 m tall, branches long, often decumbent at the base then erect, thinly greyish-white woolly, densely leafy, becoming penduncu-loid upwards. *Leaves* very variable, mostly 8–15 (–45) × 2–4 (–15) mm, becoming smaller and more distant upwards, elliptic-oblong or linear-oblong, apex acute, sometimes acuminate, mucronate, slightly narrowed to a half-clasping, subdecurrent base, margins flat or subrevolute, upper surface clothed in thin silvery grey sericeous indumentum that will strip like a skin (and sometimes does), rarely woolly-felted, lower surface closely white-woolly. *Heads* heterogamous or sometimes homogamous, cylindric, mostly c. 3 × 1 mm, sometimes up to 4,5 × 3 mm, many in compact cymes corymbose-paniculately arranged at the branch tips. *Involucral bracts* in c. 4 series, graded, closely imbricate, inner about equalling flowers, not radiating, translucent, glossy, bright canary-yellow, tips generally smooth. *Receptacle* with fimbrils exceeding the ovaries. *Flowers* 6–12 (–20), 0–4 ♀, 3–12 (–20) ☿, yellow. *Achenes* 0,75 mm, broadly cylindric, glabrous. *Pappus* bristles many, scabrid, bases cohering strongly by patent cilia. Fig. 20:2.

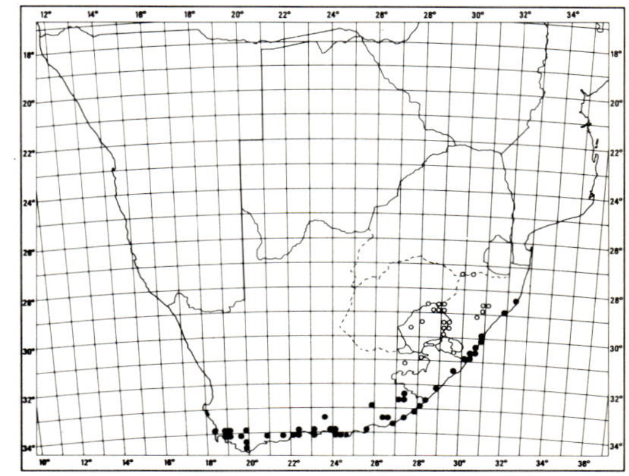

MAP 54.— ● Helichrysum cymosum subsp. **cymosum**
○ Helichrysum cymosum subsp. **calvum**

FIG. 20.—1, **Helichrysum kraussii,** flowering branch, × 1; 1a, leaf, margins revolute, × 3,3; 1b, head, × 13 (*Tosh* NU 30706). 2, **H. cymosum** subsp. **cymosum,** flowering branch, × 1; 2a, leaf, × 3,3; 2b, head, × 13 (*Hilliard & Burtt* 12449). 3, **H. melanacme,** flowering branch, × 1; 3a, leaf, × 2; 3b, head, × 13; 3c, hermaphrodite flower, × 17; 3d, female flower, × 17; 3e, pappus bristle, × 17 (*Hilliard & Burtt* 11240). 4, **H. simillimum,** part of plant, × 1; 4a, leaf, × 2; 4b, head, × 10 (*Rennie* 358).

H. cymosum subsp. *cymosum* ranges from the SW. Cape, including the Peninsula, eastwards along the coast and the coastal mountain ranges, then north-east both along the coast and inland to Albany district, Katberg and the Amatola Mountains, and as far as Lake St Lucia in Natal. Occurs up to c. 1 500 m above sea level in the Cape mountains, but not above 600 m in Natal. Also recorded from Rustenburg Kloof in the south-western Transvaal, but this needs confirmation; a pencil note on the sheet suggests the label is wrong. Grows in big straggling clumps, often in moist places, such as hollows between the coastal dunes, in shrub communities, both in the Cape scrub and on forest margins. Flowers between September and April. Map 54.

Vouchers: *Acocks* 11109 and 22975 (PRE); *Esterhuysen* 20764 (BOL; PRE); *Flanagan* 417 (PRE); *Hilliard & Burtt* 6061 (E; K; MO; NU; S); *Wright* 976 (E; K; NH; NU; M).

H. cymosum subsp. *cymosum* shows considerable variation in head size. Heads are commonly c. 3 × 1 mm but larger-headed plants occur erratically from the Cape Peninsula to the Umzimkulu River on Natal's southern border, and, from the herbarium records, appear often to grow with the typical small-headed plant. This increase in head size is due, not to an increase in flower number, but to an increase in flower size: corolla length ranges from 2−3,5 mm, and there is a corresponding increase in corolla width.

H. cymosum can be recognized by its cylindric heads with canary-yellow bracts, and leaf shape and indumentum; the leaves are generally elliptic-oblong to linear-oblong with acute tips, the upper surface covered with paper-like indumentum. Only rarely are the leaves (apart from the uppermost ones) acuminate or the upper leaf surface cobwebby or thinly woolly. *Salter* 9036 (BOL; PRE; SAM), 9065 (SAM) and 9789 (BM), collected at Kirstenbosch, are anomalous in that the heads are like those of *H. cymosum* but the leaves resemble those of *H. tenuiculum* (no. 28); *H. tenuiculum* has not been recorded from the Peninsula, so it is unlikely that crossing has taken place. The specimens possibly represent no more than a shade form of *H. cymosum*. *Taylor* 4610 (PRE) from Jonkershoek Forestry Reserve, Stellenbosch, also has exceptionally large acuminate leaves.

H. melanacme (no. 29) is sometimes confused with *H. cymosum*. The leaves of *H. melanacme* are lanceolate, acuminate, and usually thinly woolly above, therefore leaf shape and indumentum will distinguish most specimens of *H. cymosum* from *H. melanacme*. Furthermore, the involucral bracts of *H. melanacme* are tawny or sometimes straw-coloured, the tips crisped, not smooth or nearly so as they usually are in *H. cymosum*. Confusion can also arise with *H. tenuiculum* (no. 28); see under that species.

Hedberg (Symb. bot. Upsal. 15, 1: 203, 1957) reduced the tropical African '*H. fruticosum* (Forsk.) Vatke' (that is, *Gnaphalium forskahlii* J.F. Gmel.) to subspecific rank under *H. cymosum*, but the relationship of this species lies with *H. melanacme* not with *H. cymosum* (see Hilliard & Burtt in Notes R. bot. Gdn Edinb. 38, 1: 145, 1980).

(b) subsp. **calvum** *Hilliard* in Notes R.

bot. Gdn Edinb. 32,3: 346 (1973) Compositae in Natal 169 (1977). Type: Natal-Lesotho border, summit of Drakensberg in vicinity of Bushman's River pass, c. 3050 m, *Wright* 447 (E, holo.!; NU, iso.!).

H. infaustum var. *discolor* Moeser in Bot. Jb. 44: 246 (1910). Types: Natal, *Cooper* 2592 (K!); Greytown, *Wood* 1013 (BM! E!) and 4323 (G! K! Z!); Transvaal, Sandspruit, *Rehmann* 6872 (K! Z!).

Differs from typical *H. cymosum* in its generally more crowded and often narrower leaves, smaller heads (4−7 flowers in a head, against 6−12), generally shorter fimbrils (about as long as the ovary, against about two and a half times as long), and lack of pappus bristles. It can be confused with *H. griseolanatum* (no. 11) and *H infaustum* (no. 10), but is easily distinguished by its glabrous ovaries.

Subsp. *calvum* ranges along the Drakensberg from about Saalboom Nek, Barkly Pass and Maclear in the E. Cape, along the Natal-Lesotho border to the low Berg near Volksrust and Wakkerstroom in the SE. Transvaal and around Ficksburg and Harrismith in the Orange Free State, and Maseru, Berea and Butha Buthe districts in Lesotho; also on Ngeli Mt on the Transkei-Natal border and the mountain spurs running out to Greytown and Nkandhla in Natal. Flowers from December to March. Forms low, spreading clumps in stony grassland or on steep turf slopes, between 1 200 and 3 170 m above sea level. At high altitudes the plants are, not unexpectedly, much more compact than typical *H. cymosum* from the coast, but at lower altitudes the two look so much alike that dissection is needed to separate them. Map 54.

Vouchers: *Acocks* 20176 (PRE); *Devenish* 829 (PRE); *Hilliard & Burtt* 6678 (E; K; MO; NU; PRE; S).

27. **Helichrysum aureonitens** *Sch. Bip.* in Flora, Regensburg 27: 680 (1844); Harv. in F.C. 3: 247 (1865); Moeser in Bot. Jb. 44: 285 (1910); Compton, Fl. Swaziland 628 (1976); Hilliard, Compositae in Natal 171 (1977). Type: Natal Bay, July, *Krauss* 280 (P, holo.!; G; K; M; TCD; iso.!).

Gnaphalium aureonitens (Sch. Bip.) Sch. Bip. in Bot. Ztg 3: 173 (1845).

Helichrysum helodes Hiern, Cat. Afr. Welw. Pl. 3: 561 (1898). Types: Angola, Huilla, Morro de Lopollo, *Welwitsch* 3502 (BM!) and 3503 (BM! K!).

Tufted perennial herb, whole plant greyish-white appressed woolly, stems up to c. 300 mm tall, slender, simple or sparingly branched from a creeping stock, leafy throughout. *Leaves* up to 20 (−30) × 3 (−6) mm, slightly smaller upwards, oblong or oblong-spathulate, apex subacute or obtuse, base broad, half-clasping, margins flat or subrevolute. *Heads* heterogamous, campanulate, c. 4 × 3 mm, up to c. 30 in compact corymbose clusters at the branch tips. *Involucral bracts* in c. 3 series, graded, imbricate, mostly obtuse, inner about equalling flowers, not radiating, outer shorter, woolly at base, all pellucid, yellow often washed pale brown. *Receptacle* with fimbrils about equalling ovaries. *Flowers* 37—57, 13—23 ♀, 21—37 ☿. *Achenes* less than 0,5 mm, glabrous. *Pappus* bristles many, equalling corolla, scabrid, bases cohering by patent cilia.

Widespread, from Huila in Angola to the Transvaal Highveld and eastern highlands, Swaziland, western Lesotho, Orange Free State, Natal and the eastern Cape about as far south as King William's Town and the Amatola Mts. Also recorded from southernmost Mozambique. Common, often forming extensive colonies visible from afar as grey patches in grassland, flowering between September and February. Map 55.

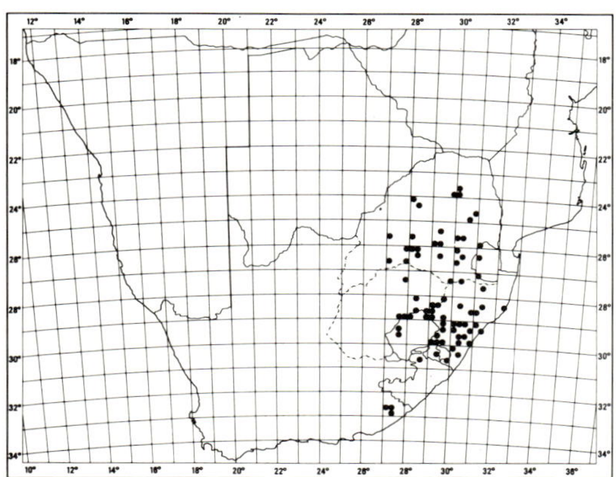

MAP 55.— **Helichrysum aureonitens**

Vouchers: *Codd 3180* (NU; PRE); *Devenish 1603* (E; MO; NU); *Moll 2547* (NU; PRE).

28. **Helichrysum tenuiculum** *DC.*, Prodr. 6: 203 (1838); Harv. in F.C. 3: 246 (1865). Type: Cape, Drakensteensberg, Du Toit's Kloof, *Drège 5747* (G-DC, holo.!; BM; E; TCD; S, iso.!).

A tufted perennial herb or subshrub, stems wiry, branching mostly near the base, decumbent and rooting, then more or less erect to c. 500 mm, thinly greyish-white woolly, closely leafy, more distantly so under the heads. *Leaves* mostly 8—35 × 1—5 mm, smaller and more distant towards the heads, lanceolate or linear-lanceolate, apex acuminate, base broad, ear-clasping, subdecurrent, margins somewhat revolute, glabrous or thinly cobwebby above, thinly white-woolly below. *Heads* heterogamous, or rarely homogamous, cylindric or narrowly campanulate, c. 3 × 2 mm, many in cymose clusters arranged in a usually compact, sometimes lax, terminal corymbose-panicle. *Involucral bracts* in c. 4 series, graded, closely imbricate, inner nearly equalling the flowers, tips obtuse, smooth or nearly so, scarcely radiating, translucent, straw-coloured or the inner yellow. *Receptacle* with fimbrils equalling the ovaries. *Flowers* 8—14, (0—) 1—3 ♀, 7—11 ☿. *Achenes* c. 0.5 mm long, barrel-shaped, glabrous. *Pappus* bristles many, about equalling the corolla, scabrid, bases cohering by patent cilia.

Recorded only from the Cape, on the mountains near Paarl (Drakensteen Mountains and Seven Sisters Mountains) to the Koudebokkeveld Mountains near Citrusdal, and the mountains about Murraysburg and Graaff-Reinet. Rarely collected. Grows on slopes, in moist places near streams, flowering in January and February. Map 56.

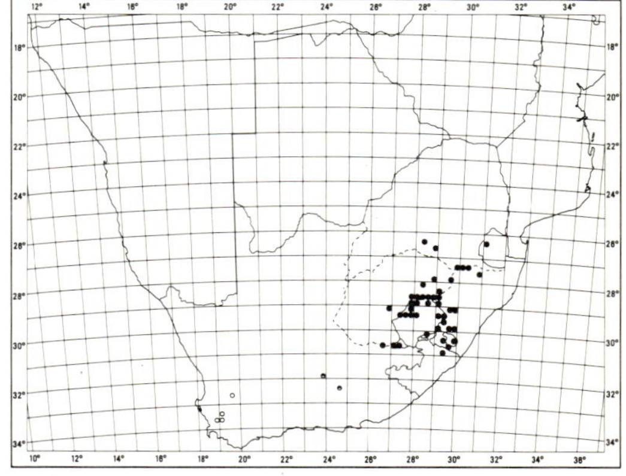

MAP 56.— ○ **Helichrysum tenuiculum**
● **Helichrysum melanacme**
◐ sympatric

H. tenuiculum is sometimes confused with *H. cymosum* (no. 26) but is distinguished by its generally differently shaped leaves with different indumentum, and heads with bracts either tawny, or the outer tawny or brownish, the inner yellow. See also *H. melanacme*, below.

Vouchers: *Esterhuysen* 21113 (BOL; NBG; NU) and 21180 (BOL; NBG; PRE); *Tyson* 161 (BM); *Wall* 245 (S).

Esterhuysen 29573 (BOL; K; NU; Oudtshoorn distr., peaklet at the top of the Swartberg Pass, gully on south side, 2 vi 1962) is close to *H. tenuiculum*, which it resembles in facies and leaf indumentum, but the heads are larger and broadly campanulate (flowers c. 18−21, 2−3 ♀, 16−18 ☿) and the leaves are narrowed to the base and are not ear-clasping. Its status will remain uncertain until these southern Cape mountains are better collected.

29. **Helichrysum melanacme** *DC.*, Prodr. 6: 203 (1838); Harv. in F.C. 3: 246 (1865); Hilliard & Burtt in Notes R. bot. Gdn Edinb. 38: 145 (1980). Type: Cape Wittebergen, *Drège* 5767 (G-DC, holo.!).

Gnaphalium melanacme (DC.) Sch. Bip. in Bot. Ztg 3: 172 (1845).

Gnaphalium kuntzei O. Kuntze, Rev. Gen. Pl. 3,2: 152 (1898). *Helichrysum kuntzei* (O. Kuntze) Moeser in Bot. Jb. 44: 279 (1910). Type: Natal, Charlestown, *O. Kuntze* (K, iso.!).

Helichrysum pullulum Burtt Davy in Jl S. Afr. Bot. 1: 109 (1935). Type: Cape [Natal], Umzimkulu River at Handcock's Drift, 2 500 ft., Dec. 1885, *Tyson* 3086 (K, holo.!; Z, iso.! Note: the SAM specimen under this number is *H. cymosum*).

H. tenuiculum sensu Hilliard, Compositae in Natal 169 (1977); non DC.

H. melanacme is closely allied to *H. tenuiculum* (no. 28) but is distinguished by its *leaves*, which are nearly always woolly or cobwebby above, and its *heads* with *involucral bracts* tawny (or very rarely the innermost yellowish), tips always decidedly crisped, and more numerous *flowers*, (15−) 20−45, 6−18 ♀, 12−32 ☿. Fig. 20:3.

Ranges from the mountains about Murraysburg and Graaff-Reinet to the Witteberg near Lady Grey, the Drakensberg and its outliers in East Griqualand, Transkei, Lesotho, Orange Free State and Natal, the low Drakensberg on the Natal-Transvaal border, and the mountains of western Swaziland; also recorded from Suikerbosrand and Greylingstad on the Transvaal Highveld. Map 56.

Forms large tangled clumps on grassy mountain slopes or near forest margins; flowering from December to April. Plants in moist shady places, such as forest margins, have larger and less woolly leaves and are of laxer habit than plants out in the open.

The areas of *H. melanacme* and *H. tenuiculum*

overlap in the Murraysburg-Graaff-Reinet area: Tyson collected both species near Murraysburg (*Tyson* 162, BM, BOL, is *H. melanacme*, while *Tyson* 161, BM, is *H. tenuiculum* and is in very young bud, while *Tyson* 162 is in full flower). Drège collected a specimen from the Camdeboo Mountains that is possibly *H. melanacme*, but it is only in very young bud.

Vouchers: *Acocks* 23609 (PRE); *Compton* 24827 (NBG; PRE); *Devenish* 573 (PRE); *Hilliard & Burtt* 7670 (E; K; MO; NU; S); *Schlechter* 6501 (BOL).

30. **Helichrysum interjacens** *Hilliard* in Notes R. bot. Gdn·Edinb. 40: 256 (1982). Type: Natal, Ngotshe distr., Itala Nature Reserve, c. 1 525 m, in damp cracks in rock domes, 4 iv 1977, *Hilliard & Burtt* 10021 (NU holo.!; E; K; MO; S, iso.!).

Bushy perennial herb or subshrub c. 600 mm tall, much branched from the base, branches long, thin, brittle, thinly white-woolly, closely leafy. *Leaves* mostly 15−20 × 2−2,5 mm, diminishing upwards and becoming more distant below the heads, linear-lanceolate, acuminate, base broad, half-clasping, eared, margins more or less revolute, upper surface cobwebby-woolly, lower more densely so. *Heads* heterogamous, campanulate, 3 × 1,5−2 mm, many in compact corymbose panicles. *Involucral bracts* in 5 series, graded, imbricate, inner about equalling flowers, light straw colour, tips obtuse, semi-pellucid, crisped, minutely radiating. *Receptacle* with fimbrils much exceeding ovaries. *Flowers* 13−25, 9−17 ♀, 3−8 ☿. *Achenes* 0,75 mm long, cylindric, obscurely ribbed, glabrous. *Pappus* bristles

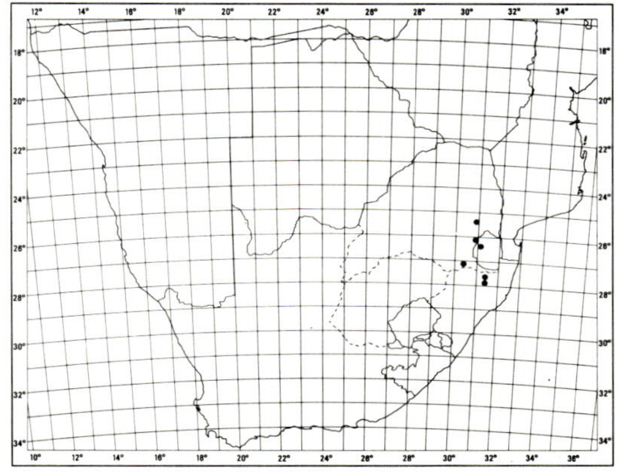

MAP 57. — **Helichrysum interjacens**

many, equalling corolla, scabridulous, bases cohering by patent cilia.

Recorded from the E. and S.E. Transvaal on Long Tom Pass (2530 CA), at Lochiel (2630 BB) and the farm Rusfontein, Wakkerstroom district (2730 AB), Mbabane in Swaziland, Itala and Ngome in Northern Natal (2731 CB and CD). Favours crevices in rock outcrops; flowering in March and April. Map 57.

H. interjacens may be of hybrid origin between *H. melanacme* (above) and *H. polycladum* (below): it is known from only a small geographical area where the ranges of these two species overlap, and may grow in close proximity to one or both of its putative parents. It closely resembles *H. melanacme* in general facies, but is readily distinguished by ♀ flowers in a head far outnumbering ☿; also, where these two species have been recorded together, *H. interjacens* has decidedly paler involucral bracts' than *H. melanacme*. *H. polycladum* is easily distinguished from both by its much narrower leaves and white involucral bracts.

Vouchers: *Compton* 27730 (BM; M; NBG; NU; PRE); 25786 (NBG; PRE); *Devenish* 1696 (E; NU); *Hilliard & Burtt* 9852 (E; K; NU; PRE; S).

31. **Helichrysum polycladum** *Klatt* in Bull. Herb. Boissier 4: 837 (1896); Moeser in Bot. Jb. 44: 281 (1910); Compton, Fl. Swaziland 634 (1976). Type: Transvaal, Mpome [Mparneberg fide Jl. S. Afr. Bot. 30: 135, 1964], *Schlechter* 4725 (Z, holo.!; BM; BOL; G; GRA; PRE; S; iso.!).

Bushy herb forming thick rounded clumps up to 1 m tall and as much across. *Stems* long, thin, sometimes tangled, woody, brittle, thinly greyish-white woolly, gland-dotted, closely leafy. *Leaves* mostly 7−22 × 1 (−2) mm or less, diminishing slightly upwards, linear-lanceolate, apex acuminate, mucronate, base broad, somewhat eared, margins revolute, upper surface more or less loosely white-woolly, often glabrescent, lower persistently loosely white-woolly, gland-dotted. *Heads* heterogamous, c. 3 × 1,5 mm, cylindric, some woolly hairs at base, many in compact corymbose panicles. *Involucral bracts* in 5 series, imbricate, graded, inner about equalling flowers, oblong-lanceolate, pellucid, tips subacute, opaque, dirty white, crisped, radiating. *Receptacle* with fimbrils much exceeding the ovaries. *Flowers* 9−15, 6−10 ♀, 1−5 ☿. *Achenes* 0,5 mm, cylindric, obscurely ribbed, glabrous. *Pappus* bristles many, equalling corolla, scabridulous, bases cohering by patent cilia.

Recorded from high ground in the Transvaal (from the Blouberg and Soutpansberg south to the low

Drakensberg on the Transvaal-Natal border, and from Heidelberg, Johannesburg and Rustenburg on the Highveld), from the environs of Mbabane in Swaziland, and Itala near Louwsburg in northern Natal. Also in the eastern highlands of Zimbabwe. Favours rocky sites, flowering in March, April and May. Easily recognized by its narrow leaves, small heads with white, crisped involucral bracts, and ♀ versus ☿ flowers in the proportion 2−10: 1. Map 58.

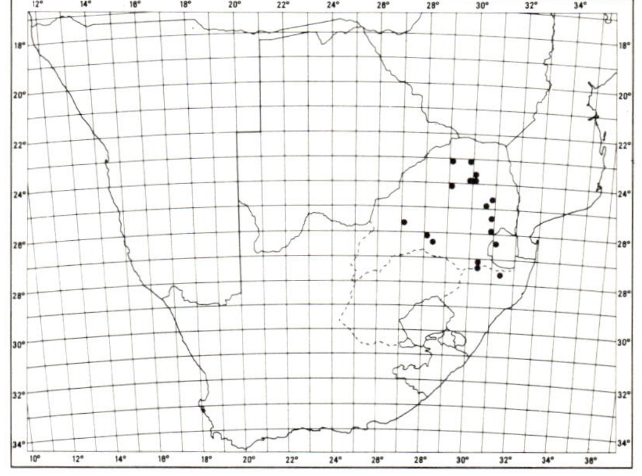

MAP 58. — **Helichrysum polycladum**

Vouchers: *Compton* 27731 (NBG; PRE); *Devenish* 1695 (E; K; MO; NU); *Esterhuysen* 21449 (BOL; E; K; PRE); *Hilliard & Burtt* 10020 (E; K; M; MO; NU; PRE; S).

32. **Helichrysum simillimum** *DC.*, Prodr. 6: 203 (1838); Harv. in F.C. 3: 246 (1865); Moeser in Bot. Jb. 44: 281 (1910); Hilliard, Compositae in Natal 196 (1977). Lectotype: Cape, Kaffraria, Katberg, *Ecklon* 1893 (G-DC, holo.!; S, iso.!).

Gnaphalium simillimum (DC.) Sch. Bip. in Bot. Ztg 3: 172 (1845).

Perennial herb, rootstock slender, creeping, branching, stems slender, tufted, simple or branching mostly near the base, erect to c. 400 mm, thinly grey woolly-felted, closely leafy. *Leaves* imbricate, up to c. 17 × 5 mm, scarcely decreasing in size upwards, lanceolate or ovate-lanceolate, apex acuminate, base cordate-clasping, shortly decurrent, both surfaces grey woolly-felted. *Heads* heterogamous, oblong-campanulate, c. 3−4 × 2−3 mm, many in compact corymbose panicles up to c. 40 mm across terminating the branches.

Involucral bracts in c. 6 series, graded, closely imbricate, inner about equalling flowers, tips opaque, pale yellow or rarely whitish, crisped, more or less squarrose. *Receptacle* with fimbrils exceeding the ovaries. *Flowers* 15−27, 5−9 ♀, 10−18 ☿. *Achenes* less than 0,5 mm long, glabrous. *Pappus* bristles many, equalling corolla, scabrid, bases cohering lightly by patent cilia. Fig. 20:4.

Ranges from Estcourt and Pietermaritzburg districts in the Natal Midlands south through East Griqualand and the Transkei to the E. Cape, from the mountainous country about Dordrecht south to the Great Winterberg, Katberg, Amatola Mountains, Stutterheim and Komgha, then a disjunction to George on the S. Cape coast. Grows in large tangled clumps, particularly near forest margins; flowering mainly between January and April. Map 59.

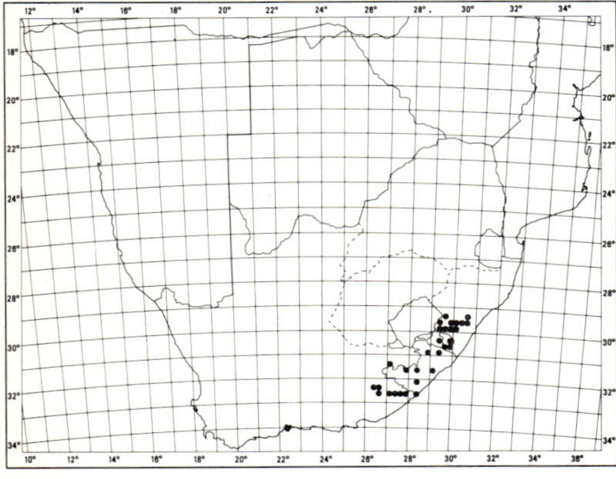

MAP 59.— **Helichrysum simillimum**

Vouchers: *Acocks* 13455 (PRE); *Flanagan* 514 (SAM); *Hilliard & Burtt* 8032 (E; K; MO; NU; S); *Tyson* 2731 (SAM).

33. **Helichrysum helianthemifolium** (*L.*) *D. Don* in Sweet, Hort. Brit. 223 (1826), in Loudon, Hort. Brit. 342 (1830); Hilliard & Burtt in Notes R. bot. Gdn Edinb. 32: 350 (1973). Type: *Gnaphalium africanum floribus minimis albicantibus* Volckamer, Fl. Noribergensis 194 cum tab. (1700)!

Gnaphalium helianthemifolium L., Sp. Pl. 851 (1753); edn 2, 1218 (1763), excl. descr. add.

G. serratum L., Pl. Rar. Afr. (1760), Amoen. Acad.

6: 98 (1763); Burm., Rar. Afr. Pl. t. 76, fig 3 (1738). Type: Cape of Good Hope, *Oldenland* in herb. Burmann (G, holo.!).

G. capitellatum Thunb., Prodr. 150 (1800), Fl. Cap. 653 (1823). *Helichrysum capitellatum* (Thunb.) Less., Syn. Comp. 305 (1832); DC., Prodr. 6: 206 (1838); Harv. in F.C. 3: 250 (1865); Moeser in Bot. Jb. 44: 281 (1910); Levyns in Adamson & Salter, Fl. Cape Penins. 784 (1950). Type: Cape, French Hoek, *Thunberg* (sheet 19110, UPS, holo.!).

H. capitellatum var. *mollius* DC., Prodr. 6: 206 (1838). Lectotype: Cape, Uniondale, between Haarlem and Avontuur, *Burchell* 5061 (G-DC!).

Diffuse herb up to 1 m high. *Stems* long, thin, woody, brittle, thinly greyish-white woolly, closely leafy. *Leaves* mostly 10−20 × 3−5 mm, smaller upwards, lanceolate or oblong-lanceolate, spreading or later deflexed, apex subacute to very acute, mucronate, base broad, subcordate-clasping, margins often somewhat reflexed, crisped or not, both surfaces thinly greyish-white woolly, upper sometimes more or less glabrescent. *Heads* heterogamous, c. 2,5 × 2 mm, cylindric, base slightly woolly, many in compact terminal corymbose panicles. *Involucral bracts* in 4 series, graded, imbricate, inner about equalling flowers, oblong, silvery pellucid, tips rounded, opaque milk-white, crisped, minutely radiating. *Receptacle* with fimbrils exceeding the ovaries. *Flowers* 12−21, 3−7 ♀, 8−14 ☿. *Achenes* 0,5 mm long, cylindric, glabrous. *Pappus* bristles many, equalling corolla, scabrid, bases cohering by patent cilia.

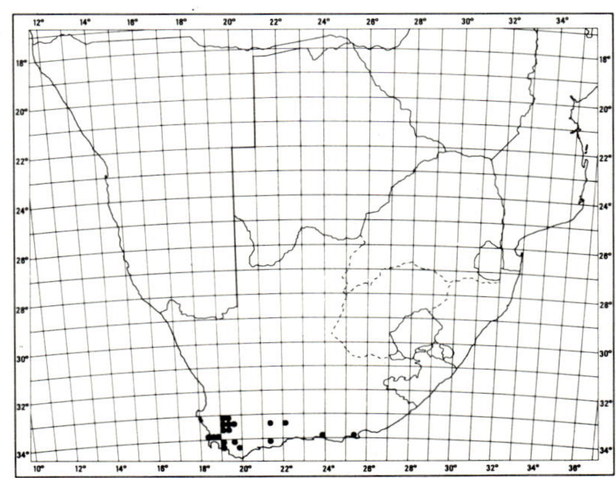

MAP 60.— **Helichrysum helianthemifolium**

On the Cape mountains from the Cedarberg south to the Peninsula and Caledon division and east to the Swartberg and the coastal ranges as far as Uitenhage. Favours streambanks or other damp places in mountain kloofs or on slopes; flowering in December and January. Map 60.

Vouchers: *Esterhuysen* 12405 (BOL; PRE); *Nordenstam* 7097 (E; S); *Rogers* 29237 (SAM); *Taylor* 4573 (PRE); *Tyson* 820 (SAM).

Group 9

Shrubs or subshrubs, rarely tufted perennial herbs; *leaves* small, generally linear to lanceolate, margins revolute; *heads* homogamous or heterogamous (sometimes in the same species), 3−5 × 2−5 mm, solitary or few to many in compact corymbose clusters or panicles; *involucral bracts* either not radiating or minutely radiating, tawny, golden-brown or white, often with purple or rose overlays, rarely yellow; *receptacle* honeycombed or fimbrilliferous; *flowers* 16−50, 0−19 ♀, ♀ sometimes more or less equalling ☿, corolla of ☿ flowers campanulate above, of ♀ flowers narrowly tubular with a conspicuous limb; *ovaries* usually hairy, sometimes glabrous; *pappus* bristles scabrid, bases cohering strongly by patent cilia, occasionally some basal fusion as well, or rarely pappus reduced *(H. anomalum)*.

Species 34−45, differing from group 8 in the shape of the ♀ flowers and in the hairy ovaries. Mainly Southern Africa; *H. dregeanum* and *H. revolutum* also recorded from S.W.A./Namibia. Mostly on poor stony or sandy soils.

1a Tips of inner involucral bracts bright canary-yellow ....36. *H. anomalum*

1b Tips of involucral bracts white, cream, rosy, straw-coloured, tawny, golden-brown or purplish brown, but never bright yellow:

  2a Tips of at least the inner involucral bracts white, cream or rosy:

    3a Tips of bracts crisped:

      4a Heads 2−3 mm broad, 2−5 clustered at the branch tips......................................41. *H. dregeanum*

      4b Heads 3−5 mm broad, usually several to many in corymbose panicles, if few, then heads c. 5 mm broad:

        5a Stems simple or subsimple from a creeping stock ....40. *H. rugulosum*

        5b Well-branched shrubs or subshrubs:

          6a Heads c. 5 mm broad, campanulate ....35. *H. teretifolium*

          6b Heads c. 3 mm broad, oblong-campanulate ....37. *H. athrixiifolium*

    3b Tips of bracts smooth or very nearly so:

      7a Heads oblong-campanulate, bracts oblong .... 38. *H. rosum*

      7b Heads campanulate, bracts broadly ovate .... 39. *H. bachmannii*

  2b Tips of at least inner involucral bracts straw-coloured, tawny, golden-brown or purplish-brown:

    8a Leaves linear, linear-lanceolate or lanceolate, margins revolute but not repand, glandular hairs, if present, small:

      9a Perennial herb with tufts of stems from a creeping stock, stems simple or branching near the base, leaves linear, heads c. 2 mm broad, many in a congested terminal cluster .... 42. *H. ammitophilum*

      9b Well-branched shrubs or subshrubs, stock creeping or not:

        10a Ovaries glabrous...34. *H. interzonale*

        10b Ovaries hairy:

          11a Tips of involucral bracts crisped:

            12a Leaves c. 4−14 × 0,5−2 mm, margins strongly revolute, heads 2−5 clustered at the branchlet tips...41. *H. dregeanum*

            12b Leaves c. 18−50 × 2−4 mm, margins weakly revolute, heads many in congested clusters at the branch tips .... 43. *H. montis-cati*

          11b Tips of involucral bracts smooth .... 44. *H. revolutum*

    8b Leaves more or less oblong, margins repand-revolute, both surfaces rough with coarse glandular hairs ....45. *H. scabrum*

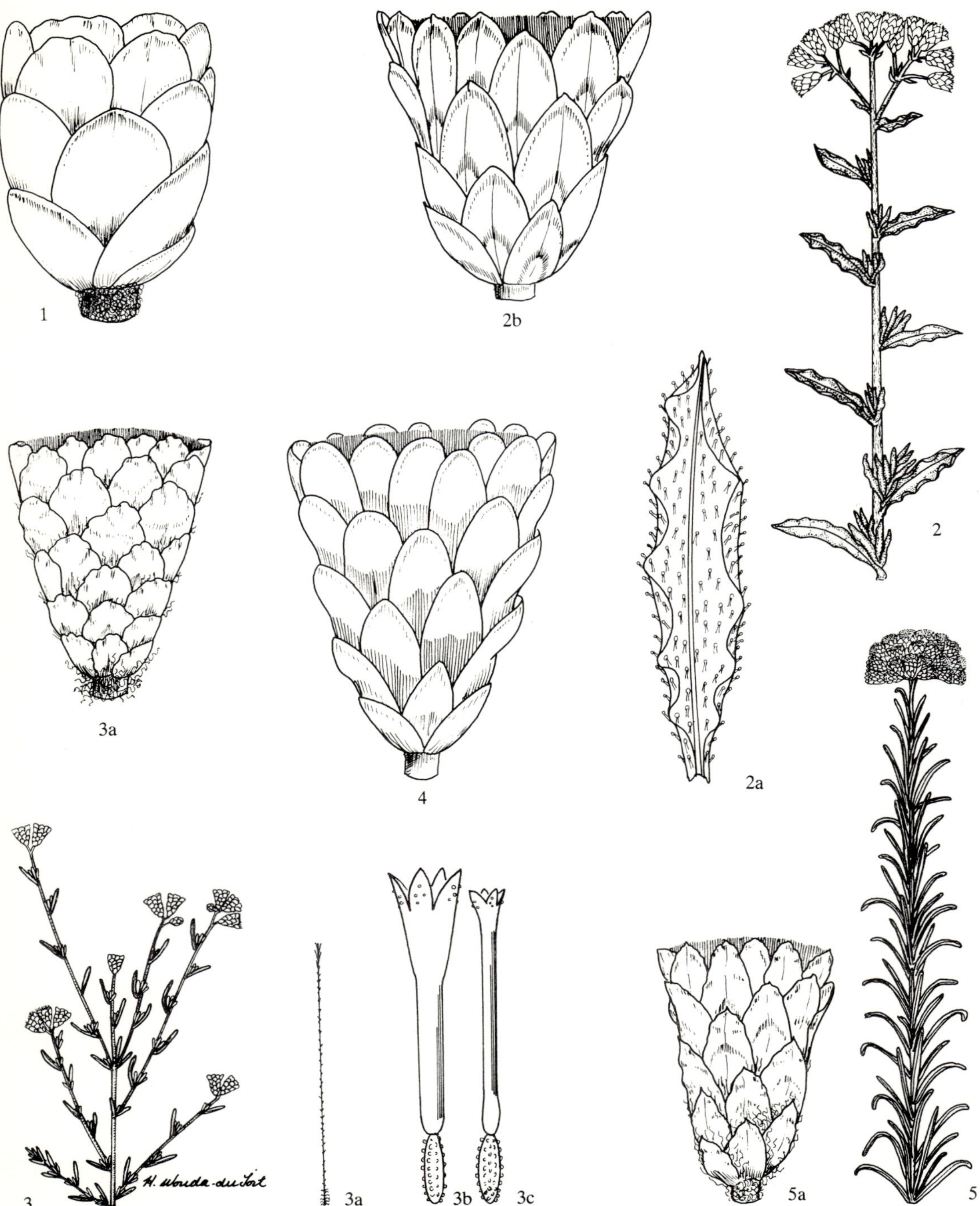

1
2b
3a
4
2a
2
3
H. Wouda-du Toit
3a
3b
3c
5a
5

34. **Helichrysum interzonale** *Compton*

in Jl S. Afr. Bot. 9: 116, fig. 4 (1943). Type: Cape, Witteberg near Bantams, 1 500 m alt., 27 x 1941, *Compton* 12164 (NBG, holo.!).

Well-branched shrublet, branches short or elongate and then decumbent and shooting at the nodes or tips, white-felted when young, closely leafy. *Leaves* mostly 2–6 (–8) × 0,5–1 mm, diminishing slightly upwards, appressed, later spreading, ascending, linear or linear-lanceolate, apex acute, mucronate, slightly hooked, base broad, clasping, margins strongly revolute, white-woolly below, thinly woolly above, soon glabrescent. *Heads* heterogamous or rarely homogamous, campanulate, c. 5 × 4 mm, solitary or up to 12 corymbosely arranged at the branch tips, subsessile or rarely peduncles reaching 8 mm. *Involucral bracts* in c. 5 series, graded, imbricate, inner about equalling flowers, semi-pellucid, tawny, tips obtuse or subacute, somewhat crisped, scarcely radiating, backs thinly woolly with orange or red gland dots. *Receptacle* with fimbrils about equalling ovaries. *Flowers* 16–29, (0–) 1–6 ♀, 13–28 ⚥, staminodes sometimes present. *Achenes* not seen, ovaries glabrous. *Pappus* bristles many, about equalling corolla, barbellate, tips barbellate to subplumose, bases cohering by patent cilia.

Along the Cape mountains from the vicinity of Ceres across the Witteberg, Klein Swartberg and Langeberg to the Groot Swartberg and Suurberg near Uniondale. Grows 'on the slopes of relatively arid quartzite mountains with a "Cape" flora bordering the Karoo' (Compton, l. c.) and 'Mountain Renosterveld. frequent on crags on S. aspect' (*Acocks* 23732). Flowers between September and November. Map 61.

Vouchers: *Acocks* 23732 (PRE); *Compton* 11837 (NBG) and 18341 (NBG); *Esterhuysen* 6276 (BOL; K); *Fourcade* 4402 (K; PRE; SAM; STE).

35. **Helichrysum teretifolium** *(L.) D. Don*

in Sweet, Hort. Brit. 223 (1826); Less., Syn. Comp. 312 (1832); DC., Prodr. 6: 205 (1838); Harv. in F.C. 3: 250 (1865); Wood & Evans, Natal Plants 4: 327 (1906); Moeser in Bot. Jb. 44: 283 (1910); Levyns in Adamson & Salter, Fl. Cape Penins. 784 (1950); Batten & Bokelmann, Wild Flow. E. Cape Prov. 160, plate 127,7 (1966); Hilliard, Compositae in Natal 197 (1977). Type: *Gnaphalium* no. 2 in herb. Cliff. (BM!).

Gnaphalium teretifolium L., Sp. Pl. 854 (1753).

G. scoparium Schrank in Denkschr. K. Akad. Wiss. Münch. 8: 160 (1824). Type: Cape of Good Hope, *Brehm* s.n. (M, holo.!; S, iso.!).

Helichrysum teretifolium var. *congestum* DC., Prodr. 6: 205 (1838). Lectotype: Cape, mouth of Great Fish River, W. side, *Burchell* 3731 (G-DC! misquoted by DC., l. c. as 373). *H. teretifolium* var. *rufescens* DC., l. c. Type: Cape, Mossel Bay, seashore, *Burchell* 6252 (G-DC!). *H. teretifolium* var. *umbelliforme* DC., l. c. Lectotype: Cape, Riversdale div., Soetemelks (sphalm. Troetemelks) River, *Burchell* 6660 (G-DC!). *H. teretifolium* var. *natalense* Harv. in F.C. 3: 250 (1865). Lectotype: Natal, *in dunis prope litus*, Natal point, July 1839, *Krauss* 291 (TCD!; G; M, isolecto.!).

Much-branched compact or straggling subshrub up to c. 300 mm tall, branchlets thinly white-felted, closely leafy to summit. *Leaves* c. 3–15 × 1–1,5 mm, not diminishing upwards, linear, rigid, spreading, margins revolute, tip uncinate, base shortly decurrent, upper surface soon glabrous, lower white-woolly. *Heads* heterogamous or very rarely homogamous, campanulate, c. 5

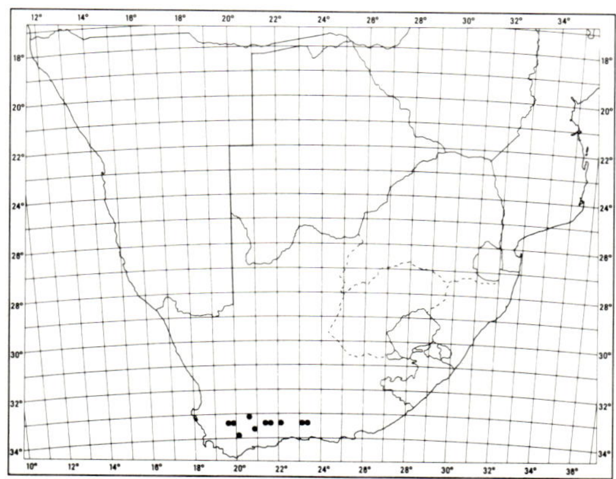

MAP 61.— **Helichrysum interzonale**

FIG. 21.–1, **Helichrysum bachmannii,** head, × 10 (*Acocks* 19685). 2, **H. scabrum,** flowering branch, × 1; 2a, leaf, to show repand, undulate margins and glandular hairs, × 4,6; 2b, head, × 10 (*Bolus* 9021). 3, **H. dregeanum,** flowering branch, × 1; 3a, head, × 10; 3b, hermaphrodite flower, × 13; 3c, female flower, × 13; 3d, pappus bristle, × 13 (*Armour* NU 42993). 4, **H. rosum,** head, × 10 (*MacOwan* 593). 5, **H. ammitophilum,** part of plant, × 1; 5a, head, × 10 (*Hilliard* 4954).

× 5 mm, few to several in compact terminal corymbs. *Involucral bracts* in 6—8 series, graded, imbricate, inner slightly exceeding flowers, tips very obtuse, crisped, opaque creamy-white or outer sometimes rosy, minutely radiating. *Receptacle* with fimbrils exceeding ovaries. *Flowers* 17—60, (0—) 2—14 ♀, 10—49 ☿. *Achenes* c. 1 mm long, glabrous or with myxogenic duplex hairs. *Pappus* bristles many, about equalling corolla, tips barbellate, bases cohering by patent cilia.

Ranges from Piquetberg and the Great Winterhoek Mountains through the SW. and southern Cape to the E. Cape as far inland as Grahamstown and Dohne, thence along the coast to Natal, where it reaches the northern limit of its distribution at Durban Bluff. Grows in shrub communities, often on sand, on stabilized dunes near the sea or on mountain slopes, often in dense stands. Flowers between July and November. Map 62.

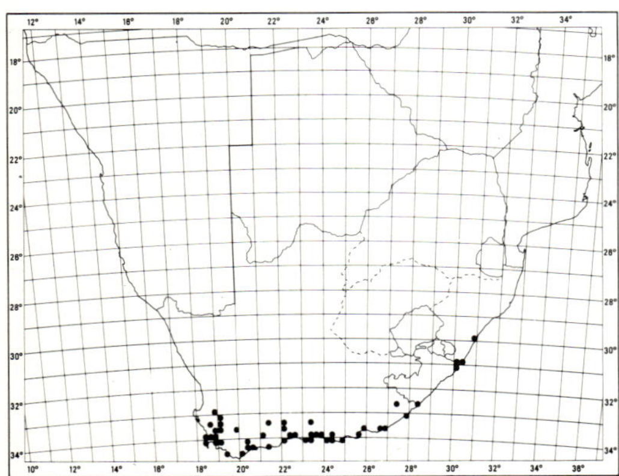

MAP 62.— **Helichrysum teretifolium**

Vouchers: *Acocks* 9141 (PRE); *Compton* 7594 (NBG); *Esterhuysen* 10681 (BOL; NBG; PRE); *Hilliard* 3995 (E; K; NH; NU); *MacOwan* 215 (SAM).

36. **Helichrysum anomalum** *Less.*, Syn. Comp. 303 (1832); DC., Prodr. 6: 204 (1838); Harv. in F.C. 3: 249 (1865); Moeser in Bot. Jb. 44: 255 (1910); Levyns in Adamson & Salter, Fl. Cape Penins. 784 (1950); Hilliard, Compositae in Natal 167 (1977). Type: Cape, *Mund & Maire* (B†).

Manopappus anomalus (Less.) Sch. Bip. in Flora, Regensburg 27: 677 (1844). *H. anomalum* var. *turbinatum* Harv. in F.C. 3: 249 (1865). *Gnaphalium*

anomalum (Less.) O. Kuntze, Rev. Gen. 3,2: 150 (1898).

Tanacetum gnaphaloides Steud. in Flora, Regensburg 13: 542 (1830), *nomen*.

Helichrysum anomalum var. *rufescens* DC., Prodr. 6: 204 (1838). Type: Uitenhage div., Zwartkops River, *Ecklon* 1577 (G-DC, holo.!; SAM, iso.!).

H. anomalum var. *brevifolium* DC., l. c. Type: Uitenhage div., Eland River, *Ecklon* 1868 (G-DC, holo.!; SAM, iso.!).

H. anomalum var. *ovatum* Harv., l. c. Type: nothing cited, nothing annotated in TCD!

Bushy dwarf shrublet up to 300 mm tall and as much across, branches closely grey-woolly, densely leafy. *Leaves* 3—14 × 1—3 mm, diminishing in size upwards, much reduced and distant below the flowering heads, linear, lanceolate or oblong, apex acute or obtuse, mucronate, sometimes recurved, base broad, half-clasping, margins revolute, both surfaces closely grey-woolly. *Heads* heterogamous, cylindric-campanulate, c. 4 × 2,5 mm, few to many in dense cymose clusters terminating the branches. *Involucral bracts* in c. 6 series, graded, closely imbricate, very obtuse, outer pale brown, pellucid, inner about equalling flowers, tips opaque or pellucid, bright canary yellow, minutely radiating. *Receptacle* with fimbrils 2 to 3 times as long as the ovaries. *Flowers* 17—50, 4—12 ♀, 13—40 ☿. *Achenes* c. 0,75 mm long, glabrous. *Pappus* bristles very few, about equalling the *ovary* in length.

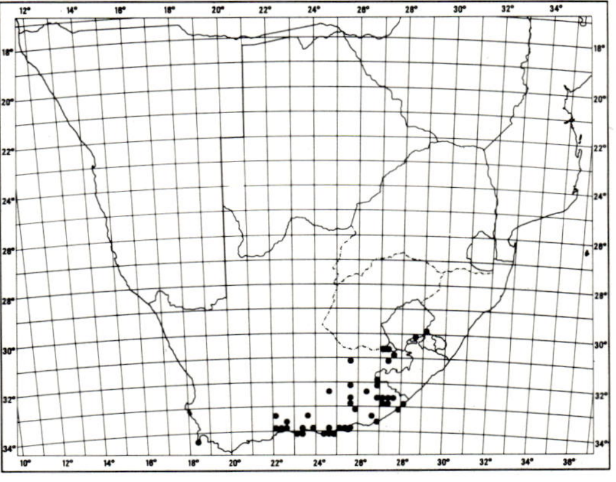

MAP 63.— **Helichrysum anomalum**

Recorded from the Cape of Good Hope Nature Reserve, then a disjunction to Swartberg Pass and Robinson Pass, thence eastwards across the mountains and through the coastal districts to Grahamstown, King William's Town and Haga-Haga near East London, and north and east across the mountains to the Graaff-Reinet and Middelburg area, to the Witteberg, and the Drakensberg from about Barkly Pass to Qacha's Nek and Sehlabathebe in Lesotho. Very variable in leaf size and to some degree in indumentum; distinctive by virtue of its extraordinarily long fimbrils and reduced pappus. Grows gregariously and with other dwarf shrubs on bare, often rocky, areas, or in rough grassland. Flowers between September and May. Map 63.

Vouchers: *Acocks* 17650 (PRE); *Compton* 10753 (NBG); *Esterhuysen* 13651 (PRE); *Hilliard & Burtt* 6659 (E; K; MO; NU; S).

37. **Helichrysum athrixiifolium** *(O. Kuntze) Moeser* in Bot. Jb. 44: 281 (1910); Hilliard, Compositae in Natal 199 (1977). Type: Natal, Colenso, *O. Kuntze* 1337 (W, holo.!; K, iso.!).

Gnaphalium athrixiifolium [O. Hoffm. ex] O. Kuntze, Rev. Gen. 3: 150 (1898).

Bushy perennial herb up to c. 450 mm high, stems woody, young parts thinly appressed greyish-white woolly, very leafy, primary leaves often subtending very dwarf lateral shoots, making the leaves look fascicled. *Leaves* mostly 10−40 × 1−5 mm, those on the dwarf shoots sometimes smaller and usually very narrow, linear, linear-lanceolate or oblong-lanceolate, apex acute, base half-clasping, very shortly decurrent, margins more or less revolute, upper surface thinly cobwebby, soon glabrescent, lower surface thinly white tomentose. *Heads* heterogamous, oblong-campanulate, c. 4 × 3 mm, few to many in compact corymbose panicles, terminating the branches. *Involucral bracts* in 5−6 series, graded, loosely imbricate, tips subacute, opaque white in the inner series, crisped, innermost about equalling the flowers, scarcely radiating. *Receptacle* shortly honeycombed. *Flowers* 16−30, 6−13 ♀, 8−17 ☿. *Achenes* 0,75 mm long, with myxogenic duplex hairs. *Pappus* bristles many, equalling corolla, scabrid, bases cohering by patent cilia.

Ranges from the Lowveld in the E. Transvaal and Swaziland to Tongaland and coastal Zululand, penetrating westwards in Natal in the basins of the Pongola and Tugela Rivers. Grows in grassland or open woodland on sandy or shaly soils, and will invade

overgrazed or eroded areas, and roadsides. Flowering recorded in all months. Map 64.

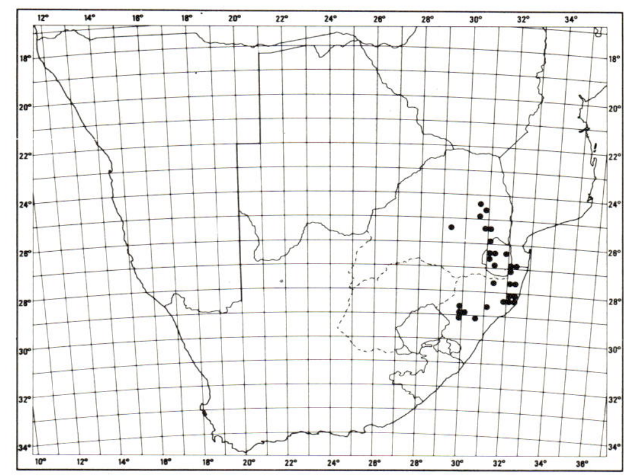

MAP 64.— **Helichrysum athrixiifolium**

Vouchers: *Codd* 4751 (NU; PRE); *Compton* 28988 (PRE); *Hilliard & Burtt* 8586 (E; K; MO; NU; S); *Strey* 5439 (NU, PRE); *Wright* 151 (E; NH; NU).

H. athrixiifolium is frequently confused with *H. rugulosum* (no. 40), but that species produces simple or subsimple stems from a creeping stock, rarely has dwarf axillary shoots, and the fimbrils on the receptacle much exceed the ovaries. However, *H. athrixiifolium* is closely allied to *H. rosum* (below), from which it is scarcely distinguishable in habit, foliage and size and shape of the heads. But the tips of the involucral bracts are always much crisped in *H. athrixiifolium*, while they are nearly always smooth, rarely slightly crisped, in *H. rosum*.

38. **Helichrysum rosum** *(Berg.) Less.*, Syn. Comp. 306 (1832); DC., Prodr. 6: 205 (1838); Harv. in F.C. 3: 251 (1865) as *H. erosum*; Moeser in Bot. Jb. 44: 282 (1910). Type: Cape of Good Hope, *Grubb* (STB, holo.!).

Gnaphalium rosum Berg., Descr. Pl. Cap. 260 (1767). *Helichrysum erosum* var. *angustifolium* Harv. in F.C. 3: 251 (1865).

G. heterophyllum Thunb., Prodr. 149 (1800), Fl. Cap. 648 (1823). *Helichrysum teretifolium* var. *heterophyllum* (Thunb.) DC., Prodr. 6: 205 (1838). Type: Cape of Good Hope, *Thunberg* (sheet 19174, UPS, holo.!).

G. erosum Thunb., Mus. Upsal. Append. VIII: 133 (1800), nomen. (*Thunberg*, sheet 19151, UPS!; sheet 19152, also written up as *erosum*, is *H. rugulosum*).

Helichrysum concolorum DC., Prodr. 6: 206 (1838). *Gnaphalium concolorum* (DC.) Sch. Bip. in Bot. Ztg

3: 172 (1845). *H. erosum* var. *concolorum* (DC.) Harv. in F.C. 3: 251 (1865). *H. rosum* var. *concolorum* (DC.) Moeser in Bot. Jb. 44: 282 (1910). Lectotype: Cape, George, *Ecklon* 665 (G-DC!).

H. concolorum var. *brevifolium* DC., Prodr. 6: 206 (1838). Type: Cape, Van Staden's River and Algoa Bay, *Drège* 5777 (G-DC, holo.!; TCD, iso.!).

H. erosum var. *latifolium* Harv. in F.C. 3: 251 (1865). No specimen cited, nor annotated in TCD!

Two varieties are recognized:

(a) var. **rosum.**

Well-branched or densely tufted sub-shrub up to c. 1 m tall, young stems thinly white-felted, closely leafy, at least some leaves with dwarf axillary shoots making the leaves look fascicled. *Leaves* mostly 8—25 × 1—3 (—5) mm, linear, linear-lanceolate or oblong-lanceolate, apex acute or subacute, base very shortly decurrent, margins revolute or subrevolute, often undulate-dentate, upper surface glabrous to thinly cobwebby-felted, often wrinkled, lower white-felted, the hairs often stringy, or occasionally loosely woolly above and below. *Heads* heterogamous, oblong-campanulate, c. 4 × 3 mm, few to many in compact corymbose panicles terminating the branchlets. *Involucral bracts* in 5—6 series, graded, inner about equalling the flowers, imbricate, bases pellucid, pale straw-coloured or tawny, tips blunt, opaque, milk-white, smooth or rarely slightly crisped, not radiating. *Receptacle* honeycombed. *Flow-*

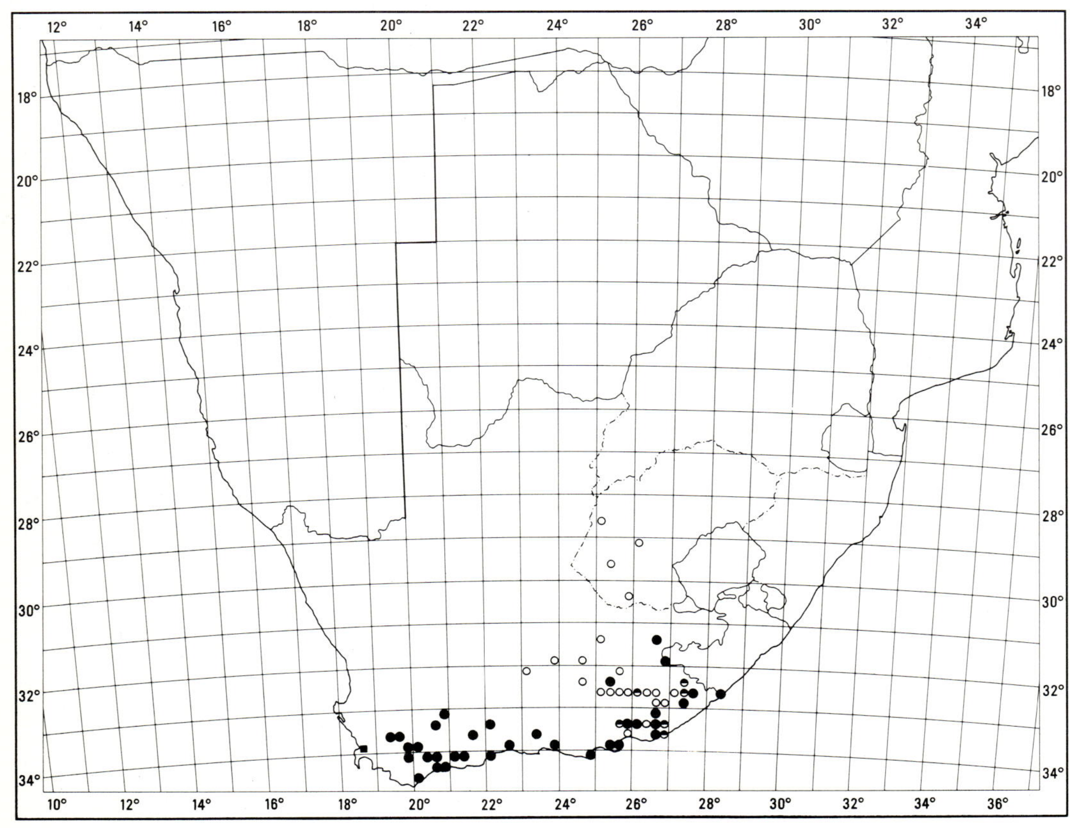

MAP 65.— ● **Helichrysum rosum** var **rosum**
　　　　　○ **Helichrysum rosum** var. **arcuatum**
　　　　　◑ sympatric

ers 15−28, 4−11 ♀, 8−21 ☿. *Achenes* c. 0,75 mm, broadly cylindric, with myxogenic duplex hairs. *Pappus* bristles many, about equalling corolla, scabridulous, bases cohering by patent cilia. Fig. 21:4.

Ranges from about Stellenbosch (Visserhok) eastwards over the mountains flanking the Karoo to about Queenstown and through the coastal districts to about Kentani in the Transkei, on the flats or on the flanks of the mountains, in shrub communities, coastal renosterveld, valley bushveld or karroid scrub, or in grassland. Flowers mainly from September to December, but as late as March. Map 65.

Vouchers: *Acocks* 9466 (PRE); *Esterhuysen* 13590 (PRE); *MacOwan* 593 (PRE; SAM); *Pegler* 1491 (BOL; PRE).

H. rosum var. *rosum* is characterized by its bushy growth, 'fascicled' leaves, and white-tipped involucral bracts that are usually smooth and concave; when the bract tips are crisped, as they occasionally are, confusion can arise with *H. athrixiifolium* (above) which is indeed very closely allied to *H. rosum* and is distinguished morphologically by this bract character, and perhaps by a difference in habit that cannot be detected in the herbarium.

De Candolle singled out specimens with loosely woolly leaves as *H. concolorum*. However, variation in degree of woolliness is very common in *Helichrysum;* in *H. rosum* plants with the more usual type of leaf indumentum can grow with woolly-leaved specimens, and all degrees of woolliness are found over the range of the species. It is scarcely worth upholding the epithet *concolorum* even at varietal level, but the combination is available for use if desired.

H. rosum is much confused with *H. rugulosum* (no. 40), but that species has simple or subsimple stems, rarely produces dwarf axillary shoots, and the teeth on the receptacle are longer than the ovaries. Also frequently confused with *H. dregeanum* (no. 41); see under that species, and below.

(b) var. **arcuatum** *Hilliard*, var. nov. *ramis arcuatis patentibus saepe ad nodos radicantibus a typo differt.*

Type: Cape, Graaff-Reinet distr., 3124 DD, farm Blaauwater, N. of Bethesda Road, c. 5 000 ft., 25 xi 1977, *Hilliard & Burtt* 10661 (NU, holo.!; E; K; PRE; S, iso.!).

Cape—Somerset East distr., 3225 CB, just off Graaff-Reinet-Somerset East road, below Bruintjeshoogte, 30 xi 1977, *Hilliard & Burtt* 10780 (E; K; M; MO; NU; PRE; S). 3225 DA, Auret Drive on S. face of Boschberg, 30 xi 1977, *Hilliard & Burtt* 10790 (E; K; M; MO; NU; PRE; S). Bedford distr., near Bedford, 1 xii 1950, *Maguire* 679 (NBG). Victoria East distr., near Alice, 29 xi 1941, *Barker* 1383 (NBG). Beaufort West distr., Nelspoort, Courland's Kloof, *Pearson* 1329 (SAM). Keiskammahoek distr., Keiskammahoek, scrub forest in Boma Pass, 20 ix 1942, *Acocks* 9117.

Murraysburg distr., Murraysburg, 1879, *Tyson* 38 (PRE). Middelburg distr., Grootfontein, 18 i 1935, *Verdoorn* 1509 (PRE). Alexandria distr., Addo Elephant Park, 22 viii 1959, *Barnard* 561 (PRE).

Orange Free State.—Fauresmith Veld Reserve, 2 x 1929, *Henrici* 1826 (PRE). Boshof, ix 1947, *Brueckner* 896 (PRE).

Differs from the typical plant in habit, namely, decumbent stems radiating from the stock, these often arcuate and producing short (less than 50 mm), simple, erect flowering shoots at the nodes, sometimes rooting where in contact with the ground; and in the involucre, the bracts being mostly pellucid, the tips of only the inner few series often opaque or semi-opaque and white or whitish, occasionally all pellucid.

Ranges from about Nelspoort and Murraysburg in the Great Karoo east to Keiskammahoek, south to Addo and NE. to the southern Orange Free State. Sympatric with the typical plant only in the E. Cape from about Addo and Grahamstown NE. to Keiskammahoek and NW. to Bedford. Grows in stony places, sometimes in karroid scrub, or along roadsides; flowering between August and April. Map 65.

Possibly because of its spreading growth, this is the variety of *H. rosum* that is most frequently confused with *H. dregeanum;* see under that species (no. 41). *H. rosum* var. *arcuatum* and *H. dregeanum* are sympatric mainly in the southern half of the Orange Free State.

39. **Helichrysum bachmannii** *Klatt* in Bull. Herb. Boissier 4: 459 (1896). Type: Cape, south of Hopefield, Nov. 1886, *Bachmann* 1202 (Z, holo.!).

Well-branched, scrambling shrub up to 2 m tall, young branches greyish-white tomentose, leafy, more distantly so below the inflorescences. *Leaves* mostly 13−35 × 2−3 mm, scarcely diminishing upwards, linear-lanceolate or linear, apex acute, mucronate, base slightly narrowed, half-clasping, margins strongly revolute, both surfaces greyish tomentose. *Heads* heterogamous, campanulate, c. 4 × 3 mm, many in corymbose clusters corymbose-paniculately arranged. *Involucral bracts* in 4 series, graded, loosely imbricate, inner about equalling flowers, not radiating, tips very obtuse, opaque white. *Receptacle* with fimbrils about equalling ovaries. *Flowers* c. 13−20, 4−6 ♀, 9−15 ☿. *Achenes* not seen, ovaries with highly myxogenic duplex hairs. *Pappus* bristles many, scabrid, bases cohering strongly by patent cilia. Fig. 21:1.

Recorded from the environs of Velddrif, Vredenburg and Hopefield in the Swartland of the SW. Cape. Grows on sand or rock outcrops, flowering between August and November. Very rarely collected. Map 66.

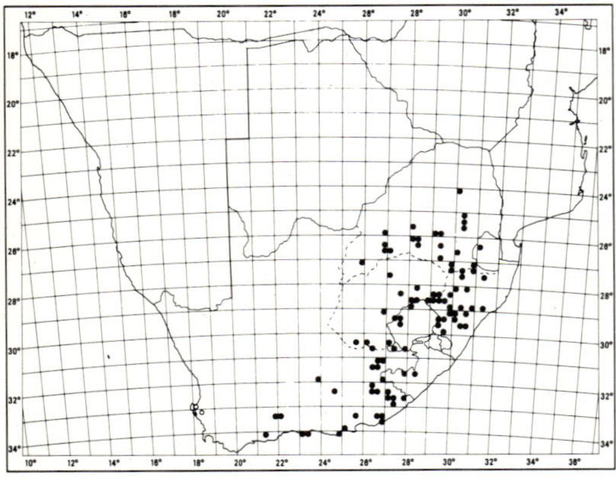

MAP 66.— ○ **Helichrysum bachmannii**
       ● **Helichrysum rugulosum**

Vouchers: *Acocks* 19685 (PRE); *Boucher* 2910 (STE); *Marsh* 1274 (K, mixed with *H. revolutum*).

40. **Helichrysum rugulosum** *Less.*, Syn. Comp. 307 (1832); DC., Prodr. 6: 205 (1838); Harv. in F.C. 3: 250 (1865); Moeser in Bot. Jb. 44: 282 (1910); Hilliard, Compositae in Natal 198 (1977). Lectotype: Cape, Plettenberg Bay, *Mundt & Maire* s.n. (G!).

H. rugulosum var. *angustifolium* DC., Prodr. 6: 205 (1838). Lectotype: Cape, Colesberg div., Naauwpoort, *Burchell* 2764 (G-DC!, not 276A as transcribed by DC.).

H. rugulosum var. *latifolium* DC., l. c. Lectotype: Cape, Albany, *Drège* 5782 (G-DC!).

Gnaphalium rugulosum (Less.) Sch. Bip. in Bot. Ztg 3: 172 (1845).

Perennial herb, rootstock creeping, branching, flowering stems tufted, simple or branching near the base, erect to c. 300 mm, thinly white-felted, closely leafy. *Leaves* sessile, up to 25 × 5 mm, rarely a little larger, often only c. 15 × 2−3 mm, diminishing upwards, somewhat spreading or ascending, lower oblong-lanceolate, acute, upper lanceolate, acuminate, sometimes very narrow, margins subrevolute, upper surface thinly cobwebby-felted, rugose, lower white-felted, the hairs often

stringy. *Heads* heterogamous, campanulate, c. 5 × 4 mm, few to many in compact corymbose panicles. *Involucral bracts* in c. 4−5 series, graded, inner about equalling flowers, loosely imbricate, tips opaque, crisped-dentate, often purplish or pink initially, later creamy, or the inner creamy, or all creamy, minutely radiating. *Receptacle* with fimbrils exceeding ovaries. *Flowers* 17−42, 8−19 ♀, 9−30 ☿, ♀ flowers sometimes equalling or slightly outnumbering ☿. *Achenes* 0,75−1 mm long, broadly cylindric, with myxogenic duplex hairs. *Pappus* bristles many, about equalling corolla, scabridulous, bases cohering by patent cilia.

Ranges from the Transvaal Highveld and eastern highlands through western Swaziland, the eastern half of the Orange Free State, the Midlands and Uplands of Natal, Transkei and E. Cape to the southern Cape as far west as the Swartberg, and Aasvogelkop in Riversdale district. Map 66.

Grows in poor stony or sandy grassland, and readily invades overgrazed areas and roadsides. Flowers mainly between December and March.

The well crisped, denticulate, somewhat squarrose involucral bracts and long fimbrils on the receptacle, together with generally simple stems without axillary leaf tufts, characterize the species. Specimens from Katberg and the Amatola Mountains (e.g. *Galpin* 2398, PRE), often have numerous short branches arising near the base of the flowering stems and very narrow acuminate leaves, but do not differ otherwise.

Vouchers: *Acocks* 9245 (PRE); *Bolus* 419 (BM; BOL); *Flanagan* 264 (NBG); *Hilliard & Burtt* 6646 (E; K; MO; NU; PRE); *Story* 1470 (PRE).

41. **Helichrysum dregeanum** *Sond. & Harv.* in F.C. 3: 251 (1865); Moeser in Bot. Jb. 44: 279 (1910); Hilliard, Compositae in Natal 200 (1977). Lectotype: Cape, Stormberg, 5 000−6 000 ft. *Drège* 5769 (S!; G-DC; K; SAM; TCD, isolecto.!).

Dwarf twiggy mat-forming subshrub, branches diffuse or erect to c. 150 mm, very slender but stiff, thinly white-woolly, closely leafy, often with dwarf axillary shoots. *Leaves* up to 14 × 2 mm but often only c. 4 × 0,5 mm, linear or linear-lanceolate, margins strongly revolute, apex subacute, base broad, very shortly decurrent, upper surface thinly and loosely woolly, lower white-woolly. *Heads* heterogamous, campanulate, c. 3,5−4 (−5) × 2−3 (−4) mm, 2−5, rarely more, clustered at the tips of the branchlets. *Involucral bracts* in c. 4 series, graded, closely imbricate, inner about equalling flowers, tips obtuse, crisped, pellucid, whitish, straw-coloured, tawny or sometimes reddish, not radiating. *Receptacle* shortly toothed. *Flowers* 18—36, 5—10 ☿, 14—29 ♀. *Achenes* c. 1 mm long, with myxogenic duplex hairs. *Pappus* bristles many, about equalling corolla, scabridulous, bases cohering by patent cilia. Fig. 21:3.

Recorded principally from the Witwatersrand and the SW. Transvaal, NE. Cape (Vryburg, Barkly West, Kimberley), Orange Free State and western Lesotho, in the mountainous parts of the E. Cape from the Stormberg and Queenstown to Rhodes, East Griqualand, and the foothills of the Drakensberg in southernmost Natal. There are isolated records from the Bokkeveld Mountains (19°E, 31°—31°50'S), Leeuwpoort near Concordia in Namaqualand (*Schlechter* 11361, BOL), the Khamiesberg (*Esterhuysen* 23693, BOL; K; S; and *Nordenstam* 1337, NU; S), and from Aus in South West Africa/Namibia (*Marloth* 5908, PRE). These far western specimens are less closely branched, and there may be more heads clustered at the branchlet tips, than specimens from elsewhere, but there seems little else to distinguish them. Map 67.

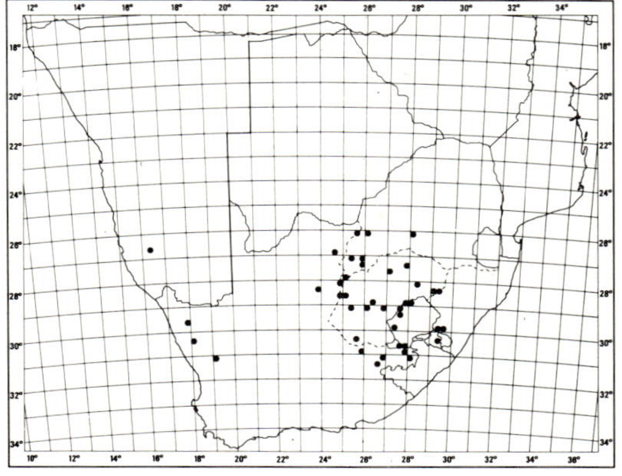

MAP 67.— **Helichrysum dregeanum**

Grows in dry sandy or stony grassland, or in dwarf shrub communities, and readily becomes a weed along roadsides and in overgrazed places. Flowers between September and February.

Much confused with *H. rosum* (no. 38), but distinguished by its few (mostly 2—5) heads in subumbellate clusters, involucral bracts ranging from almost colourless to tawny, tips all crisped, inner ones pellucid or subopaque, ranging from pale to deep tawny through pink or red tints to dingy white, and

receptacles only very shortly toothed. In *H. rosum,* the heads are few to many, often in corymbose panicles, but sometimes clustered, especially in dwarfed specimens, bract tips smooth, all but the outermost opaque milk-white (but see var. *arcuatum*) and the receptacle has teeth shorter than the ovaries or about equalling them. See also notes under *H. ammitophilum* (below).

Vouchers: *Acocks* 1430 (PRE); *Burtt Davy* 5530 (NBG; PRE); *Flanagan* 1910 (PRE, SAM); *Galpin* 6538 (NBG; PRE); *Hilliard & Burtt* 6648 (E; K; MO; NBG; NU; PRE; S).

42. **Helichrysum ammitophilum** *Hilliard* in Notes R. bot. Gdn Edinb. 40: 250 (1982). Type: Orange Free State, Harrismith, Queen's Hill, 5 500 ft, 20 ii 1970, *Hilliard* 4954 (NU, holo.!; E; K; MO; M; PRE; S, iso.!).

Perennial herb with a slender woody creeping rootstock, stems tufted, simple or branching at or near the base, erect to c. 150—200 mm, white-felted, closely leafy. *Leaves* mostly 8—15 × 0,5—1 mm, diminishing slightly upwards, suberect, imbricate, linear, apex acute, mucronate, sometimes strongly recurved, base broad, margins revolute, lower surface white-felted, upper with thin 'tissue-paper' indumentum of woolly hairs. *Heads* heterogamous, oblong-campanulate, c. 3 × 2 mm, many in compact terminal clusters 10—15 (—30) mm across. *Involucral bracts* in c. 4 series, thinly webbed together with wool, graded, closely imbricate, inner nearly equalling flowers, pellucid, straw-coloured, tips tawny, occasionally purplish, blunt, crisped-denticulate, somewhat squarrose. *Receptacle* shortly toothed. *Flowers* 18—31, 8—15 ♀, 9—18 ☿, ♀ sometimes equalling or outnumbering ☿, bright yellow. *Achenes* not seen, ovaries with myxogenic duplex hairs. *Pappus* bristles many, nearly equalling corolla. scabridulous, bases cohering by patent cilia. Fig. 21:5.

Recorded from the Harrismith-Bethlehem-Ficksburg area of the Orange Free State, Leribe in Lesotho, and the Witteberg and near Dordrecht in the E. Cape. Grows in stony or sandy turf on hilltops and slopes at c. 1 675 m; flowering in February. Map 68.

Can be confused with *H. dregeanum* (above), but is distinguished by its simple stems terminating in many-headed, clusters.

Vouchers: *Acocks* 11200 (PRE); *Cooper* 617 (BM; E; TCD; Z); *Dieterlen* 502 (K; PRE; SAM; Z; NH and BM mixed with *H. rugulosum*); *Scheepers* 1382 (PRE); *Story* 913 (PRE).

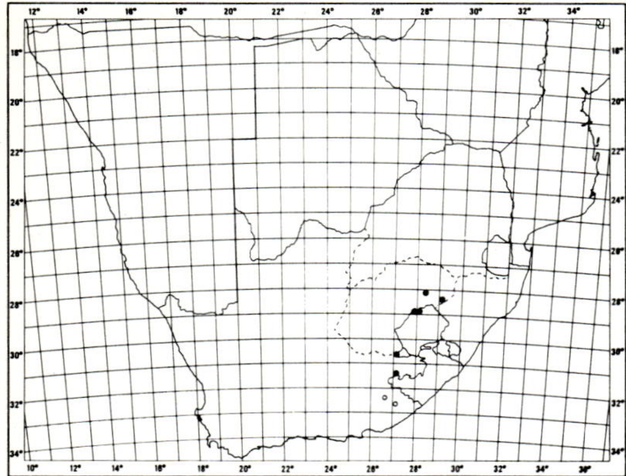

MAP 68.— ● **Helichrysum ammitophilum**
○ **Helichrysum montis-cati**

43. **Helichrysum montis-cati** *Hilliard* in Notes R. bot. Gdn Edinb. 40: 262 (1982). Type: Cape, Stockenstrom div., Katberg Pass (3226 BC), c. 5 000 ft. 28 x 1980, *Hilliard & Burtt* 13261 (NU, holo.!; E; K; MO; PRE; S, iso.!).

Bushy halfshrub, sprawling or erect to c. 1−3 m, branches long and slender, greyish-white woolly-felted, closely leafy, leaves more distant on flowering twigs, all but the reduced leaves spreading, reflexed with age. *Leaves* 18−60 × 2−6 mm, reduced leaves below flowering heads c. 8−15 × 1−2 mm, linear, acute, mucronate, passing into lanceolate, acuminate, scale-tipped bracts, base broad, half-clasping, margins more or less revolute, upper surface cobwebby at first, glabrescent, rugose, gland-dotted, lower surface greyish-white woolly-felted. *Heads* heterogamous, campanulate, c. 4 × 3 mm, many in congested glomerules 15−25 mm across at the branch tips. *Involucral bracts* in c. 5 series, imbricate, outer shorter, inner subequal, equalling flowers, tips subacute or obtuse, somewhat erose and crisped, recurved but scarcely radiating, translucent, pale golden-brown. *Receptacle* shortly honeycombed. *Flowers* 21−31, 2−7 ♀, 18−27 ☿, bright yellow. *Achenes* 1 mm long, with myxogenic duplex hairs. *Pappus* bristles many, equalling corolla, scabrid, bases cohering strongly by patent cilia.

Known only from Katberg Pass, and Cata Forest Reserve and Hogsback in the nearby Amatola mountains of the E. Cape. Grows in rough herbage near forest; flowering in October and November. Map 68.

Vouchers: *Esterhuysen* 13252 (BOL); *Hutchinson* 1672 (BM; BOL; K: PRE mixed with *H. odoratissimum*); *Sidey* 3747 (K; PRE; S); *Story* 3189 (GRA; PRE).

44. **Helichrysum revolutum** (*Thunb.*) *Less.*, Syn. Comp. 305 (1832); DC., Prodr. 6: 206 (1838); Harv. in F.C. 3: 252 (1865); Moeser in Bot. Jb. 44: 282 (1910); Levyns in Adamson & Salter, Fl. Cape Penins. 784 (1950); Merxm. & Roessl. in Mitt. bot. StSamml., Münch. 15: 367 (1979). Lectotype: Cape of Good Hope, *Thunberg* (sheet 19245, UPS!).

Gnaphalium revolutum Thunb., Prodr. 150 (1800), Fl. Cap. 652 (1823).

Helichrysum leiolepis DC., Prodr. 6: 187 (1838). *Gnaphalium leiolepis* (DC.) Sch. Bip. in Bot. Ztg 3: 171 (1845). Type: Cape, Clanwilliam div., Olifants River and at Villa Brakfontein, *Ecklon* (G-DC, holo.!; S, iso.!).

Bushy half-shrub up to 2 m tall, stems woody, young ones thinly greyish-white woolly, closely leafy. *Leaves* often with axillary tufts of reduced leaves, main stem leaves 10−40 × 1−5 mm, diminishing slightly upwards, erect or spreading, linear, linear-lanceolate to lanceolate, apex acute, mucronate, base somewhat eared, slightly decurrent, margins revolute, upper surface thinly greyish-white woolly, often glabrescent, irregularly wrinkled, often tuberculate, lower persistently greyish-white woolly. *Heads* heterogamous, campanulate, c. 4,5−5 × 4 mm, few to many in terminal compact corymbose panicles. *Involucral bracts* in 4−5 series, imbricate, graded, inner equalling flowers, oblong, tips rounded, not radiating, pellucid, glossy, pale to deep straw-coloured. *Receptacle* shortly toothed. *Flowers* 20−37, (1−) 3−8 ♀, 17−32 ☿. *Achenes* 1 mm, barrel-shaped, with myxogenic duplex hairs. *Pappus* bristles many, about equalling corolla, barbellate, bases cohering strongly by patent cilia.

One record from the Aurusberge in southern S.W.A./Namibia, then widespread in the western Cape from Namaqualand south through Vanrhynsdorp, Calvinia, Clanwilliam, Sutherland, Montagu, Worcester, Cape Town and Simonstown degree squares, from near sea level to c. 900 m, in rocky or sandy places. Flowers between July and October. Map 69.

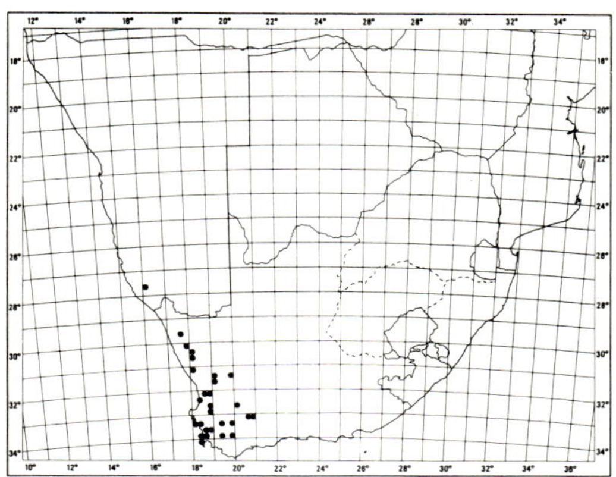

MAP 69.— **Helichrysum revolutum**

Vouchers: *Acocks* 19407 (M; PRE); *Barker* 9604 (NBG); *Compton* 9387 (NBG); *Esterhuysen* 23693 (PRE); *Nordenstam* 1339 (M; NU).

45. Helichrysum scabrum *Less.*, Syn. Comp. 312 (1832), non DC.; Harv. in F.C. 3: 252 (1865); Moeser in Bot. Jb. 44: 283 (1910). Type: Cape of Good Hope, *Thunberg* (sheet 19250, UPS, holo.!).

Gnaphalium scabrum Thunb., Prodr. 150 (1800), Fl. Cap. 655 (1823), non L. (1753).

Helichrysum repandum DC., Prodr. 6: 203 (1838). *Gnaphalium repandum* (DC.) Sch. Bip. in Bot. Ztg 3: 172 (1845). Lectotype: Clanwilliam div., Olifants River and Villa Brakfontein, *Ecklon* 780 (G-DC!).

H. repandum var. *microphyllum* DC., l. c. *H. scabrum* var. *microphyllum* (DC.) Harv., l. c. Type: Khamiesberg, *Drège* 2851 (G-DC, holo.!; SAM, iso.!).

H. scabrum var. *scaberrimum* Harv., l.c. Type: Caledon, *Ecklon* s.n.

Well-branched shrub up to 1,5 m tall, branches tangled, young stems thinly white-woolly, glandular, leafy. *Leaves* often with axillary tufts of reduced leaves, main stem leaves 6−14 × 2−4 mm, diminishing slightly upwards, spreading, oblong, oblong-lanceolate or oblanceolate, apex acute, stoutly mucronate, base somewhat eared, slightly decurrent, margins revolute, bluntly repand, undulate, both surfaces with rough coarse glandular hairs, sometimes cobwebby as well. *Heads* heterogamous, campanulate, 4−5 × 3−4 mm, few to several in terminal compact corymbose panicles. *Involucral bracts* in 4−5 series, imbricate, graded, inner about equalling flowers, oblong, tips rounded, not radiating, pellucid, glossy, straw-coloured. *Receptacle* with fimbrils about equalling ovaries. *Flowers* 17−24, 4−6 ♀, 14−18 ☿. *Achenes* not seen, ovaries with myxogenic duplex hairs. *Pappus* bristles many, nearly equalling corolla, barbellate, bases cohering by patent cilia. Fig. 21:2.

Endemic to the western Cape, and ranging from Springbok in Namaqualand south to Calvinia, Clanwilliam, and Wuppertal degree squares, and reaching Ceres. Grows on rocky mountain slopes, in karroid scrub; flowering between August and December, but mainly in September. Easily recognized by its more or less oblong leaves with revolute repand margins and rough glandular hairs. Map 70.

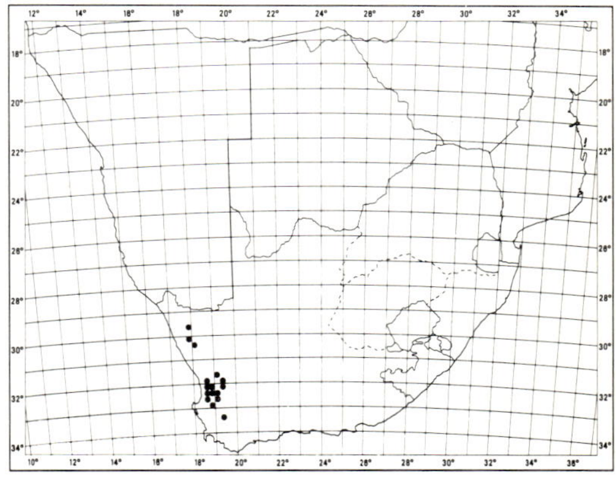

MAP 70.— **Helichrysum scabrum**

Vouchers: *Bolus* 9021 (BOL; NU); *Compton* 20778 (NBG); *Hutchinson* 825 (BOL); *Pearson & Pillans* 5875 (BOL; NBG).

Group 10

Shrubs or woody perennial herbs; *leaves* small, mostly oblong, elliptic, ovate or obovate; *heads* usually heterogamous, rarely homogamous and then within the same

1a
1d
1b
1c
2
3
4
4a
H. Wanda-du Toit
1

species, 4—6 × 2—5 mm, in congested rounded terminal clusters; *involucral bracts* either not radiating or minutely radiating, straw-coloured or tawny, often villous on the backs, stereome undivided; *receptacle* smooth, honeycombed or fimbrilliferous; *flowers* 12—32, (0—) 1—10 ♀, corolla of ☿ flowers broadly campanulate above, of ♀ flowers with a very well developed limb; *achenes* usually hairy, rarely glabrous (dimorphic in one species); *pappus* bristles shortly plumose, barbellate or rarely scabrid above, shaft scabrid or barbellate, bases cohering by patent cilia, or lightly fused in one species.

Species 46—52, confined to the W. and SW. Cape.

1a Backs of involucral bracts villous, only the extreme tips glabrous:

  2a Heads c. 4—5 × 4 mm, leaves up to 14 × 6 mm:

    3a Leaves mostly 11—14 × 5—6 mm, obovate to obovate-oblong, narrowed to the base .... 46. *H. capense*

    3b Leaves mostly 5—7 (—12) × 2,5—4 (—6) mm, narrowly ovate, base broad .... 47. *H. marifolium*

  2b Heads c. 6 × 5 mm, leaves mostly 13—23 × 6—10 mm .... 48. *H. rotundatum*

1b Involucral bracts hairy below with conspicuous glabrous tips:

  4a Receptacle honeycombed or fimbrilliferous:

    5a Leaves oblong, oblong-ovate, elliptic or oblong-lanceolate:

      6a Stems loosely woolly; receptacle honeycombed .... 49. *H. catipes*

      6b Stems woolly with spreading hairs as well; receptacle with fimbrils much exceeding ovaries .... 50. *H. dasyanthum*

    5b Leaves linear to linear-oblong .... 51. *H. plebeium*

  4b Receptacle smooth .... 52. *H. hebelepis*

46. **Helichrysum capense** *Hilliard* in Bot. J. Linn. Soc 82: 200 (1981). Type: Cape, Riversdale distr., on Kampsche Berg, 9 xii 1814, *Burchell* 7091 (G-DC, holo.!; K, iso.!).

Eriosphaera apiculata DC., Prodr. 6: 166 (1838), non *Helichrysum apiculatum* (Labill.) DC. Type as above.

E. oculus-cati sensu DC., Prodr. 6:166 (1838), non (L.f.) Less.

Helichrysum marifolium var. ? *oblongifolium* DC., Prodr. 6: 186 (1838). Type: Cape, E. slopes Table Mountain, near Constantia, *Ecklon* 504 (G-DC, holo.!).

H. marifolium sensu Harv. in F.C. 3: 230 (1865); Levyns in Adamson & Salter, Fl. Cape Penins. 782 (1950), non DC.

Perennial herb, stems diffuse, prostrate or scrambling, occasionally rooting at the nodes, branching, young parts greyish-white woolly-felted, closely leafy, flowering twigs with distant reduced leaves. *Leaves* mostly 11—14 × 5—6 mm, diminishing slightly upwards, then abruptly much reduced and distant below the heads, obovate to obovate-oblong, narrowed to the base, uppermost oblong, sessile, apex rounded or obtuse, mucronate, upper surface thinly greyish-white woolly, often glabrescent, lower thickly and persistently woolly. *Heads* heterogamous, campanulate, c. 4 × 4 mm, up to c. 12 in congested rounded terminal clusters. *Involucral bracts* in 4—5 series, imbricate, graded, inner equalling flowers, not radiating, lanceolate to oblong, straw-coloured, tips rounded or acute, often dark, whole bract thickly silky-villous dorsally, hairs yellowish. *Receptacle* smooth. *Flowers* 15—29, 4—6 ♀, 11—23 ☿. *Achenes* not seen, ovaries 5-ribbed, with duplex hairs, not myxogenic. *Pappus* bristles many, shaft barbellate, becoming shortly plumose above, bases cohering by patent cilia. Fig. 22:1.

Endemic to the SW. and S. Cape from the Peninsula east to Ruyter's Kop in Mossel Bay district, on the mountains where it favours damp shady places on cliffs or among rocks, c. 600—1 500 m. Flowers between September and December, mainly in November and December. Map 71.

FIG. 22.—1, **Helichrysum capense**, part of plant, × 1; 1a, head, × 6,6; 1b, hermaphrodite flower, × 10; 1c, female flower, × 10; 1d, pappus bristle, × 10 (*Stokoe* 7536). 2, **H. marifolium**, part of plant, × 1 (*Zeyher* 2891). 3, **H. rotundatum,** part of plant, × 1 (*Stokoe* 1074). 4, **H. catipes**, part of plant, × 1; 4a, head, × 6,6 (*Esterhuysen* 20951).

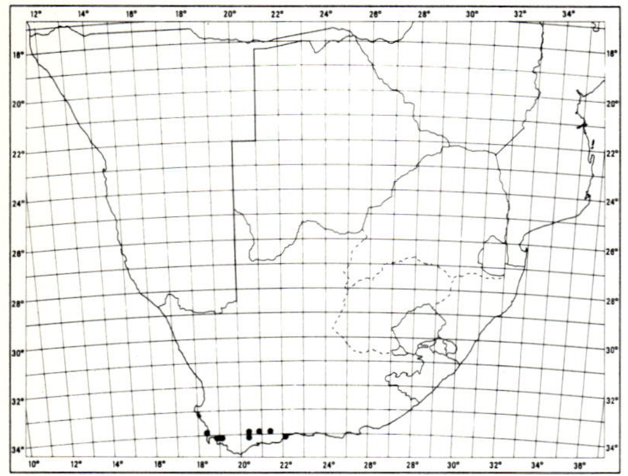

MAP 71.— **Helichrysum capense**

Vouchers: *Esterhuysen* 17592 (BOL; PRE); *Stokoe* 8128 (BOL; NBG; PRE); *Taylor* 6587 (PRE).

This is the plant that Harvey (l.c.) called *H. marifolium,* but that name should be applied to what has hitherto been called *H. umbellatum* (below).

47. **Helichrysum marifolium** *DC.,* Prodr. 6: 186 (1838). Type: Cape, Swellendam distr., mountains near Puspasvalei, Voormansbosch, Duivelsbosch and at Keurbooms River, 1 000−4 000 ft., Oct., *Ecklon* 195 (G-DC, holo.!; M, iso.!).

Gnaphalium marifolium (DC.) Sch. Bip. in Bot. Ztg 3: 171 (1845).

Eriosphaera umbellata Turcz. in Bull. Soc. Nat. Moscow 24, 2: 79 (1851). *Helichrysum umbellatum* (Turcz.) Harv. in F.C. 3: 230 (1865). Type: Cape, River Zonder Einde, *Zeyher* 2891 (K; PRE; S; SAM; TCD, iso.!).

Perennial herb, stems diffuse, probably trailing, branched, grey woolly-felted, closely leafy except near the heads. *Leaves* spreading or reflexed, mostly 5−7 (−12) × 2,5−4 (−6) mm, diminishing in size and more distant near the heads, narrowly ovate, apex subacute, mucronate, base broad, half-clasping, both surfaces greyish-white woolly-felted. *Heads* heterogamous, campanulate, c. 4−5 × 4 mm, up to c. 12 in congested, rounded terminal clusters. *Involucral bracts* in 4−5 series, graded, imbricate, inner about equalling flowers, not radiating, lanceolate to oblong, straw-coloured, tips acute to obtuse, often dark,

whole bract densely villous dorsally. *Receptacle* smooth. *Flowers* 15−20, 5 ♀, 10−15 ☿. *Achenes* not seen, ovaries with duplex hairs, not mucilaginous when wet. *Pappus* bristles many, shaft barbellate, becoming shortly plumose upwards, bases cohering by patent cilia. Fig. 22: 2.

Recorded only from the Riviersonderend Mountains, in Caledon division of the SW. Cape, on steep S. slopes and cliffs, in kloofs, and along cliffed river banks, flowering in September and October. Closely allied to *H. capense* (above) but distinguished by its differently shaped leaves, often much smaller than those of *H. capense.* Map 72.

Vouchers: *Esterhuysen* 32719A (BOL); *Stokoe* SAM 57557 (NBG; PRE; SAM).

48. **Helichrysum rotundatum** *Harv.* in F.C. 3: 230 (1865); Moeser in Bot. Jb. 44: 288 (1910). Type: Cape, Caledon div., tops of the mountains at Baviaanskloof near Genadendal, 15 Feb. 1815, *Burchell* 7726 (G-DC, holo.!; K, iso.!).

Eriosphaera rotundifolia DC., Prodr. 6: 166 (1838). Type as above.

E. coriacea DC., Prodr. 6: 167 (1838); Moeser in Bot. Jb. 44: 341 (1910); *Helichrysum coriaceum* (DC.) Harv. in F.C. 3: 230, 608 (1865), non Harv., l.c. 239; *H. fulvellum* Harv. in F.C. 3: 608 (1865). Type: Cape, Caledon div., mountains near Swellendam, Genadendal, *Drège* 1773 (G-DC, holo.!; BM; E; K; TCD, iso.!).

Small subshrub, stems loosely branched, young parts greyish-white or fulvous woolly-felted, closely leafy except towards the heads. *Leaves* spreading or reflexed, mostly 13−23 × 6−10 mm, diminishing slightly in size upwards then abruptly much reduced and distant below the heads, obovate to oblong-obovate, slightly narrowed to the base, sessile, apex rounded, mucronate, upper surface thinly greyish-white woolly, often glabrescent, lower thickly and persistently woolly. *Heads* heterogamous, campanulate, c. 6 × 5 mm, c. 7−22 in congested rounded terminal clusters. *Involucral bracts* in 4−5 series, graded, inner not quite equalling flowers, not radiating, lanceolate to oblong, straw-coloured, tips acute to rounded, whole bract thickly silky-villous dorsally. *Receptacle* smooth. *Flowers* 20−38, 6−9 ♀, 16−29 ☿. *Achenes* c. 2 mm long, barrel-shaped, ribbed, with duplex hairs, not myxogenic. *Pappus* bristles many, shaft barbellate,

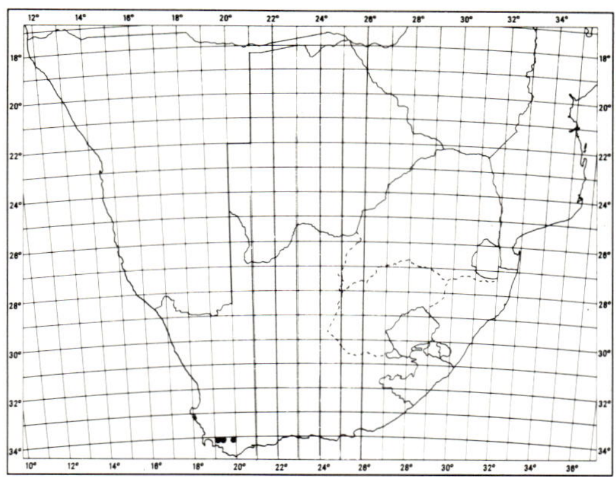

MAP 72.— **Helichrysum marifolium**

becoming shortly plumose above, bases cohering by patent cilia. Fig. 22: 3.

Apparently endemic to the western end of the Riviersonderend Mountains, above Genadendal, having been recorded from Omklaarberg, Jona's Kop and the Wildepaardeberg, flowering between October and December. Grows in moist places on south-facing cliffs; rarely collected. Map 73.

Closely allied to *H. capense* (no. 46), but a coarser plant, with larger leaves and larger heads.

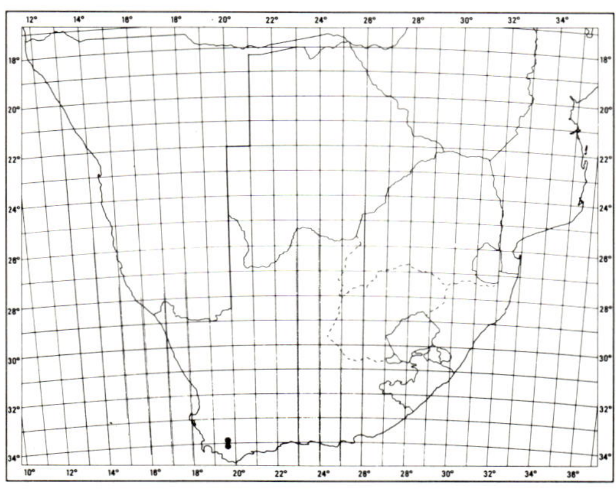

MAP 73.— **Helichrysum rotundatum**

Vouchers: *Bolus* 7391 (BOL; SAM; Z); *Schlechter* 9868 (BM; BOL; E; K; PRE; Z); *Stokoe* 6585 (BOL; K).

49. **Helichrysum catipes** *(DC.) Harv.* in F.C. 3: 223 (1865). Type: Cape, Cedarberg, *Drège* 2837 (G-DC, holo.!; G; P; S, iso.!).

Eriosphaera catipes DC., Prodr. 6: 167 (1838).

Helichrysum anaxetonoides Schltr. & Moeser in Bot. Jb. 44: 289 (1910). Type: Klein Namaland, Koude Bokkeveld, Skurfdebergen, Elandsfontein, 1 540 m, *Schlechter* 10035 (BM; BOL; G; K; PRE; Z, iso.!).

Bushy perennial herb, vegetative twigs congested, thin, woody, erect (?) or spreading, closely leafy, flowering twigs elongate, distantly leafy, all grey-woolly. *Leaves* mostly 6—13 × 3—6 mm, oblong-ovate, longer and narrower (c. 9—12 × 3 mm) on the flowering twigs, apex rounded, mucronate, base broad, clasping, both surfaces thinly greyish-woolly. *Heads* heterogamous, campanulate, c. 5 × 3 mm, c. 6—12 in congested terminal clusters surrounded by leaf-like bracts each with a scarious appendage, webbed together with wool. *Involucral bracts* in c. 4 series, scarcely graded, loosely imbricate, inner slightly shorter than flowers, oblong, tips rounded, straw-yellow, subopaque, minutely radiating, all but these tips loosely enveloped in wool. *Receptacle* honeycombed. *Flowers* c. 26—28, 5—7 ♀, 20—22 ☿. *Achenes* 1 mm long, cylindric, with duplex hairs, not myxogenic. *Pappus* bristles many, scabrid, tips barbellate, bases cohering lightly by patent cilia, some light fusion as well. Fig. 22: 4.

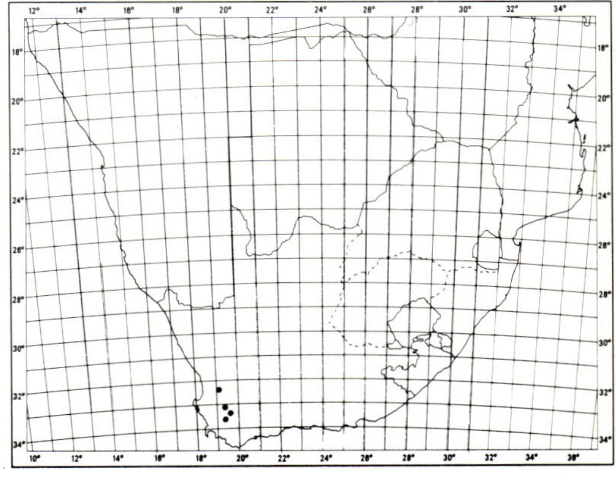

MAP 74.— **Helichrysum catipes**

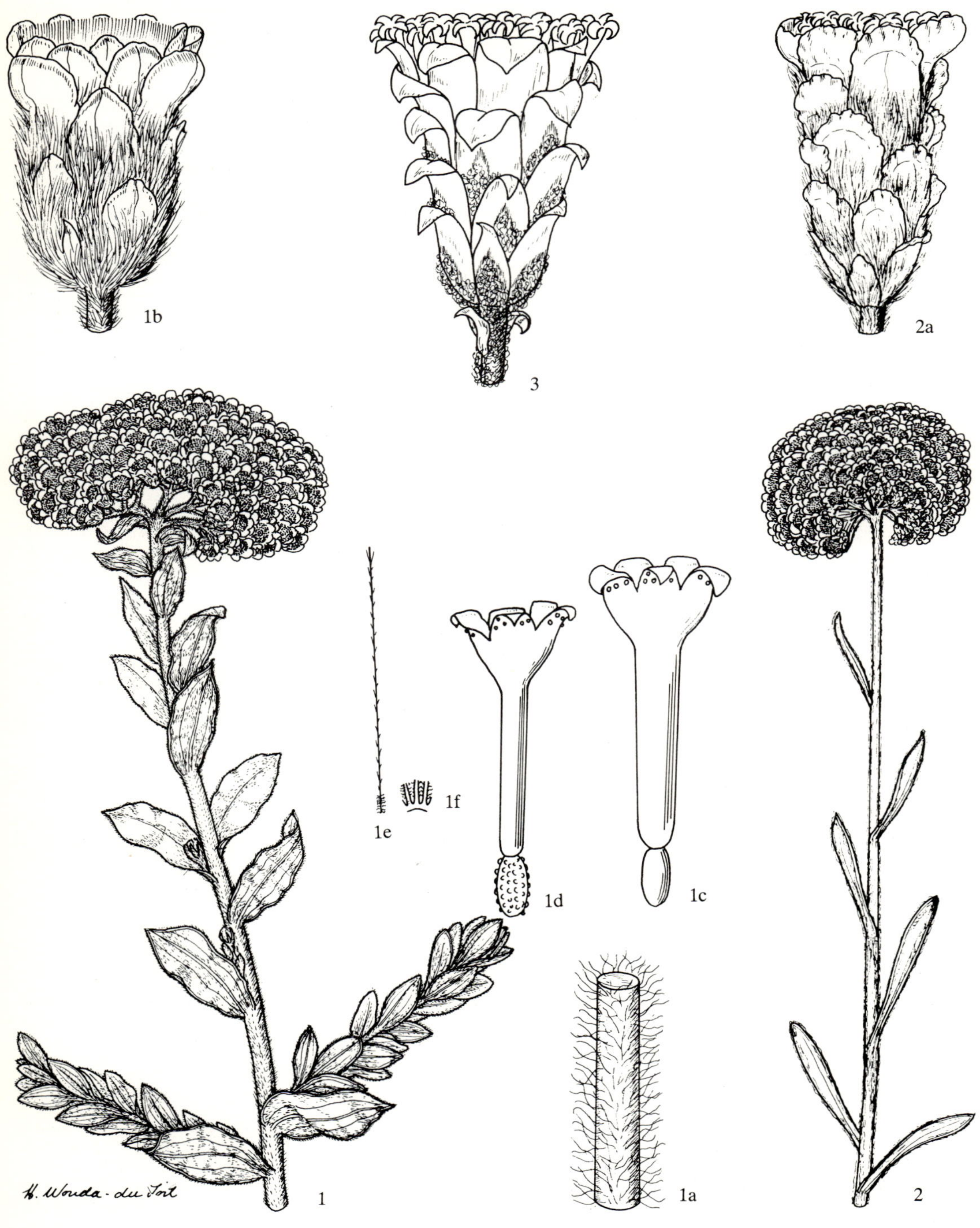

Endemic to the SW. Cape and recorded from the Cedarberg, Hex River Mountains, Skurfdeberg and Bokkeveld Sneeukop. Rarely collected. Miss Esterhuysen records 'slopes at foot of shale band' and 'shaly slopes with step ledges, in low bush' between c. 1 500 and 1 800 m, flowering in December and March. Map 74.

Vouchers: *Esterhuysen* 20951 (BOL; K; M; NBG; PRE); *Esterhuysen* 30108 (BOL; PRE).

50. **Helichrysum dasyanthum** (Willd.) Sweet, Hort. Brit. 223 (1826); Hilliard & Burtt in Bot. J. Linn. Soc. 82: 200 (1981). Type: Cape of Good Hope (sheet 15453, B-W!).

Gnaphalium dasyanthum Willd., Enum. 865 (1809).

Gnaphalium maritimum L., Mant. alt. 283 (1771), non Hill (1769). *Helichrysum maritimum* D. Don in Sweet, Hort. Brit. 223 (1826); Less., Syn. Comp. 304 (1832); DC., Prodr. 6: 204 (1838); Harv. in F.C. 3: 248 (1865); Moeser in Bot. Jb. 44: 288 (1910); Levyns in Adamson & Salter, Fl. Cape Penins. 784 (1950). Type: Cape of Good Hope, seashore, *Tulbagh* 171 (LINN 989.29!).

G. molle Thunb., Prodr. 150 (1800), Fl. Cap. 653 (1823), non Salisb. (1796). Type: Cape, French Hoek, *Thunberg* (sheet 19201, UPS, holo.!).

Helichrysum maritimum var. *microphyllum* DC., Prodr. 6: 204 (1838); Harv. in F.C. 3: 248 (1865). Lectotype: Cape, Kamiesberge, alt. 4 000—5 000 ped., *Drège* 5760 (G-DC!).

Shrub or subshrub, erect to c. 1 m, or sprawling or straggling, branches long, thin, greyish-white woolly, long spreading hairs as well, these sometimes rufous, leafy, flowering twigs more distantly so. *Leaves* mostly 6—30 × 2,5—10 mm, becoming smaller below the heads, elliptic, oblong or oblong-lanceolate, sessile, apex obtuse or subacute, mucronate, sometimes recurved, margins subrevolute, often undulate, thinly greyish-white woolly above or wool sometimes wanting, long spreading hairs as well, whitish or rufous, thickly woolly below, also with spreading hairs. *Heads* heterogamous, campanulate, c. 4×3 mm, many in compact terminal corymbose panicles. *Involucral bracts* in 4—5 series, imbricate, graded, just overtopping flowers, silky-villous dorsally, outer lanceolate, often dark-tipped, inner oblong, tips rounded, glabrous, semi-opaque, pale straw-yellow, minutely radiat-

ing. *Receptacle* with fimbrils much exceeding the ovaries. *Flowers* 17—28, 3—10 ♀, occasionally with staminodes, 12—24 ☿. *Achenes* c. 0,75 mm, those from ♀ flowers often c. one third smaller, barrel-shaped, obscurely ribbed, in ♀ flowers with duplex hairs, not myxogenic, in ☿ flowers usually glabrous, very rarely hairy. *Pappus* bristles many, equalling corolla, barbellate, bases lightly fused as well as cohering by patent cilia. Fig. 23: 1.

Confined to the W., SW. and S. Cape, from Kamieskroon and Kamiesberg in Namaqualand south through Vanrhynsdorp, Calvinia, Clanwilliam, Wuppertal and Worcester degree squares to the Peninsula, thence east through Simonstown, Caledon, Bredasdorp, Riversdale, Oudtshoorn and Mossel Bay degree squares to the environs of Willowmore, Knysna and Plettenberg Bay, along the littoral and on the mountain slopes, in fynbos and on the margins of dune scrub and forest, flowering mainly in September, October and November. Map 75.

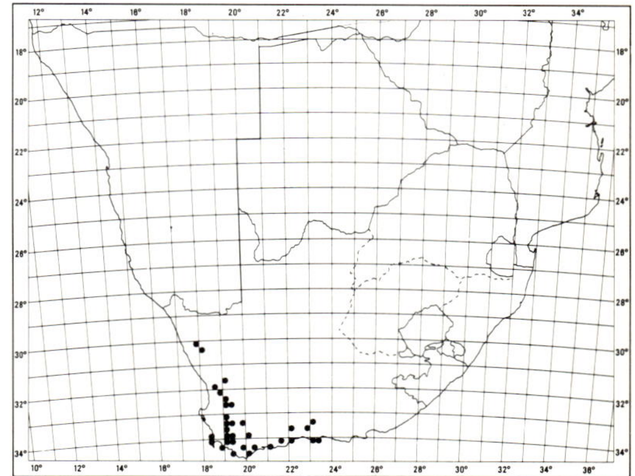

MAP 75.— **Helichrysum dasyanthum**

Straggling specimens bear a superficial resemblance to *H. capense,* but are easily distinguished by the spreading hairs on stems and leaves, and the markedly fimbrilliferous, not smooth, receptacle. Also, there are many more heads in the inflorescence, the involucral bracts have distinctly glabrous tips and the ovaries of female and hermaphrodite flowers are usually different.

FIG. 23.—1, **Helichrysum dasyanthum,** flowering branch, × 1; 1a, detail of part of stem to show spreading hairs, × 4; 1b, head, × 8; 1c, hermaphrodite flower, achene glabrous, × 13; 1d, female flower, achene larger than in hermaphrodite flower, hairy, × 13; 1e, pappus bristle, × 13; 1f, bases of pappus bristles much enlarged to show light fusion (*Acocks* 24122). 2, **H. plebeium,** flowering branch, × 1; 2a, head, × 8 (*Acocks* 22682). 3, **H. hebelepis,** head, × 8 (*Mauve & Oliver* 148).

Vouchers: *Compton* 22181 (NBG); *Esterhuysen* 21912 (BOL; PRE); *Esterhuysen* 14692 (BOL; NBG); *Tyson* 708 (BOL; SAM); *Werdermann & Oberdieck* 511 (K; PRE).

51. **Helichrysum plebeium** *DC.*, Prodr. 6: 206 (1838); Harv. in F.C. 3: 249 (1865); Moeser in Bot. Jb. 44: 287 (1910). Lectotype: Cape, Swellendam, *Ecklon* 1669 (G-DC!; SAM, isolecto.!).

Gnaphalium plebeium (DC.) Sch. Bip. in Bot. Ztg 3: 172 (1845).

Shrublet, to c. 450 mm, branches virgate, young parts greyish-white woolly-felted, leafy. *Leaves* mostly 10−20 × 1−2 mm, reduced and more distant on flowering twigs, linear to linear-oblong, apex acute, mucronate, base slightly narrowed, half-clasping, margins revolute, both surfaces greyish-white woolly, often becoming cob-webby above. *Heads* heterogamous, campanulate, c. 4 × 2−3 mm, many in compact rounded terminal clusters up to 20 mm across. *Involucral bracts* in 4−5 series, imbricate, graded, inner about equalling flowers, oblong, backs loosely silky-woolly, tips broad, rounded, pellucid, tawny, minutely radiating. *Receptacle* with fimbrils much exceeding ovaries. *Flowers* 12−18, 1−3 ♀, 9−15 ☿. *Achenes* not seen, ovaries glabrous. *Pappus* bristles many, about equalling corolla, scabrid, barbellate above, bases cohering by patent cilia. Fig. 23: 2.

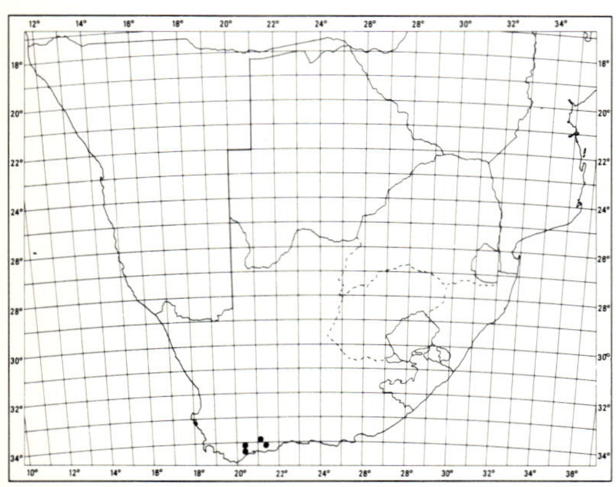

MAP 76.— **Helichrysum plebeium**

Endemic to the southern Cape, and recorded only from Swellendam and Riversdale divisions, on plateaux and hillslopes, flowering between August and October. Map 76.

Vouchers: *Acocks* 22682 (K; PRE); *Esterhuysen* 24624 (BOL); *Liebenberg* 6531 (PRE); *Wurts* 309 (NBG).

52. **Helichrysum hebelepis** *DC.*, Prodr. 6: 186 (1838); Harv. in F.C. 3: 235 (1865); Moeser in Bot. Jb. 44: 291 (1910) excl. syn. Type: Little Namaqualand, *Drège* 5753, excluding right-hand piece on sheet (G-DC!).

Gnaphalium hebelepis (DC.) Sch. Bip. in Bot. Ztg 3: 171 (1845).

H. hebelepis var. *angustius* DC., Prodr. 6: 187 (1838). Lectotype: Clanwilliam div., Olifants River and Villa Brakfontein, *Ecklon* 161 (G-DC!).

Subshrub up to 1,5 m tall, branches long, slender, young parts thinly greyish-white woolly, closely leafy, leaves becoming distant upwards. *Leaves* mostly 20−40 × 2,5−10 mm, reduced upwards, linear, linear-lanceolate or elliptic, sessile, apex subacute to acute, acuminate in uppermost leaves, mucronate, margins subrevolute, undulate in upper leaves, both surfaces thinly greyish-white woolly. *Heads* homogamous or heterogamous, turbinate-campanulate, c. 4 × 3 mm, many in dense globose terminal cymose corymbs. *Involucral bracts* in 6 series, imbricate, graded, reaching cylindric part of corolla, minutely radiating, lanceolate becoming oblong-lanceolate inwards, acute to obtuse, backs woolly, tips glabrous, subopaque, deep

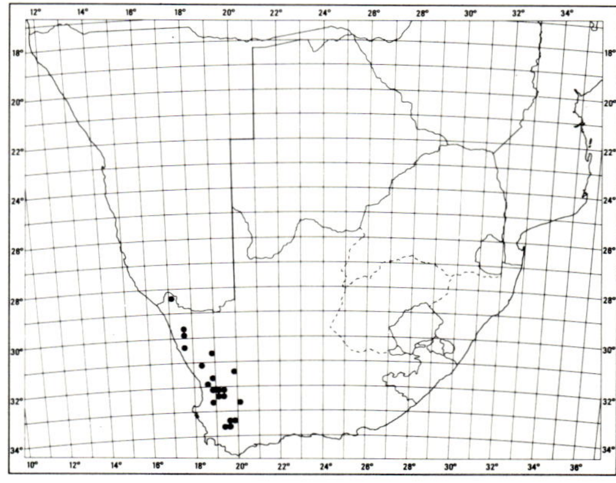

MAP 77.— **Helichrysum hebelepis**

straw-yellow sometimes tinged light golden brown. *Receptacle* smooth. *Flowers* 19—32, 0—5 ♀ (staminodes often present), 15—32 ☿. *Achenes* not seen, ovaries with duplex hairs, not myxogenic. *Pappus* bristles many, scabrid, bases cohering by patent cilia. Fig. 23: 3.

Apparently confined to the W. and SW. Cape, and ranging from the Richtersveld, just south of the Orange River, through Hondeklipbaai, Kamiesberg, Vanrhynsdorp, Calvinia, Clanwilliam, Wuppertal, Sutherland and Worcester degree squares. On stony slopes and flats, in dry scrub, flowering between July and October, mainly in August and September. Map 77.

Much confused with *H. tricostatum* : see under that species (below).

Vouchers: *Acocks* 19404 (K; PRE); *Compton* 5101 (BOL; NBG); *Esterhuysen* 5762 (BOL); *MacOwan* 1887 (K; SAM); *Nordenstam* 1011 (NU; S).

Group 11

Shrubs or subshrubs; leaves small, oblong, narrowly elliptic, spathulate or obovate; heads homogamous, 4—5 × 3—5 mm, in congested corymbose panicles; *involucral bracts* not radiating; *receptacle* smooth, honeycombed or fimbrilliferous; *flowers* 8—34, corolla broadly campanulate above; *achenes* glabrous or hairy; *pappus* bristles barbellate above, shaft scabrid, bases either cohering by patent cilia or free.

Species 53—55; 53 and 54 endemic to the W. and SW. Cape, 55 widespread in the Cape and reaching the southern Orange Free State.

1a Leaves mostly 12—30 mm long, oblanceolate, narrowly elliptic, lanceolate or linear-lanceolate:

  2a Involucral bracts glabrous or with a few woolly hairs near the base ....53. *H. tricostatum*

  2b Involucral bracts greyish-white woolly all over the backs ....54. *H. lambertianum*

1b Leaves mostly 3—10 mm long, spathulate or obovate ....55. *H. pentzioides*

53. **Helichrysum tricostatum** *(Thunb.)* Less., Syn. Comp. 310 (1832); DC., Prodr. 6: 209 (1838); Harv. in F.C. 3: 251 (1865); Moeser in Bot. Jb. 44: 290 (1910). Type: Cape, Piketberg, *Thunberg* (sheet 19277, UPS, holo.!).

Gnaphalium tricostatum Thunb., Prodr. 151 (1800), Fl. Cap. 657 (1823), non DC. (1838).

Shrub or subshrub up to 1,5 m tall, branches long, slender, young parts thinly white-woolly, leafy. *Leaves* mostly 15—30 × 5—8 mm, reduced and more distant upwards, oblanceolate to narrowly elliptic, linear-lanceolate upwards, apex acute, mucronate, base narrowed, minutely petiolate, margins subrevolute, both surfaces thinly greyish-white woolly, *Heads* homogamous, campanulate, c. 4 × 3 mm, woolly at base, many in terminal compact corymbose panicles. *Involucral bracts* in 5 series, imbricate, graded, inner about equalling flowers, oblong, translucent, tips rounded, subopaque, convex, glossy, straw-coloured, not radiating. *Receptacle* with fimbrils equalling or exceeding ovaries. *Flowers* 18—26. *Achenes* 1 mm long, cylindric, obscurely angled, glabrous. *Pappus* bristles many, about equalling corolla, scabrid, barbellate above, bases cohering lightly by patent cilia. Fig. 24: 3.

Endemic to the Sandveld of Namaqualand and the western Cape from Hondeklipbaai south to Buck Bay, N. of Cape Town. Flowers between September and December. Recorded as much eaten by stock. Map 78.

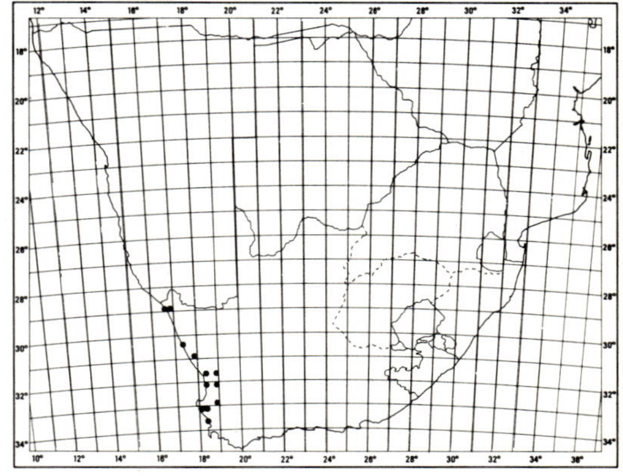

MAP 78.— **Helichrysum tricostatum**

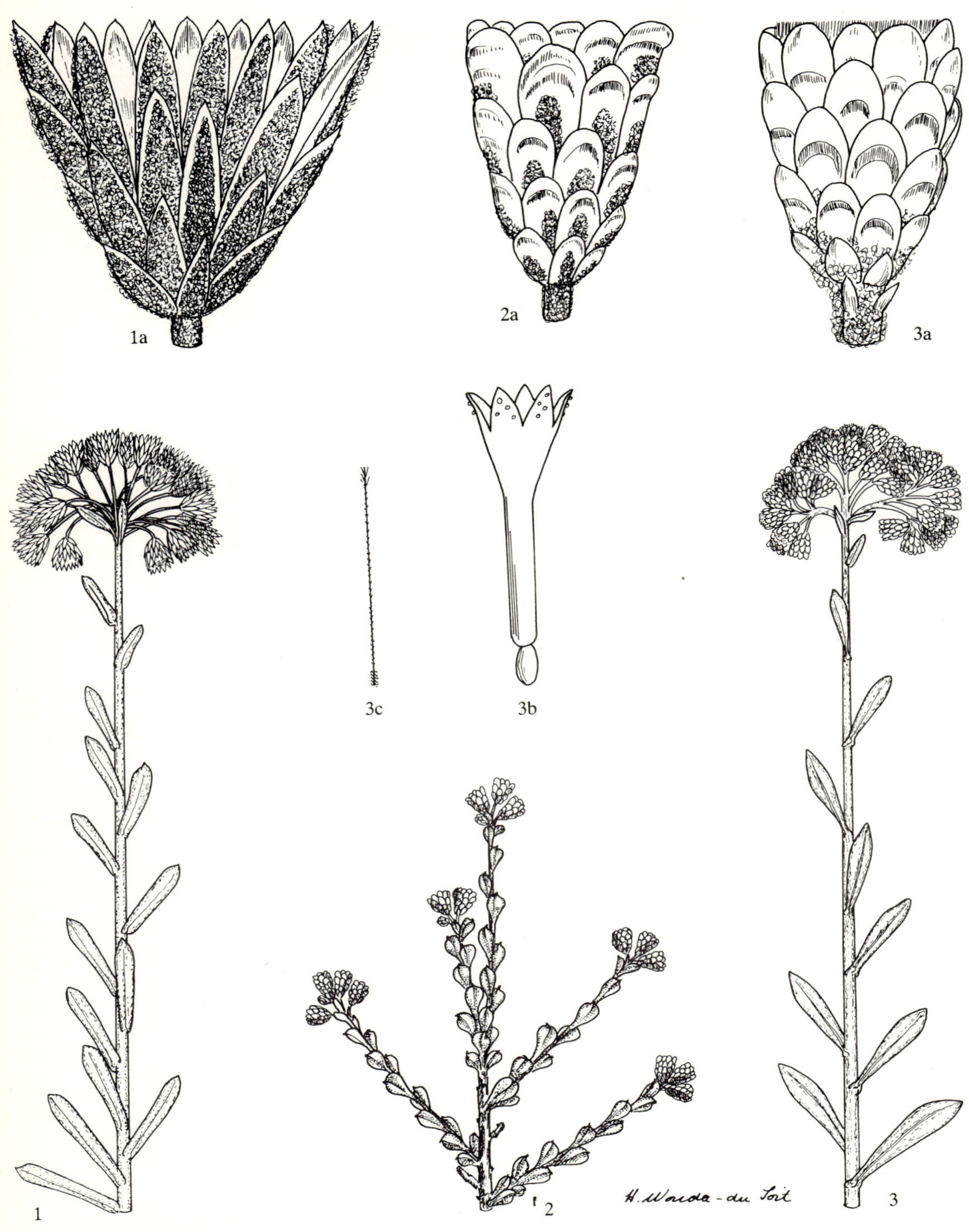
1a
2a
3a
3c
3b
1
2
3
H. Wonda-du Toit

Closely resembles *H. hebelepis* (above) in foliage but easily recognized by its different involucre, fimbrilliferous receptacle and glabrous ovaries.

Vouchers: *Acocks* 14816 (PRE); *Compton* 18906 (NBG); *Taylor* 5506 (K; PRE); *Nordenstam* 3413 (NU; S); *Pillans* 5129 (BOL; K).

54. **Helichrysum lambertianum** *DC.,*

Prodr. 6: 190 (1838); Harv. in F.C. 3: 231 (1865); Moeser in Bot. Jb. 44: 291 (1910). Lectotype: Cape, Clanwilliam, Olifants River and Villa Brakfontein, *Ecklon* 2680 (G-DC!).

Gnaphalium lambertianum (DC.) Sch. Bip. in Bot. Ztg 3: 171 (1845).

Shrubby from a stout woody caudex, up to 600 mm tall, branches long, slender, closely white-felted, young ones leafy. *Leaves* mostly 12−20 × 2−3 mm, diminishing upwards, linear-lanceolate to lanceolate, apex acute, mucronate, base subauriculate, sessile, both surfaces white-felted, minutely glandular. *Heads* homogamous, turbinate, 5 × 5 mm, few to many crowded in terminal corymbose panicles. *Involucral bracts* in 4 series, graded, imbricate, outer lanceolate, inner oblong, obtuse, apiculate, about equalling the flowers, not radiating, all greyish-white woolly dorsally, only the extreme tip glabrous. *Receptacle* smooth. *Flowers* 19−34, yellow. *Achenes* not seen, ovaries with duplex hairs. *Pappus* bristles many, scabrid, tips subplumose or barbellate, bases nude, not cohering. Fig. 24: 1.

Recorded from the Giftberg (Vanrhynsdorp distr.) south through the Cedarberg and Koude Bokkeveld to the mountains about Ceres, Worcester, Hex River, Touws River, Montagu and Villiersdorp, then a disjunction, surely not real, to Georgida in Uniondale division. Grows on dry stony slopes, among rocks, or in the crevices of rocky cliffs or large boulders, flowering between October and January. Map 79.

Easily recognized by its peculiar involucral bracts.

Vouchers: *Acocks* 15200 (PRE); *Bolus* 1051 (BOL; K; SAM); *Compton* 16751 (NBG); *Esterhuysen* 22019 (BOL; K; PRE).

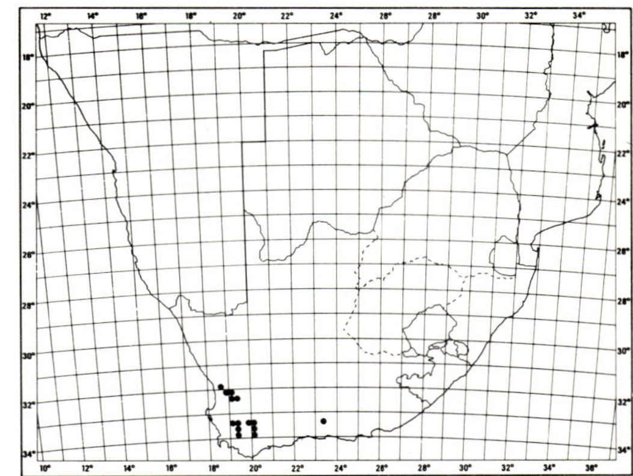

MAP 79.— **Helichrysum lambertianum**

55. **Helichrysum pentzioides** *Less.,*

Syn. Comp. 282 (1832); DC., Prodr. 6: 192 (1838); Harv. in F.C. 3: 236 (1865); Moeser in Bot. Jb. 44: 291 (1910). Type: Cape, near Mossel Bay, Kamma River, *Mundt et Maire.*

Gnaphalium pentzioides (Less.) Sch. Bip. in Bot. Ztg 3: 171 (1845).

Rounded twiggy shrub up to 1 m tall, very aromatic, branches short, stiff, divaricate, old branches bare, rough with persistent leaf bases, twigs thinly grey silky-felted, closely leafy. *Leaves* erect to spreading, mostly 3−10 × 2−4 mm, diminishing upwards, spathulate or obovate, much narrowed to base, very shortly decurrent, thickish, apex mucronate, recurved, both surfaces grey silky-felted, indumentum skin-like, glandular-punctate. *Heads* homogamous, turbinate-cylindric, c. 5 × 3 mm, few in terminal corymbose clusters. *Involucral bracts* in 4 series, graded, imbricate, inner equalling flowers, oblong, white woolly and glandular-punctate dorsally, tips rounded, slightly convex, glabrous, straw-coloured or golden brown, not radiat-

---

FIG. 24.−1, **Helichrysum lambertianum,** part of plant, × 1; 1a, head, × 8 (*Acocks* 15200). 2, **H. pentzioides,** flowering branch, × 1; 2a, head, × 10 (*Codd* 419). 3, **H. tricostatum,** part of plant, × 1; 3a, head, × 10; 3b, flower, × 13; 3c, pappus bristle, × 13 (*Acocks* 14816).

ing. *Receptacle* smooth or very shortly toothed. *Flowers* 8—15. *Achenes* not seen, ovaries with myxogenic duplex hairs. *Pappus* bristles many, nearly equalling corolla, scabrid, barbellate above, bases cohering by patent cilia. Fig. 24: 2.

Widely distributed from about Glen and Bloemfontein (Orange Free State) in the east and Prieska (N. Cape) in the west, south to Keiskammahoek, Grahamstown and Addo in the E. Cape, Montagu, Touws River and Bredasdorp in the west. Grows in karroid and other dry scrub; flowering between October and May, principally December and January. Map 80.

Vouchers: *Acocks* 8685 (PRE); *Bolus* 2200 (BM; BOL); *Esterhuysen* 19560 (BOL; NBG; PRE); *Story* 3694 (PRE).

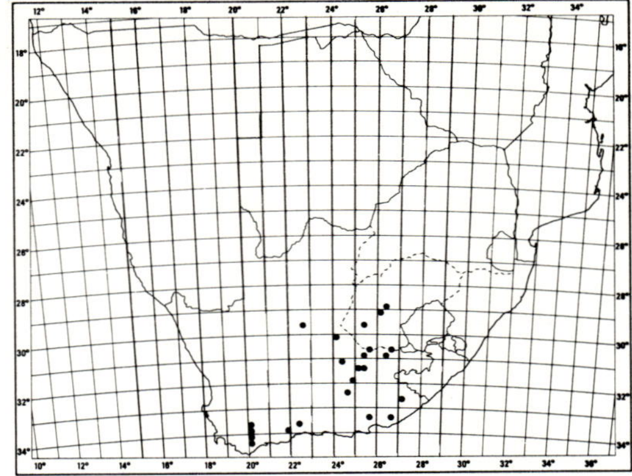

MAP 80.— **Helichrysum pentzioides**

Group 12

Shrubs, subshrubs or bushy perennial herbs, or rarely annual; *leaves* small, linear, oblong, elliptic, spathulate or obovate; *heads* homogamous, 3—7 × 2—4 mm, few to many in clusters or small corymbose panicles; *involucral bracts* not radiating, or rarely squarrose, white, pink, straw-coloured, tawny, golden-brown or yellow, often red above the stereome; *receptacle* smooth, honeycombed or fimbrilliferous; *flowers* (3—) 5—30, corolla narrowly funnel-shaped; *achenes* usually hairy, rarely glabrous and then usually in the same species; *pappus* bristles scabrid, tips sometimes barbellate, bases cohering by patent cilia, sometimes lightly fused as well, or bases free, with or without patent cilia.

Species 56—68, mainly in S.W.A./.Namibia, Botswana, Cape, Transvaal and Orange Free State, a few in Natal and southern Mozambique, often in dry or sandy places; several species narrowly endemic.

1a At least the tips of the inner involucral bracts white or rosy, or white and rose:

  2a Heads c. 4 × 4 mm; leaves elliptic-lanceolate, margins flat .... 56. *H. silvaticum*

  2b Heads c. 4—5 × 1—2 mm; leaves either linear with revolute margins or lingulate with weakly revolute margins:

    3a Heads 3—6-flowered ....58. *H. niveum*

    3b Heads 11—34-flowered:

      4a Plants erect though some stems may straggle.. 59. *H. asperum*

      4b Plants prostrate:

        5a Heads c. 4 mm long containing 8—14 flowers; branches loosely spreading...................60. *H. lineare*

        5b Heads c. 5 mm long containing 11—24 flowers; branches caespitose, forming dense mats ....61. *H. caespititium*

1b Involucral bracts various shades of yellow, buff or brown, sometimes with red patches as well, particularly above the stereome, but never with white or rosy tips:

  6a At least the tips of the inner involucral bracts clear bright yellow:

    7a Leaves linear, oblong or elliptic:

      8a Heads c. 3 × 2 mm, leaves linear or oblong, 1—2 mm broad .... 62. *H. arenicola*

      8b Heads c. 6 × 3 mm, leaves elliptic, 3—8 mm broad .... 65. *H. uninervium*

    7b Leaves spathulate, c. 5—8 × 2—3 mm ....64. *H. excisum*

6b Involucral bracts straw-coloured, buff or various shades of brown, but not clear bright yellow:

  9a Leaves linear, linear-oblong or linear-lanceolate, margins often revolute:

    10a Heads 3—6 mm long; leaves mostly concolorous, or if discolorous, heads only c. 4 mm long:

      11a Heads 5—6 mm long ....57. *H. simulans*

      11b Heads 3—4 mm long:

        12a Heads c. 3 mm long, bracts straw-coloured .... 62. *H. arenicola*

        12b Heads c. 4 mm long, bracts buff-coloured, brownish or golden brown, often with red tints as well:

          13a Heads c. 4 × 2 mm containing 7—12 flowers, bracts ± oblong-lanceolate .... 59. *H. asperum*

          13b Heads c. 4 × 3 mm containing 15—23 flowers, bracts ± ovate or subrotund and somewhat concave ....68. *H. refractum*

    10b Heads 6—7 mm long; leaves discolorous (glabrous to cobwebby above, white tomentose below)..67. *H. hamulosum*

  9b Leaves spathulate or obovate:

    14a Involucral bracts squarrose ....63. *H. fourcadei*

    14b Involucral bracts erect, not squarrose:

      15a Heads surrounded by small narrowly elliptic leaves webbed to the outer involucral bracts; pappus uniseriate..57. *H. simulans*

      15b Heads surrounded by broad leaves free from the involucral bracts, which are soon deciduous; pappus in more than one series... 66. *H. lucilioides*

56. **Helichrysum silvaticum** *Hilliard* in Notes R. Bot. Gdn Edinb. 32: 361 (1973), Compositae in Natal 192 (1977). Type: Natal, Ubombo distr., Makatini Flats, *Strey* 5287 (NU, holo.!; PRE iso.!).

Perennial herb up to c. 400 mm high, sparingly branched, branches thinly grey-woolly, leafy particularly above. *Leaves* c. 15—40 × 1,5—6 mm, slightly smaller upwards, elliptic-lanceolate, apex acute, mucronate, base narrowed, sessile, both surfaces thinly grey-woolly. *Heads* homogamous, turbinate-campanulate, c. 4 × 4 mm, many in small corymbose panicles terminating the branches. *Involucral bracts* in c. 5 series, graded, loosely imbricate, inner about equalling flowers, bases silvery, translucent, tips opaque white, acute, not radiating. *Receptacle* with fimbrils equalling ovaries. *Flowers* c. 20—30. *Achenes* 0,5 mm long, glabrous. *Pappus* bristles many, about equalling corolla, scabridulous, bases cohering strongly by patent cilia, some light fusion as well. Fig. 25: 1.

Recorded from the Mozambique coastal plain and its extension southwards into northern Zululand (Tongaland), growing in open woodland or grassland on sandy soils. Flowers in June and July; rarely collected. Map 81.

Vouchers: *Gerstner* 3467 (BOL, K); *Junod* 2912 (G).

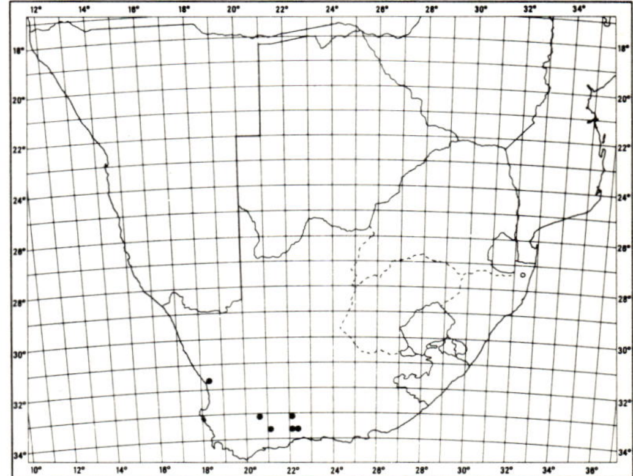

MAP 81.— ○ Helichrysum silvaticum
● Helichrysum simulans

57. **Helichrysum simulans** *Harv. & Sond.*, F.C. 3: 217 (1865). Type: Cape, George, Karroo in der Gegend von Gouritz-rivier, 1 000—3 000 Fuss, December, *Ecklon* 546 (S, holo.!).

H. silicicolum Compton in Trans. R. Soc. S. Afr. 19: 316 (1931). Type: Cape, Laingsburg distr., White Hill, Karoo Garden, 2 700 ft, on white quartzite rubble, 1 xii 1924, *Compton* 2889 (BOL, holo.!).

Dwarf twiggy shrublet c. 50—150 mm high, main stem gnarled, c. 10 mm diam.,

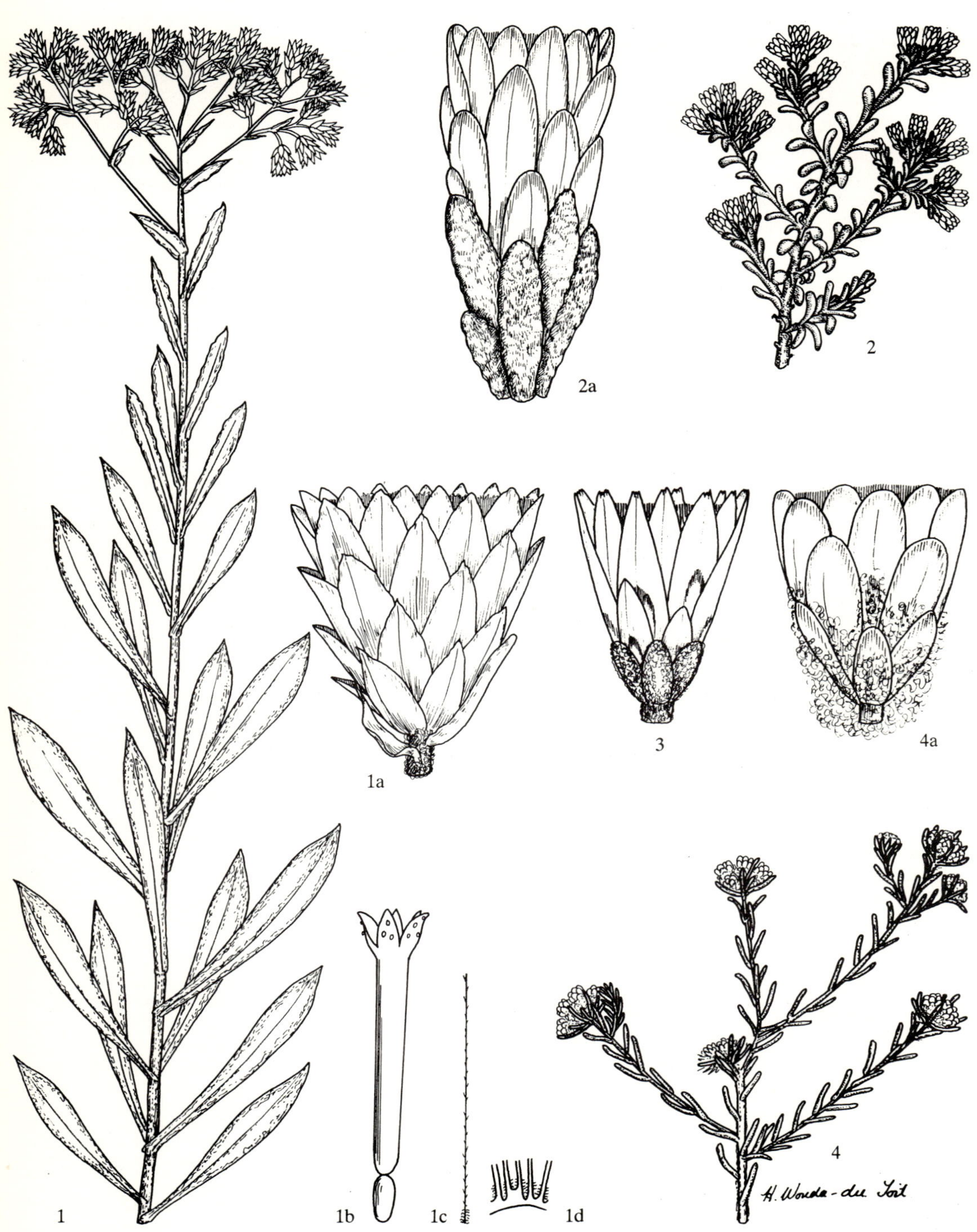
2a
2
1a
3
4a
1
1b
1c
1d
4
H. Woud-du Toit

branches stiffly erect, bare below, branch-lets grey stringy-felted, closely leafy. *Leaves* erect, imbricate, 3—6 × 1—2 mm, linear (narrowly spathulate, up to 3 mm broad, in *Hall* 177; see notes below), apex more or less obtuse, slightly recurved, base broad, shortly decurrent, both surfaces enveloped in grey stringy tissue-paper-like tomentum. *Heads* homogamous, cylindric, c. 5—6 × 2—3 mm, solitary or 2—3 together at the branchlet tips, closely surrounded by leaves webbed to outer involucral bracts. *Involucral bracts* in 5 series, graded, closely, imbricate, inner equalling flowers, not radiating, all obtuse, convex, backs some-what woolly, light or dark golden-brown, sometimes red above the stereome, semi-pellucid. *Receptacle* nearly smooth. *Flowers* 11—16. *Achenes* not seen, ovaries with myxogenic duplex hairs. *Pappus* bristles many, equalling corolla, scabrid, bases cohering lightly by patent cilia. Fig. 25: 2.

An ill-known but distinctive species recorded from the Karoo at Whitehill (Laingsburg distr.), Calitzdorp, Prince Albert, the Gouritz River, and Zebra and Moeras River on the northern flank of the Outeniqua Mountains. Grows in quartzite rubble or sand, flowering in November and December. Map 81.

Vouchers: *Compton* 12636 (NBG); *Esterhuysen* 19285 (BOL); *Stokoe* in SAM 68713 (SAM).

Hall 177 (NBG) has paler heads and broader leaves than the other specimens seen, but is perhaps scarcely distinct specifically (Vanrhynsdorp distr., Holriver, 4 miles N. of bridge, c. 200 ft. 31 x 1964, *Hall* 177, NBG).

58. Helichrysum niveum *(L.) Less.*,

Syn. Comp. 302 (1832), quoad syn. tantum; Hilliard & Burtt in Bot. J. Linn. Soc. 82: 263 (1981). Lectotype: Plukenet, Mant. 67 (1696), herb. Sloane vol. 100 folio 24 (BM!).

Gnaphalium niveum L., Sp. Pl. 852 (1753).

Helichrysum metalasioides DC., Prodr. 6: 171 (1838) p.p. excl. *Burchell* 2272 (G-DC!); Levyns in Adamson & Salter, Fl. Cape Penins. 781 (1950). *Gnaphalium metalasioides* (DC.) Sch. Bip. in Bot. Ztg 3: 169 (1845). *Helichrysum ericifolium* Less, var. *metalasioides* (DC.) Harv. in F.C. 3: 217 (1865). Type: Cape, Hout Bay, *Lambert* s.n. (G-DC, holo.!).

H. callunoides Sch. Bip. in Flora, Regensburg 27, 2:

677 (1844). *Gnaphalium callunoides* (Sch. Bip.) Sch. Bip. in Bot. Ztg. 3: 173 (1845). Type: Cape, Swellendam, Zoetendals Valley, Dec., *Krauss* 557 (P, holo.!; G; Z, iso.!).

Dwarf twiggy ericoid shrublet, branch-es erect, spreading or prostrate, c. 60—120 (—300) mm long, branchlets white-woolly-felted, closely leafy. *Leaves* 2—5 (—10) × 0,5—0,75 mm, scarcely diminishing up-wards, spreading, later reflexed, linear, obtuse, half clasping, shortly decurrent, margins strongly revolute, both surfaces grey woolly-felted, glabrescent, or wholly glabrous. *Heads* homogamous, cylindric, 4—5 × 1 mm, in small glomerules clustered at the branch tips closely surrounded by leaves. *Involucral bracts* in c. 6 series, outermost very short, webbed to the surrounding leaves, persistent, inner sub-equal, tips lanceolate, acute, milk-white or sometimes pink, slightly exceeding flowers, not radiating, soon caducous. *Receptacle* conical, honeycombed. *Flowers* (3—) 5 (—6). *Achenes* 0,75 mm, with myxogenic duplex hairs. *Pappus* bristles many, equal-ling corolla, scabrid, bases with minute patent cilia not cohering.

Along the SW. Cape coast from Saldanha Bay, the Cape Peninsula and the Cape Flats to about Still Bay in Riversdale district, often on coastal dunes, always in sandy places, not recorded above 180 m. Flowers mainly from December to February. Map 82.

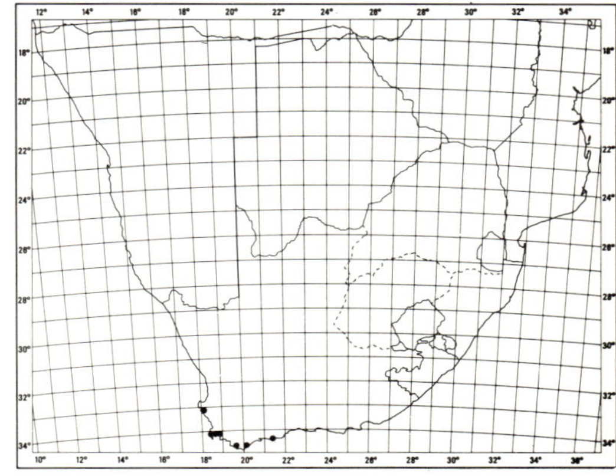

MAP 82.— **Helichrysum niveum**

FIG. 25.—1, **Helichrysum silvaticum,** part of plant, × 1; 1a, head, × 6,6; 1b, flower, × 13; 1c, pappus bristle, × 13; 1d, detail of bases of pappus bristles (*Strey* 5287). 2, **H. simulans,** flowering branch, × 1; 2a, head, × 8 (*Van Breda* 450a). 3, **H. asperum,** head, × 8 (*Hanekom* 1742). 4, **H. arenicola,** flowering branch, × 1; 4a, head, × 8 (*Leistner* 2904).

Has something of the aspect of *H. asperum* var. *glabrum* (below) but is easily distinguished by its few-flowered heads with acute white or white flushed pink involucral bracts.

Vouchers: *Esterhuysen 28761a* (BOL; PRE); *Galpin 11259* (PRE); *Taylor 5662* (PRE).

59. Helichrysum asperum *(Thunb.) Hilliard & Burtt* in Notes R. Bot. Gdn Edinb. 34: 79 (1975); Hilliard, Compositae in Natal 208 (1977). Type: Cape of Good Hope, *Thunberg* (sheet 20961, UPS, holo.!).

Stoebe aspera Thunb., Prodr. 170 (1800), Fl. Cap. 728 (1823). *Seriphium asperum* (Thunb.) Pers., Syn. Pl. 2: 501 (1807).

Gnaphalium ericoides L., Pl. Afr. Rar. 19 (1760), Sp. Pl. 2: 1193 (1763), Amoen, Acad. 6: 99 (1764), non *Helichrysum ericoides* (Lam.) Pers. (1805). *Evax ericoides* (L.) Schrank in Denkschr. K. Akad. Wiss. Münch. 8: 172 (1824). *Helichrysum ericoides* (L.) D. Don in Sweet, Hort. Brit. 293 (1830), non (Lam.) Pers. (1805). *H. ericifolium* Less., Syn. Comp. 314 (1832); DC., Prodr. 6: 172 (1838); Harv. in F.C. 3: 217 (1865) excl. var. *metalasioides;* Moeser in Bot. Jb. 44: 298 (1910) excl. vars.; Levyns in Adamson & Salter, Fl. Cape Penins. 781 (1950). Lectotype: from the Cape of Good Hope (LINN 989.17!).

Helichrysum laxum DC., Prodr. 6: 171 (1838). *Gnaphalium laxum* (DC.) Sch. Bip. in Bot. Ztg 3: 169 (1845). *Helichrysum ericifolium* var. *laxum* (DC.) Harv. in F.C. 3: 217 (1865). Lectotype: Cape, Paarlberg, *Drège* 1788 (G-DC!; BM; E; S; SAM; TCD; isolecto!).

Five varieties are recognized. Not all specimens can be assigned to a variety, and putative hybrids have been reported. See comments in text.

1a Heads mostly solitary or 2—4 at the branch tips, rarely more:

  2a Tips of involucral bracts light golden-brown, often tinged red, semipellucid .... (a) var. *asperum*

  2b Tips of involucral bracts opaque white .... (b) var. *albidulum*

1b Heads several at the branch tips:

  3a Leaves 2—5 (—10) × 0,5 mm, glabrous .... (d) var. *glabrum*

  3b Leaves either variously woolly at least when young, or if glabrous, then leaves mostly 7—10 × 1—1,5 mm:

    4a Leaves mostly 7—20 × 1—1,5 mm, woolly or glabrous; tips of involucral bracts white .... (e) var. *comosum*

    4b Leaves mostly 3—6 × 0,5—1 mm:

      5a Leaves loosely woolly; involucral bracts semipellucid, buff-coloured, sometimes with a reddish tinge .... (c) var. *appressifolium*

5b Leaves enveloped in tightly woven 'tissue-paper' indumentum (though this breaks down to wool with age); tips of involucral bracts opaque white, often reddish above the stereome .... (b) var. *albidulum*

(a) var. **asperum.**

Hard, stiff, divaricately branched, tangled, bushy shrublet mostly 15—400 mm tall, sometimes straggling, outer spreading branches then sending up numerous erect shoots, branches thinly grey-woolly, often glabrescent, rarely glandular-hispid (see note below), leafy. *Leaves* 2—6 × 0,5 mm (old primary leaves up to 20 × 1 mm), linear, acute to subacute, or obtuse, often apiculate, sessile, margins strongly revolute, both surfaces thinly white woolly-felted, often glabrescent, sometimes glandular-hispid as well. *Heads* homogamous, c. 4 × 2 mm, cylindric, solitary or 2—4, very rarely more, together at the branch tips, closely surrounded by leaves. *Involucral bracts* in 5—6 series, outermost short, webbed with wool to surrounding leaves, inner subequal, equalling flowers, erect, tips semipellucid, subacute, often erose, light golden-brown, often reddish below, not radiating, soon caducous. *Receptacle* raised, slightly tuberculate. *Flowers* 7—13. *Achenes* not seen, ovaries with myxogenic duplex hairs. *Pappus* bristles many, equalling corolla, scabrid, bases with patent cilia not cohering. Fig. 25: 3.

Ranges from the Cape Peninsula E. to Touws River and White Hill (Laingsburg distr.) and N. to Vanrhynsdorp and Calvinia. Grows in fynbos on dry stony mountain slopes or flats, or occasionally in succulent Karoo (Vanrhynsdorp), between c. 120 and 1 200 m above sea level; flowering between October and January. Map 83.

H. asperum var. *asperum* can be confused with *H. oxybelium* (no. 97), which much resembles it in habit, in disposition of the heads, and in colour of the involucral bracts, but var. *asperum* is easily recognized by its homogamous heads containing 7—13 flowers, not heterogamous and containing 19—29 flowers.

In the type specimen, which is unlocalized, only the young leaves are loosely woolly; the older leaves are merely glandular-hispid. Modern collections from the Cedarberg and from the mountains near Groot Kloof, De Doorns and Worcester tend to be glandular-hispid; elsewhere prominent glands are lacking but both glandular and eglandular plants may grow together.

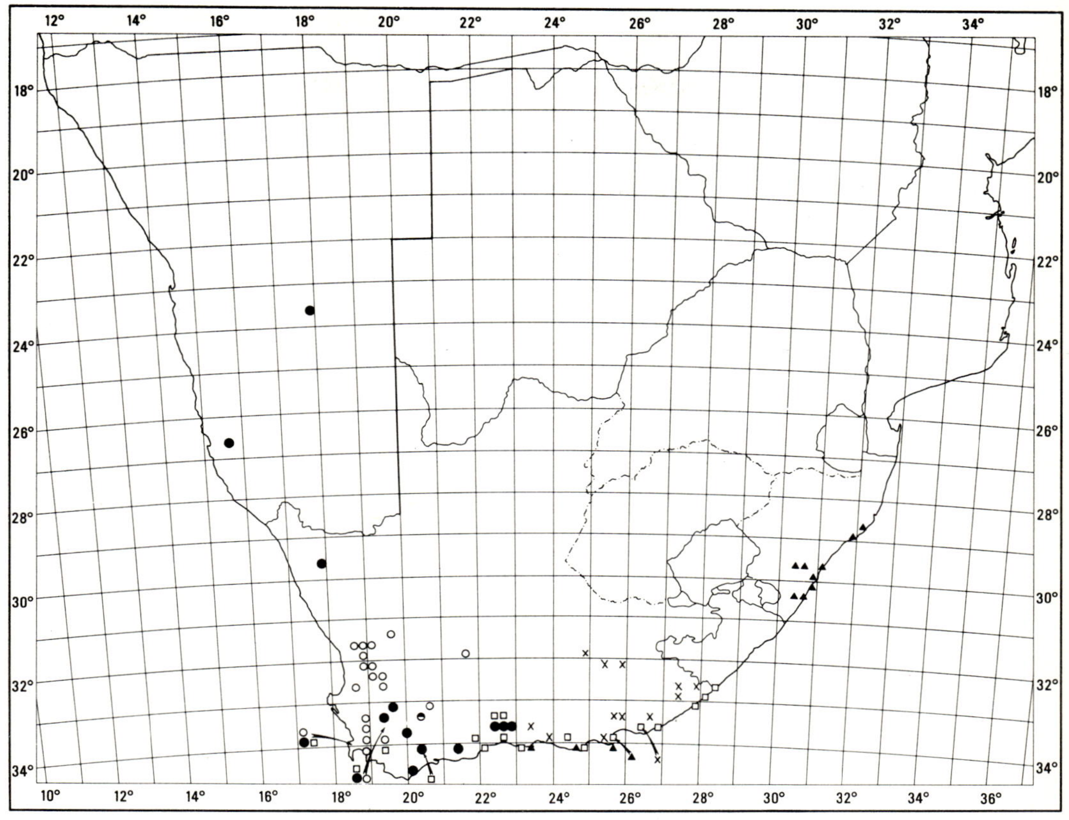

MAP 83.— ○ **Helichrysum asperum** var. **asperum**
      ● **Helichrysum asperum** var. **albidulum**
      □ **Helichrysum asperum** var. **glabrum**
      × **Helichrysum asperum** var. **appressifolium**
      ▲ **Helichrysum asperum** var. **comosum**

Vouchers: *Acocks* 15107 (PRE); *Esterhuysen* 14964 (BOL; PRE); *Galpin* 11103 (PRE); *Pillans* 5411 (NBG).

(b) var. **albidulum** *(DC.) Hilliard,* comb. nov.

Lectotype: Cape, Riversdale distr., between Soetmelks (sphalm. Troetemelks) River and Little Vette River, *Burchell* 6819 (G-DC!; K; PRE, isolecto.!).

H. ericifolium Less. var. *albidulum* DC., Prodr. 6: 172 (1838); Harv. in F.C. 3: 217 (1865); Levyns in Adamson & Salter, Fl. Cape Penins. 781 (1950).

Similar to the typical plant in habit, head size and disposition of the heads, but leaves and young stems enveloped in tightly woven, silvery-grey, papery indumentum (which may break down to wool with age) and involucral bracts with opaque white tips.

Recorded mainly from the Cape Peninsula and east through Worcester, Montagu, Bredasdorp, Riversdale and Oudtshoorn degree squares (not recorded beyond c. 23°E) with a few records from much further north, around Nababiep and Springbok in Namaqualand, and in S.W.A./Namibia in the diamond area (2615 DA Lüderitz) and Rheinpfalz (2317 DB Rehoboth). The specimen from Rheinpfalz (*Dinter* 6731, BM; E; PRE; SAM) is discussed by Merxmüller (F.S.W.A. 139: 94, 1967) under *H. fleckii.* Map 83.

Vouchers: *Esterhuysen* 417 (BOL; PRE); *Galpin* 11322 (PRE); *Schlechter* 9939 (BM; PRE).

Salter collected var. *albidulum* near Bishop's Court, Cape Peninsula, in December 1947, together with var. *glabrum* (below) and intermediates between the two varieties (*Salter* 8975, 8977, 8976, NBG). *Marais* 44 (PRE) from the Bontebok National Park, Swellendam, also appears to be intermediate (and both var. *albidulum* and var. *glabrum* occur in the park.)

(c) var. **appressifolium** *(Moeser) Hilliard,* comb. nov.

Type: Cape, Albany distr., near Grahamstown, Botha's Hill, 660 m, 21 xii 1894, *Schlechter* 6091 (BM; K; PRE; S, iso.!).

H. ericifolium Less. var. *appressifolium* Moeser in Bot. Jb. 48: 339 (1913).

Differs from the typical plant in its laxer habit, the branches tending to be corymbosely arranged rather than stiffly divaricate, and generally more heads crowded at the tips of the branchlets. The bracts are pale brownish, often with a pink tinge, usually lacking the opaque white tips of var. *albidulum*. The indumentum is woollier, less silky, than that of var. *albidulum*.

Recorded from Willowmore degree square northeast and east to Cradock and Komgha, in dry gravelly or sandy places. Map 83.

Vouchers: *Burtt Davy* 12144 (PRE); *Esterhuysen* 13634 (BOL: PRE); *Fourcade* 2949 (PRE): *Hilliard & Burtt* 10578 (E; K; NU; S).

(d) var. **glabrum** *Hilliard var. nov. a typo ramis laxis corymbose dispositis, capitulis multis ad apices ramulorum aggregatis recedit.*

Type: Cape Peninsula, Wynberg Hill, 19 i 1919, *Pillans* 3279 (PRE, holo.!).

Rounded, bushy shrublet up to c. 200 mm tall, main branches laxly corymbosely branched above, branches virgate, thinly white woolly-felted, closely leafy. *Leaves* patent or reflexed, 2–5 (−10) × 0,5 mm, scarcely smaller upwards, linear, margins strongly revolute, glabrous, glandular-punctate. *Heads* in clusters at the branchlet tips, surrounded by leaves. *Involucral bracts* either buff-coloured, sometimes with a reddish tinge, and pellucid, or subopaque to opaque white or dirty white (see notes below).

Differs from the typical plant in its lax, corymbosely arranged branches and many heads clustered at the tips of the branchlets. Plants with buff-coloured bracts have been recorded from the Cape Peninsula; along the coast, east of Swellendam, plants begin to appear with subopaque, dirty-white tips to the involucral bracts, or occasionally the tips are opaque and pure white; plants with buff-coloured bracts are also present (Albertinia, Mossel Bay, Knysna, Kareedouw, Humansdorp, Uitenhage). Still further east (at Alexandria, Port Alfred, East London, Kei Mouth and Gogwana Mouth in the Transkei) bract tips are mostly opaque white.

H. asperum var. *glabrum* grows on sandy soils, on grassy slopes or in fynbos, often near the sea, but also on the mountains of the SW. Cape up to c. 600 m. Flowers mainly between December and March. Map 83.

Vouchers: *Esterhuysen* 418 (BOL; PRE); *Flanagan* 336 (PRE); *Hilliard & Burtt* 10896 (E; K; MO; NU; PRE; S).

Specimens from Coombs Valley near Grahamstown may have woolly leaves as in var. *appressifolium* (*Hilliard & Burtt* 10852, E; K; NU; PRE) or glabrous leaves as in var. *glabrum* (*Hilliard & Burtt* 10863, E; K; NU; PRE); one plant of the latter collection has some branchlets with glabrous leaves, others with lower leaves glabrous, upper woolly.

Marais s.n. (PRE) from the Bontebok National Park at Swellendam appears to be *H. asperum* var. *glabrum* × var. *albidulum*.

(e) var. **comosum** *(Sch. Bip.) Hilliard,* comb. et stat. nov.

Type: Natal, Port Natal, *Krauss* (BM; G; K; TCD, isotypes!).

H. comosum Sch. Bip., in Flora, Regensburg 27: 678 (1844).

H. asperum (Thunb.) Hilliard & Burtt p.p.; Hilliard, Compositae in Natal 208 (1977).

H. ericifolium Less. var. *lineare* sensu Harv. in F.C. 3: 217 (1865); non *H. lineare* DC.

Similar in habit to var. *glabrum* but a coarser-looking plant, with generally larger leaves 7−20 × 1 (−1,5) mm, which may be glabrous, but are often woolly below and sometimes thinly appressed-woolly above in Natal specimens, and are always woolly in Cape specimens. The involucral bracts always have conspicuous opaque white tips.

Var. *comosum* is common on coastal sand dunes from about Plettenberg Bay to Port Elizabeth; then there is an apparent distributional gap to the Natal coast, where var. *comosum* has been recorded as far north as the Mozambique border. There are no records of any form of *H. asperum* along the Transkeian coast, apart from Kei Mouth and Gogwana Mouth (var. *glabrum*). Map 83.

In Natal, var. *comosum* has been recorded inland up to c. 900 m above sea level, always on sandy soils, but it is not as common inland as along the beaches.

Some specimens of *H. lineare* (below) from inland Natal may resemble var. *comosum* rather closely, but are distinguished by their prostrate habit. The grey-woolly Cape specimens can be confused with *H. niveum* (no. 58), but that species has fewer flowers in the head (mostly 5—6, opposed to 8—12) and shorter, narrower, more crowded leaves.

Vouchers: *Dahlstrand 20* (PRE); *Fourcade 1799* (PRE); *Hilliard & Burtt 3374* (E; NH; NU); *Rudatis 1565* (PRE); *Strey 5187* (PRE).

60. **Helichrysum lineare** *DC.,* Prodr. 6: 172 (1838). Lectotype: Cape, bank of Mashowing River between Takun and Molito, 29-30 Sept. 1812, *Burchell 2307* (G-DC, holo.!; K, iso.!).

Gnaphalium lineare (DC.) Sch. Bip. in Bot. Ztg 3: 169 (1845).

Helichrysum seineri Moeser in Bot. Jb. 44: 300 (1910); Merxm. F.S.W.A. 139: 97 (1967). Type: S.W.A./Namibia, Caprivi Strip, W. environs Sesheke, *Seiner 45* (B).

Helichrysum sp.; Hilliard, Compositae in Natal 210 (1977).

Perennial herb with a long stout taproot, woody with age, branches up to c. 200 mm long, many spreading from the crown, prostrate or rarely decumbent, simple below, corymbosely branched above especially in old plants, thinly white-woolly, closely leafy. *Leaves* 5—10 (—13) × 0,75—1 (—4) mm, more or less spreading, linear or lingulate, obtuse, sessile, margins revolute, rarely only weakly so, both surfaces white-woolly, upper often glabrescent, glandular-punctate, viscid. *Heads* homogamous, cylindric, c. 4 × 2 mm, in small glomerules at the branch tips, closely surrounded by leaves. *Involucral bracts* in c. 6 series, outermost short, webbed to the surrounding leaves with wool, persistent, inner subequal, erect, tips subacute to obtuse, often erose, usually opaque white, rarely buff or pinkish, shafts pellucid, colourless or reddish, soon caducous. *Receptacle* conical, tuberculate. *Flowers* 8—14, yellow, sometimes tipped pink. *Achenes* not seen, ovaries with myxogenic duplex hairs or hairs sometimes few or wanting. *Pappus* bristles many, equalling corolla, scabrid, bases with patent cilia not cohering.

Widespread in the dry interior of the Cape E. of 22°E., spreading eastwards to Somerset East, Stutterheim and Barkly East districts, the SW. Transvaal, Orange Free State and W. Lesotho; comparatively rare

in Natal. Also recorded along the Okavango river, and in the Caprivi Strip, Botswana, Zimbabwe and Zambia. Grows in bare places and often seasonally wet places, on sand, alluvium or limestone soils, and will colonize streamsides, around pans, and disturbed ground. Flowers mainly between August and December, sometimes as early as April to July.

H. lineare has been much confused with *H. asperum* (no. 59). However, the two species are nearly allopatric, their areas overlapping only in the eastern part of the country; it is only there that difficulty may arise in placing specimens, but *H. lineare* is always prostrate, *H. asperum* is always erect, and *H. lineare* generally flowers earlier than *H. asperum*.

Specimens from very wet ground may have large leaves in which the lingulate shape is therefore more striking, and the leaf margins may be less revolute than in specimens from drier places e.g. *Robertson & Elfers 69* and *113*, PRE, from seasonally inundated ground near the Chobe River, Botswana. Such a specimen was described as *H. seineri*. Map 84.

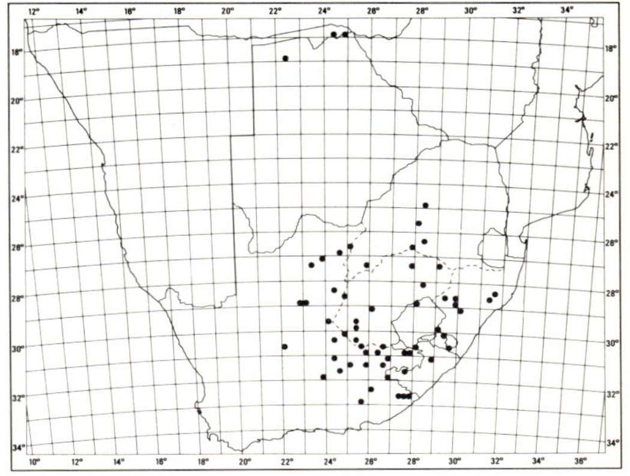

MAP 84.— **Helichrysum lineare**

Vouchers: *Acocks 9474* (PRE); *Dieterlen 522* (SAM; PRE); *Galpin 6537* (PRE); *Hilliard 5192* (E; K; NU; PRE; S); *Leistner 892* (PRE).

61. **Helichrysum caespititium** *(DC.) Harv.* in F.C. 3: 217 (1865); Moeser in Bot. Jb. 44: 299 (1910); Hilliard, Compositae in Natal 209 (1977). Type: Tambukiland, Klipplaat River, Shiloh, *Ecklon 1503* (not *Drège,* as quoted by DC.) (G-DC, holo.!; E, iso.!).

H. lineare DC. var. *caespititium* DC., Prodr. 6: 172 (1838).

H. metalasioides DC., Prodr. 6: 171 (1838) p.p. quoad *Burchell 2272* (G-DC!).

Prostrate perennial mat-forming herb, profusely branched, densely tufted, branchlets c. 10 mm tall, closely leafy. *Leaves* more or less patent, up to c. 5−10 × 0,5 mm, linear, subacute or obtuse, base broad, clasping, margins revolute, both surfaces enveloped in silvery 'tissue-paper-like' indumentum breaking down to wool, rarely loosely woolly, dotted with orange glands. *Heads* homogamous, cylindric-turbinate, c. 5 × 2 mm, solitary to several clustered at the branchlet tips. *Involucral bracts* in c. 6 series, outermost short, pellucid, webbed together and to the surrounding leaves with wool, persistent, inner series subequal, tips opaque white or sometimes pale pink, slightly exceeding the flowers, minutely radiating, soon deciduous. *Receptacle* with flattened tubercles. *Flowers* 11−24, yellow, often tipped pink. *Achenes* 0,75 mm, elliptic, with myxogenic duplex hairs. *Pappus* bristles many, equalling corolla, scabrid, bases with minute patent cilia not cohering.

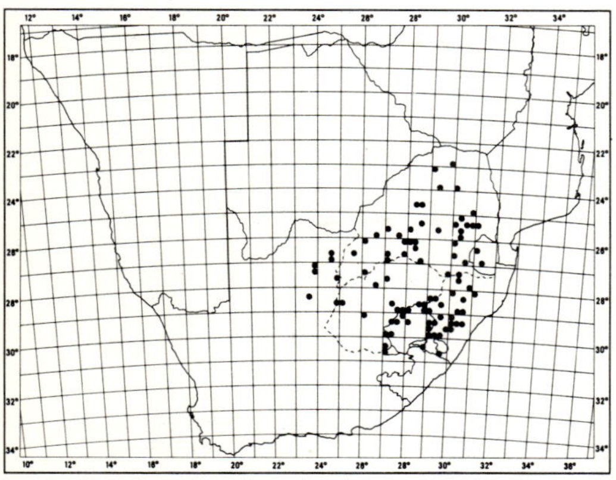

MAP 85.—— **Helichrysum caespititium**

From about Kuruman and Griqualand West in the Northern Cape to the Orange Free State, Lesotho, NE. Cape, E. Griqualand and Natal, Swaziland and the Transvaal (Highveld, Drakensberg, Waterberg, Soutpansberg) between c. 650 and 1 900 m above sea level. Also in the eastern highlands of Zimbabwe. Forms big mats on bare or sparsely grassed areas, often on disturbed sites; flowering between August and December. Closely allied to *H. asperum* (no. 59) but distinguished by its different habit, minutely radiating involucral bracts, and often more flowers in the head.

Distinguished from *H. lineare* (above) by its caespitose habit, narrower leaves, larger heads with usually more flowers but fewer heads clustered together, and minutely radiating bracts. Sometimes confused with *H. paronychioides* (no. 98). Map 85.

Vouchers: *Acocks* 10550 (PRE); *Dieterlen* 139 (PRE; SAM); *Galpin* 9460 (PRE); *Wilms* 701 (PRE).

62. **Helichrysum arenicola** M. D. Henderson in Mitt. bot. StSamml., Münch. 5: 115 (1963); Merxm., F.S.W.A. 139: 92 (1967). Type: Kimberley distr., Zandbult, 20 miles NE. Douglas, *Leistner* 1551 (PRE, holo.!).

H. rangei [Moeser ex] Dinter in Fedde, Repert. 18: 249 (1922), nomen nudum. S.W.A./Namibia, Aminuis, *Range* 779 (SAM!).

Low-growing annual or short-lived (?) perennial with a long woody taproot, branches many from the crown, mostly 50−200 mm long, prostrate or suberect, slender, subsimple or much branched in upper half, thinly white-woolly, glabrescent, leafy. *Leaves* 4−13 × 1 (−2) mm, linear-oblong, narrowed in upper two thirds, obtuse, margins revolute, lower third broadened, half-clasping, margins flat, both surfaces glandular, thinly woolly-cobwebby, glabrescent. *Heads* homogamous, campanulate, c. 3 × 2 mm, several in tight clusters up to 10 mm diam, at the branch tips closely surrounded by reduced leaves and webbed to them with wool, old clusters overtopped by several younger flowering branchlets. *Involucral bracts* in 3 series, subequal, equalling flowers, not radiating, pellucid, yellow or straw-coloured, glossy, tips obtuse, erose, soon caducous. *Receptacle* with flattened tubercles. *Flowers* 16−28, yellow. *Achenes* 0,75 mm long, elliptic, glabrous or with myxogenic duplex hairs. *Pappus* bristles many, equalling corolla, scabrid, bases with patent cilia not cohering. Fig. 25: 4.

Known from Gobabis, Sandfontein, Aminuis and Rehoboth districts in S.W.A./Namibia, Botswana, and the desert and semi-desert areas of the northern Cape and south-western Orange Free State. Grows in red Kalahari sands, on the slopes or in the depressions of dunes. Probably flowers in any month, but most records are in spring (August to November) with others in December, February, March and May. Map 86.

Vouchers: *Acocks* 18137 (PRE); *Codd* 5844 (PRE; WIND); *Leistner* 2904 (PRE); *Wilman* 1596 (SAM).

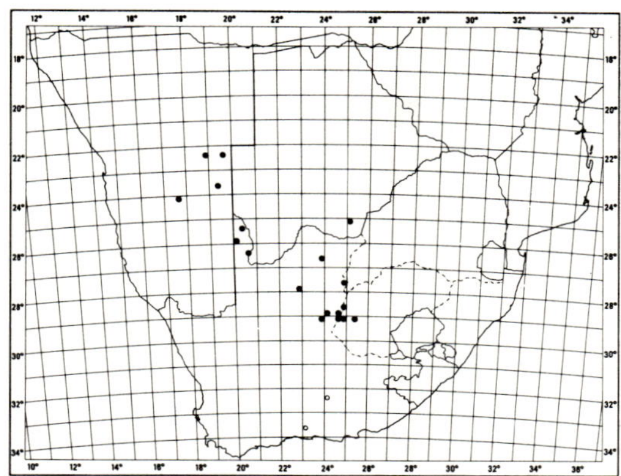

MAP 86.— ● **Helichrysum arenicola**
            ○ **Helichrysum fourcadei**

63. Helichrysum fourcadei *Hilliard, sp. nov.* H. exciso *(Thunb.) Less. affinis, sed capitulis majoribus c. 6 × 4 mm (nec 4 × 2,5 mm), plus minusve solitariis (nec aggregatis), bracteis involucralibus rubris et brunneis (nec flavis) distinguitur.*

Frutex staturae ignotae, fortasse nana; ramuli rigidi, appresse argenteo-griseo-lanati, foliati. Folia 3 —9 ×1 —3 mm, spatulata, apice obtusa (summa parva subacuta), mucronata, recurva, basi multo angustata, brevissime decurrentia, ad caulem lana contexta, utrinque indumento argenteo-griseo papyraceo induta. Capitula homogama, turbinato-cylindrica, c. 6 × 4 mm, solitaria vel 2—3 apicibus ramulorum laxe disposita. Bracteae involucrales c. 8-seriatae, gradatae, imbricatae, dorso lanatae, supra stereomate rubrescentes, apicibus rubro-brunneae, obtusae vel subacutae, squarrosae. Receptaculum plus minusve laeve. Flores c. 22—24. Achenia non visa; ovaria pilis duplicibus induta. Pappi setae multae, scabridae, apicibus barbellatae, basibus probabiliter ciliis patentibus cohaerentes (pappo immaturo).

Type: Cape, Uniondale div., 19 miles from Uniondale on Willowmore road, Karoo S. of Olifants River, 29 xi 1942, *Fourcade 5848* (BOL, holo.!; STE, iso.!).

Cape—[3224 AC] Aberdeen distr., SW. of Aberdeen, 3 xii 1950, *Maguire 737* (NBG).

Twiggy shrub, height unknown, but probably dwarf, branchlets rigid, closely appressed silvery-grey woolly, leafy. *Leaves* 3—9 × 1—3 mm, spathulate, apex obtuse, uppermost small leaves subacute, mucronate, recurved, base much narrowed, very shortly decurrent, webbed to the stems, both surfaces silvery-grey felted with a papery surface. *Heads* homogamous, turbinate-cylindric, c. 6 × 4 mm, solitary or 2—3 loosely arranged at the branchlet tips. *Involucral bracts* in c. 8 series, graded, imbricate, backs woolly, reddish above the stereome, tips reddish brown, obtuse or subacute, squarrose. *Receptacle* more or less smooth. *Flowers* c. 20—24. *Achenes* not seen, ovaries with myxogenic duplex hairs. *Pappus* bristles many, scabrid, tips barbellate, bases probably cohering strongly by patent cilia. Fig. 26:2.

Known from the type collection and one other, from near Aberdeen; flowering in December. The small silvery spathulate leaves and small more or less solitary heads with reddish brown squarrose involucral bracts combine to make this a very distinctive species. Map 86.

64. Helichrysum excisum *(Thunb.) Less.,* Syn. Comp. 282 (1832); DC., Prodr. 6: 192 (1838); Harv. in F.C. 3: 237 (1865); Moeser in Bot. Jb. 44: 291 (1910). Type: Cape of Good Hope, *Thunberg* (sheet 19153, UPS, holo.!).

Gnaphalium excisum Thunb., Prodr. 151 (1800), Fl. Cap. 655 (1823).

Densely twiggy dwarf shrub up to 450 mm tall, very aromatic, old branches bare, rough with persistent leaf bases, twigs short, stiff, thinly grey-felted, closely leafy. *Leaves* suberect to spreading, 5—8 × 2—3 mm, thickish, spathulate, much narrowed to the base, very shortly decurrent, apex mucronate, recurved, both surfaces grey-felted. *Heads* homogamous, turbinate-cylindric, c. 4 × 2,5 mm, few in terminal corymbose clusters, often further aggregated. *Involucral bracts* in 6 series, graded, imbricate, inner slightly exceeding the flowers, lanceolate, acute, tips squarrose, some woolly hairs dorsally on shafts, outer golden sometimes tinged golden-brown, inner bright canary-yellow. *Receptacle* smooth or very shortly toothed. *Flowers* 7—11. *Achenes* not seen, ovaries with myxogenic hairs. *Pappus* bristles many, scabrid, tips barbellate, bases cohering strongly by patent cilia. Fig. 26: 3.

Recorded from Montagu, Bredasdorp, Ladismith, Riversdale, Oudtshoorn and Willowmore degree squares, on hills and mountain slopes; flowering between November and February. Easily recognized by its small grey spathulate leaves and small heads with bright yellow squarrose bracts. Map 87.

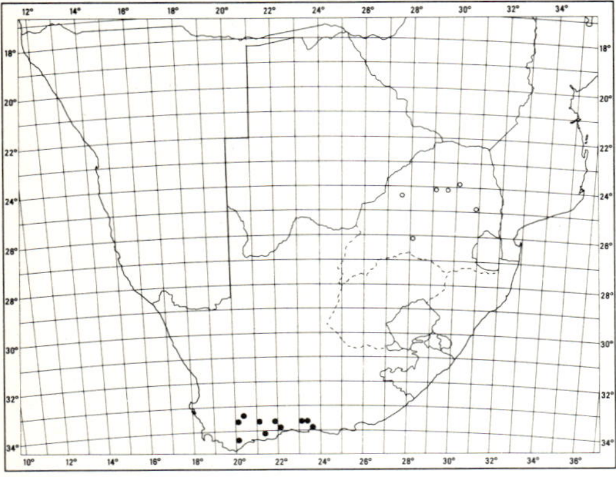

MAP 87.— ● **Helichrysum excisum**
        ○ **Helichrysum uninervium**

Vouchers: *Acocks* 23102 (PRE); *Bolus* 1053 (BM; BOL; SAM); *Compton* 10544 (NBG); *Rogers* 17893 (PRE; SAM); *Stokoe* in SAM 54699 (SAM).

65. **Helichrysum uninervium** *Burtt Davy* in Jl S. Afr. Bot. 1: 110 (1935). Type: Transvaal, Lydenburg distr., Pilgrim's Rest, July 1919, *Rogers* 18000a (K, holo.!; BM; PRE, iso.!).

Virgate subshrub, stems 300—1 000 mm tall, several from a stout woody caudex, simple or corymbosely branched above, thinly grey-felted, closely leafy but old parts often nude. *Leaves* 5—20 × 3—8 mm, smallest near the heads, elliptic, apex acute or subacute, black-mucronate, base cuneate into a short (1—2 mm) petiole, margins flat, both surfaces grey-felted with a yellowish tinge at least when dry. *Heads* homogamous, turbinate-campanulate, c. 6 × 3 mm, many on short scaly peduncles in small or large corymbose-panicles. *Involucral bracts* in c. 6 series, graded, imbricate, outer pale brown or reddish, cobwebby woolly, inner about equalling or slightly exceeding flowers, pale lemon-yellow, acute to very acute, erect or somewhat squarrose. *Recep-*

tacle smooth or shortly honeycombed. *Flowers* 6—11. *Achenes* 1 mm long, with highly myxogenic duplex hairs. *Pappus* bristles many, equalling corolla, barbellate above, bases cohering strongly by patent cilia, some light fusion in bundles as well.

A Transvaal endemic, recorded from the mountains between Potgietersrus and the Wolkberg, Pilgrim's Rest, the Waterberg and the Witwatersrand. Favours rocky sites. Flowering specimens have been collected in July, October and December, but July is probably the main month, and this would perhaps account for paucity of records. Map 87.

H. uninervium, in common with a great many other species, has triplinerved leaves, but frequently only the midvein is visible, which clearly suggested the specific epithet. A very distinctive species, but clearly allied to *H. excisum* (above).

Vouchers: *Codd* 10410 (PRE); *Dahlstrand* 1857 (PRE).

66. **Helichrysum lucilioides** *Less.*, Syn. Comp. 290 (1832); DC., Prodr. 6: 191 (1838); Harv. in F.C. 3: 236 (1865); Moeser in Bot. Jb. 44: 297 (1910); Merxm., F.S.W.A. 139: 95 (1967). Type: Cape of Good Hope, *Thunberg* (sheet 19261, UPS, holo.!).

Gnaphalium staehelinoides Thunb., Prodr. 150 (1800), Fl. Cap. 652 (1823), non *Helichrysum staehelinoides* Less. (1832). Type as above.

Leontonyx ramosissimum O. Hoffm. in Kuntze, Rev. Gen. 3,2: 162 (1898). Type: Cape, Cradock, 12 ii 1894, *O. Kuntze* (K!; fragment PRE, mounted with *Smith* 5264!).

Dwarf, hard, gnarled, intricately branched shrub 500—600 mm high, or more lax with spreading branches, main stem 20—30 mm diam., twigs thinly tomentose, closely leafy. *Leaves* mostly 3—12 × 2—5 mm, thick, obovate, obtuse, tip recurved-mucronate, base narrowed, petiole-like in larger leaves, margins somewhat crisped-undulate, both surfaces thinly greyish-white woolly (but see note below). *Heads* homogamous, cylindric, 5—7 × 2—4 mm, 2 to several clustered at the branchlet tips, surrounded by leaves. *Involucral bracts* in 4 series, soon caducous, graded, imbricate, outer lightly webbed to 1 or 2 upper leaves, inner about equalling flowers, all subpellucid, pale yellow or straw-coloured, sometimes reddish below, glossy, oblong, obtuse, somewhat concave, not radiating, soon caducous. *Receptacle* scarcely honeycombed. *Flowers* 9—18. *Achenes* not seen, ovaries with myxogenic duplex hairs. *Pappus* bristles many, in more than one series, equalling corolla, scabrid, bases cohering strongly by patent cilia. Fig. 26: 1.

Widely distributed in the dry areas from the southern part of S.W.A./Namibia and the Richtersveld in northern Namaqualand east through Griqualand West to Kimberley district and the southern Orange Free State as far east as Aliwal North degree square and south to Laingsburg district in the west, Willowmore and Cradock in the east.

Grows on rocky or stony hills or karroid plains; flowering between September and May, but mainly in October. The bushes are often grazed and may send out long decumbent new shoots that are possibly a means of vegetative propagation. Map 88.

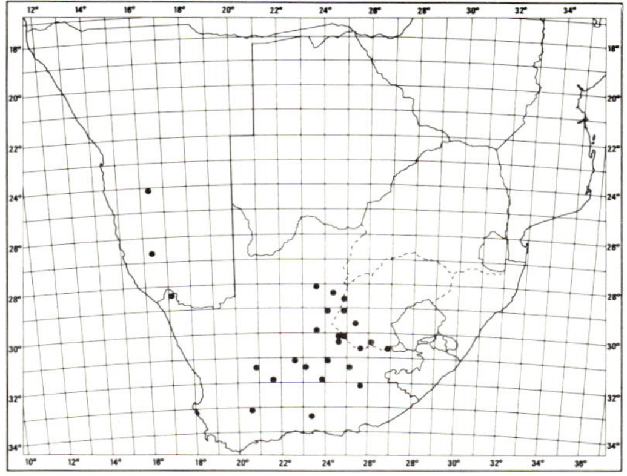

MAP 88.— **Helichrysum lucilioides**

Vouchers: *Burchell* 1690 (K, PRE); *Compton* 6739 (NBG); *Dinter* 4133 (BOL; M; PRE; SAM); *Leistner* 1522 (M; PRE); *Tyson* 367 (BOL; SAM).

Acocks 15147 (Fraserburg distr., Qaggaskop Hills, rocky ridge on top of koppie, Arid Karoo, frequent, 1 — 2 ft, dense yellowish shrubs, 21 x 1948) seems to be a variant of *H. lucilioides* with leaves densely glandular as well as lightly woolly.

67. **Helichrysum hamulosum** *[E. Mey. ex] DC.*, Prodr. 6: 192 (1838); Harv. in F.C. 3: 249 (1865); Moeser in Bot. Jb. 44: 251 (1910). Type: South Africa, *Drège* 932 (G-DC, holo.!).

Well-branched ericoid shrublet up to 600 mm tall, old branches nude, rough with old leaf bases, branchlets white-tomentose,

closely leafy. *Leaves* suberect, imbricate, 6—25 × 1 mm, linear, apex acute, mucronate, hooked, base broad, sessile, margins strongly revolute, upper surface glabrous or with a few woolly hairs, lower white-tomentose. *Heads* homogamous, cylindric, c. 6—7 × 2 mm, many in dense terminal corymbose panicles. *Involucral bracts* in 5—6 series, graded, imbricate, a little shorter than the flowers, ovate-lanceolate, acute or subacute, straw-coloured, glossy, erect. *Receptacle* with fimbrils up to half as long as ovaries. *Flowers* 5—14. *Achenes* 1 mm long, with myxogenic duplex hairs. *Pappus* bristles many, nearly equalling corolla, scabrid, bases nude, not cohering. Fig. 26: 5.

Widespread in the dry interior of the Cape, from Kamiesberg, Calvinia, Wuppertal and Worcester degree squares east through the Great and Little Karoos to Hanover, Graaff-Reinet and Oudtshoorn. A constituent of shrub communities; flowering from December to March. Map 89.

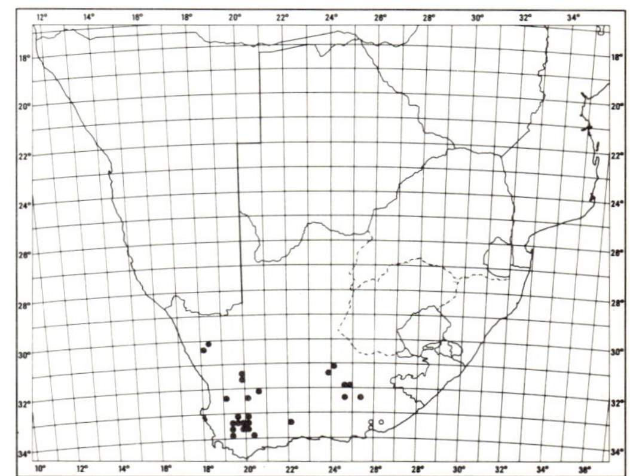

MAP 89.— ● **Helichrysum hamulosum**
      ○ **Helichrysum refractum**

Vouchers: *Acocks* 18619 (PRE); *Bolus* 1052 (BM; BOL; NBG; NU); *Compton* 12907 (NBG); *Hilliard & Burtt* 10718 (E; K; MO; NU; PRE; S); *Schlechter* 9908 (BM; BOL; E).

68. **Helichrysum refractum** *Hilliard, sp. nov.* H. praecincto *Klatt affinis, sed foliis lineari-lanceolatis marginibus valde revolutis (nec spatulato-oblongis marginibus vix revo-*

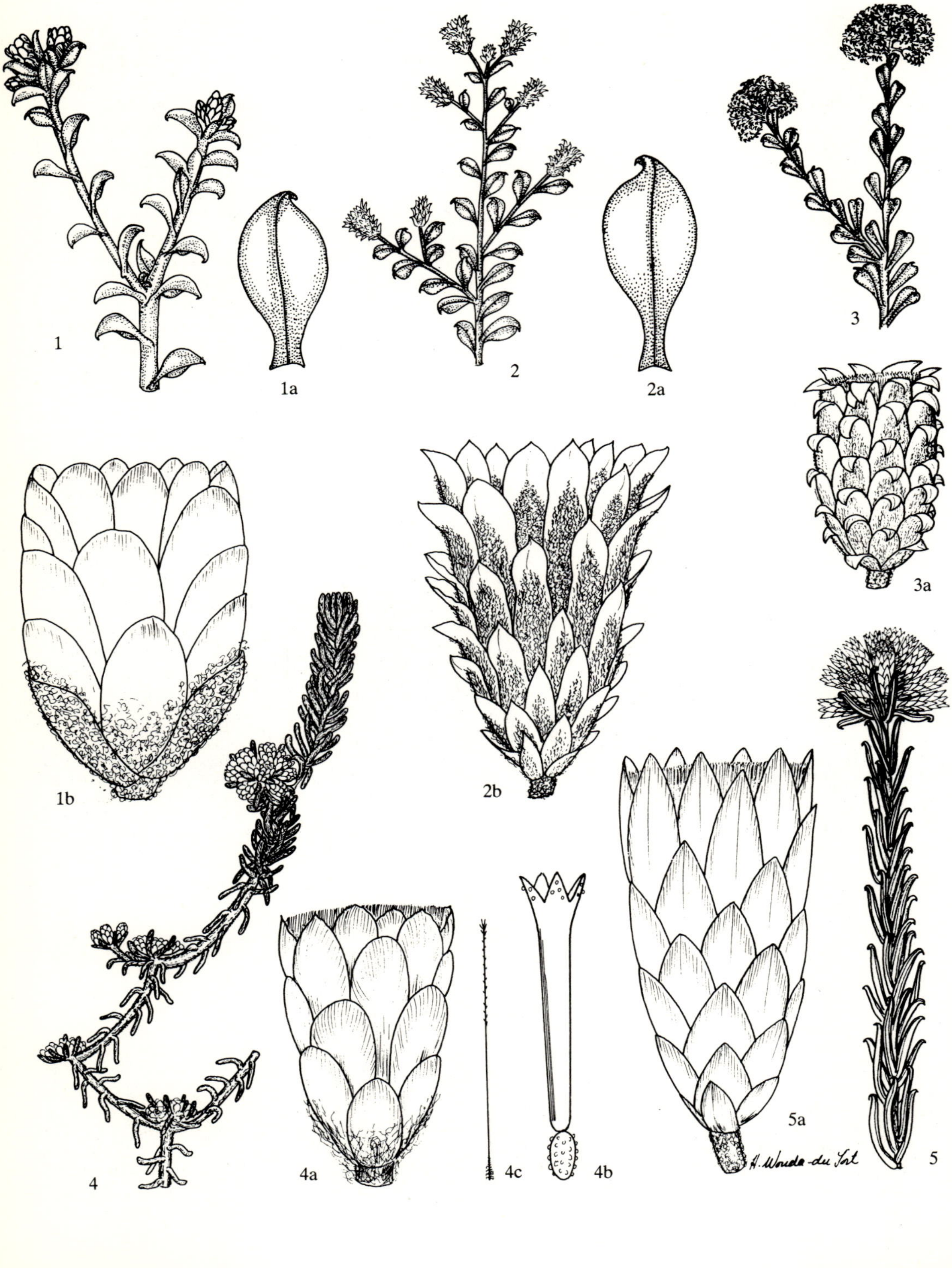

1
1a
1b
2
2a
2b
3
3a
4
4a
4b
4c
5
5a
A. Wouda-du Toit

lutis), capitulis 15 − 23-floris (nec 11 − 17-floris), acheniis pilosis (nec glabris) differt.

Frutex multiramosus, vel effusus vel densus et rotundatus, vel tegetiformis, caulibus primariis 3 −5 mm diam., ramulis ad 450 mm longis, partibus vetustioribus nudis, ramulis juvenilibus griseo-lanato-pannosis, omnino dense foliatis. Folia *patentia demum reflexa, 4 −10 × 1 mm, lineari-lanceolata, apice obtuso, basi paulo dilatata semi-amplectente, marginibus valde revolutis, supra tenuiter lanata, pagina inferiore vix visibili cano-lanato-pannosa.* Capitula *homogama, campanulata, c. 4 × 3 mm, c. 10 −15 in glomerulos terminales densos 10 mm diametro ramulis juvenilibus mox superantes disposita.* Bracteae involucrales *c. 5-seriatae, mox caducae, gradatae, imbricatae, extimae redactae, acutae, inter se et ad capitula proxima lana contextae, intimae flores subaequantes, non radiantes, glabrae, apicibus obtusissimis erosis paulo concavis translucidis pallide aureo-brunneis.* Receptaculum *breviter favosum.* Flores *c. 15 −23.* Achenia *non visa; ovaria pilis duplicibus myxogenis praedita.* Pappi setae *permultae, scabridae, basibus ciliis patentibus cohaerentibus.*

Type: Cape, Alexandria distr. (3325 BD), Olifantskop Pass, N. side, c. 2 400 ft, False Fynbos, rare by roadside, dense hemispherical bush, 8 ix 1960, *Acocks* 21419 (PRE, holo.!).

Cape−[3325 BD] Sandflats, inland dunes, 1 000 ft, 15 ix 1955, *Johnson* 1307 (PRE). [3326 BD] Albany distr., near Grahamstown, Atherstone, 1 800 ft, *Rogers* 3285 (Z); ibidem, March 1908, *Rogers* 181 (K; BOL 10639 is almost certainly the same collection). [3325 DA] Port Elizabeth distr., near Addo, on inland sand dunes around vlei, wet for part of year, forms carpet c. 10″ high, 7 April 1954, *Noel,* s.n., RUH 7743 (BOL).

The specific epithet refers to the leaves, which become more or less reflexed with age.

Shrubby, much-branched perennial, either sprawling or forming a dense rounded bush or mat, main stems 3−5 mm diam., branches up to 450 mm long, old parts becoming nude, young branches grey-woolly-felted, closely leafy throughout. *Leaves* spreading, eventually reflexed, 4−10 × 1 mm, linear-lanceolate, apex obtuse, base slightly broadened, half-clasping, margins strongly revolute, upper surface thinly woolly, lower surface scarcely visible, greyish-white woolly-felted. *Heads* homogamous, campanulate, c. 4 × 3 mm, c. 10−15 in dense glomerules 10 mm across at the branch tips, soon overtopped by young growth. *Involucral bracts* in c. 5 series, soon caducous, imbricate, graded, outermost reduced, acute, webbed together and to adjoining heads, innermost about equalling flowers, not radiating, glabrous, very obtuse, erose, somewhat concave, translucent, pale golden-brown. *Receptacle* shortly honeycombed. *Flowers* c. 15−23. *Achenes* not seen, ovaries with myxogenic duplex hairs. *Pappus* bristles very many, scabrid, bases cohering lightly by patent cilia. Fig. 26: 4.

Recorded only from the eastern Cape, in Albany, Port Elizabeth and Alexandria districts, in shrub communities and on inland sand dunes; flowering in September. Easily recognized by its narrow leaves with strongly revolute margins, reflexed in age, and the tight clusters of heads overtopped by new growth. Rarely collected. Map 89.

FIG. 26.−1, **Helichrysum lucilioides,** part of plant, × 1; 1a, leaf, showing recurved tip, × 2,6; 1b, head, × 6,6 (*Story* 4263). 2, **H. fourcadei,** flowering branch, × 1; 2a, leaf, showing recurved tip, × 2,6; 2b, head, × 6,6 (*Fourcade* 5848). 3, **H. excisum,** flowering branch, × 1; 3a, head, × 6,6 (*Fourcade* 4513). 4, **H. refractum,** flowering branch, × 1; 4a, head, × 8; 4b, flower, × 13; 4c, pappus bristle, × 13 (*Acocks* 21419). 5, **H. hamulosum,** flowering branch, × 1; 5a, head, × 6,6 (*Hilliard & Burtt* 10718).

Group 13

Subshrubs or perennial herbs, either erect, or prostrate or decumbent and mat-forming; *leaves* small or medium-sized, linear, oblong, oblanceolate, spathulate or obovate; *heads* homogamous or heterogamous, 3−6 × 3−6 mm, usually in congested clusters, rarely solitary or subsolitary; *involucral bracts* not radiating, straw-coloured, tawny or golden brown; *receptacle* nearly smooth to honeycombed or shortly fimbrilliferous; *flowers* 11−70, 8−38 ♀, ♀ occasionally outnumbering ☿, corolla of ☿ flowers campanulate above, of ♀ flowers narrowly cylindric with conspicuous limb; *achenes* glabrous; *pappus* bristles with scabrid or barbellate tips, bases with patent cilia, cohering or not.

Species 69−74, 2 endemic to the high Drakensberg, 2 to the mountains of the E. Transvaal and Swaziland, 1 to the E. Cape, 1 confined to the Transvaal and Natal.

1a Heads homogamous:

  2a Well-branched shrubs or shrublets:

    3a Clusters of heads closely surrounded and partly obscured by leaves and soon overtopped by new shoots arising below the clusters of heads. E. Cape coast .... 69. *H. praecinctum*

    3b Clusters of heads neither partly obscured by leaves nor overtopped by lateral shoots. Natal and Transvaal:

      4a One or two small bracts present below the clustered heads, the individual heads more or less free from one another; receptacle shortly honeycombed ....70. *H. galpinii*

      4b Several reduced leaves present below the clustered or solitary heads and webbed to the outer involucral bracts; receptacle with fimbrils equalling the ovaries............................. 71. *H. obductum*

  2b Tufted perennial herb, forming small mats .... 72. *H. truncatum*

1b Heads heterogamous:

  5a Many-stemmed mat-forming perennial herb, leaves up to 4 mm broad .... 73. *H. flanaganii*

  5b Perennial herb with one or several leaf rosettes crowded on the crown and several lateral simple or sparingly branched flowering stems; radical leaves up to 15 mm broad.......................... 74. *H. basalticum*

69. Helichrysum praecintum *Klatt* in Bull. Herb. Boissier 4: 838 (1896); Moeser in Bot. Jb. 44: 299 (1910). Type: Cape, Kei Mouth, 7 i 1895, *Schlechter* 6199 (Z, holo.!; BM (fragment); BOL; PRE, iso.!).

H. rogersii S. Moore in J. Bot., Lond. 46: 41 (1908); Moeser in Bot. Jb. 44: 287 (1910). Type: Cape, Port Alfred, *Rogers* 944 (BM, holo.!).

Dwarf shrublet branching from the base, taproot woody, deeply penetrative, stems up to 200 mm long, nude at the base and rooting there, laxly branched above, thinly white-felted, leafy throughout. *Leaves* mostly 5−20 × 1,5−5 mm, diminishing upwards, spathulate-oblong to oblong, apex rounded, base narrowed, half-clasping, both surfaces thickly white woolly-felted. *Heads* homogamous, cylindric, c. 5× 3 mm, several clustered in glomerules 10 − 15 mm across at the branch tips, surrounded by leaves. *Involucre* surrounded by 2 − 3 reduced leaves webbed together with wool, bracts in 3 − 4 series, soon caducous, subequal, equalling flowers, loosely imbricate, not radiating, pale straw-coloured to light golden-brown, obtuse, apiculate, translucent, tips opaque or subopaque. *Receptacle* nearly smooth. *Flowers* 11−17. *Achenes* 0,75 mm long, glabrous. *Pappus* bristles many, delicate, scabrid, equalling corolla, bases cohering lightly by patent cilia. Fig. 27: 1.

Recorded along the E. Cape coast from Port Alfred to Kei Mouth, growing on bare, well compacted and often stony sand dunes and slopes on the foreshore and flowering between December and March. Map 90.

Confused in herbaria with *Vellereophyton vellereum* (p. 7,2: 36), which also grows along the coast, but in damp dune slacks or in salt marshes at the river mouths, and is easily distinguished by its heterogamous heads with very different white-tipped involucral bracts.

Vouchers: *Batten* 324 (E; K; M; MO; NU; S); *Flanagan* 335 (PRE; SAM); *Hilliard & Burtt* 11108 (E; K; NU; S); *Tyson* s.n., Kowie (NU; PRE; SAM).

70. Helichrysum galpinii *N.E. Br.* in Kew Bull. 1906: 22 (1906); Compton, Fl.

Swaziland 630 (1976). Type: Transvaal, near Barberton, summit Saddleback Mtn, 5 000 ft, *Galpin* 544 (K, holo.!; BOL; PRE, iso.!).

H. galpinii var. *tenuis* Burtt Davy in Jl S. Afr. Bot. 1: 108 (1935). Type Transvaal, Lydenburg distr., Belfast, *Bolus* 11970 (K, holo.!; BM; BOL; PRE, iso.!).

Dwarf shrub, main stems up to 10 mm diam., woody, gnarled, lateral branches prostrate, spreading, intricately branched, ultimate branchlets erect to c. 30 mm, densely leafy. *Leaves* spreading, mostly 5−15 × 0,75−1 (−2) mm, linear, apex subacute, base broad, clasping, both surfaces enveloped in thick silvery grey, closely woven, slightly glossy, skin-like indumentum. *Heads* homogamous, campanulate, c. 4 × 4 mm, up to 10 in a tight globose cluster surrounded by a few linear bracts, the heads more or less free from each other, each cluster at the tip of an erect terminal remotely leafy silky-felted peduncle 10 − 50 mm tall. *Involucral bracts* in c. 5 series, graded, imbricate, about equalling flowers, tawny or straw-coloured, tips obtuse or subacute, crisped, minutely radiating. *Receptacle* shortly honeycombed. *Flowers* 15−36. *Achenes* not seen, ovaries glabrous. *Pappus* bristles many, about equalling corolla, tips barbellate, bases nude, not cohering. Fig 27: 2.

In the E. Transvaal and Swaziland from the mountains at Haenertsburg south along the escarpment and over the Steenskampsberge to the Barberton Mountains and Mbabane district in Swaziland. Forms mats spreading over rock outcrops and hanging from rocky cliffs (sandstone, granite, quartzite); flowering between September and December, principally November and December. Map 90.

Var. *tenuis* is no more than a lax specimen produced under favourable growing conditions.

Vouchers: *Codd* 8208 (PRE); *Compton* 30254 (NBG, PRE); *Galpin* 14635 (NU; PRE); *Hilliard* 2973 (NU); *Junod* 972 (G).

71. **Helichrysum obductum** *H. Bol.* in Trans. S. Afr. phil. Soc. 18: 391 (1907); Moeser in Bot. Jb. 44: 285 (1910); Hilliard, Compositae in Natal 171 (1977). Lectotype: Natal, Great Noodsberg, 2 000 − 3 000 ft, *Wood* 5292 (BOL!; E; NH, isolecto.!).

H. obductum var. *laxior* H. Bol., l.c. Type: Transvaal, Drakensberg, Devil's Kantoor, Sept. 1886, *Bolus* 7806 (BOL, holo.!; K; PRE, iso.!).

A well-branched shrub, branches either prostrate or erect to c. 600 mm, branchlets silky grey-felted, closely leafy. *Leaves* erect or spreading, imbricate, c. 5−20 × 1−2 (−3) mm, linear, acute or subacute, mucronate, slightly hooked, base half-clasping, slightly decurrent, margins revolute, both surfaces enveloped in silvery tissue-paper-like indumentum that hides dense coarse hairs below. *Heads* homogamous, campanulate, 4−6 (−7) × 3−6 (−7) mm, solitary or 2−3−c. 20 in very compact corymbose clusters, often webbed to surrounding reduced leaves, at the branchlet tips. *Involucral bracts* in 4−5 series, outermost shorter, inner subequal, loosely imbricate, slightly exceeding flowers, scarcely radiating, tips obtuse or acute, tawny, glossy. *Receptacle* with fimbrils equalling the ovaries. *Flowers* 15−44. *Achenes* not seen, ovaries, glabrous. *Pappus* bristles many, scabrid, about equalling corolla, bases cohering by patent cilia. Fig. 27: 3.

Ranges from the Wolkberg south through the highlands of the E. Transvaal to the mountains of N. Natal and across the high ground to the Noodsberg, then to Kloof, near Durban, and Ismont, near Mid Illovo, south of Durban. Grows in rocky places, sometimes prostrate on rock sheets, often erect among boulders and outcrops on rocky ridges. Flowers can be found in any month, but the main season seems to be between April and September. Map 91.

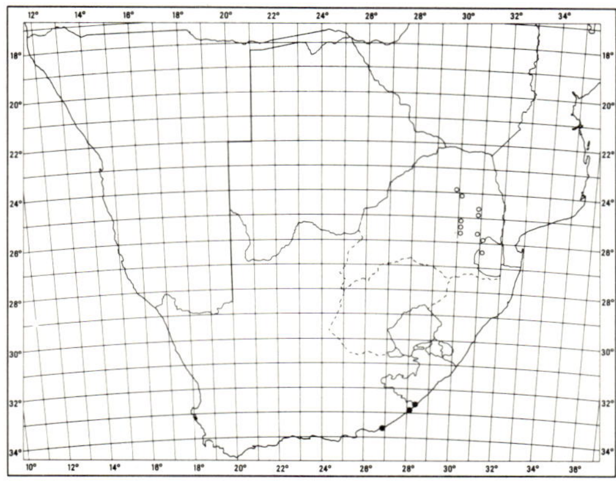

MAP 90.— ● **Helichrysum praecinctum**
○ **Helichrysum galpinii**

1
1a
1b
1c
2
2a
3
3a
3b
4
4a
H. Wouda-du Toit

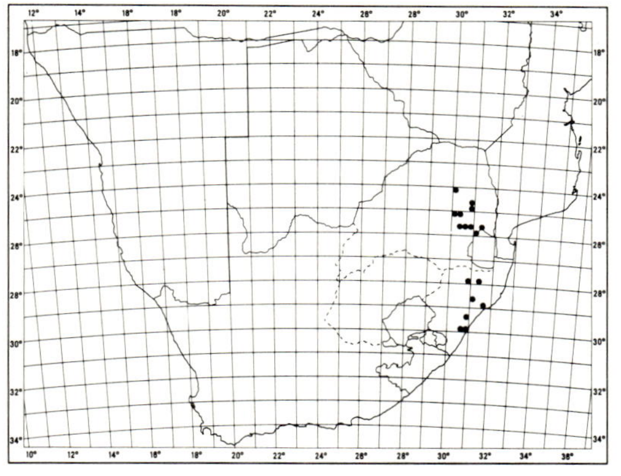

MAP 91.— **Helichrysum obductum**

Twiggy shrubs with solitary heads and short more or less appressed leaves look very different from loosely branched specimens with long lax leaves and heads in small glomerules, but many intermediate states occur, making the recognition of infraspecific categories impossible. Just such an intermediate plant with heads solitary or few in small glomerules and short but lax leaves, was described as var. *laxior,* now relegated to synonymy.

Vouchers: *Codd* 7602 (K; PRE, solitary heads, short appressed leaves); *Hilliard & Burtt* 9949 (E; K; MO; NU; PRE; S, several heads in glomerules, long, lax leaves); *Rogers* 19507 (PRE, few heads in glomerules, short lax leaves).

72. **Helichrysum truncatum** *Burtt Davy*

in Jl S. Afr. Bot. 1: 110 (1935); Compton, Fl. Swaziland 636 (1976). Type: Transvaal, Lydenburg distr., Devil's Knuckles, between Lydenburg and Spitzkop, *Wilms* 739 (K, holo.!; BM, iso.!).

Mat-forming perennial herb, stems decumbent and often rooting at the base, then erect to c. 150 mm, simple or branching from the base, leafy. *Leaves* rosetted initially, oblanceolate, uppermost lanceolate, up to 50 × 5 (−7) mm, becoming shorter and narrower upwards and passing into bracts with brown scarious tips, apex subacute to acute to acuminate, apiculate, base broad, clasping, both surfaces and the

stems enveloped in silvery silky skin-like indumentum. *Heads* homogamous, cylindric, c. 3−4 × 1,5−2 mm, many in dense, flat-topped clusters 10−20 mm across, solitary at the branch tips, surrounded by brown papery bracts and webbed together with grey silky-woolly hairs. *Involucral bracts* in 3 series, graded, equal, equalling the flowers, very loosely imbricate, silvery, pellucid, at least the outer ones golden-brown above, tips blunt, lacerate, crisped, not radiating. *Receptacle* shortly honeycombed. *Flowers* 7−14. *Achenes* not seen, ovaries glabrous. *Pappus* bristles very few, (often some flowers epappose, others with 1 or 2 bristles), shorter than corolla, subplumose above, smooth below, bases not cohering. Fig 27: 4.

Rarely collected, and recorded only from the Wolkberg, NE. Transvaal, Devil's Knuckles near Lydenburg, Long Tom Pass over Mount Anderson, MacMac, the Machadodorp area, and near Forbes Reef and the Palwane Hills in western Swaziland. Grows in marshy grassland or sedge mats over rock flushes, flowering between January and April. Map 92.

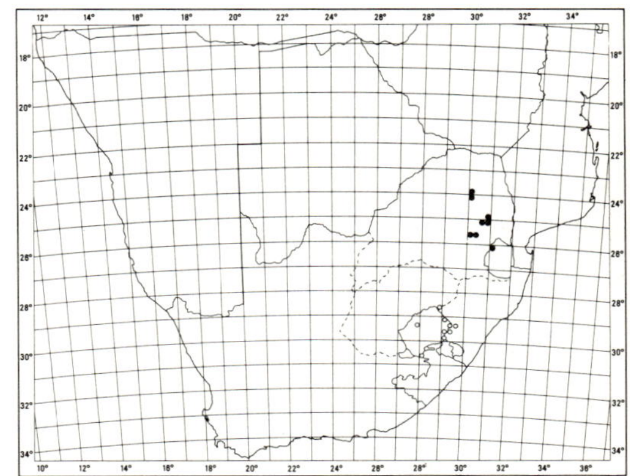

MAP 92.— ● **Helichrysum truncatum**
○ **Helichrysum flanaganii**

Vouchers: *Compton* 25901 (NBG; PRE); *Culverwell* 0775 (E; NU); *Hilliard & Burtt* 6015 (E; K; M; NH; NU; PRE; S; Z); *Muller & Scheepers* 160 (PRE).

FIG. 27.−1, **Helichrysum praecinctum,** part of plant, × 1; 1a, head, × 8; 1b, flower, × 6,6; 1c, pappus bristle, × 6,6 *(Batten 324).* 2, **H. galpinii,** flowering branch, × 1; 2a, head, × 8 *(Hilliard 2946).* 3, **H. obductum,** part of lax, loosely branched plant with heads in glomerules, × 1 *(Hilliard & Burtt 9949);* 3a, part of plant with heads mostly solitary, × 1 *(Nixon s.n.);* 3b, head, × 10 *(Hilliard & Burtt 9949).* 4, **H. truncatum,** part of plant, × 1; 4a, head, × 8 *(Hilliard & Burtt 6015).*

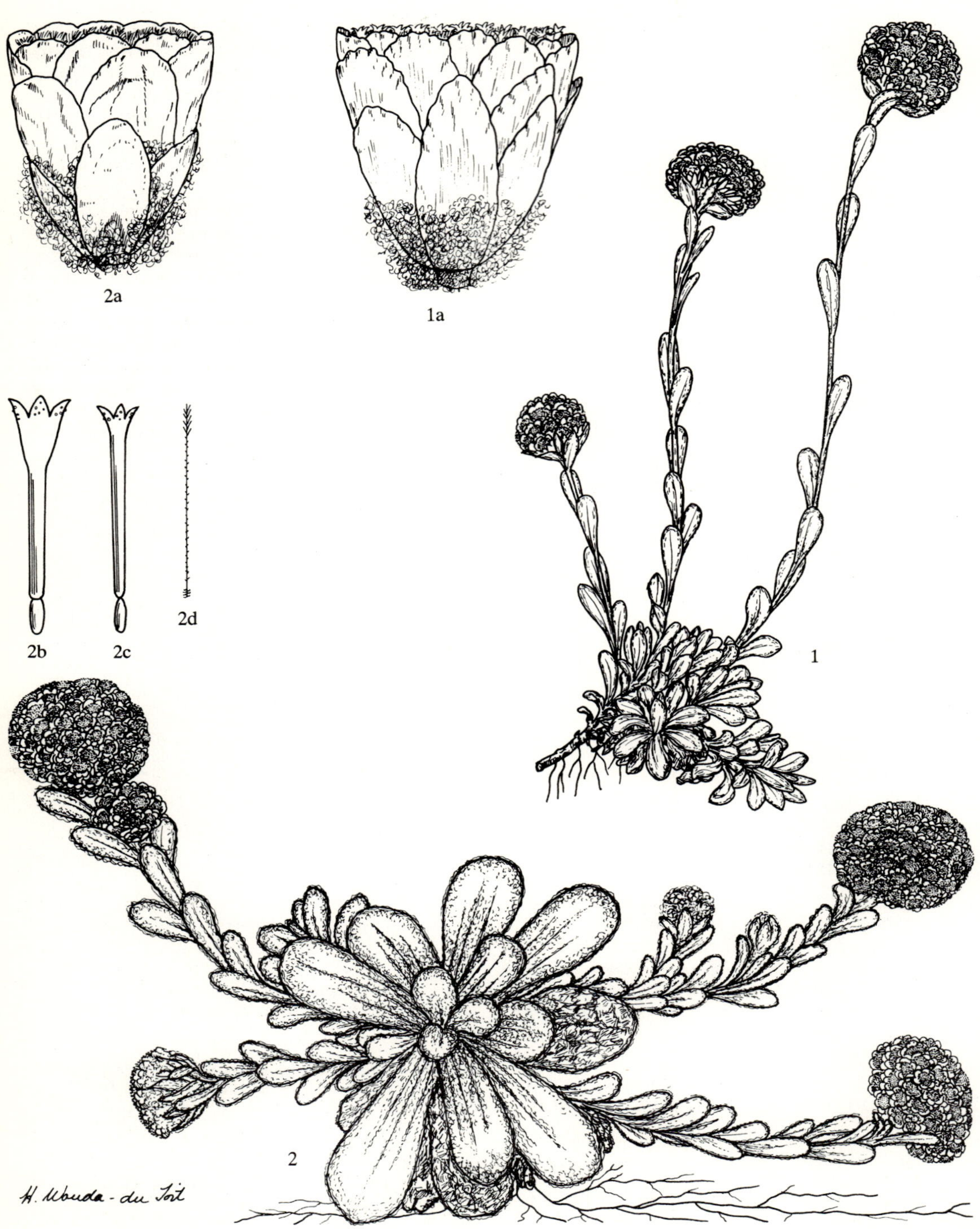

2a
1a
2b
2c
2d
1
2
H. Mbuda-du Toit

73. **Helichrysum flanaganii** *H. Bol.* in Trans. S. Afr. phil. Soc. 18: 385 (1907); Hilliard, Compositae in Natal 163 (1977). Lectotype: Orange Free State, summit Mont aux Sources, *Flanagan* 1964 (BOL!; K, isolecto.!)

Profusely branched mat-forming perennial herb, main branches prostrate, rooting, soon nude, ultimate branchlets elongating at flowering, ascending, c. 40—200 mm long, greyish-white woolly, closely leafy. *Leaves* up to 8 (—20) × 4 mm, obovate or oblong-obovate, narrowed to the base and half-clasping, apex obtuse, apiculate, both surfaces glandular-punctate and closely greyish-white woolly, or without wool and with stalked glands. *Heads* heterogamous, sub-cylindric, c. 3 × 3 mm, many felted together at the base in a congested, rounded, terminal corymbose cluster up to 18 mm across. *Involucral bracts* in c. 4 series, graded, imbricate, inner about equalling flowers, not radiating, tawny-yellow, tipped golden-brown. *Receptacle* smooth or shortly honeycombed. *Flowers* c. 35—55, 8—15 ♀, 22—40 ☿, yellow. *Achenes* 0,75 mm, glabrous. *Pappus* bristles many, delicate, tips barbellate, bases cohering lightly by patent cilia. Fig. 28:1.

On the high mountains of Lesotho and along the Drakensberg from Mont aux Sources (Bergville district, Natal) to Naudé's Nek, north of Maclear in the E. Cape. Forms thick mats over damp rock sheets, on bare earth or in short damp turf, between c. 1 650 and 3 200 m. Flowers from October to December. The typical form has grey-woolly leaves, but green-leaved plants are not uncommon, and the two may grow together. Map 92.

Vouchers: *Hilliard* 5391 (E; K; MO; NU; PRE; S); *Hilliard & Burtt* 7085 (E; K; MO; NU; S); *Ruch* 2435 (PRE); *Wright* 329 (E; K; NU; S).

74. **Helichrysum basalticum** *Hilliard* in Notes R. bot. Gdn Edinb. 32: 344 (1973), Compositae in Natal 164 (1977). Type: Lesotho, Phutha, near Mokhotlong, c. 10 000 ft. alt., *Compton* 21570 (NBG, holo.!).

Gnaphalium alticolum Compton in Jl S. Afr. Bot. 19: 113 (1953), non *Helichrysum alticolum* H. Bol., 1907. Type as above.

Perennial herb with a cluster of very short branches at the crown, each producing a rosette of leaves, flowering stems several, lateral, simple or branched, prostrate, 40—120 (—400) mm long, grey-woolly, leafy. *Radical leaves* up to 40 × 15 mm, obovate-oblong, apex obtuse, base slightly narrowed, half-clasping, both surfaces thickly grey-woolly; *cauline leaves* similar but commonly smaller and narrowly oblong. *Heads* heterogamous, campanulate, c. 4 × 4 mm, many felted together at the base in a congested, rounded, terminal, corymbose cluster 15—20 mm across, occasionally up to 4 of these clusters in turn closely agglomerated. *Involucral bracts* in 3 series, loosely imbricate, about equalling the flowers, not radiating, translucent, tawny-yellow, mostly tipped golden brown. *Receptacle* honeycombed. *Flowers* c. 40—70, c. 27—38 ♀, 22—35 ☿, yellow. *Achene* 0,75 mm, glabrous. *Pappus* bristles many, delicate, tips barbellate, bases with patent cilia, not cohering. Fig. 28: 2.

Along the high Drakensberg, from Mont aux Sources (Bergville district, Natal) to Naudés Nek in the Cape Drakensberg north of Maclear. Forms small mats on basalt sheets and bare stony ground: flowering between February and April. Rarely collected. Map 93.

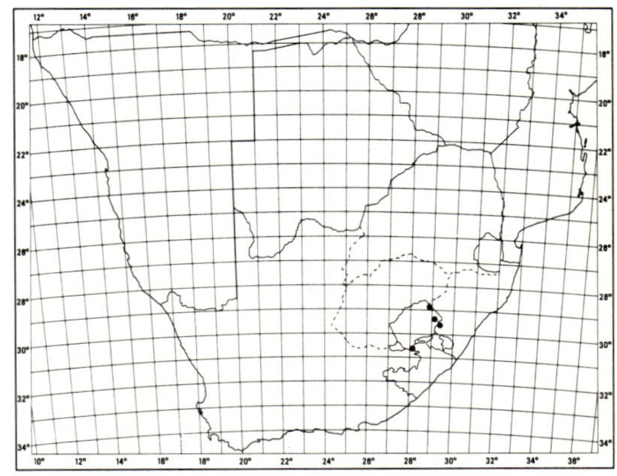

MAP 93.— **Helichrysum basalticum**

Vouchers: *Galpin* 14102 (NH; S); *Hilliard* 5340 (E; K; M; NBG; NU; PRE; S); *Hilliard & Burtt* 6725 (E; K; M; MO; NBG; NU; PRE; S); *Nordenstam* 2136 (S).

FIG. 28.—1, **Helichrysum flanaganii,** part of plant, × 1; 1a, head, × 10 (*Hilliard* 5391). 2, **H. basalticum,** whole plant, × 1; 2a, head, × 10; 2b, hermaphrodite flower, × 10; 2c, female flower, × 10; 2d, pappus bristle, × 10 (*Hilliard & Burtt* 6725).

Group 14 (includes *Leontonyx*)

Annual or perennial herbs, often more or less prostrate, forming small mats; *leaves* small, elliptic-oblong to spathulate, obovate or subrotund; *heads* homogamous or heterogamous, often in the same species, $3-6 \times 2-4$ mm, usually in congested terminal clusters, rarely solitary or subsolitary; *involucral bracts* not radiating, sometimes squarrose, whitish, creamy, yellow, orange, red, pink, purple or golden-brown, colours often combined, lamina characteristically with a median, opaque patch often drawn out into an acute to acuminate, squarrose tip, inner 2 series falling quickly, outer 2 series webbed together with wool, sometimes persistent; *receptacle* smooth; *flowers* $9-45$, $0-8$ ♀; corolla of ☿ flowers funnel-shaped, of ♀ flowers narrowly cylindric; *achenes* glabrous or hairy, often in the same species; *pappus* bristles sometimes in more than one series, scabrid, bases cohering by patent cilia.

Species 75—84, 6 species endemic to the Cape, 2 to S.W.A./Namibia and the Cape, 1 to the high Drakensberg, 1 widespread, from the E. highlands of Zimbabwe to the Cape Peninsula, frequently in dry sandy places.

(Not all specimens will key out, especially of no's 78, 79 and 80; see notes after species descriptions).

1a Heads either solitary or in glomerules, either terminal or corymbosely arranged:

  2a Branches erect or decumbent and then erect ....77. *H. spiralepis*

  2b Branches prostrate or diffuse:

    3a Involucral bracts either obtuse and then often emarginate, or subacute to acute, not recurved or very weakly so:

      4a Heads c. 5 mm long:

        5a Plants either closely branched and forming dense mats or, if more loosely branched, then plants lacking long simple leafy runners, and pappus in more than one series:

          6a Plants closely branched and forming small dense mats; found generally between 1 200 and 3 300 m above sea level:

            7a Inner involucral bracts at least 1 mm broad immediately below the opaque apical patch; on the high Lesotho plateau ....75. *H. lineatum*

            7b Inner involucral bracts less than 1 mm broad immediately below the opaque apical patch; on the mountains in the SW. and S. Cape ....79. *H. zwartbergense*

          6b Plants loosely branched; generally along the coast, often on the seashore ....78. *H. litorale*

        5b Plants loosely branched, branches sometimes forming long simple leafy runners, pappus uniseriate ....76. *H. dunense*

      4b Heads c. 4 mm long:

        8a Plants forming small dense mats ....83. *H. tysonii*

        8b Plants loosely branched:

          9a Heads solitary ....81. *H. solitarium*

          9b Heads crowded in glomerules:

            10a Heads homogamous, containing c. 9—12 flowers, ovaries glabrous ....76. *H. dunense*

            10b Heads heterogamous, containing c. 23—29 flowers, ovaries hairy ....82. *H. leontonyx*

    3b Involucral bracts very acute to acuminate, recurved or hooked:

      11a Heads c. 5—6 mm long, pappus bristles in more than one series ....80. *H. tinctum*

      11b Heads 3—4 mm long, pappus uniseriate:

        12a Plant with a loose open habit; tips of involucral bracts long-acuminate ....82. *H. leontonyx*

        12b Plant a small dense mat; tips of involucral bracts very acute ....83. *H. tysonii*

1b Heads either solitary or in glomerules, crowded towards the tips of the branchlets in a leafy, racemose, compound inflorescence ....84. *H. micropoides*

75. Helichrysum lineatum *H. Bol.* in Trans. S. Afr. phil. Soc. 18, 3: 386 (1907). Type: Cape, Barkly East distr., Drakensberg, Doodman's Krans Mtn., 9 000 ft, 8 iii 1904, *Galpin* 6669 (BOL, holo.!; K; NH, iso.!).

Profusely branched perennial (?) herb forming mats up to 200 mm across, or tufted, stems prostrate or decumbent, very slender, greyish-white felted, closely leafy. *Leaves* imbricate with woolly-backed oblong bases as long as the blade webbed to the stem, blades mostly 4−10 × 3−7 mm, elliptic-oblong to oblong-spathulate, apex rounded, both surfaces closely greyish-white woolly-felted. *Heads* homogamous, very rarely heterogamous, cylindric, c. 5 × 2 mm, many clustered at the branch tips closely surrounded by leaves. *Involucral bracts* in 4 series, soon caducous, outermost short, webbed with wool to 2−3 reduced surrounding leaves, others subequal, about equalling flowers, not radiating, all translucent, upper part light golden-brown, obtuse, extreme tip opaque or subopaque along the midline. *Receptacle* nearly smooth. *Flowers* 18−25, very rarely 1 ♀. *Achenes* spindle-shaped, c. 1 mm long, glabrous in type material, with duplex hairs in other collections. *Pappus* bristles many, about equalling corolla, delicate, scabrid, bases cohering lightly by patent cilia. Fig. 29: 1.

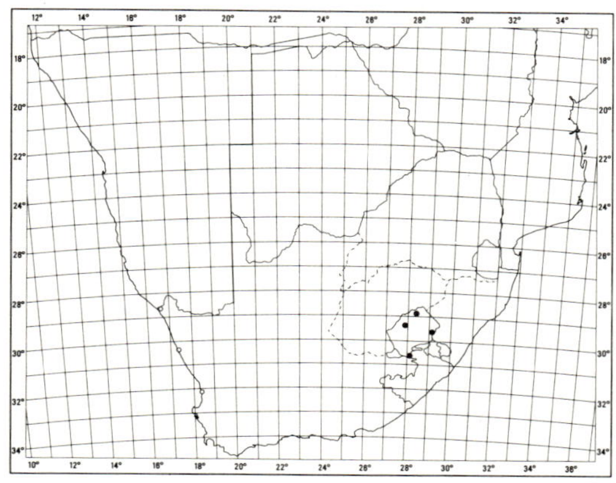

MAP 94.— ● **Helichrysum lineatum**
       ○ **Helichrysum dunense**

Endemic to the high mountains in Lesotho and the Drakensberg on the Cape-Lesotho and Natal-Lesotho borders, c. 2 700−3 200 m. Grows in short grassland, then tufted, or on bare earth, then forming small mats. Rarely collected. Map 94.

Vouchers: *Jacot Guillarmod* 2083 (PRE); *Hilliard & Burtt* 8780 (E; NU).

76. Helichrysum dunense *Hilliard, sp. nov.* H. lineato *H. Bol. affinis sed habitu diffuso, capitulis 10−12-floris (nec 18−25-floris) distinguenda.*

Herba perennis, radice palari lignosa in caudicem 10 mm diametro incrassata; caules multi, ad 600 mm longi, simplices vel laxe ramosi, plus minusve prostrati, tenuiter albo-lanati, crebre foliati. Folia plerumque 6−10 × 2,5−4,5 mm, spatulata, apice rotundato, basi lata semi-amplexicauli, utrinque dense cano-lanata. Capitula homogama cylindrico-campanulata, c. 4−5 × 3 mm, pluria ad apices ramorum aggregata, foliis arcte circumcincta. Bracteae involucrales 4-seriatae, mox caducae, extimae breves, ad folia contigua lana contextae, ceterae subaequales, quasi flores aequantes, non radiantes, translucentes, saepe purpureo-tinctae, apicibus abrupte acutis opacis dilute aureo-brunneis. Receptaculum fere laeve. Flores c. 9−12. Achenia 0,75 mm longa, glabra. Pappi setae multae, quasi corollam aequantes, scabridae, basibus ciliis patentibus obsoletis praeditis.

Type: Cape, Clanwilliam div., 1 mile S. of Steenbokfontein (S. of Lambert's Bay), coastal sand dunes, 11 xii 1963, *Nordenstam* 3412 (E, holo.!; S, iso.!).

Cape.—[3218 AB] 6½ miles S. of Lambert's Bay, coastal dunes, frequent, 16 x 1958, *Acocks* 19898 (PRE); Lambert's Bay, Nortier Experimental Farm, littoral dunes, 6 xi 1954, *Boucher* 2603 (STE). Namaqualand, Hondeklipbaai, plentiful on sandy slopes round bay, Oct. 1924, *Pillans* 18066 (BOL).

Perennial (?) herb, taproot woody, stems many from the crown, up to 600 mm long, loosely branched, more or less prostrate, thinly white or greyish-white woolly, closely leafy. *Leaves* mostly 6−10 × 2,5−4,5 mm, spathulate, apex rounded, base broad, half-clasping, both surfaces closely greyish-white woolly. *Heads* homogamous, cylindric-campanulate, c. 4−5 × 3 mm, several clustered at the branch tips closely surrounded by leaves. *Involucral bracts* in 4 series, soon caducous, outer short, webbed with wool to surrounding

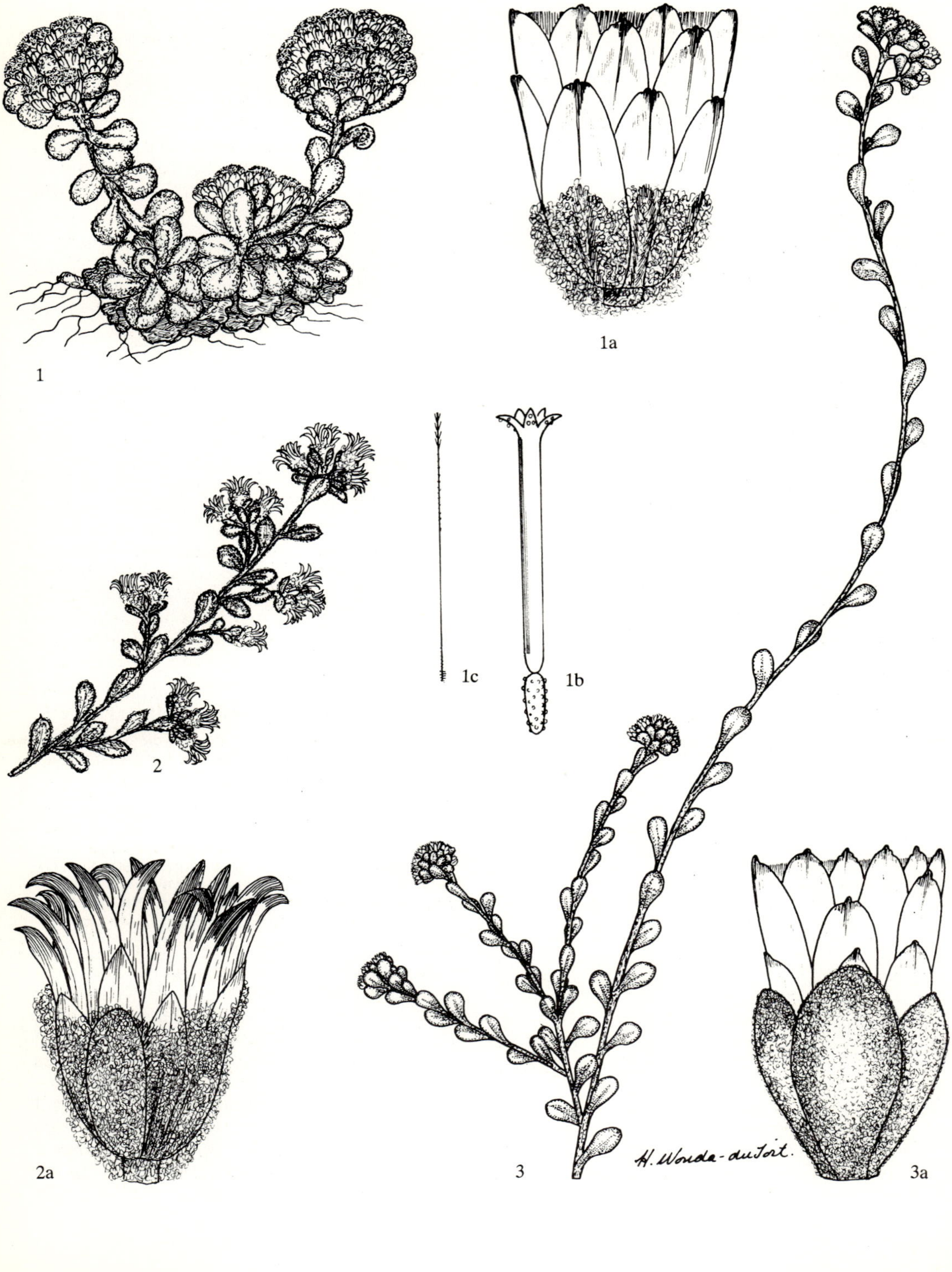

1

1a

1c 1b

2

2a

3

3a

leaves, others subequal, about equalling the flowers, not radiating, translucent, often tinged purple, tips abruptly acute, opaque, light golden-brown. *Receptacle* nearly smooth. *Flowers* c. 9—12. *Achenes* 0,75 mm long, glabrous. *Pappus* bristles many, about equalling the corolla, scabrid, bases with obsolete patent cilia. Fig. 29: 3.

Known from a few collections on the western Cape coast in the vicinity of Lambert's Bay and at Hondeklipbaai and Grootderm in Namaqualand, on sand dunes; flowering between October and December. Map 94.

Not unlike *H. litorale* (no. 78) in facies, but more loosely branched and pappus uniseriate.

77. **Helichrysum spiralepis** *Hilliard & Burtt* in Bot. J. Linn. Soc. 82: 199 (1981). Type: Herb. Van Royen (L, sheet 900, 346—99!).

Gnaphalium scabrum L., Sp. Pl. 855 (1753), non *Helichrysum scabrum* Less. *G. squarrosum* L., Sp. Pl. edn. 2: 1197 (1763). *Leontonyx tomentosus* Cass. in Dict. sci. Nat. 25: 466 (1822). *Spiralepis squarrosa* D. Don in Mem. Wern. nat. Hist. Soc. 5: 551 (1826). *Leontonyx squarrosus* DC., Prodr. 6: 167 (1838); Harv. in F.C. 3: 205 (1865); Levyns in Adamson & Salter, Fl. Cape Penins. 786 (1950); Hilliard, Compositae in Natal 255 (1977). Type as above.

Gnaphalium connatum Spreng. in Flora, Regensburg 12, Beilage 1: 7 (1829), nomen.

Leontonyx squarrosus var. *pallidus* DC., Prodr. 6: 167 (1838). Lectotype: Cape, Stellenbosch, *Ecklon* (G-DC!).

L. squarrosus var. *longifolius* DC., l.c. Lectotype: Cape, *Burchell* 4144 (G-DC!).

Eriosphaera dubia DC., l.c., non *Helichrysum dubium* Cass. Type: Cape, Lady Grey distr., Witteberg, *Drège* 5807 (G-DC, holo.!; K; S, iso.!).

Leontonyx squarrosus var. *discretus* Harv. in F.C. 3: 205 (1865). Type: Natal, *Gerrard & McKen* 272 (TCD, holo.!).

Helichrysum involucratum Klatt in Bull. Herb. Boissier 4: 461 (1896), non F. Muell. (1863). Type: [Transkei], East Griqualand, Mt Malowe, 1 675 m, *Tyson* 3095 (Z, holo.!; K; SAM, iso.!).

A tufted perennial (?) herb, all parts grey-woolly, stems annual, often numerous, up to c. 300 mm tall, erect or decumbent then erect, simple or branched from near the base into corymbose panicles, leafy. *Radical leaves* rosetted, up to c. 70 × 15 mm, mostly oblong-spathulate, sometimes linear-oblong, stem leaves smaller. *Heads* heterogamous or rarely homogamous, c. 5 × 3 mm, narrowly campanulate, congested in small terminal glomerules surrounded by leafy bracts, glomerules rarely solitary, usually in contracted or spreading corymbose panicles. *Involucral bracts* in 4 series, 2 outer shorter, webbed to subtending leaves with wool, 2 inner longer, equalling or slightly exceeding the flowers, nearly glabrous, margins hyaline, tips opaque, ranging from a small apiculus to a long acuminate squarrose point, whitish, creamy, pink or red-purple, all bracts soon caducous, the whole head disintegrating. *Receptacle* nearly smooth. *Flowers* 12—36 (0—) 1—8 ♀, 8—27 ☿. *Achenes* c. 0,75 mm long, obscurely angled, with myxogenic duplex hairs or glabrous. *Pappus* bristles copious, in several series, minutely scabrid, bases cohering lightly by patent cilia.

Ranges from Table Mountain and Constantia Berg and the mountains E. and S. of Worcester along the coastal mountains to the E. Cape, then NE. both at the coast and inland through Natal, Lesotho, the NE corner of the O.F.S. to western Swaziland and the eastern highlands of the Transvaal; also in the eastern highlands of Zimbabwe. Grows in grassland. Map 95.

In the Cape, *H. spiralepis* always has acuminate hooked involucral bracts and hairy ovaries. In Natal, this is the form that is commonest at the coast, and the only form recorded from around Ficksburg and Golden Gate in the O.F.S. and nearby Leribe in Lesotho. It flowers between September and December and is mostly fading in January. In the Cape and Natal Drakensberg and neighbouring Lesotho, the Natal Midlands and Uplands, northern Natal and northwards, plants with glabrous ovaries predominate. Bract tips may be acuminate to more or less acute. Flowering is mainly between January and March. The name *Eriosphaera dubia* was given to a dwarf form with more or less simple flowering stems, acuminate bracts and glabrous ovaries. This has been recorded from the Cape Drakensberg, nearby Witteberg and Lesotho and southernmost Natal Drakensberg, (e.g. Lesotho, Sehlabathebe, *Beverly & Hoener* 591, NU; Maclear distr., Naude's Nek, *Hilliard & Burtt* 6628, E; K; NU; S). However, it grades imperceptibly into taller, more branched plants. Plants with acuminate to more or less acute involucral bracts are found along the high Natal Drakensberg. In the Natal midlands, only plants with more or less acute bracts are known; this form was

---

FIG. 29.—1, **Helichrysum lineatum,** whole plant, × 1; 1a, head, × 6,6; 1b, flower, × 10; 1c, pappus bristle, × 10 (*Hilliard & Burtt* 12032). 2, **H. tinctum,** part of plant, × 1; 2a, head, × 6,6 (*Esterhuysen* 13076). 3, **H. dunense,** part of plant, × 1; 3a, head, × 10 (*Nordenstam* 3412).

given the illegitimate name *Helichrysum involucratum*. It has not proved possible to partition the variation satisfactorily on a geographic or any other basis, and *H. spiralepis* is therefore treated as a single polymorphic species.

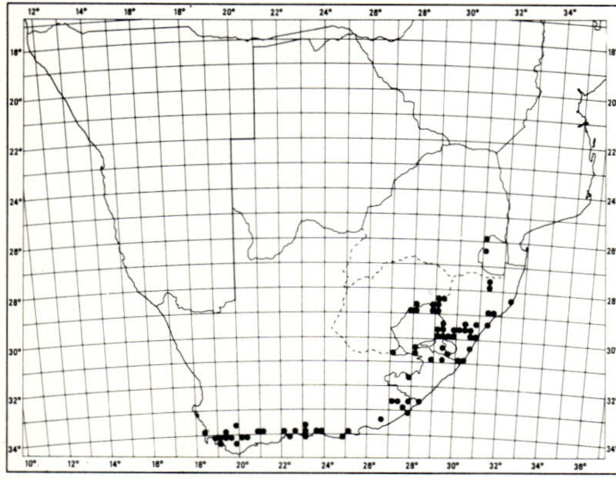

MAP 95.— **Helichrysum spiralepis**

Vouchers: *Compton* 21693 (NBG); *Lewis* 5333 (NBG); *Tyson* 2948 (NBG); (all with acuminate squarrose bracts and hairy ovaries); *Hilliard* 4909 (E; COI; K; NH; NU; S; SRGH); *Meeuse* 10076 (PRE) (with acute bracts and glabrous ovaries).

78—80. Helichrysum agg. tinctum.

The boundaries between the components of this aggregate, *H. tinctum*, *H. zwartbergense* (no. 79) and *H. litorale* (no. 78), are not always clearcut. However, the retention of three binomials seems justified when distribution patterns are considered in conjunction with particular combinations of characters. Thus *H. litorale*, characterized by obtuse involucral bracts, is common along the sea coast, and inland is nearly confined to low-lying sandy areas. *H. tinctum*, characterized by acuminate squarrose involucral bracts, occurs on higher ground (mostly between 250 and 1 350 m above sea level) but overlaps in distribution with *H. litorale* in its lower sites. There are problems at higher altitudes as well; see notes after the species descriptions. *H. zwartbergense*, which typically has obtuse involucral bracts, may be only a variant of *H. tinctum* confined to high altitudes. Field studies are needed.

78. Helichrysum litorale *H. Bol.* in Trans. S. Afr. phil. Soc. 18: 381 (1907);

Hilliard & Burtt in Bot. J. Linn. Soc. 82: 198 (1981). Lectotype: Cape, East London distr., sandhills on beach, Gooda River Mouth, alt. 10 ft, 24 xii 1905, *Galpin* 7342 (BOL!; BM; PRE, isolecto.!).

Gnaphalium spathulatum Thunb., Prodr. 151 (1800), Fl. Cap. 656 (1823), non Burm. f. (1768). *Leontonyx spathulatus* Less., Syn. Comp. 327 (1832), excl. syn. *G. spadiceum* Lam. (1788); Harv. in F.C. 3: 207 (1865); Levyns in Adamson & Salter, Fl. Cape Penins. 786 (1950). *L. spathulatus* var. *hirsutus* Harv., l.c. Type: Cape of Good Hope, Saldanha Bay, seashore, *Thunberg* (sheet 19255, UPS, holo.!).

Leontonyx angustifolius DC., Prodr. 6: 168 (1838), non *Helichrysum angustifolium* (Lam.) Pers. Lectotype: Cape, George, *Ecklon* 265 (G-DC!; S, isolecto.!).

L. candidissimus DC., l.c. non *Helichrysum candidissimum* Don. *L. spathulatus* Less. var. *candidissimus* (DC.) Harv. in F.C. 3: 207 (1865). Lectotype: Cape of Good Hope, *Ecklon* 1017 (G-DC!).

Prostrate or diffuse grey- or white-woolly herb, possibly perennial, root woody, branches several from the crown, up to 450 mm long, sometimes rooting in the lower parts, loosely branched, branchlets thinly greyish-white woolly, leafy. *Leaves* mostly 8—20 × (2—) 5—12 mm, considerably smaller immediately below the heads, spathulate to obovate, apex mostly rounded, tip or sometimes whole leaf folded and recurved, base narrowed, half-clasping, both surfaces greyish-white woolly, the wool sometimes closely woven. *Heads* homogamous or heterogamous, cylindric, c. 5 × 2—3 mm, solitary or aggregated in small glomerules c. 10 mm diam. terminating the branches and short lateral branchlets in the upper axils, surrounded by reduced leaves webbed together with wool. *Involucral bracts* in 4 series, outermost short, lanceolate, surrounded by 1 or 2 leaves webbed to them with wool, inner about equalling the flowers, subequal, loosely imbricate, soon caducous, tips subpellucid, or opaque at the extreme tips, mostly golden-brown, sometimes crimson, subacute to more or less truncate, then often emarginate, not radiating. *Receptacle* scarcely honeycombed. *Flowers* 9—26, 0—5 ♀, 9—26 ☿. *Achenes* 0,75 mm long, glabrous or with myxogenic hairs. *Pappus* bristles in 2—3 series, very many, very delicate, equalling corolla, scabrid, bases cohering lightly by patent cilia.

Common along the coast, often on the seashore dunes, from Britannia Bay north of Cape Town to Gonubie River mouth north of East London, in the SW. ranging inland as far as Tulbagh and Verkeerde Vlei, and in the E. recorded from Swartwatersberg (Albany district), and Fort Cunynghame and Dohne near Stutterheim. Flowers mainly between November and January. Map 96.

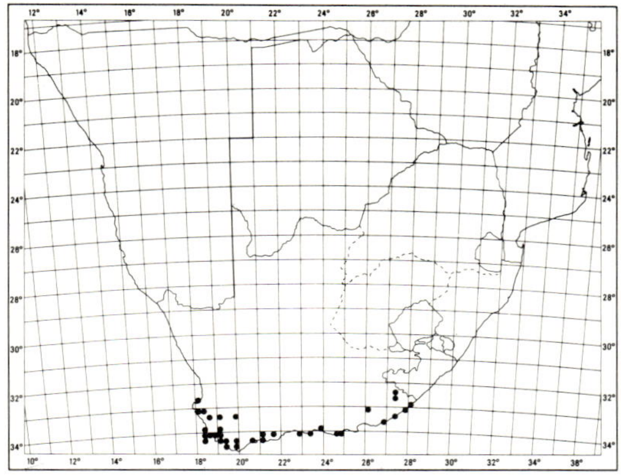

MAP 96.— **Helichrysum litorale**

Nearly all the records are from very low altitudes; Swartwatersberg (850 m), Dohne (1 500 m), Table Mountain (760 m) appear to be exceptional.

There is no doubt that *H. litorale* is very close to *H. tinctum* (no. 80 — see notes there) and some specimens are difficult to place; but the fact that plants with more or less obtuse bracts are very nearly confined to low altitudes, plants with acuminate bracts to higher altitudes, seems sufficient reason to keep them apart. Also, although both species appear to be at the peak of their flowering in November, *H. tinctum* begins flowering in September and is fading in December, while early flowering records are few in *H. litorale*, which may flower as late as February.

Vouchers: *Acocks* 23201 (PRE); *Dyer* 3360 (PRE); *Esterhuysen* 4944 (NBG); *Hilliard & Burtt* 12460 (E; NU).

79. **Helichrysum zwartbergense** *H. Bol.* in Trans. S. Afr. phil. Soc. 18: 382 (1907). Type: Cape, Prince Albert distr., rocky places on summit of Zwartberg range, c. 5 500 ft, Dec. 1904, *Bolus* 11532 (BOL, holo.!; BM; K, iso.!).

Grey-woolly herb, probably perennial, woody at the root, stems many from the crown, 40—150 mm long, prostrate, profusely branched, forming dense little mats, branchlets slender, white-woolly, closely leafy. *Leaves* 3—8 (—15) × 1,5—5 (—7) mm, spathulate, obovate or subrotund, tips rounded, base somewhat narrowed, half-clasping, both surfaces densely greyish-white woolly. *Heads* homogamous, or very rarely one flower ♀, cylindric, c. 5 × 3 mm, solitary or several clustered at the tips of the branchlets, surrounded by leaves. *Involucral bracts* in 4 series, soon caducous, outer 2 series shorter, webbed with wool to 3—4 reduced surrounding leaves, inner series subequal, about equalling flowers, not radiating, all light golden-brown, sometimes red above the stereome, semi-pellucid, extreme tip opaque, straw-coloured, obtuse and then usually emarginate, or sometimes subacute to shortly acuminate. *Receptacle* nearly smooth. *Flowers* 15—44. *Achenes* 1 mm, glabrous or hairy and then myxogenic. *Pappus* bristles very many, in more than 1 series, equalling corolla, delicate, scabrid, bases cohering by patent cilia.

On the high mountains of the SW. Cape, from the Cedarberg near Clanwilliam to the Tsitsikama Mountains, Uniondale district, and Cockscomb in the Great Winterhoek Mountains, Uitenhage district, generally between 1 200 and 2 200 m above sea level. Flowers mainly in December and January. Map 97.

H. zwartbergense is characterized by its caespitose habit and mostly obtuse involucral bracts. However, the bract tips may be acute or shortly acuminate. The obtuse tips are themselves interesting: many are bidentate, and look exactly as though they had been nipped off. It seems likely that only a small and easily varied genetic factor is responsible for the degree of development of the tips.

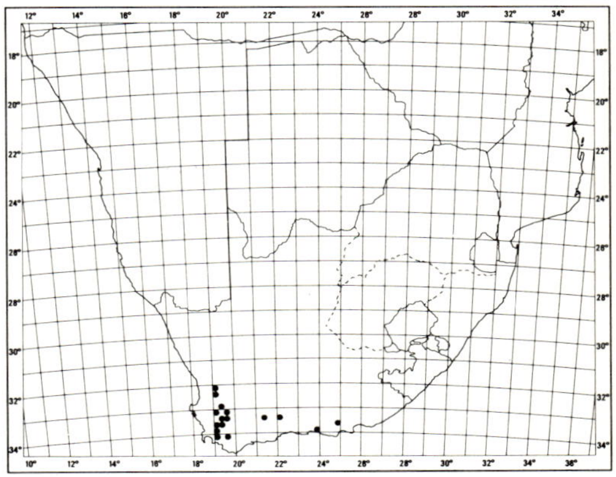

MAP 97.— **Helichrysum zwartbergense**

Vouchers: *Compton* 12852 (NBG); *Esterhuysen* 134491 (BOL; PRE); *Lewis* 1118 (SAM); *Stokoe* in SAM 51421 (PRE).

It may well be that *H. zwartbergense* is only a high altitude, more compact, form of *H. tinctum* (no. 80 — see notes there); there is certainly no clearcut distinction between them, either in growth habit or in degree of development of the bract tips. However, over most of its range, *H. zwartbergense* is easily recognized; difficulty arises in the mountains between Wellington, Ceres and Worcester. A number of collections made over the years by Miss Esterhuysen are particularly interesting and point the need for further study in the field. The following are some examples; there are others in BOL. *Esterhuysen* 12466 (BOL; NBG), Lower Wellington, Sneeukop, N. aspect, 1 200—1 370 m, bracts ranging from obtuse to acuminate and squarrose, compact habit; *Esterhuysen* 12467 (BOL; PRE), same locality, but S. aspect at 1 585 m, obtuse to acute bracts, plant open and loosely branched; *Esterhuysen* 22582 (BOL; PRE) and 9917 (BOL), Waaihoek Mountain, Worcester, 1 500—1 800 m, obtuse to subacute bracts, compact habit; 9918 (BOL) same locality, 1 675 m, acuminate squarrose bracts, compact habit; *Esterhuysen* 14278 (BOL), Milner Peak, Hex River Mountains, 1525 m, acute to acuminate squarrose bracts, compact habit; *Esterhuysen* 14255 (BOL) and 14871 (BOL, PRE), same locality, 1 830 m, obtuse bracts, compact habit; *Esterhuysen* 14773 (BOL, NBG), Michell's Peak, Ceres, 1 525 m, bracts obtuse or acuminate, squarrose, compact habit; *Esterhuysen* 19754 (NBG), Sneeuwgat Peak, Tulbagh, 1 675—1 830 m, acute to acuminate squarrose bracts, compact habit; *Esterhuysen* 16547 (BOL; NBG), Du Toit's Kloof, above Mountain Club Hut, obtuse to acute to acuminate squarrose bracts, compact habit. It is clear then that typical plants of *H. zwartbergense* with obtuse bracts can grow intermingled with plants with acute to acuminate bracts that therefore accord better with *H. tinctum*. The only record of loosely branched *H. tinctum* from this critical area is *Esterhuysen 12467*, quoted above, and *Esterhuysen* 13547 (NBG), Haalhoek Spitzkop, 900—1 200 m, which is a mixture of *H. tinctum* and *H. zwartbergense*.

80. **Helichrysum tinctum** *(Thunb.) Hilliard & Burtt* in Bot. J. Linn. Soc. 82, 3: 200 (1981). Type: Cape of Good Hope, *Thunberg* (sheet 19276, UPS, holo.!).

Gnaphalium tinctum Thunb., Prodr. 151 (1800), Fl. Cap. 656 (1823). *Leontonyx coloratus* Cass. in Dict. Sci. nat. 25: 467 (1822). *Spiralepis tincta* (Thunb.) D. Don, l.c. *L. coloratus* var. *gracilis* Less., Syn. Comp. 326 (1823).

G. glomeratum L., Pl. Rar. Afr. 20 (1760), Amoen. Acad. 6: 99 (1764), non *Helichrysum glomeratum* Klatt. *Spiralepis glomerata* (L.) D. Don in Mem. Wern. nat. Hist. Soc. 5: 552 (1826). *Leontonyx glomeratus* (L.) DC., Prodr. 6: 168 (1838); Harv. in F.C. 3: 206 (1865); Levyns in Adamson & Salter, Fl. Cape Penins. 786 (1950). *L. glomeratus* var. *verus* Harv., l.c. Lectotype: South Africa (LINN 989.85!).

Leontonyx stramineus DC., Prodr. 6: 168 (1838), non *Helichrysum stramineum* Hiern. *L. glomeratus* var. *stramineus* (DC.) Harv., l.c. Type: Cape, Piquetberg, *Drège* 6338 (G-DC, holo.!).

L. bicolor DC., l.c., non *Helichrysum bicolor* Lindl. Type: Kamiesberg, *Drège* 2838 (G-DC, holo.!; S, iso.!).

L. angustifolius DC. var. *diffusus* DC., l.c. Type: Cape, Drakensteenberg, *Drège* 1790 (G-DC, holo.!).

L. glomeratus var. *intermedius* Harv., l.c. Type: nothing cited.

Prostrate or diffuse grey- or sometimes white-woolly herb, possibly annual, main branches 40—450 mm long, many from the crown, congested or lax, much branched above, loosely woolly, leafy. *Leaves* mostly 7—20 × 4—8 mm, spathulate to elliptic-spathulate, obtuse, base narrowed, margins flat, both surfaces woolly-felted, uppermost leaves, particularly those surrounding the heads, often folded lengthwise and recurved. *Heads* homogamous or rarely heterogamous, campanulate, 5—6 × 3—4 mm, sometimes solitary, mostly c. 3—12 crowded in small glomerules terminating the branchlets. *Involucral* bracts in 4 series, outer short, webbed with wool to surrounding leaves, inner subequal, exceeding flowers, soon caducous, colour very variable, pale straw to deep straw to golden brown or crimson, or combination of these colours, tips very acute to acuminate, opaque,

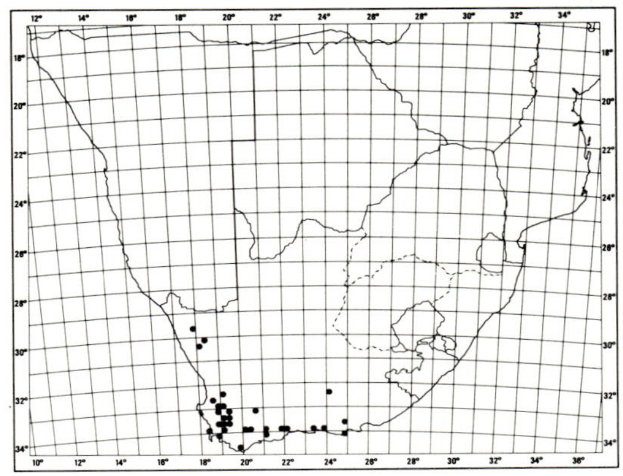

MAP 98.— **Helichrysum tinctum**

strongly hooked, concolorous or colour contrasting with that of shaft. *Receptacle* nearly smooth. *Flowers* 13—27, 2—3 occasionally ♀. *Achenes* not seen, ovaries with myxogenic duplex hairs, or very rarely glabrous. *Pappus* bristles many, in several series, equalling corolla, delicate, scabrid, bases cohering by patent cilia. Fig. 29: 2.

Ranges from Springbok, the Kamiesberg and the mountains about Piquetberg and the Cedarberg across the high ground and mountains, including Table Mountain, to the Great Winterhoek Mountains and Kruisfontein Mountains near Humansdorp, mainly between 250 and 1 350 m above sea level, with one isolated record (*Acocks* 21780) from the Camdeboo plateau near Aberdeen at 1 675 m. Flowers mainly between September and November. Map 98.

H. tinctum is characterized by its acuminate, hooked involucral bracts and hairy ovaries; only very rarely are the bracts emarginate. The type of *Leontonyx bicolor* seems to be no more than an extreme variant with solitary or 2−3 heads at the branch tips.

Vouchers: *Acocks* 19854 (PRE); *Barker* 7698 (NBG); *Compton* 6460 (NBG); *Esterhuysen* 19319 (BOL; PRE); *Galpin* 4146 (PRE).

The Drège specimens from near the Olifants River and the Orange (Gariep), cited by De Candolle and Harvey as *Leontonyx glomeratus*, are *Helichrysum leontonyx* (no. 82), which is readily distinguished by its smaller heads with uniseriate pappus.

81. **Helichrysum solitarium** *Hilliard, sp. nov.* H. litorali *H. Bol. affinis sed habitu multo graciliore, lana tenuiore laxiore, foliis minoribus (plerumque 5 −7 mm longis, nec 8 −20 mm) planis (numquam complanatis nec apicibus recurvis), capitulis minoribus 4 mm longa (nec 5 mm) semper solitariis (numquam coacervatis nec ad apices ramulorum arcte aggregatis).*

Herba perennis gracilis, verosimiliter tegetem formans; rami filiformes, diffusi, 8 −100 mm longi, tenuiter et laxe lanati, partibus juvenilibus foliatis. Folia 5 −7 × 1,5 −2 mm, spatulata, apice subacuto, basi multo angustata semi-amplexicauli, utrinque laxe cano-lanata. Capitula homogama, cylindrica, 4 × 3 mm, apicibus ramulorum solitaria, foliis superata. Bracteae involucrales 4-seriatae, gradatae, dorso lanatae, exteriores ad folia duo proxima laxe contextae, interiores flores vix superantes, non radiantes, apicibus truncatis emarginatis opacis pallide bubalinis, area opaca in linea media breviter descendente. Receptaculum breviter favosum. Flores c. 18 −20. Achenia non visa; ovaria pilis duplicibus myxogenis praedita. Pappi setae multae, verosimiliter biseriatae (specimine juvenili), corollam quasi aequantes, scabridae, basibus ciliis minutis patentibus vix cohaerentibus.

Type: Cape, Ceres div., Bokkeveld Tafelberg, SE. slopes, in shade, 6 000 ft, 8

xii 1940, *Esterhuysen* 3934 (BOL, holo.!).

Delicate perennial herb, probably mat-forming, branches filiform, prostrate or diffuse, 80−100 mm long, thinly and loosely woolly, young parts leafy. *Leaves* 5−7 × 1,5−2 mm, spathulate, apex subacute, base much narrowed, half-clasping, both surfaces loosely greyish-white woolly. *Heads* homogamous, cylindric, 4 × 3 mm, solitary at the branchlet tips, overtopped by leaves. *Involucral bracts* in 4 series, graded, woolly on backs, outer loosely webbed to 2 adjacent leaves, inner scarcely exceeding flowers, not radiating, tips truncate, emarginate, opaque, pale buff, opaque patch extending briefly down midline. *Receptacle* shortly honeycombed. *Flowers* c. 18−20. *Achenes* not seen, ovaries with myxogenic duplex hairs. *Pappus* bristles many, probably biseriate (material young), about equalling corolla, scabrid, bases scarcely cohering by minute patent cilia. Fig. 30: 1.

Known only from the type collection. Map 99.

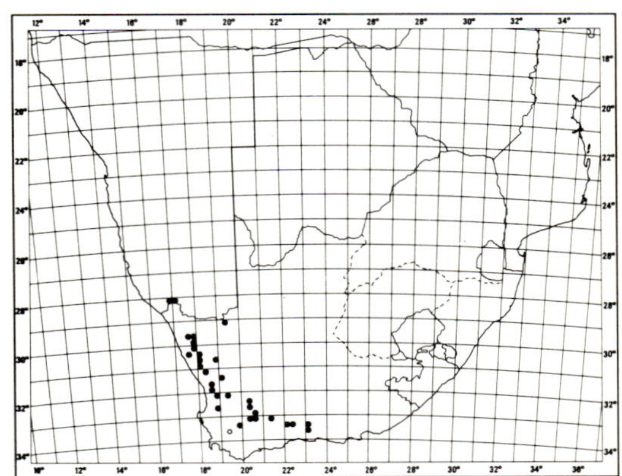

MAP 99.— ● **Helichrysum leontonyx**
○ **Helichrysum solitarium**

82. **Helichrysum leontonyx** *DC.*, Prodr. 6: 169 (1838); Harv. in F.C. 3: 214 (1865); Moeser in Bot. Jb. 44: 296 (1910). Type: Cape, Little Namaqualand, Springbok distr., between Kaus Mountains and Gariep, *Drège* 2843 (G-DC, holo.!).

Gnaphalium leontonyx (DC.) Sch. Bip. in Bot. Ztg 3: 169 (1845).

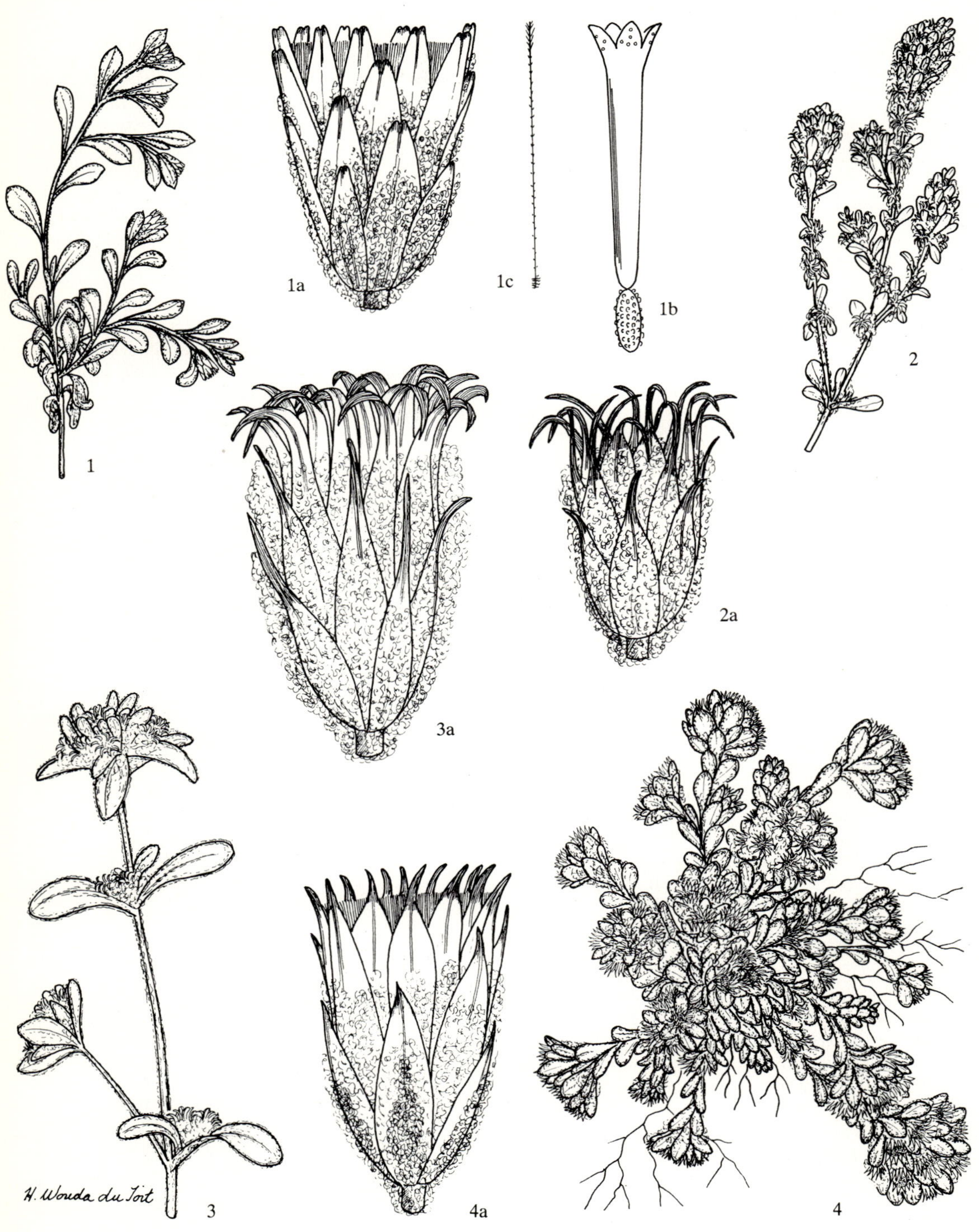

Grey-woolly annual herb with a slender taproot, branches 15−150 mm long, many from the crown, simple or much branched, slender, prostrate, often reddish, loosely woolly, distantly leafy but leaves crowded under the heads and at the crown. *Radical leaves* often withered at flowering, mostly 15−30 × 5−10 mm, oblong-elliptic to spathulate, apex rounded, base narrowed, margins flat, cauline leaves similar, but often smaller, mostly 5−15 × 1,5−6 mm, oblong-elliptic to spathulate, often folded lengthwise, apex recurved, all with both surfaces loosely grey-woolly. *Heads* heterogamous, campanulate, c. 3−4 × 2−2,5 mm, sessile or shortly pedunculate, crowded in terminal glomerules c. 5−15 mm across. *Involucral bracts* in 3−4 series, outermost short, webbed with wool to leaves, inner subequal, exceeding flowers, backs woolly, tips usually acuminate, often markedly squarrose, rarely only acute, dull yellow, orange or red, sometimes red above the stereome. *Receptacle* nearly smooth. *Flowers* 20−37, (1−) 2−7 ♀, 18−30 ☿, yellow, sometimes tipped red. *Achenes* 0,75 mm long, with myxogenic duplex hairs, or rarely glabrous. *Pappus* bristles equalling corolla, barbellate above, scabrid below, bases cohering lightly by patent cilia. Fig. 30: 3.

Recorded from the Obibberge in southernmost S.W.A./Namibia near the Orange River, the Springbok area in Namaqualand and the environs of Pofadder further east, south to the Roggeveld escarpment, Ceres and the Little Karoo from Laingsburg east to Willowmore and Uniondale. Map 99.

Grows in sand, and will colonize roadsides and old fields. Flowers from July to October, mainly in September.

Being a desert annual dependent on erratic rainfall, plants of *H. leontonyx* are very variable in size, and stems less than 20 mm long will bear clusters of heads. There is also considerable variation in degree of branching and size of glomerules. However, the involucral bracts are generally acuminate with recurved tips; the only exceptions seen are *Compton* 21147 (NBG) with more or less acute bracts, and *Goldblatt* 5905A (E; MO; NU) in which the bracts vary from acute to shortly acuminate.

H. leontonyx is frequently found in herbaria under *H. tinctum* (*Leontonyx glomeratus*) (no. 80) from which it is most easily distinguished by its small heads and uniseriate pappus, a character calling for careful observation.

Vouchers: *Acocks* 16946 (PRE); *Compton* 11257 (NBG); *Esterhuysen* 5817 (PRE); *Maguire* 999 (NBG, STE).

Acocks 15588 (PRE) from Aroab, cited in F.S.W.A. 139: 95 as *H. leontonyx*, is *H. micropoides* (no. 84; see under that species for distinguishing characters). *Merxmüller & Giess* 28606 (M; WIND) from the Obibberge is *H. leontonyx*, and the only specimen I have seen from S.W.A./Namibia.

83. **Helichrysum tysonii** *Hilliard* in Notes R. bot. Gdn Edinb. 40: 269 (1982). Type: Cape, Graaff-Reinet district, Renosterberg above Lootsberg railway halt, c. 1 800 m, farm Blaauwater, 24 xi 1977, *Hilliard & Burtt* 10626 (NU, holo.!; E; K; MO; PRE; S, iso.!).

Annual herb forming small dense mats up to c. 200 mm across, stems many from the crown, well branched, prostrate, rooting, branchlets grey-tomentose, leafy. *Leaves* 4−10 × 1,5−6 mm, narrowly to broadly spathulate, tips rounded, extreme tip folded and slightly recurved, base broad, clasping, margins flat, both surfaces grey-woolly. *Heads* homogamous or heterogamous, campanulate, c. 4 × 2 mm, many crowded in small terminal glomerules, closely surrounded by leaves, the innermost leaves webbed to the outer involucral bracts. *Involucral bracts* in c. 3 series, outermost short, inner subequal, slightly exceeding flowers, tips very acute, slightly recurved, opaque straw-coloured or reddish. *Receptacle* nearly smooth. *Flowers* 19−32, 0−3 ♀, 19−31 ☿. *Achenes* 0,75 mm long, with myxogenic duplex hairs. *Pappus* bristles many, equalling corolla, scabrid, bases scarcely cohering by patent cilia. Fig. 30: 4.

Recorded from the mountains of the E. central Cape: Koudeveldberg, Sneeuberg, Compassberg, Renosterberg and Wapadsberg. Grows on bare stony ground, and will colonize eroded places and hard gravelly road verges. Flowers between September and November. Map 100.

Distinguished from *H. leontonyx* (above) by its congested habit and acute not acuminate, involucral bracts.

FIG. 30−1, **Helichrysum solitarium,** part of plant, × 1; 1a, head, × 10; 1b, flower, × 13; 1c, pappus bristle, × 13 (*Esterhuysen* 3934). 2, **H. micropoides,** part of plant, × 1; 2a, head, × 10 (*Leistner* 2905). 3, **H. leontonyx,** part of plant, × 1; 3a, head, × 10 (*Acocks* 19547). 4, **H. tysonii,** plant, × 1; 4a, head, × 10 (*Hilliard & Burtt* 10618).

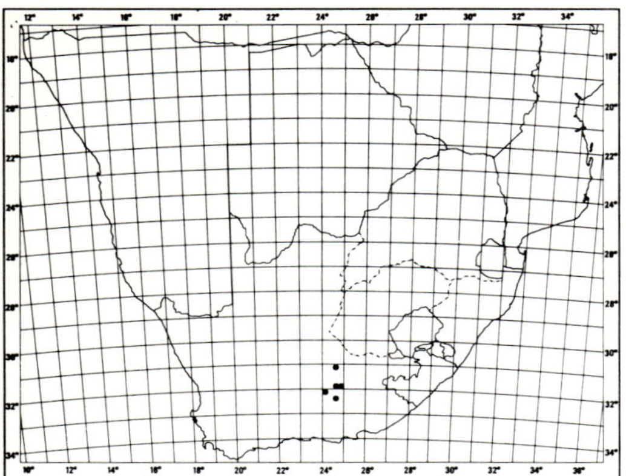

MAP 100.— **Helichrysum tysonii**

Vouchers: *Acocks* 16514 (PRE); *Acocks* 24580 (PRE); *Bolus* 509 (BOL, K); *Hilliard & Burtt* 10618 (E; K; M; NU; S); *Tyson* 417 (PRE).

84. **Helichrysum micropoides** *DC.*, Prodr. 6: 170 (1838); Harv. in F.C. 3: 215 (1865); Merxm., F.S.W.A. 139: 96 (1967). Type: South Africa, *Drège* 5811 (G-DC, holo.!).

Gnaphalium micropoides (DC.) Sch. Bip. in Bot. Ztg 3: 169 (1845).

G. prostratum Thunb., Prodr. 150 (1800), Fl. Cap. 652 (1823); Less., Syn. Comp. 330 (1832); Harv. in F.C. 3: 262 (1865); non *Helichrysum prostratum* Hook. f. (1844). Type: Cape, Onder Bokkeveld, *Thunberg* (sheet 19234, UPS, holo.!).

Helichrysum filagineum DC., l.c.; Harv., l.c. *Gnaphalium araneosum* Sch. Bip. in Bot. Ztg 3: 169 (1845). Type: South Africa, *Drège* 6339 (G-DC, holo.!).

H. namaquense Schltr. & Moeser in Bot. Jb. 44: 295 (1910). Type: Namaqualand, Zabies, *Schlechter* 11224 (Z, holo.!; BM; E; G; K; PRE; S; fragment NU, iso.!).

Herb with a slender taproot, probably often annual, occasional specimens woody, branches few to many from the crown, prostrate or decumbent, mostly 20—150 mm long, very slender, simple or sparingly branched, loosely woolly, glabrescent, often reddish, distantly leafy except at the crown and under the heads. *Radical leaves* up to 15 × 6 mm, stem leaves sometimes smaller, upper leaves mostly 3—8 × 1,5—3 mm, obovate, spathulate or oblong-spathulate, apex obtuse, often folded lengthwise and

strongly recurved, base narrowed, half-clasping, margins somewhat undulate, both surfaces glandular, loosely grey-woolly. *Heads* homogamous or heterogamous, narrowly campanulate, c. 3 × 2 mm, each opposite a leaf, solitary or few in small glomerules crowded towards the tips of the branchlets forming a leafy racemose compound inflorescence. *Involucral bracts* in 3—4 series, outermost short, webbed to surrounding leaves with wool, inner subequal, exceeding flowers, woolly on the backs, pellucid, sometimes red above the stereome, tips opaque, subacute to acuminate, mostly squarrose, straw-coloured to reddish. *Receptacle* nearly smooth. *Flowers* 8—23, 0—2 (—3) ♀, 8—22 ☿, yellow, sometimes tipped red. *Achenes* 0,75 mm long, glabrous or with myxogenic duplex hairs. *Pappus* bristles many, equalling corolla, tips barbellate, bases scarcely cohering by patent cilia. Fig. 30: 2.

Found in the southern part of S.W.A./Namibia and the arid parts of the Cape from Lüderitz E. and SE. to the Kalahari Gemsbok National Park, Hay and De Aar, and S. through Namaqualand and the W. Karoo to Ceres. Grows in sand, sometimes forming a dense cover on sand dunes, or on hard-packed sand. Flowers between July and November. Map 101.

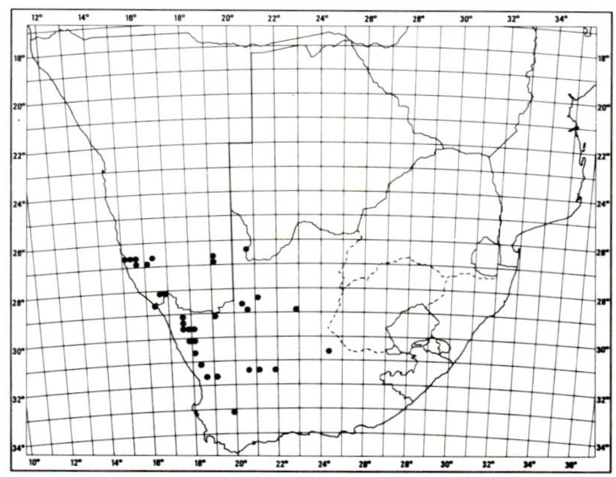

MAP 101.— **Helichrysum micropoides**

H. micropoides is very variable in the degree to which the tips of the involucral bracts are produced; there is a tendency for plants in the E. half of the range to have bracts more markedly acuminate than in plants from the W. part, but there is neither discontinuity in bract length nor absolute geographical separation. Similarly, plants with homogamous heads are commoner in the W. than in the E.

The species can usually be recognized by the peculiar arrangement of the heads in racemose compound inflorescences, but this is not always clearly visible in young specimens. However, it is a character that serves to distinguish *H. micropoides* immediately from its close ally, *H. leontonyx* (above), which has heads crowded in terminal glomerules. Also, in *H. leontonyx* the heads are always heterogamous with generally more flowers and more female flowers than in *H. micropoides*.

H. micropoides is also much confused with *H. tinctum* (no 80), which it resembles in the involucre. However, the pappus is clearly uniseriate, a character requiring careful observation; also, the heads are much smaller.

Vouchers: *Acocks* 15003 (PRE); *Bolus* 7441 (BOL; K); *Compton* 22008 (M; NBG); *Merxmüller & Giess* 2993 (M; WIND); *Wasserfall* 1102 (K; M; NBG; PRE).

Group 15

Shrublets, subshrubs, perennial or annual herbs; *leaves* small, mostly more or less spathulate or obovate, sometimes linear, linear-lanceolate or elliptic; *heads* homogamous or heterogamous, 3−7 (−10) × (1,5−) 2−7 (−10) mm, either solitary or few subcorymbosely arranged, or in congested clusters; *involucral bracts* radiating or sometimes squarrose, white, pink, creamy, yellow, rufous or golden-brown, often parti-coloured or flushed red above the stereome; *receptacle* shortly honeycombed; *flowers* 7−290, (0−) 1−50 ♀, corolla of ☿ flowers funnel-shaped, of ♀ flowers narrowly cylindric; achenes usually hairy, sometimes glabrous, often in the same species; *pappus* bristles scabrid, tips scabrid or barbellate, bases free or cohering lightly by patent cilia.

Species 85−111, mostly concentrated in S.W.A/Namibia and the W. and SW. Cape, several narrowly endemic; three species ranging widely from Zimbabwe and Angola southwards, always in warm, dry, sandy places.

1a Involucral bracts predominantly various shades of brown, buff, gold or yellow (tips of inner bracts may be white or pink), or combinations of these colours:

  2a Heads subglobose, bracts brown, concave, innermost often tipped opaque white .... 85. *H. cochleariforme*

  2b Heads campanulate or cylindric:

    3a Heads solitary or 2−3 together, frequently further arranged in corymbose panicles:

      4a Heads c. 7 mm long, outer bracts pale or dark brown, inner with opaque milk-white tips ....92. *H. stellatum*

      4b Heads 3−6 mm long:

        5a Stiff divaricately branched shrublet, heads solitary ....97. *H. oxybelium*

        5b Annual or perennial, if shrubby, branches not stiffly divaricate:

          6a Plants woody at least at base:

            7a Tips of involucral bracts concave:

              8a Heads containing c. 15−25 flowers, tips of involucral bracts mostly recurved, those of inner bracts often differently coloured from outer.............................86. *H. cylindriflorum*

              8b Heads containing c. 40−70 flowers, tips of involucral bracts not recurved, uniformly coloured ....88. *H. pulchellum*

            7b Tips of involucral bracts not concave:

              9a Tips of involucral bracts acute to acuminate:

                10a Leaf bases more or less minutely ear-clasping:

                  11a Involucral bracts uniformly light golden-brown , tips squarrose.........87. *H. aureofolium*

                  11b Outer involucral bracts brown, inner pink, whitish, or clear bright yellow .... 91. *H. incarnatum*

              10b Leaf bases oblong, very shortly decurrent or not, but not ear-clasping .... 94. *H. pumilio*

            9b Tips of involucral bracts obtuse ....95. *H. obtusum*

6b Annuals:

    12a Involucral bracts uniformly light golden-brown ....89. *H. alsinoides*

    12b Involucral bracts brown flushed with crimson, inner generally tipped opaque white ....90. *H. leptorhizum*

3b Heads crowded in terminal glomerules surrounded by leaves:

    13a Heads 5—6 mm long, tips of involucral bracts very acute to acuminate and recurved .... 94. *H. pumilio*

    13b Heads 3—5 mm long, tips of bracts not recurved:

        14a Shrublet, tips of bracts obtuse .... 95. *H. obtusum*

        14b Herbaceous, annual or perennial:

            15a Involucral bracts more or less uniformly pale golden-brown:

                16a Stems simple or subsimple, heads many in terminal glomerules.................. 108. *H. albertense*

                16b Stems loosely branched, heads few in glomerules further arranged in panicles ....89. *H. alsinoides*

            15b Tips of inner involucral bracts white ....96. *H. deserticola*

1b Involucral bracts predominantly white or white and crimson, rose or pink, sometimes wholly crimson or pink:

    17a Heads solitary at the tips of very dwarf lateral shoots (they may appear to be merely surrounded by leaves) these producing very floriferous, raceme-like twigs:

        18a Heads c. 6 mm long, outer involucral bracts white, inner rosy............................98. *H. paronychioides*

        18b Heads c. 7—9 mm long, all the bracts white:

            19a Heads c. 7—8 mm long, containing c. 20—30 flowers, all leaves persistently grey-woolly ....99. *H. spiciforme*

            19b Heads c. 8—9 mm long, containing c. 40—55 flowers, leaves on side branches soon losing their wool and contrasting with the persistently woolly leaves on the main stems............. 100. *H. amboense*

    17b Heads solitary, clustered, or arranged in corymbose panicles, not racemosely arranged:

        20a Heads subglobose, 7—10 mm long ....101. *H. argyrosphaerum*

        20b Heads cylindric or campanulate:

            21a Shrublet with stiffly divaricate branches, heads solitary, 4—5 mm long.....................97. *H. oxybelium*

            21b Annual or perennial, branches not stiffly divaricate, heads mostly in clusters or corymbosely arranged:

                22a Heads 3—4 mm long, involucral bracts pellucid, tips rounded, not radiating...........89. *H. alsinoides*

                22b Heads mostly at least 5 mm long, involucral bracts radiating or squarrose; or if heads only 3—4 mm, then tips of involucral bracts either acute or opaque, or both:

                    23a Heads 4—10 mm long, tips of involucral bracts acute to acuminate, or if obtuse, then heads at least 6 mm long:

                        24a Heads heterogamous:

                            25a Heads 4—5 mm long, in congested terminal globose clusters .... 107. *H. herniarioides*

                            25b Heads at least 6 mm long:

                                26a Bases of at least the larger leaves minutely ear-clasping ....92. *H. stellatum*

                                26b Bases of leaves not ear-clasping:

                                    27a Leaves mostly obovate or broadly elliptic, base much narrowed and petiole-like... 106. *H. gariepinum*

                                    27b Leaves linear, linear-lanceolate, oblanceolate or oblong, base more or less oblong, very shortly decurrent:

                                        28a Heads 7—10 mm long, involucral bracts generally white, with some rose in var. *aurosicum* ....93. *H. cerastioides*

        (*Note*: *H. cerastioides* and *H. pumilio* subsp. *fleckii* present acute taxonomic problems in S.W.A./Namibia; not all specimens will key out; consult the notes following the species descriptions).

85. **Helichrysum cochleariforme** *DC.*,
Prodr. 6: 185 (1838); Hilliard & Burtt in Bot. J. Linn. Soc. 82: 262 (1981). Type: Cape, Mossel Bay, dry branch of Gouritz River, 6 Nov. 1814, *Burchell* 6491 (G-DC, holo.!; BOL; K, iso.!).

Gnaphalium cochleariforme (DC.) Sch. Bip in Bot. Ztg 3: 171 (1845).

G. imbricatum L., Sp. Pl. 855 (1753). *Helichrysum imbricatum* (L.) Less., Syn. Comp. 279 (1832), non (L.) Thunb. (1823); Harv. in F.C. 3: 221 (1865); Moeser in Bot. Jb. 44: 293 (1910). Lectotype: specimen illustrated in Burm., Rar. Afr. Pl. t. 80, fig. 2 (G!).

G. discolorum sensu Thunb., Prodr. 149 (1800), non L.

Helichrysum stellatum var. *globiferum* Harv. in F.C. 3: 221 (1865). Type: Cape, Breede River, Port Beaufort, *Mundt* 21 (TCD, holo.!; K, iso.!).

Bushy half-shrub 50—200 mm high, woody at the base and much branched there, branches erect or spreading, simple or forked, often with dwarf axillary shoots, loosely grey-woolly, leafy throughout. *Leaves* mostly 5—30 × 2—5 mm. diminishing slightly upwards, spathulate to oblong- or linear-spathulate, apex obtuse, apiculate, base narrowed, half-clasping, both surfaces grey-woolly. *Heads* heterogamous, subglobose, c. 5—7 × 5—8 mm, solitary or few together at the tips of the twiglets, these corymbose-paniculately arranged. *Involucral bracts* in 6—8 series, loosely imbricate, graded, inner slightly exceeding flowers, all bracts obtuse, concave, smooth, reddish- or golden-brown, innermost often tipped opaque milk-white, minutely radiating. *Receptacle* shortly honeycombed. *Flowers* 56—120, 13—26 ♀, 45—102 ☿. *Achenes* 0,75 mm long, glabrous or with myxogenic duplex hairs. *Pappus* bristles many, scabrid, about equalling corolla, bases nude, not cohering. Fig. 32:4.

Endemic to the SW. and S. Cape, from the sand flats about Piquetberg south and then east along the coast to the Gouritz River, Mossel Bay district. Grows in open sandy places in coastal scrub or sandy hollows between the dunes; flowering between September and December, chiefly in October. Map 102.

A very distinctive species, easily recognized by its subglobose heads with golden-brown, concave bracts, the innermost bracts often tipped opaque white and minutely radiating.

Vouchers: *Acocks* 15207 (K; PRE); *Compton* 10216 (NBG); *Esterhuysen* 19566 (BOL); *Oliver* 4054 (K; PRE); *Taylor* 4281 (PRE).

86. **Helichrysum cylindriflorum** *(L.)* Hilliard & Burtt in Bot. J. Linn. Soc. 82: 261 (1981). Lectotype: Herb. Plukenet (Herb. Sloane vol. 100 folio 104, BM!).

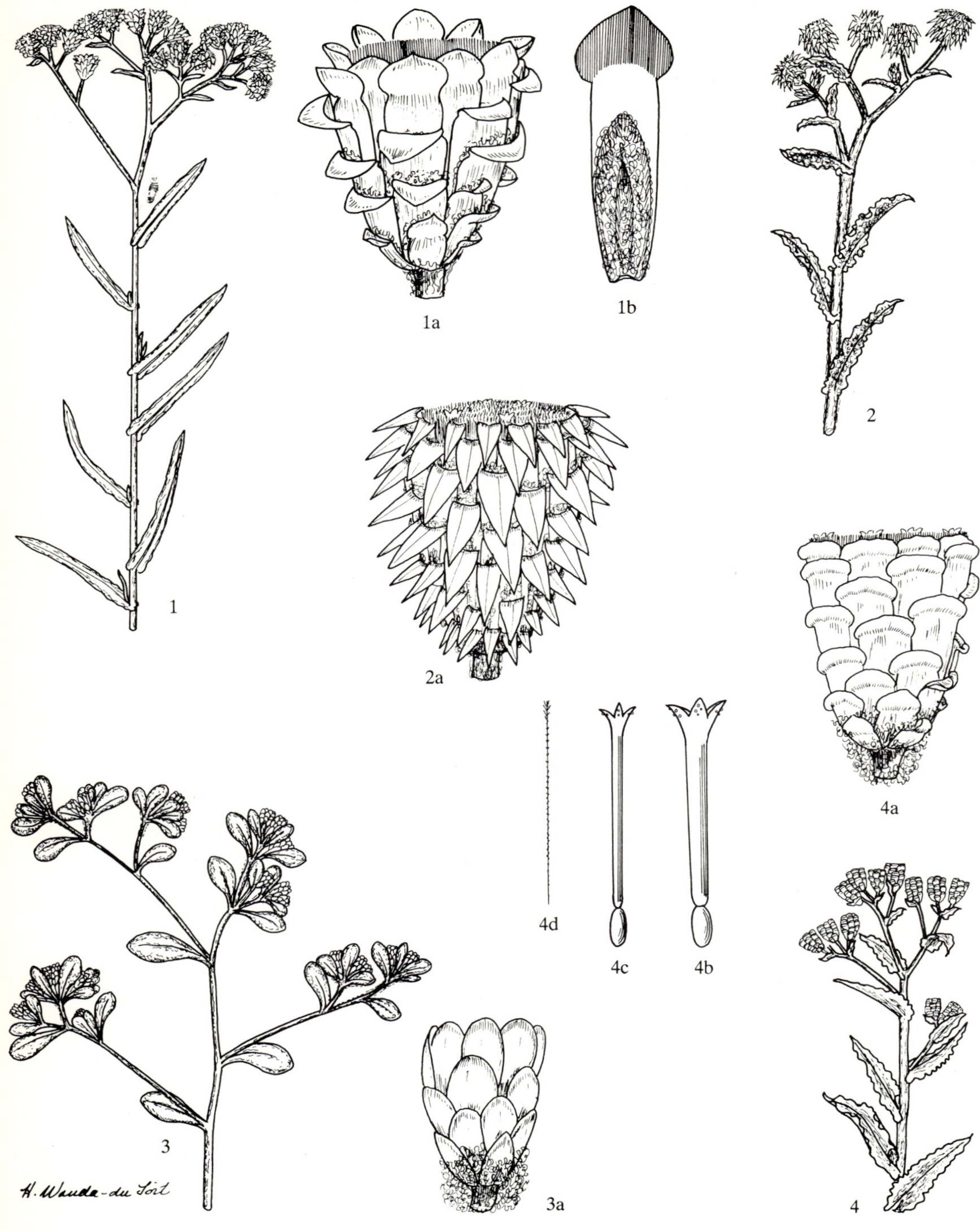

1a · 1b · 1 · 2 · 2a · 4a · 4d · 4c · 4b · 3 · 3a · 4

H. Wanda-du Toit

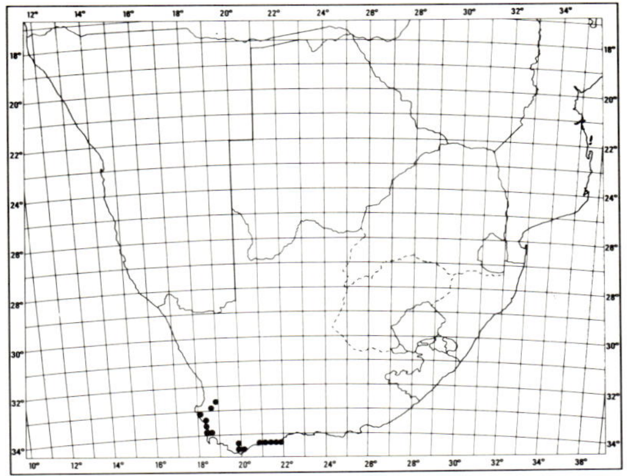

MAP 102.— **Helichrysum cochleariforme**

Gnaphalium cylindriflorum L., Pl. Rar. Afr. 19 (1760), Amoen. Acad. 6: 99 (1763), edn 2,6: 58 (1789). *G. cylindricum* L., Sp. Pl. edn 2: 1194 (1763), nom. illegit. *Helichrysum cylindricum* D. Don. in Sweet, Hort. Brit. 223 (1826); Less., Syn. Comp. 281 (1832); DC., Prodr. 6: 190 (1838); Harv. in F.C. 3: 220 (1865); Moeser in Bot. Jb. 44: 293 (1910); Levyns in Adamson & Salter, Fl. Cape Penins. 781 (1950).

Gnaphalium aethiopicum minus ramosum capitulis coccineis Plukenet, Almagestum Bot. t. 298, fig. 4 (1694), p. 172 (1696). *Note*: The figure is wrongly quoted in the text as t. 198 fig. 4 and by Linnaeus as 293 fig. 4.

G. sessile Burm. f., Prodr. Fl. Cap. 25 (1768); Burm., Rar. Afr. Pl. t. 80 fig. 3 (1739). Type: specimen in herb. Burm. (G!).

G. conyzoides Thunb., Prodr. 149 (1800), nom. illegit.

G. rubellum Thunb., Prodr. 150 (1800), Fl. Cap. 653 (1823). *Helichrysum rubellum* (Thunb.) Less., Syn., Comp. 280 (1832); DC., Prodr. 6: 191 (1838); Harv. in F.C. 3: 220 (1865), incl. vars. Type: Cape, Hex River Mountains, *Thunberg* (sheet 19247, UPS!).

G. paniculatum Thunb., Fl. Cap. 647 (1823), non Berg. (1767). Type: Cape of Good Hope, *Thunberg* (sheet 19225, UPS!).

Helichrysum imbricatum sensu DC., Prodr. 6: 191 (1838), incl. vars, non (L.) Less.

H. fastigiatum Harv. in F.C. 3: 219 (1865). Lectotype: Cape, near Rivier Zonder Einde and Buffeljagd River, *Zeyher* 2859 (TCD!; BOL; K; PRE; S, isolecto.!).

H. leipoldtii H. Bol. in Trans. S. Afr. phil. Soc. 18: 382 (1907). Type: Cape, common in sandy fields

around Clanwilliam, c. 120 m, also on Ram's Kop near Clanwilliam, *Leipoldt* 510 (BOL, holo.!; K; NBG; SAM, iso.!).

H. leipoldtii var. *parvifolia* H. Bol., l.c. Type: Cape, Clanwilliam, Wuppertal, in rocky places, *Bolus* 9022 (BOL, holo.!).

Bushy half-shrub c. 50—300 mm high, woody at the base and much branched there, branches intricate or more lax, erect or spreading, young parts thinly grey-woolly, leafy, often with dwarf axillary shoots. *Leaves* mostly 5—30 (—40) × 1—5 mm, diminishing slightly upwards, lanceolate, oblong-lanceolate or subspathulate, apex acute to obtuse, apiculate, base broad, half-clasping, often subauriculate, margins flat or undulate, both surfaces thinly greyish-white woolly, sometimes glabrescent with age. *Heads* heterogamous, rarely homogamous, cylindric-campanulate, c. 4—5 × 2,5—3 mm, few clustered at the tips of the branchlets, these in turn corymbose-paniculately arranged. *Involucral bracts* in c. 7 series, graded, loosely imbricate, outer membranous, backs woolly, tips glabrous, acute, golden-brown or rufous, squarrose, inner with tips slightly exceeding the flowers, these somewhat concave, acute to obtuse, mostly opaque (see notes below), white, creamy or yellow, very rarely pinkish, minutely radiating. *Receptacle* nearly smooth to shortly honeycombed. *Flowers* (10—) 16—33, (0—) 1—4 ♀, 13—30 ☿. *Achenes* not seen, ovaries usually with myxogenic duplex hairs, sometimes glabrous, very rarely ♀ hairy, ☿ glabrous. *Pappus* bristles many, equalling corolla, scabrid, bases not cohering. Fig. 31:1.

Widespread in the SW. districts from Vanrhynsdorp and Calvinia to the Langekloof near Uniondale, with an outlier on the Kamiesberg, from c. 30—1 350 m above sea level, but commoner on mountain and hill slopes than on flats, in rocky or sandy places. Flowers between September and December. Map 103.

Vouchers: *Acocks* 23681 (PRE); *Compton* 16775 (NBG); *Galpin* 4141 (PRE); *Schlechter* 9050 (PRE).

H. cylindriflorum is very variable in stature and in degree of branching, dense twiggy little rounded bushes grading imperceptibly to taller, more sparingly branched and more open plants in which the leaves are often larger, particularly longer, than in the more dwarf forms. Characteristically, the small cylindric-

---

FIG. 31.—1, **Helichrysum cylindriflorum,** part of plant, × 1; 1a, head, × 8; 1b, involucral bract, × 13 (*Acocks* 17348). 2, **H. aureofolium,** part of plant, × 1; 2a, head, × 8 (*Goldblatt* 4072). 3, **H. alsinoides,** part of plant, × 1; 3a, head, × 8 (*Schlechter* 8170). 4, **H. pulchellum,** part of plant, × 1; 4a, head, × 8; 4b, hermaphrodite flower, × 10; 4c, female flower, × 10; 4d, pappus bristle, × 10 (*Acocks* 19057).

campanulate heads are clustered at the branch tips and have involucral bracts with oblong shafts loosely webbed together with wool, the tips of the outer 4−5 series membranous, warm shades of brown and squarrose, the tips of the inner series opaque or subopaque, somewhat concave, radiating, white, creamy or yellow. These inner tips are mostly less than 1 mm long; very rarely do they approach 2 mm. Specimens with bright yellow bracts were distinguished as *H. leipoldtii*, and appear to predominate in parts of the range, for example in the Hex River Mountains, the Witteberg, and around Touws River, but in the Cedarberg, and in Vanrhynsdorp and Clanwilliam districts, both yellow and creamy bracts are common, while elsewhere pale bracts predominate.

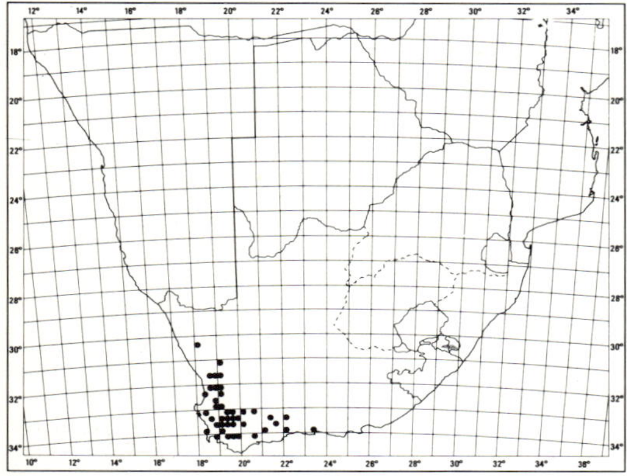

MAP 103.— **Helichrysum cylindriflorum**

One extreme of the variation range, designated *H. fastigiatum* by Harvey, is a dense twiggy dwarf bush in which the heads are aggregated as in typical *H. cylindriflorum* but all the bracts are brown, subpellucid, the tips somewhat concave and slightly crisped. The other extreme is a bush reaching 600 mm in height, also richly branched and tangled, but the heads are borne singly at the tips of leafy stalks (e.g. *Acocks* 17438, PRE; *Esterhuysen* 17719a, BOL; NBG; PRE).

The same range, from dwarf compact bushes to tall ones, with heads usually on long stalks, or sometimes with shorter stalks and the heads therefore more closely aggregated, is found in specimens with acute, opaque tips ranging in colour from white through cream to yellow (e.g. *Esterhuysen* 18101, BOL; PRE; *Pillans* 8592, BOL; PRE; *Pillans* 7349, BOL).

Taylor 5541, PRE, from the Kamiesberg, seems to be an extreme variant: a densely twiggy dwarf shrublet with heads borne singly, or sometimes paired, on long filiform leafy twigs. The tips of the bracts are buff-coloured, either opaque or semipellucid.

Yet another variant has heads on very long stalks and semi-opaque bract tips ranging from buff to yellowish, but the leaves are remarkably glandular-setose. However, glandular hairs are usually present on

the leaves of all specimens, hidden under the wool (*Esterhuysen* 1668, BOL; K; PRE; SAM; *Esterhuysen* 14859, BOL; *Esterhuysen* 12070, BOL; from the Cedarberg and the mountains above Worcester).

Too broad a species concept may have been adopted, but no more can be done in the herbarium.

87. **Helichrysum aureofolium** *Hilliard, species nova* H. cylindrifloro *(L.) Hilliard & Burtt affinis, sed capitulis late campanulatis, (nec cylindrico-campanulatis), floribus c. 60−75 (nec 15−35), apicibus bractearum involucralium c. 2 mm longis (nec 1 mm) recedit.*

Herba caespitosa perennis (?), radice palari lignosa; caules plures, ad c. 130 mm, erecti, plerumque laxe ramosi, interdum subsimplices, appresse albo-lanati, foliati. Folia 8−30 × 2−5 mm, in ramis florentibus minora, oblanceolata, sursum lanceolata, apice acuto, basi inferiorum angustata plus minusve auriculata superiorum lata semi-amplexicauli, marginibus crispatis, utrinque tenuiter appresse cano-lanata. Capitula hetero-gama, campanulata, c. 5 × 5 mm, 7 mm trans bracteas radiantes diametro, solitaria vel 2−3 in pedunculis brevibus ad apices ramulorum in paniculas corymbosas laxas disposita. Bracteae involucrales c. 7-seriatae, gradatae, imbricatae, dorso tenuiter lanatae, apicibus glabris acutis subopacis dilute aureo-brunneis squarrosis, intimae flores paulo superantes. Receptaculum favosum. Flores c. 62−77, 3−5 ♀, 59 − 72 ☿. Achenia non visa; ovaria vel glabra vel ea florum femineorum pilis myxogenis praedita. Pappi setae multae, corollam aequantes, apicibus barbellatis, inferne scabridae, basibus ciliis minutis patentibus praeditis non cohaerentibus.

Type: Cape, 15 km east of Matjiesrivier in Agter Cedarberg, E. facing slopes of rocky sandstone, 7 ix 1976, *Goldblatt* 4072 (NU, holo.!; E; MO; NBG, iso.!).

Tufted perennial (?) herb, taproot woody, stems several from the crown, erect to c. 130 mm, mostly loosely branched, white appressed-woolly, leafy. *Leaves* 8−30 × 2−5 mm, smaller on the inflorescence branches, oblanceolate becoming lanceolate upwards, apex acute, base of lower leaves narrowed, more or less ear-clasping, of upper broad, half-clasping, margins crisped, both surfaces thinly greyish-white appressed-woolly. *Heads* heterogamous, broadly campanulate, c. 5 × 5 mm, 7 mm across the radiant bracts, solitary or 2−3 on short peduncles at the branchlet tips, arranged in loose corymbose panicles. *Involucral bracts* in c. 7 series, graded, imbricate, backs lightly woolly, tips glabrous, acute, subopaque, light golden-brown, squarrose, the innermost slightly exceeding the flowers. *Receptacle* honeycombed. *Flowers* c. 62−77, 3−5 ♀, 59−72 ☿. *Achenes* not seen, ovaries either glabrous, or those of ♀ flowers with myxogenic hairs. *Pappus* bristles many, equalling corolla, tips barbellate, shaft scabrid, bases with minute patent cilia, not cohering. Fig. 31:2.

Only twice recorded: from Groot Rivier Pass in Clanwilliam district and near Matjiesrivier in the Agter Cedarberg, on rocky sandstone slopes; flowering in September. Map 104.

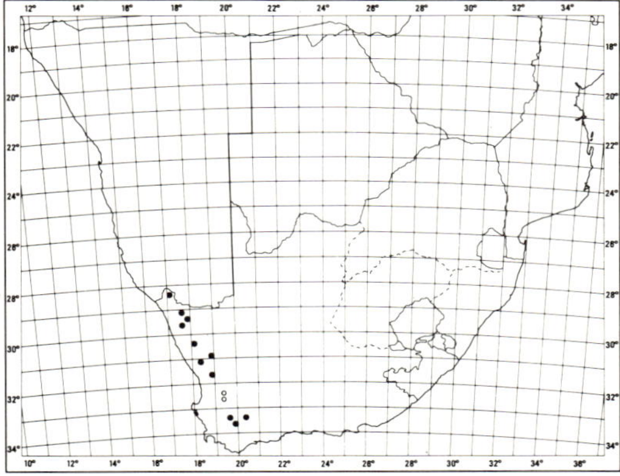

MAP 104.— ○ **Helichrysum aureofolium**
     ● **Helichrysum pulchellum**

H. aureofolium is allied to *H. cylindriflorum* (no. 86) but can be distinguished by its broadly campanulate, not cylindric campanulate, heads containing c. 60−75 flowers, not 15−35, and involucral bracts with tips c. 2 mm long, not 1 mm.

Voucher: *Compton 5096* (NBG).

The specific epithet honours Dr Peter Goldblatt who collected the type material, and also draws attention to the colour of the involucral bracts.

88. **Helichrysum pulchellum** *DC.*,
Prodr. 6: 190 (1838); Harv. in F.C. 3: 220 (1865); Moeser in Bot. Jb. 44: 293 (1910); Hilliard & Burtt in Bot. J. Linn. Soc.

82: 262 (1981). Type: Cape, Ceres div., between Hex River Mountains and Warm Bokkeveld, Sept., *Drège* 614 (G-DC, holo.!; K; S, iso.!).

H. cochleariforme sensu Harv. in F.C. 3: 220 (1865); Moeser in Bot. Jb. 44: 293 (1910), non DC.

H. concinnum N.E. Br. in Kew Bull. 1897: 269 (1897). Type: Namaqualand, near Ezel's Fontein and Rood Berg, 3 500−4 000 ft, *Drège* (K, holo.!).

Perennial herb 50−200 mm tall, much branched from the base and woody there, stems simple or subsimple branching above into the compound inflorescence, grey-woolly, leafy. *Leaves* up to 22 × 5 mm, diminishing upwards, linear, linear-oblong, or lanceolate, acute or subacute, mucronate, base broad, half-clasping, often auriculate, margins crisped undulate, both surfaces grey-woolly or cobwebby, sometimes glabrescent. *Heads* heterogamous, campanulate, c. 4−5 × 3−4 mm, solitary or few in terminal clusters, these arranged in a leafy corymbose panicle. *Involucral bracts* in c. 6−7 series, graded, loosely imbricate, innermost just overtopping flowers, all tips subacute to obtuse, somewhat concave, light golden-brown, innermost minutely radiating. *Receptacle* shortly honeycombed. *Flowers* 40−66, 6−13 ♀, 30−55 ☿. *Achenes* not seen, ovaries glabrous, or rarely with myxogenic duplex hairs. *Pappus* bristles many, equalling corolla, tips shortly subplumose or barbellate, bases nude, not cohering. Fig. 31:4.

Ranges from the Richtersveld south of the Orange River through Namaqualand to Koo and Karoopoort near Ceres and Pieter Meintjies near Laingsburg, in dry sandy or rocky places; flowering between September and November. Map 104.

Sometimes confused with *H. cylindriflorum* (above) but distinguished by its involucral bracts, all with broad golden-brown subpellucid tips, not squarrose as in *H. cylindriflorum* (though the innermost are minutely radiating), and the consistently larger number of flowers in the head.

Vouchers: *Acocks* 14800 (K; PRE); *Nordenstam* 1642 (S); *Rogers* 17894 (K; PRE; Z); *Salter* 1582 (BM; BOL; K); *Scully* 1183 (BM; BOL; E; K; SAM).

89. **Helichrysum alsinoides** *DC.*,
Prodr. 6: 169 (1838); Harv. in F.C. 3: 214 (1865); Moeser in Bot. Jb. 44: 296 (1910); Merxm. & Roessl. in Mitt. bot. StSamml., Münch. 15: 365 (1979). Type: South Africa, *Drège* 5811 (G-DC, holo.!).

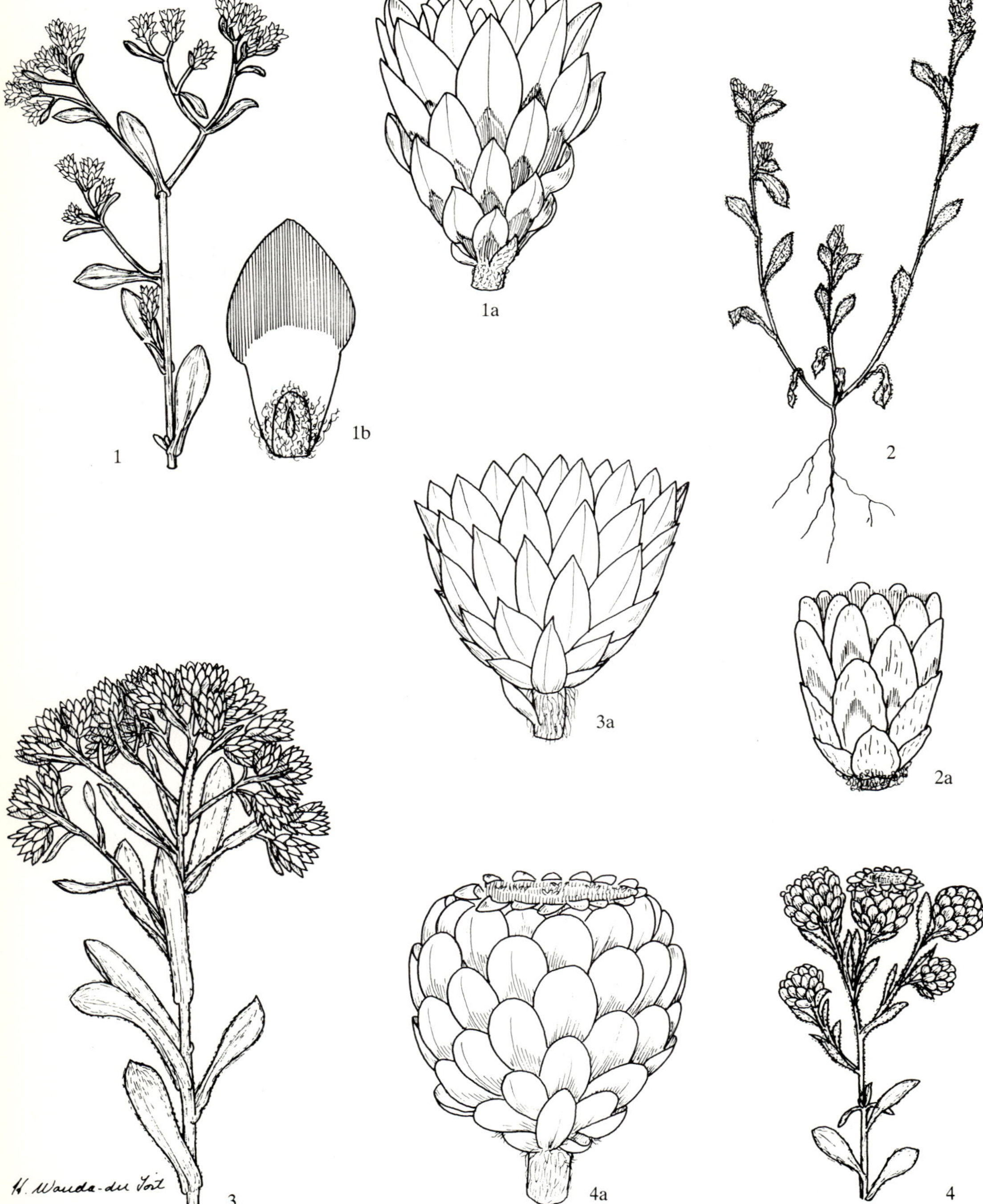

1
1a
1b
2
2a
3
3a
4
4a
H. Wanda-du Toit

Gnaphalium alsinoides (DC.) Sch. Bip. in Bot. Ztg 3: 169 (1845).

Annual herb with a slender taproot, branches 30—150 mm long, many from the crown, slender, prostrate, loosely branched, reddish, loosely woolly, distantly leafy except at the crown and under the heads. *Leaves* 5—23 × 1—9 mm, spathulate to oblong-spathulate, apex obtuse, base narrowed, half-clasping, margins flat, both surfaces loosely grey-woolly. *Heads* homogamous or heterogamous, campanulate, c. 3—4 × 2,5 mm, in small leafy clusters at the branch tips. *Involucral bracts* in 4—5 series, graded, outer lightly webbed with wool to surrounding leaves, inner about equalling flowers, not radiating, tips very obtuse, somewhat concave, semipellucid, whitish to light golden-brown. *Receptacle* nearly smooth. *Flowers* 14—38, 0—3 ♀, 14—37 ♀. *Achenes* 0,75 mm long, with myxogenic duplex hairs. *Pappus* bristles many, equalling corolla, tips scabrid, bases not cohering. Fig. 31:3.

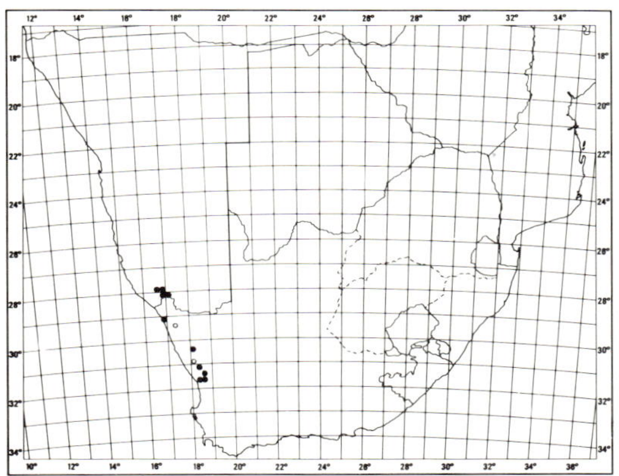

MAP 105.— ● **Helichrysum alsinoides**
○ **Helichrysum leptorhizum**

Recorded from southernmost S.W.A./Namibia south to Vanrhynsdorp, probably always in seasonally damp places and flowering between July and October. Map 105.

Specimens from north of the Orange River and near Port Nolloth have whitish or pale involucral bracts, only the innermost sometimes tinged light brown; further south, bracts are light golden-brown.

Vouchers: *Compton* 11554 (M; NBG); *Esterhuysen* 1364 (BOL); *Giess* 13835 (M; PRE); *Merxmüller & Giess* 28601 (M).

90. **Helichrysum leptorhizum** *DC.*, Prodr. 6: 169 (1838); Harv. in F.C. 3: 214 (1865); Moeser in Bot. Jb. 44: 296 (1910). Type: Little Namaqualand, Springbok distr., between Kaus Mountains and Gariep, *Drège* (G-DC, holo.!; K; P, iso.!).

Gnaphalium leptorhizum (DC.) Sch. Bip. in Bot. Ztg 3: 169 (1845).

Delicate annual herb c. 60 mm tall, stems several from the base, weakly erect, filiform, simple below, weakly branched above, thinly and loosely greyish-white woolly, glabrescent, distantly leafy. *Leaves* 5—15 × 2,5—4 mm, lower oblanceolate, much narrowed to the base, becoming spathulate to elliptic upwards, apex subacute to obtuse, apiculate, both surfaces thinly greyish white-woolly cobwebby. *Heads* heterogamous, campanulate, c. 3 × 1,5—2 mm, 2 or 3 together on minute branchlets in the upper leaf axils and at the branchlet tips, shortly pedunculate. *Involucral bracts* in c. 5 series, graded, outer pale golden-brown, pellucid, inner about equalling flowers, brown flushed crimson, tips obtuse, more or less opaque whitish, not radiating. *Receptacle* scarcely honeycombed. *Flowers* c. 19—22, 10—11 ♀, 9—11 ♀. *Achenes* 0,75 mm long, with myxogenic duplex hairs. *Pappus* bristles about equalling corolla, scabrid, bases with minute patent cilia, not cohering. Fig. 32:2.

Collected by Drège in northern Namaqualand and by Schlechter at Eenkokerboom, south of Garies, in September 1897, otherwise unknown. Map 105.

Voucher: *Schlechter* 11055 (BM; BOL; G; K; PRE).

91. **Helichrysum incarnatum** *DC.*, Prodr. 6: 191 (1838). Type: Cape, Breede River, *Drège* 5775 (G-DC, holo.!; K; fragment S, iso.!).

FIG. 32.—1, **Helichrysum incarnatum**, flowering twig, × 1; 1a, head, × 8; 1b, involucral bract, × 13 (*Hilliard & Burtt* 13059). 2, **H. leptorhizum**, whole plant, × 1; 2a, head, × 10 (*Schlechter* 11055). 3, **H. stellatum**, flowering twig, × 1; 3a, head, × 5,3 (*Howes* 215). 4, **H. cochleariforme**, flowering twig, × 1; 4a, head, × 4,6 (*Acocks* 24098).

Gnaphalium incarnatum (DC.) Sch. Bip. in Bot. Ztg 3: 171 (1845). *Helichrysum rubellum* (Thunb.) Less. var. *incarnatum* (DC.) Harv. in F.C. 3: 220 (1865).

H. ramulosum DC., Prodr. 6: 176 (1838). *Gnaphalium ramulosum* (DC.) Sch. Bip. in Bot. Ztg 3: 169 (1845). Type: Cape, Hassaqua's Kloof, 28 i 1815, *Burchell* 7530 (G-DC, holo.!; K, fragment BOL, iso.!).

Bushy half-shrub c. 100−200 mm high, branching from the base and woody there, branches often simple or subsimple, branching only above into the compound inflorescence, thinly greyish-white woolly, leafy, leaves often with dwarf axillary shoots, these sometimes terminating in a solitary head. *Leaves* mostly 12−20 × 3−5 mm, diminishing in size upwards, subspathulate, uppermost lanceolate or oblong, apex obtuse to subacute, mucronate, base broad, half-clasping, sometimes auriculate, margins often slightly crisped, both surfaces greyish-white woolly. *Heads* c. 5 × 4 mm, turbinate-campanulate, 2 or 3 together at the tips of the branchlets, these arranged in a large leafy corymbose panicle. *Involucral bracts* in c. 6 series, graded, imbricate, all lanceolate, acute, outer somewhat woolly on the backs below, tips membranous, light golden-brown sometimes suffused pink, acute, not squarrose, becoming opaque inwards, pink to light scarlet, innermost sometimes whitish, overtopping flowers, radiating. *Receptacle* scarcely honeycombed. *Flowers* 27−48, 4−8 ♀, 21−44 ☿. *Achenes* not seen, ovaries with myxogenic duplex hairs. *Pappus* bristles many, equalling corolla, scabrid, bases not cohering. Fig. 32:1.

Endemic to the SW. Cape, rarely collected, and recorded only from sandy flats in Worcester, Bredasdorp and Riversdale degree squares; flowering between September and November. Map 106.

Vouchers: *Acocks* 1670 (S); *Barker* 7510 (NBG); *Hilliard & Burtt* 13059 (E; K; MO; NU; S); *Muir* 674 (PRE).

H. incarnatum is allied to *H. cylindriflorum* (no. 86) but is distinguished by subtle differences in the involucre. The bracts are decidedly lanceolate, not mostly oblong as they are in *H. cylindriflorum*, and although there is some wool on the backs of the shafts, it is not so conspicuous as it is in *H. cylindriflorum* and does not web the bracts together. Many of the bracts have a rosy cast (rare in *H. cylindriflorum*) though the tips of the inner bracts may be creamy white: these tips are more acute than they generally are in *H. cylindriflorum*, longer (the longest c. 2−2,5 mm), flatter, and always opaque, and the outer bracts are not squarrose. There appear to be generally more flowers in the head too, though counts are too few to be certain.

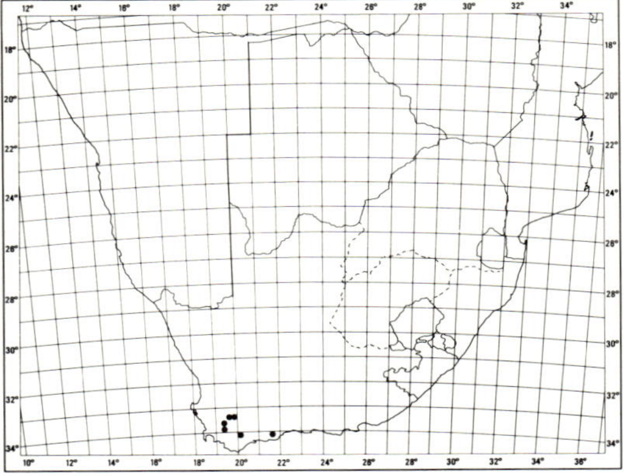

MAP 106.— **Helichrysum incarnatum**

Bolus 9020 (BOL; K; PRE) may be a variant of *H. incarnatum*: the leaves are slightly larger (up to 35 × 7 mm), the heads slightly larger (c. 7 × 7 mm), containing c. 60 flowers, and the tips of the inner involucral bracts are bright yellow, not pink. The specimen was collected between Pakhuis and Biedouw at an altitude of 360 m, in October 1897 (when the flowers were only in bud); this is well north of the known area of *H. incarnatum*. Specific rank may be appropriate.

92. **Helichrysum stellatum** *(L.) Less.*, Syn. Comp. 279 (1832); DC., Prodr. 6: 174 (1838); Harv. in F.C. 3: 220 (1865); Moeser in Bot. Jb. 44: 292 (1910). Lectotype: specimen from which Burmann, Rar. Afr. Pl. t. 80, fig. 1 was prepared (G-DC!).

Gnaphalium stellatum L., Pl. Rar. Afr. 19 (1760), Amoen. Acad. 6: 98 (1763).

G. pyramidale Berg., Descr. Pl. Cap. 255 (1767), non Thou. (1811). Type: Cape of Good Hope, *Grubb* (STB, holo.!).

G. plenum Burm. f., Prodr. Fl. Cap. 25 (1768). Type: Specimen in herb. Burm. (G!).

G. carneum Lam., Encycl. 2: 741 (1788). Type: Cape of Good Hope (P-LAM!).

G. congestum Lam., l.c., non *Helichrysum congestum* Moench (1802). *H. congestum* (Lam.) D. Don in Sweet, Hort. Brit. 223 (1826), nom. illegit. Type: Cape of Good Hope (P-LAM!).

G. fulvum Lam., Encycl. 2: 743 (1788). Type: Cape of Good Hope (P-LAM!).

Xeranthemum fragrans Andr., Bot. Rep. 9, t. 561 (1809). *Helichrysum fragrans* (Andr.) Lindl. in Loudon, Encycl. 702 (1829). Type: cult. Clapham 1803, no specimen found.

H. stellatum var. *laxum* DC., Prodr. 6: 174 (1838). Lectotype: Cape, Stellenbosch div., Zwartland, Riebekkasteel and Paardeberg, *Ecklon* 444 (G-DC!).

Bushy half-shrub 100−450 mm tall, woody at the base and much branched there, branches erect or spreading, simple or forked, thinly greyish-white woolly, leafy throughout. *Leaves* mostly 10−40 × 2−6 mm, diminishing slightly upwards, linear-oblong to spathulate, apex acute to obtuse, apiculate, base narrowed, half-clasping, often auriculate, margins flat or crisped, both surfaces grey-woolly. *Heads* hetero-gamous, broadly campanulate, c. 7 × 7 mm, many in terminal corymbose panicles. *Involucral bracts* in 7−8 series, graded, loosely imbricate, inner exceeding flowers, radiating, all lanceolate, acute or subacute, flat, outer pale to dark golden-brown, inner opaque milk-white, occasionally whole head suffused rose pink. *Receptacle* very shortly honeycombed. *Flowers* 59−142, 8−22 ♀, 50−117 ☿. *Achenes* 0,75 mm long, glabrous or with myxogenic duplex hairs. *Pappus* bristles many, about equalling corolla, scabrid, tips barbellate, bases nude, not cohering. Fig. 32:3.

Endemic to the SW. Cape, from Vanrhynsdorp and Calvinia districts south to the Cape Peninsula and Worcester and east to Montagu and Swellendam. Grows in sandy places; flowering mainly in September and October. Map 107.

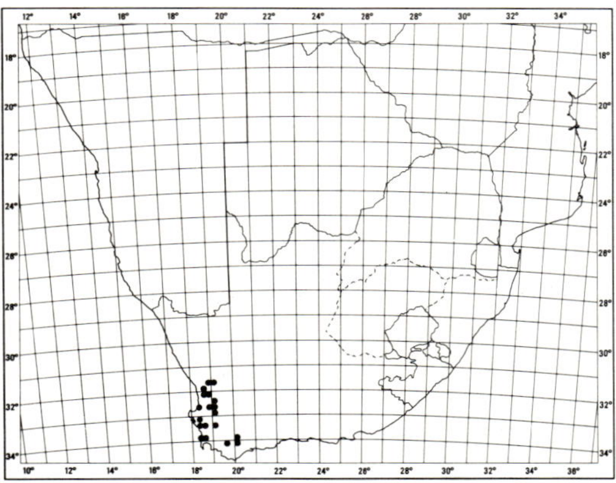

MAP 107.— **Helichrysum stellatum**

Sometimes confused with *H. cochleariforme* (no. 85) but easily distinguished by the flat, acute or subacute, not concave and obtuse, involucral bracts.

Vouchers: *Compton* 11769 (NBG); *Esterhuysen* 16005 (BOL; NBG); *Hall* 4508 (NBG; PRE); *Pillans* 8739 (NBG; PRE).

Penfold s.n. (NBG) from Strandfontein (3418 BA) may be a hybrid between *H. stellatum* and *H. indicum* (no. 110).

93. **Helichrysum cerastioides** *DC.*,
Prodr. 6: 171 (1838); Harv. in F.C. 3: 216 (1865); Moeser in Bot. Jb. 44: 302 (1910); Merxm., F.S.W.A. 139: 93 (1967). Lecto-type: Cape, beyond the Gariep [Orange], not far from Litakun, *Burchell* 2225−1 (G-DC!; K, isolecto.!).

Gnaphalium cerastioides (DC.) Sch. Bip., in Bot. Ztg 3:169 (1845).

Helichrysum obvallatum DC., Prodr. 6: 174 (1838); Harv. in F.C. 3: 222 (1865). *Gnaphalium obvallatum* (DC.) Sch. Bip. in Bot. Ztg 3: 169 (1845). Lectotype: Cape, Queenstown distr., Klipplaat River, Shiloh, Nov., *Drège* 5809 (G-DC!; BM, isolecto.!).

H. cerastioides var. *gracile* Moeser, l.c. Lectotype: S.W.A./Namibia Grootfontein, *Dinter* 666 (Z!).

H. gracile [Moeser ex] Dinter in Fedde, Repert. 18: 249 (1922), nomen.

Two varieties are recognized:

(a) var. **cerastioides.**

Bushy herb, taproot woody, main branches up to c. 200 mm, erect, decumbent or prostrate, simple or branched, grey-woolly, closely leafy throughout. *Leaves* 5−20 × 1−4 mm, scarcely decreasing in size upwards, often patent, linear, linear-lanceolate, oblanceolate or oblong, obtuse to acute, margins more or less revolute, base broad, half-clasping, shortly decurrent, both surfaces grey-woolly. *Heads* hetero-gamous, campanulate, c. 7−10 × 6−10 mm, solitary or few clustered at the tips of very short side branches at the tips of the main branches, surrounded by leaves webbed together with wool. *Involucral bracts* in c. 6 series, subequal, loosely imbricate, exceed-ing flowers, acute to acuminate, white, pellucid and glossy below, tips white sometimes tinged rose, subopaque, radiat-ing, outermost sometimes tinged pale golden-brown. *Receptacle* shortly honey-combed. *Flowers* 41−172, 10−36 ♀, 36−136 ☿, yellow, tipped pink. *Achenes* 0,75 mm long, with myxogenic duplex hairs. *Pappus* bristles many, equalling corolla, scabrid, bases with patent cilia, not coher-ing. Fig. 33:5.

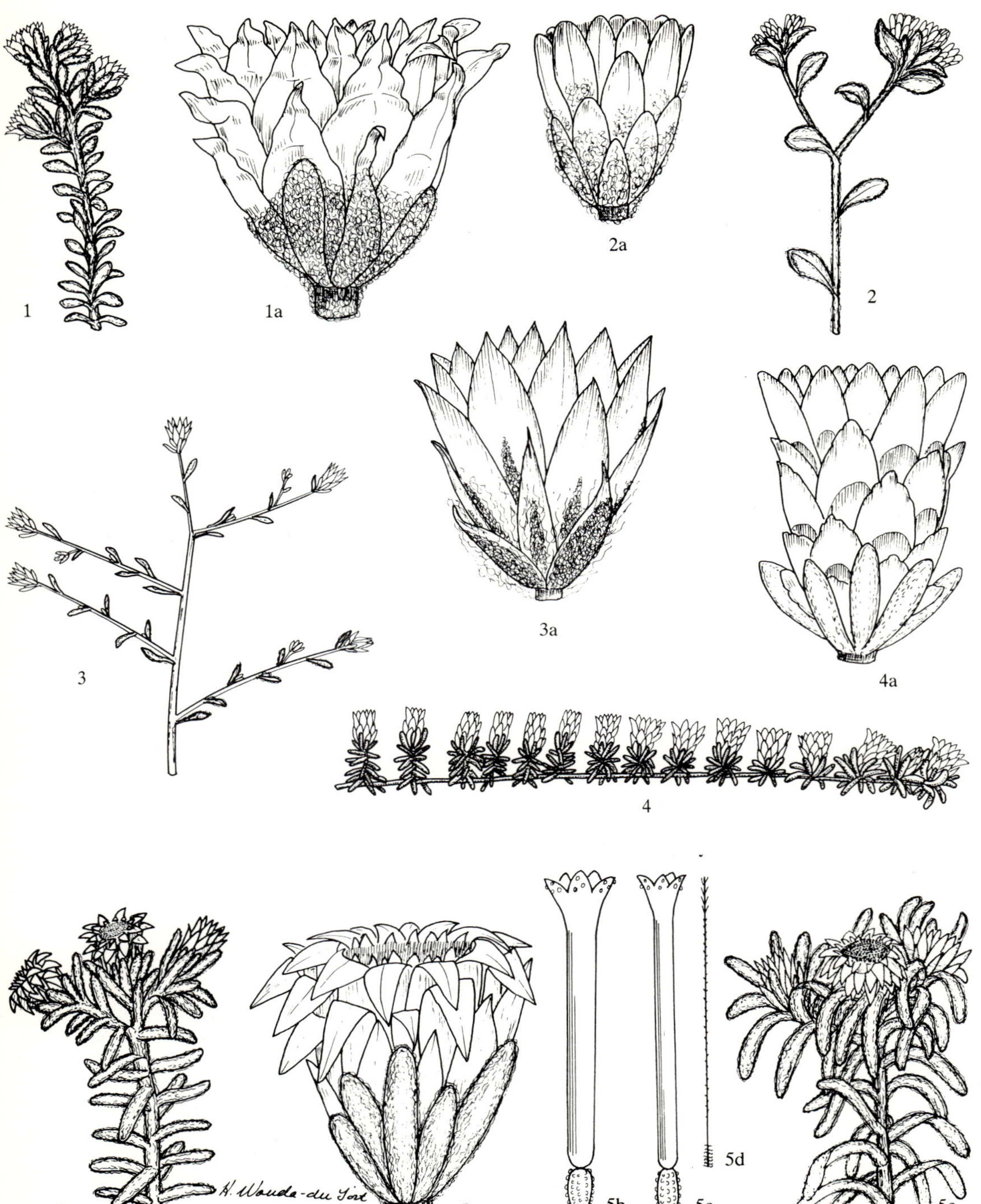

Widespread in central southern Africa from NE. S.W.A./Namibia through Botswana to Zimbabwe, the arid NE. Cape (E of 22°E), the Transvaal Highveld, the dry western Orange Free State and the E. central Cape from Victoria West to Cradock and Queenstown. Favours open flats and slopes on sand, gritty quartzites or calcareous soils; flowering mainly between June and October. Often behaves as a weed. Map 108.

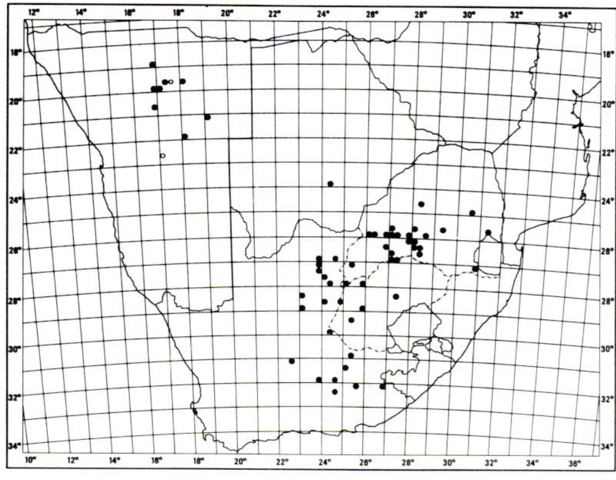

MAP 108.— ● **Helichrysum cerastioides** var. **cerastioides**
○ **Helichrysum cerastioides** var. **aurosicum**

Over the eastern part of the range, specimens resemble the type of *H. obvallatum* and have large heads (both in overall size and in the number of flowers in a head) and large leaves with margins scarcely revolute. North-westwards, from the N. central Cape to Zimbabwe, Botswana and Namibia, both heads and leaves tend to become smaller, and the leaves are more spreading with margins more strongly revolute. The types of both *H. cerastioides* and var. *gracile* match this form. In S.W.A./Namibia, *H. cerastioides* is sometimes confused with pale-headed specimens of *H. pumilio* subsp. *fleckii*, but the narrow, more or less linear, patent leaves of *H. cerastioides* in this region tend to give it a characteristic facies. Also *H. cerastioides* grows on the flats, while *H. pumilio* subsp. *fleckii* is on the mountains. However, var. *aurosicum* is problematical (see below).

Vouchers: *Acocks* 11720 (PRE); *Dinter* 7671 (M; PRE; WIND); *Flanagan* 1422 (NBG; PRE); *Leach & Bayliss* 12975 (M; PRE; WIND); *Tyson* 261 (SAM).

(b) var. **aurosicum** *Merxm. & Schreiber*

in Mitt. bot. StSamml. Münch. 2: 328 (1957); Merxm., F.S.W.A. 139: 93 (1967). Type: S.W.A./Namibia, Grootfontein, Gaub near Auros, *Rehm* s.n. (M, holo.!).

H. engelianum Dinter in Fedde, Repert 18: 249 (1922). Type: S.W.A./Namibia, Windhoek, Moltke-blick, 6 ix 1909, *Engel* in herb. Dinter 1566 (K, SAM, iso.!).

Distinguished from var. *cerastioides* by its rosy heads. Specimens from Auros, whence came the type, have the more or less linear spreading leaves typical of *H. cerastioides* in that region. But other specimens, for example *Giess* 13673 (M; PRE) and *Merxmüller & Giess* 28030, (M; PRE) from the Auasberge in Windhoek district have spathulate leaves more like those of *H. pumilio* and indeed it is difficult to decide whether these and other specimens are better placed under *H. cerastioides* or under *H. pumilio* subsp. *fleckii*. The problem demands a field study, which should be extended to the serpentine rocks of the Great Dyke in Zimbabwe whence came *Wild* 6380 (K; M) which is scarcely to be distinguished from specimens from Auros. Map 108.

94. **Helichrysum pumilio** *(O. Hoffm.) Hilliard & Burtt* in Bot. J. Linn. Soc. 82: 198 (1981). Type: Cape, Beaufort West, 950 m, 6 ii 1894, *Kuntze* (NY, holo.; K, iso.!).

Leontonyx pumilio O. Hoffm. in O. Kuntze, Rev. Gen. 3,2: 162 (1898).

Gnaphalium pusillum Thunb., Prodr. 149 (1800), Fl. Cap. 651 (1823), non Haenke (1791). *G. nanum* Willd., Sp. Pl. 3: 1898 (1803), non *Helichrysum nanum* Klatt. *Leontonyx pusillus* Less., Syn. Comp. 327 (1832). *Helichrysum pachyrhizum* Harv. var. β *thunbergii* Harv. in F.C. 3: 222 (1865). Type: Cape of Good Hope, *Thunberg* (sheet 19238, UPS, holo.!).

H. bolusianum Moeser in Bot. Jb. 48: 338 (1913). Type: Cape, Prince Albert Road railway station, 600 m, Dec. 1905, *Bolus* 11972 (BOL, iso.!).

H. hutchinsonii Phill. in Ann. S. Afr. Mus. 9: 343 (1917). Type: Cape, Namaqualand, between Steinkopf

---

FIG. 33.—1, **Helichrysum pumilio** subsp. **pumilio,** flowering branchlet, × 1; 1a, head, × 8 (*Acocks* 24611). 2, **H. obtusum,** flowering branchlet, × 1; 2a, head, × 6,6 (*Van der Schijff* 8182). 3, **H. oxybelium,** flowering branch, × 1; 3a, head, × 8 (*Acocks* 19059). 4, **H. paronychioides,** part of plant, × 1; 4a, head, × 8 (*Prosser* 1010). 5, **H. cerastioides** var. **cerastioides,** flowering branchlet of plant from western part of range, × 1; 5a, head, × 5,3; 5b, hermaphrodite flower, × 10; 5c, female flower, × 10; 5d, pappus bristle, × 10 (*Leach & Bayliss* 12975); 5e, flowering branchlet of plant from eastern part of range, × 1 (*Smith* 313).

and the Orange River, near Henkries, Oct. 1911, *Phillips* 1612 (SAM, holo.!).

H. laneum S. Moore in J. Bot., Lond. 56: 6 (1918). Type: Cape, Laingsburg, Aug. 1915, *Rogers* 16760 (BM, holo.!; BOL; G; K; P; PRE; Z, iso.!).

Two subspecies are recognized:

(a) subsp. **pumilio.**

Closely branched rounded dwarf shrublet c. 20—200 mm high or sometimes lax and open, taproot thick, woody, main stem gnarled, short, eventually up to 20 mm diam., branches erect or spreading, grey-woolly, densely leafy or sometimes more distantly so. *Leaves* mostly 4—20 × 1—4 mm, linear-spathulate, apex subacute, apiculate, base narrowed, petiole-like, very shortly decurrent, margins subrevolute, both surfaces thinly to thickly glandular pubescent, glandular hairs usually hidden by persistent grey wool. *Heads* heterogamous, campanulate, c. 5—6 × (3—) 5—7 mm, sessile, solitary and terminal, often overtopped by younger growth, or solitary on dwarf lateral branchlets, or leaf-opposed and crowded at branchlet tips, or in small terminal corymbose clusters. *Involucral bracts* in c. 5 series, graded, loosely imbricate, inner about equalling flowers, outermost webbed to surrounding leaves, all semipellucid, often flushed red above the stereome, tips rich golden-brown to palest buff, or whitish, subopaque or innermost pellucid, very acute to acuminate or rarely more abruptly contracted to an acute tip, somewhat squarrose. *Receptacle* nearly smooth. *Flowers* 30—96, 6—30 (—51) ♀, 25—84 ♀, yellow, sometimes tipped pink. *Achenes* 0,75 mm, with myxogenic duplex hairs. *Pappus* bristles many, equalling corolla, scabridulous, bases cohering lightly by patent cilia. Fig. 33:1.

Recorded mainly from the arid parts of the Cape, from Namaqualand west to Prieska and extending into the Orange Free State, and south to Laingsburg, Prince Albert Road, Graaff-Reinet and Pearston in the Karoo. Favours gravelly or rocky places; flowering mainly between August and March. A 'Karoo bush', tufted and twiggy or laxer under more favourable conditions, palatable to stock. Map 109.

Despite differences in facies produced by growing conditions or grazing, the species is easily recognized by its linear-spathulate leaves allied to very acute, more or less squarrose involucral bracts. A specimen collected in the mountains near Murraysburg (*Tyson*

sub *MacOwan* 255, SAM) is remarkable in that ♀ flowers outnumber ♀ (42—51 ♀, 36—37 ♀).

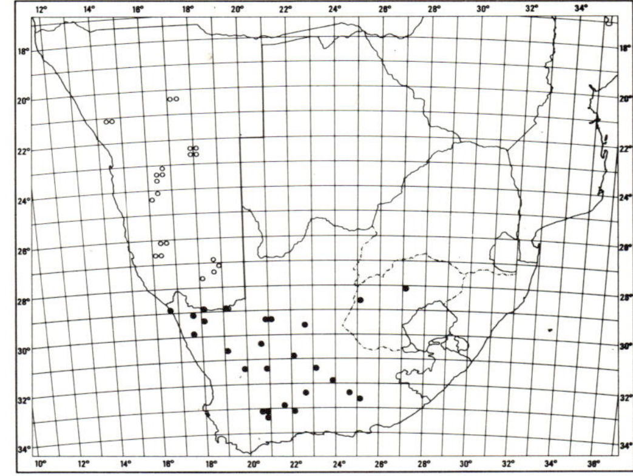

MAP 109.— ● **Helichrysum pumilio** subsp. **pumilio**
          ○ **Helichrysum pumilio** subsp. **fleckii**

Vouchers: *Acocks* 17708 (K; M; PRE); *Barker* 7101 (NBG); *Galpin* 9998 (PRE); *Hutchinson* 973 (K, PRE); *Pillans* 14149 (BOL).

(b) subsp. **fleckii** *(S. Moore) Hilliard,* comb. et stat. nov. Type: S.W.A./Namibia, Great Namaqualand, Nauchas [?], April 1891, *Fleck* s.n. (Z, holo.!; BM, iso.!).

H. fleckii S. Moore in Bull. Herb. Boissier sér. 2, 4: 1017 (1904); Moeser in Bot. Jb. 44: 303 (1910); Merxm., F.S.W.A. 139: 93 (1967).

H. dinteri S. Moore in Bull. Herb. Boissier sér 2, 4: 1016 (1904). *H. fleckii* var. *dinteri* (S. Moore) Merxm. & Schreiber in Mitt. bot. StSamml. Münch. 2: 329 (1957); Merxm., F.S.W.A. 139: 934 (1967). Type: Hereroland, Waterberg plateau, *Dinter* 387 (Z, holo.!; BM, iso.!).

H. viscidissimum Hutch. in Ann. Bolus Herb. 3: 7 (1920); Merxm., F.S.W.A. 139: 98 (1967). Type: S.W.A./Namibia, Naukluft Mts below Goas, *Pearson* 9066 (K, holo.!; BOL; SAM, iso.!).

H. viscidissimum Hutch. var. *volkii* Merxm. in Mitt. bot. StSamml. Münch. 2: 331 (1957), F.S.W.A. 139: 99 (1967). Type: S.W.A./Namibia, Otjiwarongo distr., Little Waterberg, *Volk* 2332 (M, holo.!).

Differs from the typical plant principally by its white and rose, crimson or almost scarlet involucral bracts, the rosy tinge sometimes rather pale. In the southernmost part of the range, some specimens are not or scarcely to be distinguished from typical *H.*

pumilio. Heads range from c. 5—7 mm long; leaves are generally thinly grey-woolly, but there appear to be either local races or sporadic occurrences of plants distinguished by absence of wool (*H. viscidissimum*) or by the possession of very densely woolly leaves (*H. dinteri*), or older leaves can be green and glandular, young leaves grey-woolly on the same plant; phenomena all paralleled in several other species of *Helichrysum*.

Subsp. *fleckii* has been recorded only in the highlands of S.W.A/Namibia, from the Waterberg and Brandberg, the mountains about Windhoek, the Khomas Hochland, the Gamsberg and Naukluft mountains, south to the heights near Aus and the Little and Great Karasberge. Grows in rocky or gravelly places, often in clefts in rocks, mainly on mica schist, quartz and granite. Flowers mainly between July and December. Can be confused with *H. cerastioides*, particularly var. *aurosicum*: see under that species (no. 93). Map 109.

Vouchers: *Dinter* 853 (SAM); *Merxmüller & Giess* 28262 (M; PRE; WIND); *Örtendahl* 515 (BOL; K; PRE); *Merxmüller & Giess* 28143 (M; PRE), all typical *H. fleckii*; *Meyer* 1183 (M; WIND), typical *H. dinteri*; *Strey* 2165 (BOL; NBG; PRE), typical *H. viscidissimum*; *Acocks* 18037 (PRE), *Kinges* 2378 (M; PRE), pale heads, scarcely to be distinguished from subsp. *pumilio*.

95. **Helichrysum obtusum** (*S. Moore*) *Moeser* in Bot. Jb. 44: 297 (1910); Merxm. & Schreiber in Mitt. bot. StSamml. Münch. 2: 329 (1957); Merxm., F.S.W.A. 139: 96 (1967). Type: S.W.A./Namibia, Great Namaqualand, Gubub [Garub?], July 1897, *Dinter* 1212 (Z, holo.!; fragment BM, iso.!).

H. dinteri var. *obtusum* S. Moore in Bull. Herb. Boissier 2, sér. 4: 1016 (1904).

H. obtusum var. *namibense* Merxm. & Schreiber in Mitt. bot. StSamml., Münch. 2: 329 (1957). Type: S.W.A./Namibia, Lüderitzbucht, *Dinter* 6007 (M, holo.!; BM; BOL; E; G; K; fragment NU; PRE; S; SAM; Z, iso.!).

H. obtusum var. *microphyllum* Merxm. & Schreiber in Mitt. bot. StSamml. Münch. 2: 329 (1957). Type: S.W.A./Namibia, Windhoek distr., between Kapps Farm and Windhoek, *Walter* 134 (M, holo.!; PRE; WIND, iso.!).

Shrublet, main stem very short, 5—50 × 5—10 mm, bare, woody, gnarled, branches numerous, intricate, spreading, forming a dense cushion up to 150 × 300 mm, often much smaller, white silky-woolly-felted, leafy. *Leaves* mostly 4—10 × 2—5 mm, oblong-lanceolate to broadly elliptic or suborbicular, narrowed below in broader leaves, half-clasping, apex sub-acute, recurved, margins undulate, both surfaces densely white-woolly-felted. *Heads* heterogamous, very rarely homogamous, campanulate, c. 4 × 4 mm, solitary or few clustered at the tips of the branchlets, closely surrounded by leaves. *Involucral bracts* in 3—4 series, subequal, loosely imbricate, about equalling the flowers, not radiating, obtuse, often emarginate, golden-brown to buff or whitish, often red-purple above stereome, stereome woolly outside, outer bracts and surrounding leaves webbed together with wool. *Receptacle* nearly smooth. *Flowers* (12—) 20—42, (0—) 2—9 ♀, 12—33 ☿. *Achenes* not seen, ovaries with myxogenic duplex hairs, or very rarely glabrous. *Pappus* bristles many, about equalling corolla, scabrid, bases with patent cilia, not cohering. Fig. 33:2.

Recorded mainly from the southern part of S.W.A./Namibia, from Windhoek southwards to the Orange River, and from Alexander Bay in Namaqualand eastwards to Pofadder, Riemvasmaak and Kakamas. Six records from much further east (Cape, Middelburg, *Acocks* 19133, PRE; Barkly West division, Boetsap, *Brueckner* 138, PRE; Griqualand West, Campbell, *Wilman* 1436, BOL; Griqualand West, *Cook* 11, BOL; Douglas, Mazelsfontein, *Anderson* 580 (BOL); O.F.S., Luckhoff, *Verdoorn* 2162, PRE) are included with slight reservation, but I can detect no real distinguishing feature, though the heads are homogamous, while generally, but not always, heterogamous in the typical plant. *H. obtusum* grows in sand or gravel among rocks on the desert hills; four of the six doubtful specimens were recorded on limestone; the other two collectors did not record soil type. Flowering noted in all months, but chiefly in September. Map 110.

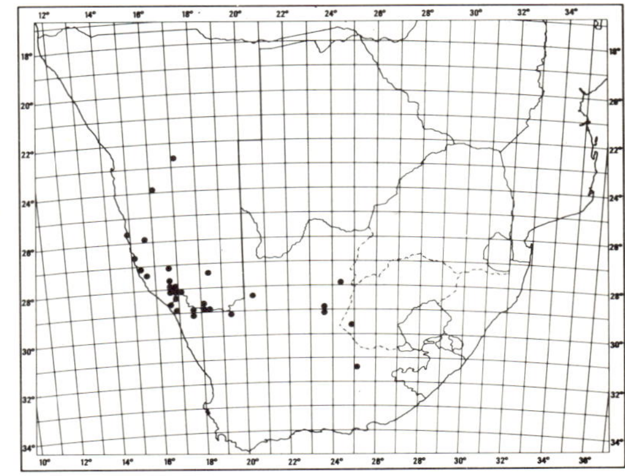

MAP 110.— **Helichrysum obtusum**

Specimens can be much dwarfed with densely tufted branches and small leaves, or laxer with relatively large leaves. Variation, which is probably induced by growing conditions, is continuous and there is no geographical patterning to it; it is not possible to uphold the two varieties described, purportedly distinguishable in characters of leaf size, shape and degree of reflexion of the tips.

Sometimes confused with *H. litorale* (no. 78), but the crisped-undulate leaf margins of *H. obtusum* are a good distinguishing feature, and so is the uniseriate pappus.

Vouchers: *Acocks* 14400 (PRE); *Giess* 12983 (M; PRE; WIND); *Merxmüller & Giess* 28279 (M; PRE; WIND); *Pillans* 5575 (BOL); *Range* 495 (SAM).

96. **Helichrysum deserticola** Hilliard, sp. nov.

H. obtuso *(S. Moore) Moeser affinis sed bracteis involucralibus opacis albis et abrupte apiculatis (nec translucentibus obtusis) et acheniis glabris (nec acheniis plerumque pilis myxogenis praeditis) distinguitur.*

Fortasse perennis, radice palari incrassata (4 mm diam.) et lignosa; caules e caudice multi dense caespitosi c. 20–60 mm longi, simplices vel subsimplices, griseolanati, foliati. Folia plerumque 4–10 × 1–4 mm, elliptico-spatulata, apice subacuto, interdum longitudinaliter plicata et recurvata, basi angustata semi-amplectente, marginibus paulo undulatis, utrinque dense albolanata. Capitula heterogama vel raro homogama, c. 3,5–4 × 2–2,5 mm, campanulata, pauca vel multa in glomerulos terminales condensata, foliis ad bracteas involucrales exteriores lana contextis circumcincta. Bracteae involucrales 3–4- seriatae, subaequales, flores aequantes, non radiantes, apicibus exteriorum obtusis interiorum abrupte acutis et apiculatis, opacae, albae, saltem exteriores aureo-brunneo-tinctae. Receptaculum brevissime favosum. Flores 14–38, (0–) 1–3 ♀, 14–36 ☿, flavi roseo-apiculati. Achenia 0,75 mm longa, glabra. Pappi setae multae, corollam aequantes, apicibus scabris, basibus non cohaerentibus.

Type: S.W.A./Namibia, Gubub [Garub?], sandflats, 18 x 1922, *Dinter* 4115 (SAM, holo.!; BM; G; Z, iso.!).

S.W.A./Namibia.—Lüderitz Süd, Rote Kuppe, 19 vii 1922, *Dinter* 3814 (BOL; PRE); [2615 DC], Namib, Kaukausib, rocky ground, x 1911, *Range* 1130 (SAM).

Probably perennial, taproot becoming thick (4 mm diam.) and woody, stems many, densely tufted from the crown, c. 20–60 mm long, simple or subsimple, grey-woolly, leafy. *Leaves* mostly 4–10 × 1–4 mm, elliptic-spathulate, apex subacute, sometimes folded lengthwise and recurved, base narrowed, half-clasping, margins somewhat undulate, both surfaces thickly white-woolly. *Heads* heterogamous or rarely homogamous, c. 3,5–4 × 2–2,5 mm, campanulate, few to many crowded in terminal glomerules 5–10 mm across, surrounded by leaves webbed to the outer involucral bracts. *Involucral bracts* in 3–4 series, subequal, equalling flowers, not radiating, tips of outer obtuse, inner abruptly acute and apiculate, opaque, white, outer at least tinged golden-brown. *Receptacle* very shortly honeycombed. *Flowers* 14–38, (0–) 1–3 ♀, 14–36 ☿, yellow, tipped pink. *Achenes* 0,75 mm, glabrous. *Pappus* bristles many, equalling corolla, tips scabrid, bases nude, not cohering.

Recorded only from southernmost S.W.A./Namibia (Kaukausib, Rote Kuppe, Garub). Grows on sandy or rocky flats; flowering between July and October. Rarely collected. Map 111.

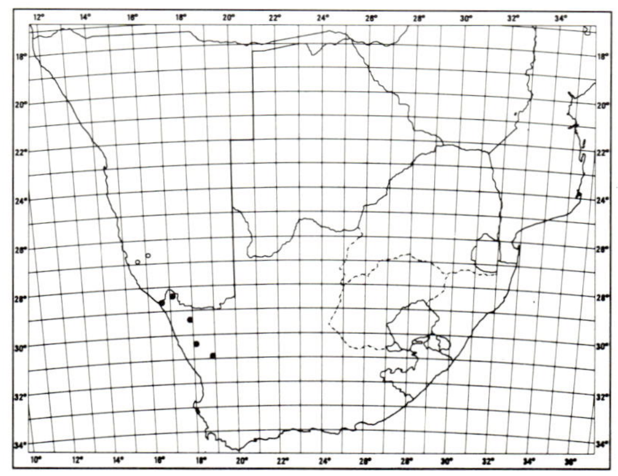

MAP 111.— ○ **Helichrysum deserticola**
● **Helichrysum oxybelium**

Closely allied to *H. obtusum* (above) but distinguished by the opaque white tips to the involucral bracts, the innermost of which are abruptly contracted into a sharp apiculus.

97. **Helichrysum oxybelium** DC., Prodr. 6: 171 (1838); Harv. in F.C. 3: 216

(1865); Moeser in Bot. Jb. 44: 299 (1910). Lectotype: Cape, Namaqualand, Kamiesberg, Boschmanland and mouth of the Gariep, *Ecklon* 1670 (G-DC.!).

Gnaphalium oxybelium (DC.) Sch. Bip. in Bot. Ztg 3: 169 (1845).

Rounded, stiff, twiggy, divaricately branched shrublet up to 300 mm tall, old twigs bare, reddish brown, subspinescent, young twigs glandular-pubescent, distantly leafy. *Leaves* 4–10 × 0,75–1 mm, linear, sessile, acute, apiculate, hooked, margins revolute, both surfaces thickly and loosely greyish-white woolly when very young, later only glandular-pubescent with some woolly hairs clinging particularly in the leaf axils and on the lower surface. *Heads* heterogamous, c. 4,5 × 3 mm, turbinate-campanulate, solitary at the tips of the twigs, surrounded by reduced leaves. *Involucral bracts* in 4–6 series, outermost short, glandular-pubescent below, thinly woolly, inner subequal, lanceolate, acuminate, about equalling flowers, erect, pellucid, glossy, purplish, tips golden-brown. *Receptacle* flat, honeycombed. *Flowers* 19–29, 7–13 ♀, 11–16 ☿. *Achenes* not seen, ovaries with myxogenic duplex hairs. *Pappus* bristles many, equalling corolla, scabrid, bases nude, not cohering. Fig. 33:3.

Recorded from Grootklip in Vanrhynsdorp district and Khamiesberg, Steinkopf, Ratelpoort Mountain, Corkscrew Mountain and the mouth of the Orange River in Namaqualand. Grows in 'Namaqualand Broken Veld of low granite domes' fide Acocks. Flowers in September and October. Map 111.

H. oxybelium is often confused with some forms of *H. asperum* sens. lat. (no. 59), but is most readily distinguished by its heterogamous heads containing more flowers than the homogamous heads of *H. asperum*. The Drège specimen cited by De Candolle as *H. oxybelium* proves to be *H. asperum* (Karroo, Konstapel, June, *Drège* G-DC!; BM!; E!).

Vouchers: *Acocks* 19059 (PRE); *Pearson* 2967 (K) and 5734 (BOL; K); *Schlechter* 11483 (PRE); *Schlechter* 11482 (BOL).

A specimen collected by Nordenstam in the Richtersveld differs from typical *H. oxybelium* in its lack of stalked glands, persistently woolly leaves with scarcely revolute margins and obtuse rather than very acute involucral bracts (Richtersveld, Granite Boss, S. of Kuboos, W. slope, granite rocks and coarse sand, 5 xi 1962, *Nordenstam* 1778, E; S!). However, *H. oxybelium* is known from so few collections that I hesitate to give formal recognition to this variant.

98. **Helichrysum paronychioides** *DC.*, Prodr. 6: 171 (1838); Harv. in F.C. 3: 216 (1865); Moeser in Bot. Jb. 44: 303 (1910). Type: Cape, beyond the Gariep [Orange], Pellat Plains, 10–28 August, *Burchell* 2234 (G-DC, holo.! BOL; E; M; S, iso.!).

Gnaphalium paronychioides (DC.) Sch. Bip. in Bot. Ztg 3: 169 (1845).

Mat-forming subshrub, stock woody, up to 10 mm diam., main branches up to 300 mm long, many, prostrate, radiating from the crown, branching and giving rise to numerous very short (up to 5 mm) erect flowering branchlets, very closely felted, densely leafy, the whole forming a flowery mat. *Leaves* 3–8 × 0,5 (–1,75) mm, not diminishing in size upwards, imbricate, linear, acute, base broad, half-clasping, margins generally revolute, both surfaces closely enveloped in closely woven, silky grey tissue-paper-like indumentum, glandular as well, rarely woolly. *Heads* heterogamous, narrowly campanulate, c. 6 × 4 mm, sessile, solitary at the tips of the branchlets, these racemosely arranged. *Involucral bracts* in c. 6 series, graded, loosely imbricate, inner exceeding flowers, all radiating, semi-pellucid, glossy, outer white, inner rose-red, tips more or less opaque, subacute, mostly white. *Receptacle* shortly honeycombed. *Flowers* 19–31, 3–10 ♀, 12–25 ☿, yellow, tipped pink. *Achenes* not seen, ovaries with myxogenic duplex hairs. *Pappus* bristles many, equalling corolla, scabridulous, bases cohering strongly by patent cilia. Fig. 33: 4.

Ranges from the central Transvaal (Springbok Flats) through the Bushveld and Lowveld east to the southern part of Kruger National Park, south and west to the NE. Cape about Vryburg, Fourteen Streams and Hopetown, the dry parts of the Orange Free State, and W. Lesotho. Also recorded from Lobatsi in SE. Botswana. Grows in dry sandy open places; flowering between June and November, mainly August and September. Map 112.

Sometimes confused with *H. caespititium* (no. 61), but distinguished, among other characters, by its solitary heterogamous heads on very short lateral branchlets.

Vouchers: *Acocks* 11715 (PRE); *Fawkes* 207 (NBG); *Galpin* 11560 (PRE); *Martin* 1021 (NBG); *Pegler* 931 (BOL).

De Candolle (l.c.) commented that his new species was remarkable for its glabrous foliage, but this is not

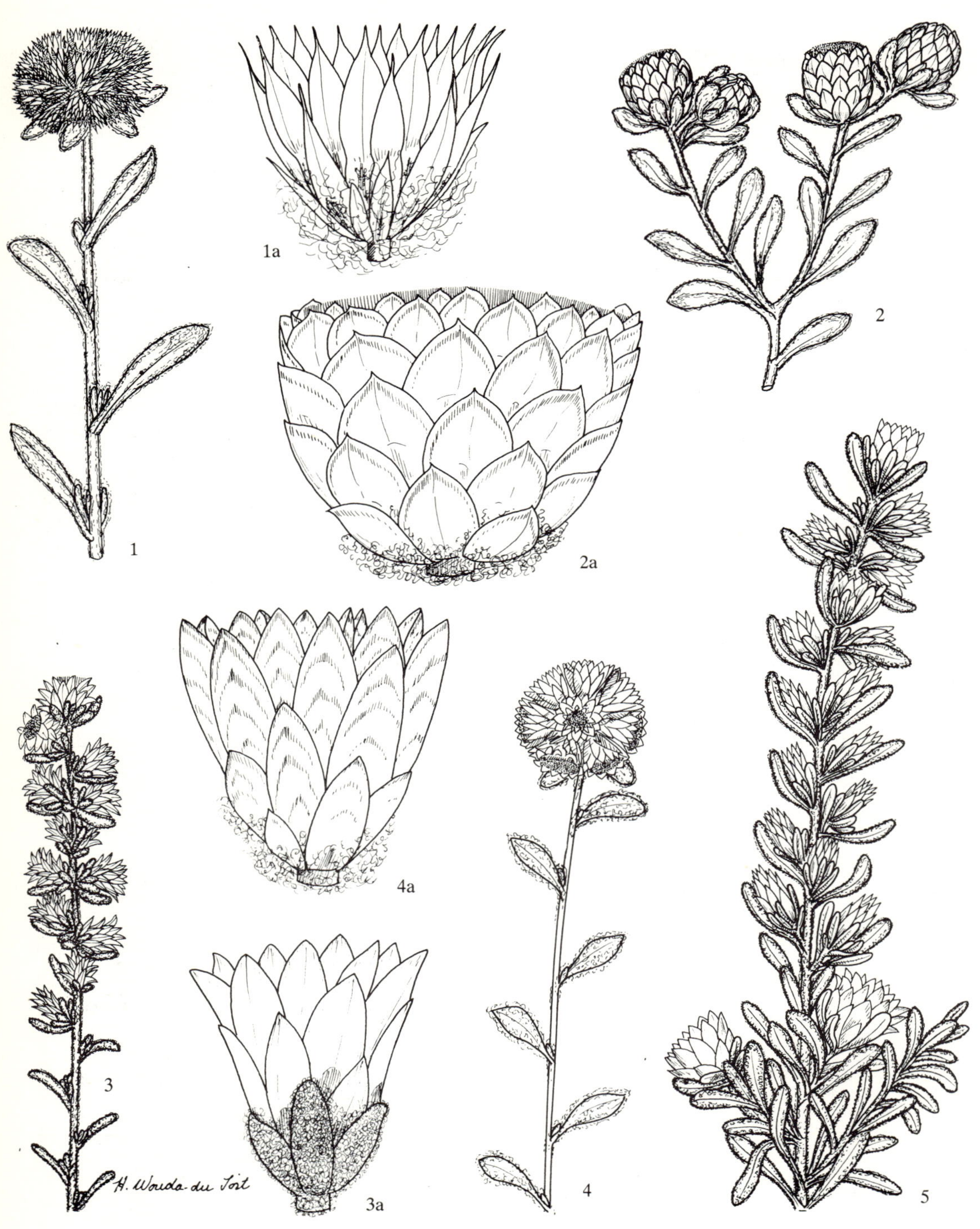
H. Wouda-du Toit

so; the leaves are clothed in very closely felted indumentum.

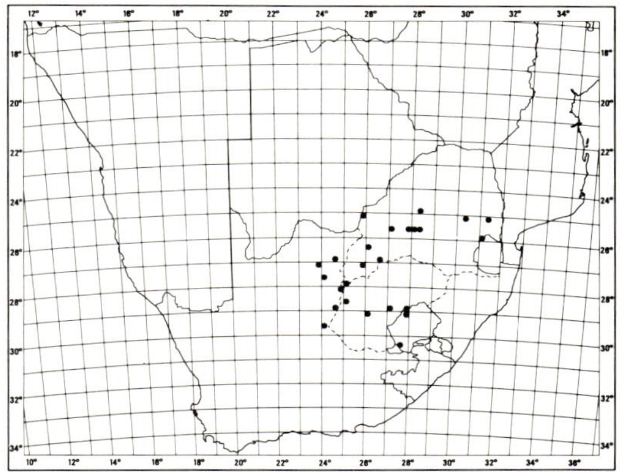

MAP 112.— **Helichrysum paronychioides**

99. Helichrysum spiciforme *DC.*, Prodr. 6: 171 (1838); Harv. in F. C. 3: 216 (1865); Moeser in Bot. Jb. 44: 302 (1910); Merxm., F. S. W. A. 139: 97 (1967). Type: Cape, near Takun, Pellat Plains at Jabiru Fountain, *Burchell* 2247 (G-DC, holo.!; BM; BOL; M; P; S; SAM; Z, iso.!).

Gnaphalium spiciforme (DC.) Sch. Bip. in Bot. Ztg 3: 169 (1845).

Woody perennial, main branches up to 400 mm long, apparently erect or decumbent, branching above, grey woolly-felted, glandular, leafy, each leaf with a very short (up to 10 mm) closely leafy axillary shoot. *Leaves* mostly 3−10 (−23) × 1−1,5 (−4) mm, the smallest on the axillary shoots, linear or the largest oblanceolate, obtuse, base broad, half-clasping, margins revolute, both surfaces thickly grey-woolly, glandular. *Heads* heterogamous, campanulate, c. 7−8 × 5 mm, 10 mm when expanded, solitary, sessile at the tip of each axillary shoot, these racemosely arranged and the whole very floriferous. *Involucral bracts* in

5−6 series, graded, loosely imbricate, inner exceeding flowers, radiating, tips subopaque, white, very acute. *Receptacle* nearly smooth. *Flowers* c. 23−30, 7−11 ♀, 16−21 ☿, yellow, tipped pink. *Achenes* not seen, ovaries with myxogenic duplex hairs. *Pappus* bristles very many, equalling corolla, scabridulous, bases cohering by patent cilia. Fig. 34: 3.

Recorded from Gordonia in the northern Cape (Van Zyl's Rus, Korannaberg, Kuie and Kgop Pans) east to Litakun and SE. to Du Toit's Pan near Kimberley, on deep loose sand. Also in Gobabis district, S.W.A./Namibia (fide Merxmüller, l.c.) and in Botswana. Map 113.

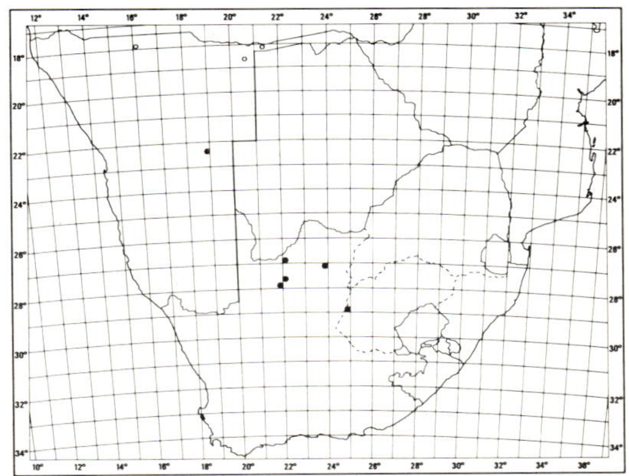

MAP 113.— ● **Helichrysum spiciforme**
         ○ **Helichrysum amboense**

Rarely collected; probably flowers between July and October.

Vouchers: *Drake* 2247 (PRE); *Leistner* 2086 (PRE); *Tuck* SAM 16303 (SAM).

100. Helichrysum amboense *Schinz* in Bull. Herb. Boissier 6: 561 (1898). Type: S.W.A./Namibia, Amboland, Oshiheke near Olukonda, August 1894, *Rautanen* 162 (Z, holo.!; BM; G, iso.!).

FIG. 34.−1, **Helichrysum candolleanum,** part of plant, × 1; 1a, head, × 5,3 (*Bagot-Smith* 5). 2, **H. argyrosphaerum,** part of plant, × 1,3; 2a, head, × 5,3 (*Hilliard* 4686). 3, **H. spiciforme,** part of plant, × 1; 3a, head, × 5,3 (*Drake* 2247). 4, **H. erubescens,** part of plant, × 1; 4a, head, × 6,6 (*Hall* 368). 5, **H. amboense,** part of plant, × 1,3 (*Maguire* 1629).

H. spiciforme DC. var. *amboense* (Schinz) Moeser in Bot. Jb. 44: 302 (1910); *H. spiciforme* subsp. *amboense* (Schinz) Merxm. in Mitt. bot. StSamml. Münch. 2: 330 (1957), F.S.W.A. 139: 97 (1967).

Perennial herb, main stems several from the crown, up to 600 mm long, erect (?) or prostrate (?), woody, up to 3 mm diam., glandular-hispid, thinly woolly, flowering branches stiffly divaricate, up to 70 mm long, slender (c. 1 mm diam.), glandular-hispid. *Leaves* on main stems up to 35 × 10 mm, oblanceolate, apex obtuse or subacute, base much narrowed, both surfaces greyish-white woolly, on flowering twigs up to 8 × 2 mm, linear, apex obtuse or subacute, base broad, half-clasping, margins revolute, both surfaces glandular-hispid, woolly only in extreme youth. *Heads* heterogamous, campanulate, c. 8–9 × 8 mm, c. 12 mm when expanded, solitary, sessile at the tips of much abbreviated dwarf shoots, appearing merely surrounded by leaves, shoots racemosely arranged, the whole twig very floriferous. *Involucral bracts* in c. 5 series, graded, loosely imbricate, inner exceeding flowers, tips radiating, subopaque white, very acute. *Receptacle* nearly smooth. *Flowers* c. 36–55, 7–15 ♀, 29–40 ☿. *Achenes* not seen, ovaries with myxogenic duplex hairs. *Pappus* bristles many, equalling corolla, scabridulous, bases cohering strongly by patent cilia. Fig. 34: 5.

Recorded only from northern S.W.A./Namibia, from the environs of Olukonda in the west to Tamsoe and Andara (Okavango) in the east. Very rarely collected; the description above is based almost entirely on the isotype in Geneva. Grows in sand; flowering in July and August. Map 113.

Distinguished from *H. spiciforme* (above) by its more branching habit (dwarf axillary shoots bearing solitary heads on side branches, not directly on the main stems except at their tips), leaves on the side branches soon losing their wool and then contrasting with the persistently woolly leaves on the main stems, and possibly by larger (longer) heads.

Voucher: *Maguire* 1629 (BOL; NBG; PRE).

101. **Helichrysum argyrosphaerum** *DC.*, Prodr. 6: 174 (1838); Harv. in F. C. 3: 222 (1865); Oliv. & Hiern in F. T. A. 3: 351 (1877); Moeser in Bot. Jb. 44: 302 (1910); Merxm., F.S.W.A. 139: 92 (1967); Hilliard, Compositae in Natal 211 (1977). Lectotype:

Cape, Prieska distr., near Orange River, 'between Gariep Station and shallow ford', *Burchell* 1645 (G-DC!; K; PRE, isolecto.!).

Gnaphalium argyrosphaerum (DC.) Sch. Bip. in Bot. Ztg 3: 170 (1845).

Herb, possibly annual, taproot woody, branches up to 300 mm long, many, radiating, prostrate or decumbent, sub-simple to profusely branched, thinly woolly, leafy throughout. *Leaves* up to 25 × 7 mm, scarcely smaller upwards, spathulate to oblanceolate, subacute, mucronate, narrowed to a flat petiole-like base, both surfaces thinly grey-woolly. *Heads* heterogamous, subglobose, c. 7–10 × 7–10 mm, solitary at the tips of the branchlets, surrounded by leaves. *Involucral bracts* in c. 9 series, graded, imbricate, inner about equalling flowers, pellucid, shining, silvery becoming increasingly pink-tinged inwards, fading again at maturity, tips mostly obtuse, radiating. *Receptacle* very shortly honeycombed. *Flowers* 166–290, 19–36 ♀, 142–258 ☿, yellow, tipped pink. *Achenes* 0,75 mm long, elliptic, with myxogenic duplex hairs. *Pappus* bristles many, equalling corolla, scabrid, bases cohering lightly by patent cilia. Fig. 34: 2.

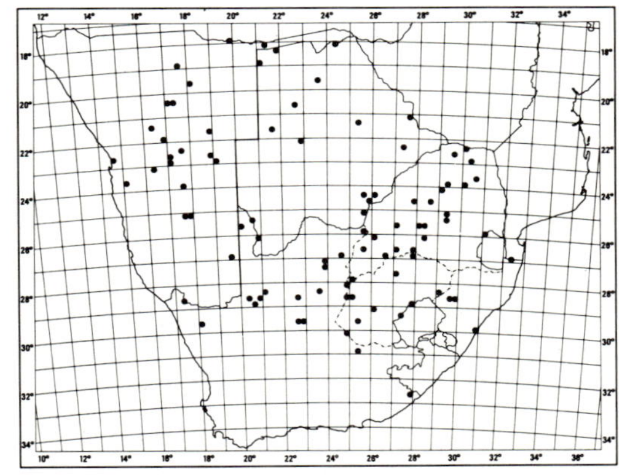

MAP 114.— **Helichrysum argyrosphaerum**

Widely distributed from Zambia, Malawi, Zimbabwe and Mozambique to Botswana, Angola, S.W.A./Namibia, Transvaal (excluding the E. highlands), N. Cape, Orange Free State, the Natal coastal plain and Tugela basin, with a single record from Komgha in the E. Cape. Grows in hot, dry, sandy places, and readily becomes a weed. Flowers between June and December. There are several records of its being regarded as excellent forage for both wild and domestic animals, but blindness and encephalopathy in sheep and cattle result from ingestion of large quantities (Onderstepoort J. Vet. Sci. 42,4: 135-148, 1975). Map 114.

Vouchers: *Acocks* 20983 (PRE); *Gerstner* 6196 (NBG; PRE); *Merxmüller & Giess* 3590 (PRE; WIND); *Pooley* 655 (E; NH; NU).

102. **Helichrysum candolleanum** *Buek,* Index DC. Prodr. 2: vi (Oct. 1840); Hilliard, Compositae in Natal 211 (1977). Type: *Bojer* 15 (G-DC, holo.!; P, fragment, iso.!) (cited by De Candolle as collected in Madagascar, but stated by Humbert, Fl. Madag. Comp. 565, 1962, to have come from E. Africa).

H. leptolepis DC., Prodr. 6: 170 (1838), non DC., Prodr. 6: 194 (1838); Harv. in F. C. 3: 222 (1865); Moeser in Bot. Jb. 44: 303 (1910); Merxm., F. S. W. A. 139: 95 (1967). *H. detersum* Steud., Nom. edn 2, 1: 738 (Nov. 1840). *Gnaphalium candolleanum* (Buek) Sch. Bip. in Bot. Ztg 3: 169 (1845). *G. leptolepis* (DC.) Kuntze, Rev. Gen. Pl. 3,2: 152 (1898). Type as for *H. candolleanum.*

H. pachyrhizum Harv. in F. C. 3: 222 (1865), nom. illegit., excl. var. *β thunbergii* Harv. Lectotype: Transvaal, Aapies River, *Zeyher* 895 (K! B; BM; BOL; E; SAM, isolecto.!).

H. damarense O. Hoffm. in Bot. Jb. 10: 275 (1889). Type: S.W.A./Namibia, Karibib, *Marloth* 1287 (PRE, iso.!).

H. leptolepis var. *intermedium* S. Moore in Bull. Herb. Boissier 2 sér. 4: 1016 (1904). Type: S.W.A., Otavi, *Dinter* 557 p.p. (Z, holo.!; BM, iso.!).

H. leptolepis var. *latifolium* S. Moore, l.c. Type: S.W.A./Namibia, Waterberg Plateau, *Dinter* 557 p.p. (Z, holo.!; BM, iso.!).

Perennial herb sometimes flowering in seedling stage, taproot stout, woody, crown becoming woody at base with age, main branches c. 150−450 mm long, subsimple to profusely branched, prostrate, decumbent or erect, thinly woolly becoming floccose, leafy. *Leaves* (4−) 10−35 × (1−) 1,5−8 (−12) mm, linear, oblong-lanceolate, lanceolate, oblanceolate, occasionally narrowly to broadly obovate or spathulate, acute, subacute or obtuse, mucronate, base narrowed, half-clasping, both surfaces thickly to thinly greyish-woolly, often yellowish when young. *Heads* homogamous, campanulate, c. (4−) 5−6 × 3,5−4 (−6)

mm, many in tight globose clusters c. 15−20 mm across at the tips of the branchlets, surrounded by woolly leaves mostly longer than the compound head. *Involucral bracts* in c. 6 series, subequal, slightly exceeding the flowers, loosely imbricate, very acute to acuminate, radiating, shining, pellucid, silvery white, inner in particular often suffused or tipped pink or crimson, rarely with flecks of colour, tips sometimes subopaque. *Receptacle* with pit margins shortly produced, or nearly smooth. *Flowers* 18−50, yellow, tipped red. *Achenes* 0,75 mm, elliptic, with myxogenic hairs, or rarely glabrous. *Pappus* bristles many, about equalling corolla, scabrid, bases cohering by patent cilia. Fig. 34: 1.

Widely distributed from S.W.A./Namibia to Botswana, the Transvaal Bushveld, Swaziland Lowveld, and coastal Natal as far south as Durban, penetrating inland up the dry river valleys. Also in Zimbabwe, Mozambique and, reputedly, Madagascar (see note under type specimen). Favours sandy or gravelly soils in grassland or open woodland, and open sandy places such as watercourses; sometimes a weed. Flowering recorded throughout the year, but mainly between June and September.

H. candolleanum is very variable in stature, in habit, and in leaf size, shape and degree of woolliness; some of this variation is undoubtedly environmentally controlled, but some is possibly genetic. In S.W.A./Namibia, there is some geographical partitioning of variation, but it seems best to avoid formal recognition of infraspecific groups partly because variation is by no means sharply discontinuous, partly because the situation is not understood; the problems will not yield to herbarium studies.

The commonest form of the species, which occurs throughout the major part of the geographical range, is very woolly, well-branched, the branches short, the leaves mostly 3−5 mm broad, and plants may be prostrate to erect. Plants will flower in the seedling stage and can then look deceptively different, but the small homogamous heads with very acute to acuminate bracts are diagnostic (e.g. *Acocks* 18105, PRE; *Compton* 26959, NBG; PRE).

In the Kaokoveld of NW. S.W.A./Namibia and nearby parts of Angola, plants are more or less erect and subshrubby with long slender branches and narrow (1−2 mm) leaves. Similar plants occur sporadically together with the more or less prostrate form southwards to about the 24th parallel in S.W.A./Namibia and as far east as the Transvaal and Tongaland. The 'leggy' plants have been segregated as *H. riparium* Brenan (Merxm., F.S.W.A. 139: 97, 1967). The diagnostic feature of *H. riparium*, as emphasized by Brenan, is the crisped-undulate reflexed involucral bracts: some specimens from NW. S.W.A./Namibia certainly have the inner, young, bracts somewhat crisped, and outer bracts reflexed, but not more so than many other specimens of *H. candolleanum* over its geographical range: they seem to be no more than a variant of *H. candolleanum* (e.g. *Giess* 8996, M; NBG; PRE; WIND; *Merxmüller & Giess* 30527, M; PRE; WIND; *De Winter & Leistner* 5805, K; M; PRE; WIND). Field studies are desirable to ascertain the status of these plants and of *H. riparium*.

Plants with broad, exceedingly woolly leaves are found around Otavi, in the Waterberg, and nearby; these were segregated as var. *intermedium* and var. *latifolium*, here placed in synonymy (e.g. *Giess* 12392, PRE; WIND). Map 115.

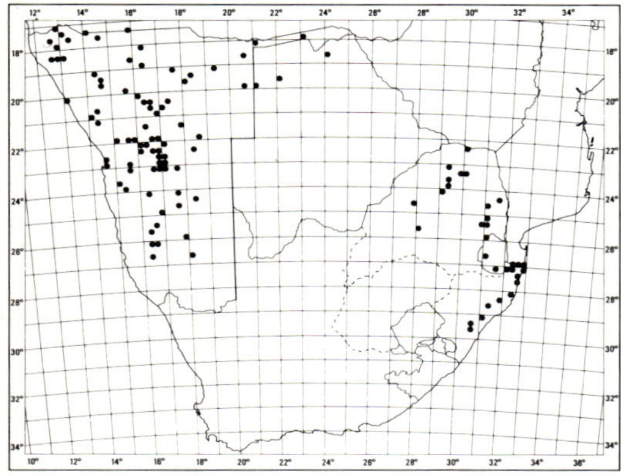

MAP 115.— **Helichrysum candolleanum**

103. **Helichrysum erubescens** *Hilliard, sp. nov.* H. candolleano *Buek affinis sed apicibus bractearum involucri obtusis vel breviter acutis (nec acutissimis nec acuminatis) differt.*

Herba verosimiliter annua vel breviter perennis, radice palari lignosa; caules 30—300 mm longi, plures, prostrati vel suberecti, rigidi, simplices vel laxe ramosi, albo-arachnoideo-lanati mox glabrescentes, remote foliati. Folia plerumque 7—22 × 3—10 mm, vel obovate vel elliptica vel spatulata, apice acuta, apiculata, basi angustata, utrinque albo-lanata. Capitula homogama, campanulata, c. 6 × 6 mm expansa 10 mm lata, usque ad 8 in glomerulo globoso ad apicem rami foliis circumcincta, omnia lana contexta. Bracteae involucrales 5—6-seriatae, laxe imbricatae, subaequales, flores superantes, semi-pellucidae, albae, apicibus radiantibus obtusis vel acutis plus minusve valde rubro-striatis demum decolorantibus. Achenia non visa; ovaria pilis duplicibus

myxogenis praedita. Pappi setae multae, corollam aequantes, scabridae, basibus ciliis patentibus leviter cohaerentibus.

Type: S.W.A./Namibia, Kaokoveld, im Rivierlauf 6 Meilen südlich Orupembe, 10 vi 1963, *Giess & Leippert* 7525 (PRE, holo.!; M; NBG, iso.!).

S.W.A./Namibia. — [1812 BA] Anabib (Orupembe) long. 1233 lat. 1810, bed of small sandy watercourse, 11 viii 1956, *Story* 5701 (M; PRE); ibid., 12 viii 1956, *Story* 5727 (PRE); ibid., 12 viii 1956, *Story* 5725 (PRE); Orupembe, 7 vi 1951, *Hall* 368 (NBG); ibid., river bed, 8 vi 1951, *Hall* 380 (NBG); im und am Rivierlauf bei Orupembe, 6 vi 1965, *Giess* 8856 (PRE); 32 km südlich Orupembe am Weg nach Sanitatas, Sechomib Rivier, 9 vi 1963, *Giess & Leippert* 7444 (M).

Herbaceous, probably annual or a short-lived perennial, taproot woody, stems 30—300 mm long, several from the crown, prostrate .or perhaps sometimes suberect, wiry, simple or loosely branched, white cobwebby-woolly, soon glabrescent, distantly leafy. *Leaves* mostly 7—22 × 3—10 mm, obovate to elliptic or spathulate, apex obtuse, apiculate, base narrowed, both surfaces white-woolly. *Heads* homogamous, campanulate, c. 6 × 6 mm, 10 mm when expanded, up to c. 8 in a globose cluster at the branch tips, surrounded by leaves, all webbed together with wool. *Involucral bracts* in 5—6 series, loosely imbricate, subequal, exceeding the flowers, semi-pellucid, white, tips radiating, obtuse to acute, more or less heavily flecked crimson producing a rosy cast, fading with age.

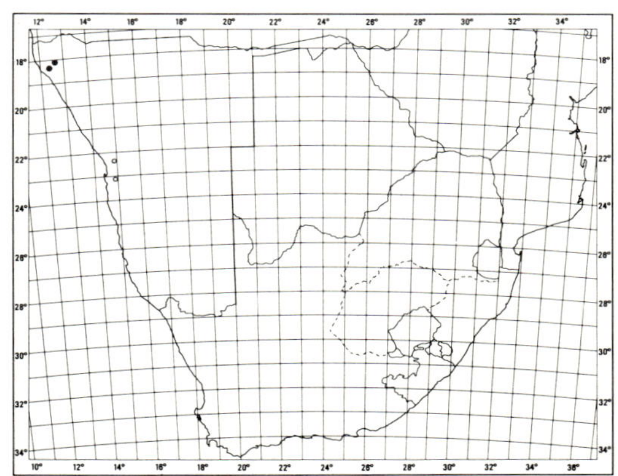

MAP 116.— ● **Helichrysum erubescens**
○ **Helichrysum marlothianum**

Receptacle nearly smooth. *Flowers* 30—50, yellow, tipped pink. *Achenes* not seen, ovaries with myxogenic duplex hairs. *Pappus* bristles many, equalling corolla, scabrid, bases cohering lightly by patent cilia. Fig. 34: 4.

Known only from the environs of Orupembe and Anabib in the Kaokoveld, where it grows in the sandy beds of watercourses; flowering recorded in June and August. Map 116.

Resembles prostrate, sparsely branched specimens of *H. candolleanum* (above) but is distinguished by its less pointed bracts flecked with crimson, not wholly coloured. Also confused with *H. gariepinum* (no. 106), which has heterogamous, not homogamous, heads, and with *H. roseo-niveum* (no. 105), which has larger heads.

104. **Helichrysum marlothianum** *O. Hoffm.* in Bot. Jb. 10:275 (1889); Moeser in Bot. Jb. 44: 304 (1910). Type: S.W.A./Namibia, Usakos, Hykamkab [Haigamkab], 400 m, April 1886, *Marloth* 1211 (BOL; K; NBG; PRE; SAM, iso.!).

Bushy annual herb with a woody taproot, profusely branched from the base, stems up to 120 mm long, apparently decumbent then erect, much branched, cobwebby, leafy. *Leaves* up to 12 (−20) × 6 (−8) mm, elliptic, subacute, much narrowed to a petiole-like base, thickly greyish-white woolly. *Heads* homogamous, campanulate, 4—6 × 3 mm, solitary, or 2—3 at the tips of the branches and dwarf axillary branchlets, closely enveloped by surrounding leaves webbed together with wool. *Involucral bracts* in c. 3—5 series, subequal, loosely imbricate, tips acute to subacuminate, opaque milk-white, sometimes reddish above the stereome, exceeding the flowers, radiating. *Receptacle* very shortly honeycombed. *Flowers* 18—39, yellow, tipped reddish. *Achenes* 1 mm long, elliptic in outline, closely ribbed, with myxogenic duplex hairs. *Pappus* bristles many, equalling corolla, scabrid, bases nude, not cohering.

Known only from a few collections, in the Namib desert, between the Kuiseb and Swartkop Rivers, growing in sandy watercourses. Flowering recorded in January, April and July. Map 116.

Sometimes confused with *H. herniarioides* (no. 107) but more closely branched than that species with woollier leaves and heads borne singly or 2—3 together at the branch tips and on short axillary shoots. Also allied to *H. roseo-niveum* (below) from which it is distinguished by its smaller heads, and to *H. gariepinum* (no. 106) which generally has more flowers in the head, a few of them female. Some specimens may be difficult to place; more collections and field studies are needed.

Vouchers: *Fock* 8395 (M; PRE); *Galpin & Pearson* 7666 (K; PRE; SAM); *Marloth* 1477 (PRE).

105. **Helichrysum roseo-niveum** *Marloth & O. Hoffm.* in Bot. Jb. 10: 275 (1888); Moeser in Bot. Jb. 44: 304 (1910); Merxm., F.S.W.A. 39: 97 (1967). Type: S.W.A./Namibia [Haigamkab] Hykamkab and Husab, *Marloth* 1212 (K; PRE; SAM; fragment BOL, iso.!).

Bushy herb, flowering when young and then single-stemmed, but main stem can become woody and nearly 10 mm in diam., taproot woody, stems c. 30—200 mm long, wiry, simple or branched, erect or decumbent, copiously white cobwebby-woolly, leafy. *Leaves* mostly 15—25 (−45) × 5—15 (−20) mm, diminishing slightly upwards, obovate or ovate, very obtuse, base narrowed, petiole-like in larger leaves, both surfaces copiously white cobwebby-woolly. *Heads* homogamous, campanulate, c. 7—10 × 8 mm, c. 15 mm across the radiating bracts, solitary or several clustered at the branch tips surrounded by green bracts webbed to outer involucral bracts. *Involucral bracts* in c. 5 series, subequal, loosely imbricate, exceeding flowers, silvery, pellucid, tips radiating, subacute, opaque white, inner often light crimson, fading to pink then white. *Receptacle* nearly smooth. *Flowers* (32−) 43—77, yellow, tipped pink. *Achenes* 1 mm, elliptic, with myxogenic duplex hairs. *Pappus* bristles many, equalling corolla, scabrid, bases nude, not cohering. Fig. 35: 4.

A most elegant plant known only from the Kaokoveld and Namib Deserts of southernmost Angola and S.W.A./Namibia and south to Rostock and Rehoboth degree squares. Grows in sandy or gravelly watercourses, around rocks, or in the crevices of rocks, possibly flowering in any month, but most records between May and July. Map 117.

Allied to *H. gariepinum* (below) from which it is distinguished by its homogamous heads; also, the two species are allopatric, *H. gariepinum* occurring only in the southern part of S.W.A./Namibia and extending into the semi-desert areas of the northern Cape. See also under *H. marlothianum* (above).

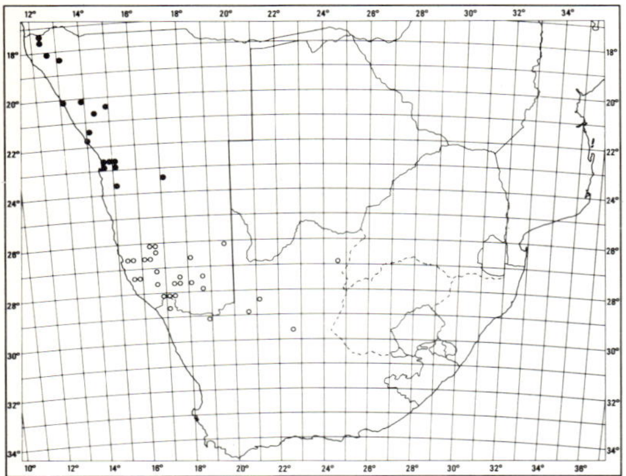

MAP 117.— ● **Helichrysum roseo-niveum**
○ **Helichrysum gariepinum**

Vouchers: *Galpin & Pearson* 7658 (K; PRE; SAM); *Hall* 382 (NBG); *Nordenstam & Lundgren* 846 (E; S); *De Winter & Leistner* 5792 (K; M; PRE; WIND).

106. **Helichrysum gariepinum** *DC.*, Prodr. 6: 174 (1838); Harv. in F. C. 3: 223 (1865); Moeser in Bot. Jb. 44: 304 (1910); Merxm., F.S.W.A. 139: 94 (1967). Lecto-type: Cape, rocky places near the Gariep [Orange], *Drège* 2841 (G-DC!; BM; PRE, isolecto.!).

Gnaphalium gariepinum (DC.) Sch. Bip. in Bot. Ztg 3: 169 (1845).

Herbaceous, taproot woody, stems several to many from the crown, 200—300 mm long, prostrate or decumbent, simple or branched, wiry, white cobwebby-woolly, often glabrescent, leafy. *Leaves* c. 7—20 × 3—12 mm, scarcely diminishing upwards, thin, spathulate or obovate to elliptic, much narrowed to the base, subpetiolate, apex obtuse, apiculate, both surfaces white cobwebby-woolly, often glabrescent. *Heads* heterogamous, campanulate, mostly c. 6—7 × 5—6 mm, c. 10 mm across the radiating bracts, solitary or up to 8 clustered at the branch tips, shortly pedunculate, closely surrounded by leaves. *Involucral bracts* in c. 3—5 series, surrounded by green lanceolate bracts webbed together with wool, sub-equal, loosely imbricate, exceeding the flowers, pellucid, silvery, tips radiating, subopaque, obtuse to acute, mostly white,

outermost sometimes tinged pale golden-brown, inner sometimes rosy to crimson, fading to white. *Receptacle* nearly smooth. *Flowers* 40—99, (1—) 3—8 ♀, 36—91 ☿, yellow, tipped pink. *Achenes* 0,75 mm with myxogenic duplex hairs or sometimes glabrous. *Pappus* bristles many, equalling corolla, scabrid, bases cohering lightly by patent cilia. Fig. 35: 5.

Found in the arid parts of the N. Cape from the environs of the Orange River east to Vryburg and Prieska, and north to Lüderitz, Aus, Keetmanshoop and Aroab in S.W.A./Namibia, in sandy, gravelly or stony places. Flowers between May and October, mainly August and September. Recorded as 'used by sparrows to line nests'! Map 117.

Can be confused with *H. roseo-niveum* (no. 105), *H. marlothianum* (no. 104) and *H. herniarioides* (below); see under those species to distinguish. An occasional specimen is difficult to place e.g. *Giess, Volk & Bleissner* 5317 (M; WIND) with heads only c. 5 mm long containing c. 35—40 flowers, 2 of them female, thus approaching *H. marlothianum; H. marlothianum* appears to differ in its homogamous heads. However, *H. marlothianum* is a poorly known species, requiring field study. See also comments under *H. herniarioides* (below).

Vouchers: *Giess* 9445 (M; PRE; WIND); *Merxmüller & Giess* 3364 (PRE; WIND); *Schlieben* 11573 (K; PRE); *Wisura* 2243 (NBG).

107. **Helichrysum herniarioides** *DC.*, Prodr. 6: 170 (1838); Harv. in F. C. 3: 215 (1865); Moeser in Bot. Jb. 44: 304 (1910); Merxm., F.S.W.A. 139: 94 (1967). Lecto-type: Cape, Silverfontein, *Drège* (G-DC!; BM; PRE; S; SAM; TCD, isolecto.!).

Gnaphalium herniarioides (DC.) Sch. Bip. in Bot. Ztg 3: 169 (1845).

G. indicum sensu Dinter in Fedde, Repert. 17: 311 (1921), non L.

Annual herb with a woody taproot, primary stem very short, erect, subsequent branches many from the base, prostrate, ascending or erect, mostly 30—200 mm long, simple or sparingly to well-branched, slender, cobwebby, distantly or closely leafy. *Leaves* mostly 4—12 × 2—7 mm, linear, oblong-obovate, obovate or spathulate, base much narrowed, apex rounded to subacute, loosely white-woolly, sometimes glabrescent, glandular. *Heads* homogamous or heterogamous, campanulate, c. 4—5 × 4—6 mm across the radiating bracts, many crowded in terminal glomerules surrounded by a few leaves. *Involucral bracts* in 5—6 series, subequal, loosely imbricate, outer pellucid, often tinged red-brown, webbed with wool, inner acute, subacute or obtuse, tips opaque or subopaque snow-white, sometimes tipped red-brown, sometimes with a reddish patch above the stereome, or sometimes the whole bract suffused pink, equalling or slightly exceeding the flowers, radiating. *Receptacle* smooth or nearly so. *Flowers* 9—42, 0—4 ♀, 9—40 ☿. *Achenes* 0,75 mm long, obscurely ribbed, glabrous or with myxogenic duplex hairs. *Pappus* bristles many, equalling corolla, scabrid, bases with minute patent cilia, not cohering. Fig. 35: 6.

In the more arid parts of the Cape from about Ceres N. and NE. to the environs of Prieska and Upington, Namaqualand and the western part of S.W.A./Namibia about as far north as the Omaruru river; common in open sandy or stony places and in open thorn scrub on sand, often in watercourses or depressions. Flowering recorded between March and December, but peak flowering from July to September. Recorded as eaten by stock. Map 118.

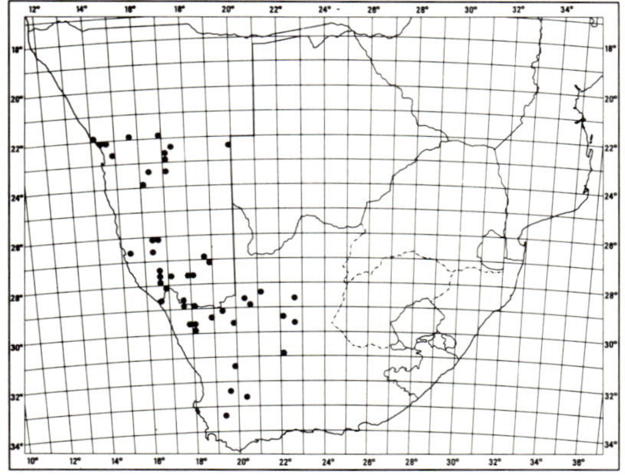

MAP 118.— **Helichrysum herniarioides**

As already noted by Merxmüller (F.S.W.A. 139: 94, 1967) three groups can be distinguished within *H. herniarioides*. Plants of the northernmost group (Okahandja, Windhoek, Rehoboth and Nauchas degree squares) are prostrate, ascending or erect, leaves 1—2 (—3) mm at their broadest, heads homogamous, 19—29 flowers, involucral bracts obovate or rotund, very obtuse, white or brown and white, sometimes innermost pink, ovaries always hairy (e.g.

Codd 5733, PRE; WIND; *Dinter* 7767, M; WIND; *Giess* 13595 PRE; WIND; *Nordenstam* 3662, M; S). Plants from Lüderitz and Aus degree squares are prostrate, leaves mostly 3—4 mm at their broadest, heads homogamous, 9—25 flowers, involucral bracts more or less lanceolate, acute to obtuse, white or brown and white, inner often pink, ovaries hairy or glabrous (*Dinter* 4096, M fragment, PRE; SAM; *Merxmüller & Giess* 2841, PRE; WIND). From roughly the 27th parallel southwards into the Cape, plants are prostrate, leaves mostly 3—7 mm broad, heads homogamous or heterogamous with 0—4 ♀ flowers, 9—40 ☿ flowers, involucral bracts often narrow, sometimes broad, acute to obtuse, mostly white or brown and white, rarely inner pink, ovaries hairy or glabrous (*Acocks* 14452, PRE; *Giess* 13752, PRE; WIND; *Giess* 9458, WIND; *Compton* 7919, NBG).

A fourth variant of *H. herniarioides* (not of *H. marlothianum*: see Merxmüller l.c. 95) comprises plants from Uis and Swakopmund degree squares (*Galpin & Pearson* 7644, K; PRE; SAM; *Giess* 3573, M, PRE; *Giess, Volk & Bleissner* 5781, M; PRE; *Marloth* 1248, M; PRE; SAM; *Merxmüller & Giess* 1754, M). They differ from the other northern group mentioned above by their more acute bracts and glabrous ovaries. A plant from further east, in Sandfontein degree square (*Wilman* 15347, BOL; SAM), also belongs here. The stems in young plants are simple or subsimple, prostrate and distantly leafy; the stems of older plants are loosely branched and are possibly erect.

The complexities of *H. herniarioides* cannot be unravelled in the herbarium.

H. herniarioides can be confused with *H. gariepinum* (no. 106): heads are 4—5 mm long with 9—40 flowers in *H. herniarioides*, 6—7 mm long with 40—100 flowers in *H. gariepinum*. Heads are always heterogamous in *H. gariepinum*, and the involucral bracts are mostly white with the inner often rosy. See also under *H. marlothianum* (no. 104).

108. **Helichrysum albertense** *Hilliard, sp. nov.* H. herniarioides *DC. affinis sed capitulis minoribus 3 mm longis (nec 4—5 mm) et bracteis involucri apicibus pellucidis vel subopacis et albidis brunneo-suffusis (nec opacis niveis nec rubro-vel brunneo-notatis).*

Herba annua, radice palari tenui; caules 10—70 mm longi, plures, prostrati, tenuissimi, simplices vel subsimplices, fusce rubrobrunnei, arachnoidei, praeter sub capitulis remote foliati. Folia plerumque 4—11 × 2—3 mm, spatulata, apice rotundato, basi angustata semi-amplexicauli, marginibus planis, utrinque laxe griseo-lanata, plus minusve glabrescentia. Capitula heterogama vel raro homogama, campanulata, 3 × 2 mm, multa in glomerulos terminales foliis circumcinctos congesta. Bracteae involucrales c. 6-seriatae, subaequales, exteriores ad folia circum-

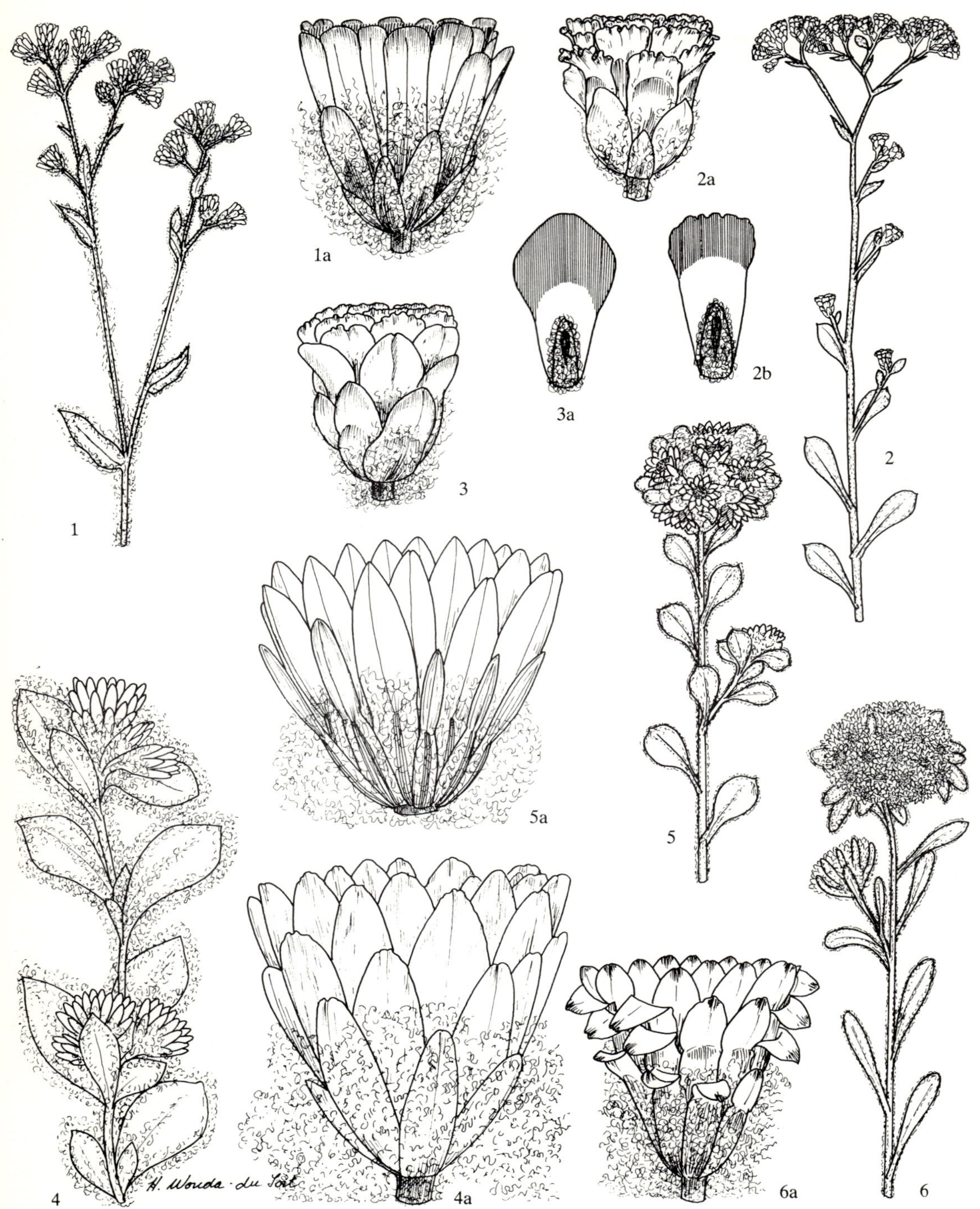

cingentia lana contextae, interiores dorso pilis paucis lanatae, apicibus rotundatis saepe emarginatis pellucidis vel subopacis albidis plerumque laete rubro- vel aureo-brunneo-suffusis, vel omnino aureo-brunneis et pellucidis vix radiantibus. Receptaculum *tuberculis planis praeditum. Flores 12—22, (0—) 1—4 ♀, 11—21 ☿. Achenia 0,75 mm longa, glabra. Pappi setae multae, corollam aequantes, apicibus scabridis, basibus non cohaerentibus.*

Type: Cape, Prince Albert div., Prince Albert Road railway station, 2 200 ft, xii 1905, *Bolus* 11989 (BOL, holo.!; PRE, iso.!).

Cape. — Prince Albert distr., [3322 AA] Prince Albert, alt. c. 2 300 ft, xii 1904, *Bolus* 11530 (BOL; K); Abrahamskraal, *Wilman* 2629 (K). Fraserburg distr., [3221 DB], Leeu-Gamka, 'Layton', 3 500 ft, very common, 22 ix 1967, *Shearing* 124 (PRE). Jansenville distr., Klipplaat, Oct. 1925, *Thode* A584 (PRE). Laingsburg distr., [3321 AB] near Ketting Station, c. 2 000 ft, Karroid Brokenveld, locally frequent on flats, 19 ix 1953, *Acocks* 17129 (PRE, mixed with *Helichrysum leontonyx* DC.).

Annual herb with a slender taproot, branches 10—70 mm long, several from the crown, prostrate, very slender, simple or subsimple, dark reddish brown, cobwebby, distantly leafy except under the heads. *Leaves* mostly 4—11 × 2—3 mm, spathulate, apex rounded, base narrowed, half-clasping, margins flat, both surfaces loosely grey-woolly, glabrescent. *Heads* heterogamous or rarely homogamous, campanulate, 3 × 2 mm, many crowded in terminal glomerules surrounded by leaves. *Involucral bracts* in c. 6 series, subequal, outer webbed with wool to surrounding leaves, inner with some woolly hairs on backs, tips rounded, often emarginate, pellucid or subopaque, whitish mostly lightly suffused reddish- or golden-brown, or wholly golden-brown and then pellucid, scarcely radiating. *Receptacle* raised, with flattened tubercles. *Flowers* 12—22, (0—) 1—4 ♀, 11—21 ☿. *Achenes* 0,75 mm long, glabrous. *Pappus* bristles many, equalling corolla, tips scabrid, bases with patent cilia not cohering.

Recorded only from the Karoo, from Leeuw-Gamka and Ketting Station (Laingsburg district) east to Klipplaat near Jansenville and south to Prince Albert in 'Karroid broken veld' and 'on flat rante', flowering between September and December. Map 119.

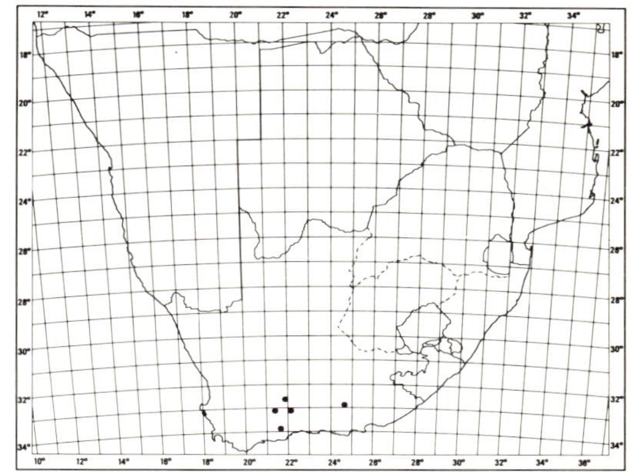

MAP 119.— **Helichrysum albertense**

Allied to *H. herniarioides* (above) but differing in its smaller heads and involucral bracts with tips pellucid or subopaque, whitish suffused with brown, not opaque snow-white nor with a red or brown patch.

109. **Helichrysum jubilatum** *Hilliard, sp. nov.* H. alsinoidei *DC. affinis, sed foliis ellipticis (nec spatulatis), inflorescentiis compositis a foliis non circumcinctis, floribus femineis numero quasi dimidium hermaphroditorum aequantibus (nec capitulis homogamis vel floribus femineis 1—3 tantum) distinguitur.*

Herba annua e basi ramosa; caules erecti vel decumbentes, 100—250 mm longi, tenuiter albo-lanati, remote foliati. Folia plerumque 8—30 × 5—15 mm, sub capitulis minora, elliptica vel oblanceolata, obtusa vel acuta, mucronata, basi semi-amplexicauli, utrinque cano-lanato-arachnoidea. Capitula heterogama, campanulata, 3,5—4 × 2,5 mm, pro parte maxima in paniculas cymosas terminales aggregata; capitula subterminalia

---

FIG. 35.—1, **Helichrysum jubilatum,** part of plant, × 1; 1a, head, × 8 (*Nordenstam* 1823). 2, **H. indicum,** part of plant, × 1; 2a, head, × 8; 2b, bract from the third series inwards, × 10 (*Orchard* 333). 3, **H. marmarolepis,** head, × 8; 3a, bract from the third series inwards, × 10 (*Ihlenfeldt* 1235). 4, **H. roseo-niveum,** part of plant, × 1; 4a, head, × 5 (*Hilliard* 4698). 5, **H. gariepinum.** part of plant, × 1; 5a, head, × 6,6 (*Giess* 14538). 6, **H. herniarioides,** part of plant, × 1; 6a, head, × 8 (*Giess* 13752).

interdum solitaria vel 2—3 ad apices ramulo-rum nudorum ad 30 mm longorum. Bracte-ae involucrales *5-seriatae, gradatae, exter-iores pellucidae, pallide stramineae, dorso lanatae, seriebus duabus interioribus sub-aequalibus et flores quasi aequantibus, apicibus obtusis opacis niveis vix radianti-bus.* Receptaculum *fere laeve.* Flores *c. 35—41, 10—14 ♀, 24—29 ☿.* Achenia *0,75 mm longa, pilis myxogenis praedita.* Pappi setae *multae, corollam aequantes, apicibus scabridis, basibus non cohaerentibus.*

Type: Cape, Namaqualand div., Rich-tersveld, c. 5 miles E. of Lekkersing on road to Stinkfontein, kloof in hill south of the road, annual, disc whitish, 7 xi 1962, *Nordenstam* 1823 (S, holo.!; E, iso.!).

Cape— [2817 DC] Doornrivier, 10 ix 1929, *Herre* STE 12205 (STE).

Annual herb, branching from the base, stems erect or decumbent, 100—250 mm long, loosely branched, thinly white-woolly, distantly leafy. *Leaves* mostly 8—30 × 5—15 mm, smallest under the heads, elliptic, obovate or oblanceolate, obtuse or acute, mucronate, base half-clasping, both surfaces greyish-white woolly-cobwebby. *Heads* heterogamous, campanulate, 3,5—4 × 2,5 mm, mostly crowded in terminal cymose panicles, subterminal heads sometimes soli-tary or 2—3 at the tips of nude branchlets up to 30 mm long. *Involucral bracts* in c. 5

series, graded, outer pellucid, pale straw-coloured, backs woolly, inner 2 series subequal, about equalling the flowers, tips obtuse, opaque snow-white, scarcely radiat-ing. *Receptacle* nearly smooth. *Flowers* c. 35—41, 10—14 ♀, 24—29 ☿. *Achenes* 0,75 mm long with myxogenic hairs. *Pappus* bristles many, equalling corolla, tips scab-rid, bases nude, not cohering. Fig. 35: 1.

Recorded only from Namaqualand, near Lekker-sing and Doornrivier; flowering between September and November. The joyous epithet commemorates the provenance of the type specimen. Map 120.

110. **Helichrysum indicum** *(L.) Grierson* in Notes R. bot. Gdn Edinb. 31:138 (1971). Lectotype: from the Cape of Good Hope, *Hermann* herb. 4: 15, no. 307 (BM!).

Gnaphalium indicum L., Sp. Pl. 852 (1753), non al.

G. paniculatum Berg., Descr. Pl. Cap. 256 (1767), non Thunb. (1823). Type: Cape of Good Hope, *Grubb* (STB, holo.!).

G. notatum Thunb., Prodr. 150 (1800), Fl. Cap. 653 (1823), nom. illegit.

G. expansum Thunb., Prodr. 149 (1800), Fl. Cap. 651 (1823). *Helichrysum expansum* (Thunb.) Less., Syn. Comp. 276 (1832); DC., Prodr. 6: 170 (1838); Harv. in F. C. 3: 216 (1865); Moeser in Bot. Jb. 44: 304 (1910); Levyns in Adamson & Salter, Fl. Cape Penins. 780 (1950). Lectotype: Cape of Good Hope, *Thunberg* (sheet 19155, UPS!).

G. drabaeforme Schrank in Denkschr. Akad. Muench. 8: 162 (1824). Type: Cape of Good Hope, *Brehm* s.n. (M, holo.!).

Helichrysum expansum var. *erectum* DC., l.c. Lectotype: ex Mus. Berol. (G-DC!).

H. expansum var. *patulum* DC., l.c. Lectotype: *Sieber* 201 (G-DC!).

Annual herb with a woody taproot, stems mostly 70—30 mm long, erect or spreading, subsimple or well-branched from the base, thinly white-woolly, leafy. *Leaves* very variable in size, lowermost up to 50 × 12 mm, upper mostly c. 10—15 × 2—5 mm, spathulate or oblong-spathulate to linear, base narrowed, apex subacute, mucronate, both surfaces greyish-white woolly. *Heads* homogamous or very rarely heterogamous, cylindric-campanulate, c. 3 × 2 mm, few to many on short stalks crowded at the branch tips, these often in loose corymbose panicles. *Involucral bracts* in 5—6 series, loosely imbricate, outer short, pellucid,

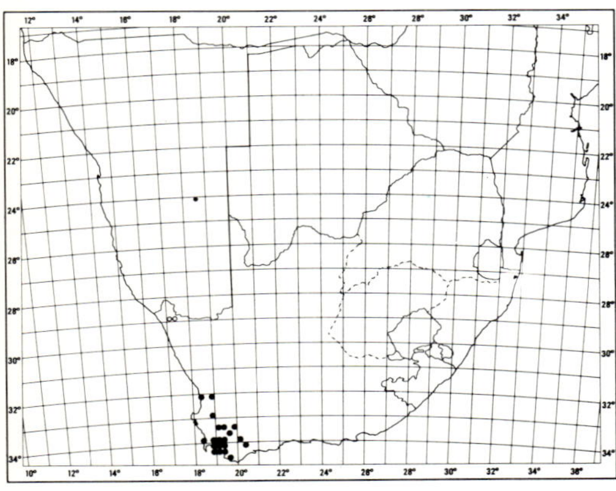

MAP 120.— ○ **Helichrysum jubilatum**
        ● **Helichrysum indicum**

acute, pale brown or occasionally reddish, bases loosely woolly, inner longer, subequal, about equalling flowers, tips opaque milk-white, sometimes suffused pink, or crimson above the stereome, or rarely creamy, very obtuse, crisped, minutely radiating. *Receptacle* smooth or nearly so. *Flowers* 14−27, occasionally 1 or 2 ♀. *Achenes* 0,75 mm long, usually with myxogenic duplex hairs or occasionally glabrous. *Pappus* bristles many, equalling corolla, scabrid, bases cohering lightly by patent cilia. Fig. 35: 2.

Endemic to the SW. Cape, from Lambert's Bay and Clanwilliam south to Betty's Bay and Swellendam. Grows on sandy flats and slopes and will become a weed in rough or cultivated ground. Flowers between October and February. Map 120.

Vouchers: *Bolus* 3817 (BOL); *Esterhuysen* 12394 (BOL); *MacOwan* 2404 (SAM); *Parker* 3454 (BOL; NBG); *Taylor* 4269 (PRE).

Levyns (in Adamson & Salter, Fl. Cape Penins. 780, 1950) says 'It seems likely that this species hybridizes with *H. rutilans*' [that is, *H. moeserianum*]. *Levyns* 8545 (BOL) has faintly yellowish, not pure white, bracts, and could be of hybrid origin as Mrs Levyns suggests; however, pollen and ovules appear to be normal.

111. **Helichrysum marmarolepis** *S. Moore* in J. Bot., Lond. 37: 370 (1890); Moeser in Bot. Jb. 44: 305 (1910). Type: Namaqualand, *Scully* 219 (BM, holo.!; K, iso.!).

Closely resembles *H. indicum* (above), but distinguished by the heads on shorter stalks and therefore borne in more con-gested clusters than in *H. indicum*, in the shape of the involucre, campanulate in *H. marmarolepis*, cylindric-campanulate in *H. indicum*, and in the involucral bracts, less woolly on the backs than in *H. indicum*, and more loosely imbricate (hence the shape of the head). Fig. 35: 3.

Endemic to Namaqualand and the W. Cape, from about the Orange River south to Heeren Logement, N. of Clanwilliam. Apparently more northerly in its distribution than *H. indicum*, which has not been recorded north of Clanwilliam and Lambert's Bay. Grows in Sandveld, flowering mainly in September and October. Map 121.

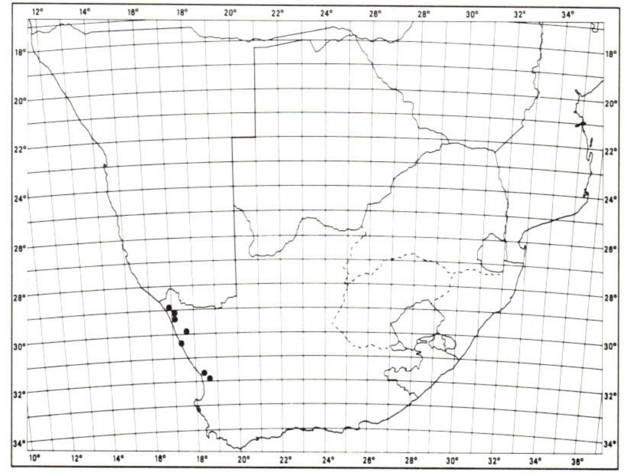

MAP 121.— **Helichrysum marmarolepis**

Vouchers: *Bolus* 9562 (BOL, PRE); *Nordenstam & Lundgren* 1647 (E; S); *Pillans* 18069 (BOL); *Schlechter* 8352 (BOL); *Steyn* 451 (NBG).

Group 16

Shrubs or half-shrubs, *leaves* large, ovate or subrotund, petiolate; *heads* homogamous or heterogamous, 2−3 × 2 mm, in congested clusters arranged in large spreading corymbose panicles; *involucral bracts* minutely radiating, white or silvery; *receptacle* smooth or shortly honeycombed; *flowers* 8−24, 0−2 ♀, corolla of ☿ flowers campanulate above, of ♀ flowers more or less narrowly funnel-shaped with a conspicuous limb; *achenes* hairy; *pappus* bristles scabrid, tips barbellate, bases cohering by patent cilia.

Species 112, 113, narrowly endemic in Natal, the Orange Free State and Transkei; taxonomically isolated.

1a A shrub; leaf blade broadly ovate or subrotund, obtuse, petiole not winged ....112. *H. populifolium*

1b A half-shrub; leaf blade ovate, acute to acuminate, base decurrent on the petiole in narrow wings...113. *H. hypoleucum*

112. **Helichrysum populifolium** *DC.*, Prodr. 6: 180 (1838); Harv. in F.C. 3: 229 (1865); Moeser in Bot. Jb. 44: 289 (1910); Hilliard, Compositae in Natal 207 (1977).

1
1a
2
2a
2b
2c
2d
H. Wonda-du Toit

Type: Cape, between the Umzimkulu and Umzimvubu Rivers, *Drège* 5010 (G-DC, holo.!; BM; P; S, iso.!).

Gnaphalium populifolium (DC.) Sch. Bip. in Bot. Ztg 3: 170 (1845).

Soft-wooded shrub up to 2 m tall and as much across, branches white-felted, leafy. *Leaves* on white-felted petioles up to 70 mm long, blade up to c. 130 × 110 mm, broadly ovate or subrotund, apex obtuse, base cordate, upper surface cobwebby, glabrescent, lower white-felted, digitately nerved. *Heads* homogamous, campanulate, c. 3 × 2 mm, in small clusters arranged in very large divaricate panicles terminating the branchlets. *Involucral bracts* in 4—5 series, graded, imbricate, outer pale brown, woolly, inner with opaque white obtuse tips equalling the flowers, minutely radiating. *Receptacle* conical, minutely tuberculate. *Flowers* 16—24. *Achenes* 0,5—0,75 mm long, with duplex hairs. *Pappus* bristles many, slightly exceeding corolla, tips barbellate, bases cohering lightly by patent cilia. Fig. 36:1.

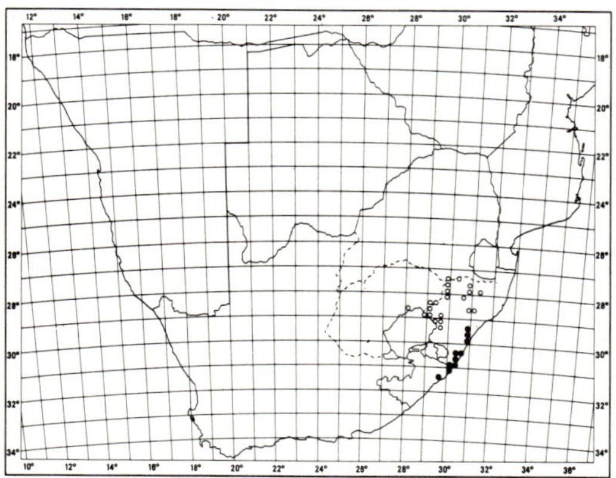

MAP 122.— ● **Helichrysum populifolium**
○ **Helichrysum hypoleucum**

Apparently confined to Table Mountain Sandstone formations from the Noodsberg NW. of Durban south to Port St Johns in the Transkei. Grows on large outcropping masses of rock or along the cliffs above gorges, up to c. 800 m; flowering between February and May. The large poplar-like leaves make this a most distinctive species. Map 122.

Vouchers: *Galpin* 2848 (PRE); *Hilliard & Burtt* 10237 (E; K; M; MO; NU; S); *Strey* 8106 (K; NU; PRE).

113. **Helichrysum hypoleucum** *Harv.* in F.C. 3: 253 (1865); Moeser in Bot. Jb. 44: 290 (1910); Hilliard, Compositae in Natal 216 (1977). Lectotype: Natal, Umgate [Enyati?] and Ingoma [Ngome?], *Gerrard & M'Ken* 1005 (TCD!; BM; K, isolecto.!).

Gnaphalium subcordatum O. Kuntze, Rev. Gen. Pl. 3,2: 154 (1898). Type as above.

Soft-wooded half-shrub, erect or scandent, climbing to c. 2 m, branches white-felted, leafy. *Leaves* on narrowly winged petioles up to 25 mm long, base clasping, auriculate, blade up to 80 × 50 mm, cordate-ovate, apex acute to acuminate, upper surface thinly cobwebby at first, soon merely minutely glandular-pubescent, lower white-felted, digitately nerved. *Heads* heterogamous, campanulate, c. 2 × 2 mm, many in very congested corymbose clusters, these arranged in large spreading corymbose panicles. *Involucral bracts* in 4 series, graded, imbricate, outer woolly, inner about equalling flowers, glossy, silvery translucent, obtuse, minutely radiating. *Receptacle* honeycombed. *Flowers* 8—13, 1—4 ♀, 7—11 ☿, bright yellow. *Achenes* c. 0,5 mm long, barrel-shaped, with myxogenic duplex hairs. *Pappus* bristles many, slightly shorter than corolla, scabrid, bases lightly cohering by patent cilia. Fig. 36:2.

Recorded from the low Drakensberg on the Transvaal-Natal-Orange Free State border, the mountains of N. Natal and the Biggarsberg, Qudeni and Nkandla in Zululand, the mountainous NE. part of the Orange Free State, and the main Drakensberg and its outliers as far south as Giant's Castle and Kamberg. Grows in tangled clumps on forest margins or on steep grassy mountain slopes, roughly 1 200—1 800 m. Flowers between February and April. Map 122.

Vouchers: *Acocks* 11433 (PRE); *Hilliard* 4927 (E; K; NH; NU; S); *Wright* 462 (E; K; M; NH; NU).

FIG. 36.—1, **Helichrysum populifolium,** part of flowering branch, × 0,7; 1a, head, × 8 (*Strey* 8648). 2, **H. hypoleucum,** part of flowering branch, × 0,7; 2a, head, × 8; 2b, hermaphrodite flower, × 13; 2c, female flower, × 13; 2d, pappus bristle, × 13 (*Wright* 462).

Group 17

Shrubs, subshrubs or perennial herbs, *leaves* small to medium-sized, ranging from linear, oblong or lanceolate to elliptic, spathulate, panduriform, ovate or subrotund, base sometimes petiole-like in broader leaves; *heads* homogamous or rarely heterogamous and then generally within the same species, 4—15 mm long, solitary or many in corymbose panicles; *involucral bracts* radiating, usually at least the tips of the inner white, occasionally rosy; *receptacle* smooth, honeycombed, or rarely fimbrilliferous; *flowers* 8—145, rarely 1—4 ♀, corolla of ♀ flowers narrowly campanulate above; *achenes* usually hairy, or rarely glabrous in the same species; *pappus* bristles with scabrid, barbellate or subplumose tips, bases usually with patent cilia, cohering or not, or rarely (*H. acrophilum*; no. 127) fused and falling in a ring, or fused in bundles (*H. rotundifolium*; no. 120).

Species 114—129, 10 endemic to the S. and SW. Cape, 9 of them confined to the mountains, 4 endemic to the extended Drakensberg Centre, 2 ranging from the Cape Peninsula to Natal, often on cliffs or rocks.

1a Compact cushion-forming shrubs, often dwarf:

  2a Leaves subrotund, ovate or elliptic, up to about twice as long as broad:

    3a Heads either few to several clustered at the branch tips, or if solitary, then c. 6—8 mm long. Drakensberg and outliers:

      4a Heads containing 9—25 flowers, flowering from mid-February to end of June ....114. *H. sutherlandii*

      4b Heads containing 19—145 flowers, flowering from July to September ....115. *H. confertum*

    3b Heads solitary at the tips of the branchlets, c. 5 mm long. S. Cape mountains .... 119. *H. saxicola*

  2b Leaves linear to narrowly oblong or narrowly elliptic, at least twice as long as broad:

    5a Leaves c. 5—40 × 1—4 mm, surface of indumentum like tissue-paper; flowers c. 20—38 ....117.*H. sessile*

    5b Leaves c. 2,5—5 × 0,5—2 mm, surface of indumentum not papery; flowers c. 15—21 ....118. *H. archeri*

1b Laxly branched shrubs or subshrubs, or perennial herbs:

  6a Bases of all but the upper reduced leaves distinctly petiole-like (although base may be broadened and ear-clasping); heads 3—5 mm long:

    7a Outer involucral bracts webbed together with wool; leaf bases not ear-clasping...........114. *H. sutherlandii*

    7b Outer involucral bracts either glabrous or thinly cobwebby, not webbed together; leaf bases distinctly ear-clasping ....116. *H. sphaeroideum*

  6b Leaf bases either not petiole-like or heads c. 6 mm or more long:

    8a Heads c. 8—10 mm long, solitary at the branch tips; surface of leaf indumentum like tissue-paper.. 121. *H. altigenum*

    8b Heads either few to many in clusters or corymbose panicles, or rarely solitary and then leaves woolly-felted:

      9a Plants with rosettes of radical leaves ....120. *H. rotundifolium*

      9b Plants without rosettes of radical leaves:

        10a Leaves lanceolate to elliptic, acute, upper surface cobwebby at first, soon coarsely and harshly pubescent ....124. *H. felinum*

        10b Leaves differing from above in either shape or indumentum:

          11a Heads c. 3—8 mm long:

            12a Main stem leaves panduriform. High Drakensberg ....128. *H. amplectens*

            12b Leaves not panduriform:

              13a Leaves up to 35 × 12 mm:

                14a Leaf tips acute or obtuse, but not more or less truncate:

                  15a Flowering stems pendunculoid and nude below the compound inflorescence ....123. *H. outeniquense*

                15b Flowering stems leafy under the compound inflorescence (although the leaves may be slightly reduced in size):

114. Helichrysum sutherlandii *Harv.* in F.C. 3: 218 (1865); Moeser in Bot. Jb. 44: 306 (1910), excl. *H. confertum* N.E. Br.; Hilliard, Compositae in Natal 212 (1977). Lectotype: [Lesotho] Basutoland, Mequatlong, *Cooper* 709 (TCD!; E; K; Z, isolecto.!).

Gnaphalium pulviniforme O. Kuntze, Rev. Gen. 3: 153 (1898). *Helichrysum pulvinatum* [O. Hoffm. ex] O. Kuntze, Rev. Gen. 3: 153 (1898) in syn. Type: Natal, Van Reenen's Pass, 1 900 m, *Kuntze* s.n. (Z, holo.!; K, iso.!).

H. sutherlandii var. *semiglabrum* N.E. Br. in Kew Bull. 1906: 21 (1906). Type: Natal, near Van Reenen's Pass, 5 000−6 000 ft, *Wood* 5702 (K, holo.!; NH, iso.!).

Well-branched shrub up to 400 mm tall, commonly lax, sometimes compact, vegetative twigs closely leafy, flowering twigs more distantly so. *Leaves* up to c. 25 × 12 mm, rhomboid or elliptic, abruptly narrowed to a flat, petiole-like, clasping base up to c. 12 mm long, apex more or less acute, upper surface thinly grey-woolly, lightly cobwebby or sometimes glabrous, lower surface white-felted, triplinerved, nerves nearly always prominent below. *Heads* homogamous, narrowly campanulate, 4−5 mm long, 5−7 mm across the radiating bracts, few to many in corymbose panicles c. 20−120 mm across. *Involucral bracts* in c. 6 series, graded, imbricate, outer pale brownish, webbed together with wool, inner much exceeding flowers, tips milk-white, opaque. *Receptacle* honeycombed. *Flowers* 9−25, yellow. *Achenes* 1−1,25 mm long, barrel-shaped, with myxogenic duplex hairs. *Pappus* bristles many, equalling corolla, scabrid, bases not cohering. Fig. 37: 1.

Ranges from the Barberton Mountains in the SE. Transvaal to the low Drakensberg on the Transvaal-Natal border, the NE. mountainous corner of the Orange Free State, Lesotho, the Natal Drakensberg and its outliers, and the more elevated parts of the Natal Midlands. It is particularly common in the Drakensberg, where it grows in clumps hanging from crevices in cliff faces and rock outcrops. Flowering begins in mid-February, is at its peak in April, and is fading by July. Map 123.

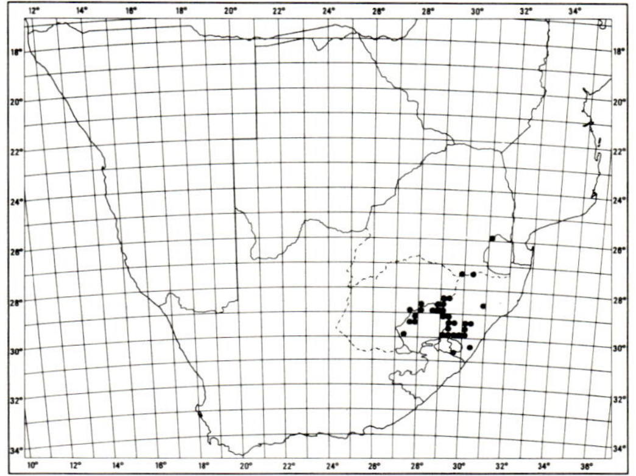

MAP 123.— **Helichrysum sutherlandii**

The commonest form of *H. sutherlandii* is a lax rounded shrublet with the flowering stems standing well away from the body of the plant. A compact, small-leaved form with very short flowering twigs occurs in parts of the Drakensberg, mainly above c. 2 600 m, and may be difficult to distinguish from small-headed plants of *H. confertum* (below), which may grow with *H. sutherlandii*, but generally flowers earlier, in spring and late winter. It is possible that there is some crossing between the species, but field and laboratory studies are needed to confirm this.

Vouchers: *Hilliard* 4994 (E; K; M; MO; NU; PRE; S) and 5010 (E; K; NU; PRE), both similar to the lectotype; *Wright* 761 (E; K; MO; NU; S), *Esterhuysen* 8842 (BOL), both the common lax plant.

115. Helichrysum confertum *N.E. Br.* in Kew Bull. 1895: 25 (1895); Hilliard, Compositae in Natal 214 (1977). Type: Natal, Drakensberg near Bushman's River, 6 000−7 000 ft, *Evans* 49 (K, holo.!; BOL; NH, iso.!).

Compact well-branched dwarf shrub, rounded in outline or depressed, old branches bare, gnarled, young ones densely tufted, closely leafy. *Leaves* appearing rosetted when viewed from above, up to 10 (−20) × 6 (−10) mm, ovate to elliptic, narrowed to a broad, flat, clasping, petiole-like base up to c. 8 (−10) mm long, apex

obtuse, upper surface greyish, lower white, woolly-felted, triplinerved, nerves not or scarcely visible. *Heads* homogamous, campanulate, very variable in size, mostly 5—8 mm long, 8—16 mm across the radiating bracts, larger heads solitary or few at the branchlet tips, smaller in several-headed corymbose clusters. *Involucral bracts* in c. 9 series, graded, imbricate, outer pale golden-brown or pinkish, bases webbed together with wool, inner much exceeding flowers, tips opaque milk-white. *Receptacle* honeycombed. *Flowers* 19—145 yellow, often tinged red. *Achenes* c. 1,25 mm long, elliptic, with myxogenic duplex hairs. *Pappus* bristles many, about equalling corolla, scabrid, tips barbellate, bases not cohering. Fig. 37: 2.

Recorded on the Natal Drakensberg from Cathedral Peak to Thamathu Pass in Underberg district, and nearby Sehlabathebe in Lesotho, and Pot River Pass north of Maclear in the E. Cape, on both basalt and Cave Sandstone, between c. 1 800 and 3 000 m. Forms compact cushions up to a metre or more across on cliff faces; flowering between July and September. Plants with small heads have been recorded from many of the high Drakensberg passes from Organ Pipes to Bushman's River Pass; elsewhere, plants tend to have large heads. Small-headed specimens may be difficult to distinguish from compact specimens of *H. sutherlandii* (above). Map 124.

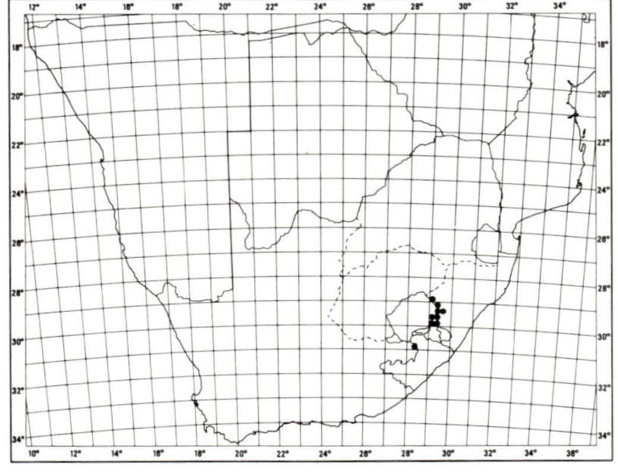

MAP 124.— **Helichrysum confertum**

Vouchers: *Beverly & Hoener* 718 (E; NU); *Esterhuysen* 17344 (BOL; NBG); *Hilliard & Burtt* 7449 (E; K; NU; PRE; S); *Wright* 186 (E; K; NH; NU).

116. **Helichrysum sphaeroideum** *Moeser* in Bot. Jb. 44: 306 (1910). Type: SW. Cape, Langeberg near Zuurbrak, *Schlechter* 2047 (Z, holo.!; BM; G; PRE; S, iso.!).

Straggling subshrub perhaps up to 600 mm tall, branching from the base, branches often tangled, long, slender, thinly white-woolly, leafy, becoming distantly so and pedunculoid below the compound inflorescence. *Leaves* 7—40 × 4—15 mm, lower with an elliptic to subrotund, acute to obtuse, mucronate blade, narrowed below and petiole-like, base expanded, ear-clasping, petiole becoming less pronounced upwards, uppermost leaves oblong, sessile, passing into distant, lanceolate-acuminate bracts, margins of all minutely crisped, upper surface coarsely hairy, lightly cobwebby as well, at least initially, lower surface thinly white-woolly-felted. *Heads* homogamous, cylindric-campanulate, 3—4 × 2—4 mm, few to many in terminal congested clusters 10—20 mm across, becoming somewhat lax with age. *Involucral bracts* in 4—5 series, more or less graded, outermost pellucid, tinged palest brown, glabrous or thinly cobwebby, inner with tips opaque milk-white, sometimes reddish above the stereome, or rarely the whole tip suffused pink, obtuse, or subacute, somewhat crisped, minutely radiating. *Receptacle* shortly honeycombed. *Flowers* 8—31. *Achenes* 0,75 mm long, ± cylindric, with peculiar thickened hairs, not myxogenic. *Pappus* bristles many, equalling corolla, scabrid, bases cohering strongly by patent cilia. Fig. 38: 1.

On the Cape mountains from about Tulbagh, Paarl and Worcester to the Langeberg between Swellendam and Riversdale, between 900 and 1 600 m above sea level. Favours damp sheltered rocky places, often on steep slopes in gullies or ravines; flowering between September and February, but mainly in December and January. Map 125.

FIG. 37.—1, **Helichrysum sutherlandii,** flowering twig, common form, × 1 (*Wright* 147); 1a, flowering branch, small-leaved form, × 1 (*Schelpe* 1309); 1b, head, × 6,6; 1c, hermaphrodite flower, × 13; 1d, pappus bristle, × 13 (*Wright* 147). 2, **H. confertum,** compact branchlet, × 1 (*Wright* 648); 2a, lax branchlet, × 1 (*Wright* 138). 3, **H. archeri,** part of plant, × 1; 3a, head, × 6,6 (*Cannon* in herb. *Marloth* 10549). 4, **H. sessile,** part of plant, × 1; 4a, head, × 5,3 (*Hilliard & Burtt* 11980).

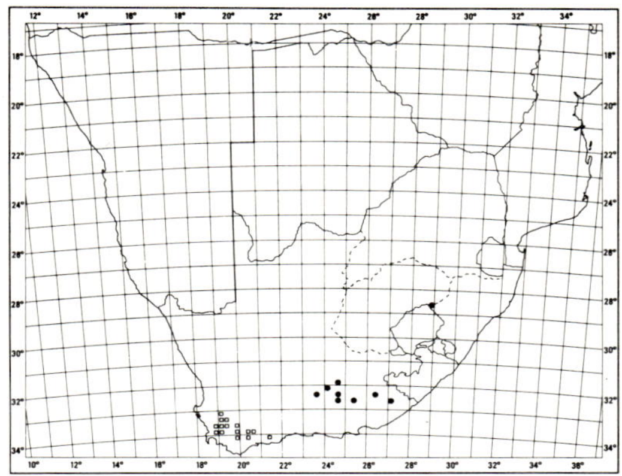

MAP 125.— □ **Helichrysum sphaeroideum**
        ● **Helichrysum sessile**

Vouchers: *Esterhuysen* 4025 (BOL; K; NBG; PRE), 22698 (BOL; K; PRE), 22532 (BOL; K; PRE), 11324 (BOL; K; NBG; PRE; SAM).

Both Burchell and Drège collected *H. sphaeroideum*, but De Candolle confused the Burchell specimen he saw (7296, G-DC, K, from a mountain near Swellendam) with *H. serpyllifolium* var. *polifolium* (that is, *Plecostachys polifolia*). The Drège specimen (at K), distributed as *H. serypyllifolium* var. *polifolium* a (from Du Toit's Kloof) is partly *H. sphaeroideum*, partly *Plecostachys polifolia*.

117. **Helichrysum sessile** *DC.*, Prodr. 6: 173 (1838); Harv. in F.C. 3: 219 (1865); Moeser in Bot. Jb. 44: 322 (1910); Hilliard, Compositae in Natal 217 (1977). Type: Cape, Sneeuwbergen, *Drège* 5738 (G-DC, holo.!; P, iso.!).

Gnaphalium evaciforme Sch. Bip. in Bot. Ztg 3: 169 (1845). Type as for *H. sessile*.

Helichrysum ernestianum DC., Prodr. 6: 173 (1838). *Gnaphalium ernestianum* (DC.) Sch. Bip. in Bot. Ztg 3: 169 (1845). Type: Cape, Camdebooberg, altd. 2 000−3 000 ped., *Drège* 863 (G-DC, holo.!).

H. laevigatum Markötter in Annale Univ. Stellenbosch 8, sect. A, no. 1: 46 (1930). Type: O.F.S., Witzieshoek, c. 1 830 m, *Thode* s.n. (STE 2794, holo.!).

A well-branched, very compact cushion-forming dwarf shrublet, 40−100 mm high, old main stems woody, gnarled, up to 15 mm diam., branches densely tufted, clothed below in dead dry leaves, upper leaves closely imbricate, appearing rosetted from above. *Leaves* c. 5−40 × 1−4 mm, narrowly elliptic or oblong, apex subacute to obtuse, both surfaces enveloped in thick white indumentum, surface silky-smooth, drying like crumpled tissue paper, breaking down with age to felt. *Heads* homogamous, campanulate, c. 5−7 × 8−12 mm across the radiating involucral bracts, solitary or 2−12 together on short or long (up to 100 mm) leafy peduncles terminating the branchlets. *Involucral bracts* in c. 4−6 series, loosely imbricate, base woolly, lamina silvery translucent, sometimes crimson above the stereome, tips opaque milk-white, deltoid, acute, radiating, *Receptacle* scarcely honeycombed. *Flowers* c. 20−38, yellow. *Achenes* 1,5 mm long, with elongated highly myxogenic duplex hairs. *Pappus* bristles many, about equalling corolla, scabrid, apical cells thickened, bases cohering by patent cilia. Fig. 37: 4.

Recorded from the mountains about Graaff-Reinet (Camdeboo Mountains, Koudeveldberge, Cave Mountain, Oudeberg, Sneeuwberg, Tandjesberg), Boschberg at Somerset East, Bruintjeshoogte Mountain near Somerset East, the Great Winterberg and the Amatola Mountains (Hogsback) between 1 370 and 2 300 m above sea level, then a great gap to the mountains near Witzieshoek in the Orange Free State where Thode, Bolus and Flanagan all collected it on the farm Bester's Vallei (Qwa Qwa Mountain) at c. 1 830 m. The small cushions grow from cracks and pockets in weathered rock faces; flowering in December and January. Map 125.

Heads may be small and crowded (described as *H. ernestianum*) or larger and then solitary or few together, on short or long peduncles, peduncle length being a response to environmental conditions. But the tufts of narrow leaves with their thick white silky covering are unmistakable, and serve to distinguish *H. sessile* from *H. sessilioides* (no. 197), which has the upper leaf surfaces clothed in thin 'tissue-paper' indumentum, the lower white-felted.

Vouchers: *Bolus* 1851 (BOL; K); *Esterhuysen* 19704 (BOL); *Marie Galpin* in herb. Galpin 2652 (K; PRE); *Hilliard & Burtt* 11980 (E; K; NU; S); *Nordenstam* 1961 (S).

118. **Helichrysum archeri** *Compton* in Trans. R. Soc. S. Afr. 19: 315 (1931). Lectotype: Laingsburg div., Whitehill, on

---

FIG. 38.−1, **Helichrysum sphaeroideum**, flowering branch, × 1; 1a, head, × 5,3 (*Esterhuysen* 11324). 2, **H. rotundifolium**, whole plant, × 1, 3; 2a, head, × 4 (*Acocks* 19984). 3, **H. altigenum**, whole plant, × 1; 3a, head, × 4 (*Esterhuysen* 27983). 4, **H. saxicola**, flowering branch, × 1,3; 4a, head, × 6,6 (*Taylor* 1095).

white quartzite patches, 13 x 1929, *Compton* 3597 (BOL!; K, isolecto.!).

Dwarf twiggy shrublet up to c. 100 mm high, main stem up to 10 mm diam., gnarled, branchlets grey-woolly, closely leafy to the tips. *Leaves* imbricate, mostly 2,5–5 × 0,5–2 mm, linear or narrowly oblong, keeled, apex obtuse, slightly recurved in older leaves, base broad, half-clasping, both surfaces grey woolly-felted. *Heads* homogamous, campanulate, c. 5 × 3 mm (6 mm across the radiating bracts), solitary or few loosely associated at the tips of the innumerable short branchlets, surrounded by a few leaves lightly webbed together. *Involucral bracts* in 4–5 series, graded, imbricate, 2 outer series brown, inner exceeding flowers, tips subacute to acute, opaque white, sometimes tinged red above the stereome and then the white tips often streaked pink, radiating. *Receptacle* shortly honeycombed, *Flowers* 15–21. *Achenes* not seen, ovaries with myxogenic duplex hairs. *Pappus* bristles many, equalling corolla, scabrid, bases cohering lightly by patent cilia. Fig 37: 3.

Recorded only from the Karoo Garden, White Hill, and the mountains near Matjiesfontein, both in Laingsburg district. Flowers in September, October and November. Map 126.

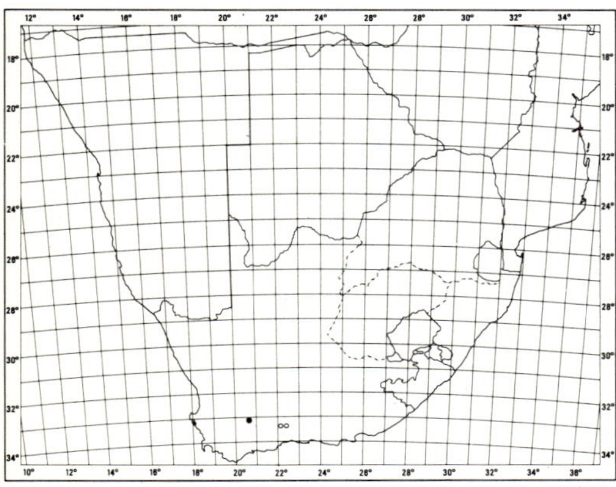

MAP 126.— ● **Helichrysum archeri**
       ○ **Helichrysum saxicola**

Vouchers: *Compton* 12633, 16283 (NBG); *Compton* 5907 (BOL); *Foley* 102 (PRE).

119. **Helichrysum saxicola** *Hilliard, sp. nov.* H. archeri *Compton et* H. sessili *DC. affinis sed ab ambobus foliis ellipticis (nec linearibus nec oblongis) et bracteis involucralibus obtusis (nec acutis) facile distinguenda.*

Fruticulus nanus pulviniformis 30–50 mm altus, caulibus vetustis ad c. 5 mm diam., ramis primariis prostratis radicantibus ramulos permultos arcte congestos erectos vel ascendentes omnino crebre foliatos emittentibus. Folia arcte imbricata, 1–3 × 0,75–1,5 mm, elliptica, obtusa, basi lata amplexicauli, utrinque dense cano-lanato-pannosa. Capitula homogama, c. 5 × 4 mm, 7 mm trans bracteas radiantes diametro, solitaria, apicibus ramulorum sessilia. Bracteae involucrales 5-seriatae, gradatae, laxe imbricatae, exteriores obtusae, albidae, inter se lana contextae, apicibus interiorum flores superantibus elliptico-oblongis obtusis opacis niveis. Receptaculum tuberculis planis praeditum. Flores c. 22–24, flavi, extimi rubro-apiculati. Achenia 1,25 mm longa, pilis duplicibus myxogenis induta. Pappi setae multae, corollam aequantes, apicibus barbellatis inferne scabridae, basibus non cohaerentibus.

Type: Cape, Prince Albert div., Swartberg, between Kliphuisvlei and Plaatsberg, 6 500 ft, Jan. 1954, *Taylor* 1095 (SAM, holo.!)

Cape—Prince Albert div., Swartberg, cracks of rocks on peaks, 6 000 ft, May 1926 [sterile], *Pocock* 62 (BOL; PRE); Swartberg Pass, 6 000 ft, Jan. 1941, *Esterhuysen* 4533 (BOL); ibidem, Dec. 1943, *Stokoe* 9081 (BOL); Great Swartberg, Jan. 1935, *Stokoe* 6772 (BOL); Swartberg, Tierberg area, 6 000 ft, grooves and crevices of rock, 1 vi 1962 [sterile], *Esterhuysen* 29567 (BOL); Oudtshoorn div., Swartberg, rock crevices, Jan. 1947, *Stokoe* SAM 62955 (SAM).

Cushion-forming dwarf shrublet 30–50 mm high, old main stem up to c. 5 mm diam., gnarled, main branches prostrate, rooting, giving rise to innumerable tightly congested erect or ascending branchlets, closely leafy throughout. *Leaves* closely imbricate, 1–3 × 0,75–1,5 mm, elliptic, obtuse, base broad, clasping, both surfaces thickly greyish-white woolly-felted. *Heads* homogamous, campanulate, c. 5 × 4 mm, 7 mm across the radiating bracts, solitary, sessile at the branchlet tips. *Involucral bracts* in 5 series, graded, loosely imbricate, outer obtuse, whitish, webbed together with wool, tips of inner exceeding flowers, elliptic-oblong, obtuse, opaque snow-white. *Receptacle* with flattened tubercles. *Flowers* c. 22−24, yellow, the outermost tipped red. *Achenes* 1,25 mm long, with myxogenic duplex hairs. *Pappus* bristles many, equalling corolla, tips barbellate, shaft scabrid, bases not cohering. Fig. 38: 4.

Recorded only from the Swartberg in Prince Albert and Oudtshoorn divisions. Grows in rock crevices, 1 500 to 2 000 m, in full flower in January. Map 126.

Closely allied to *H. sessile* (no. 117) and *H. archeri* (above) but distinguished from both by its shorter and relatively broader leaves and by its blunter involucral bracts.

120. **Helichrysum rotundifolium** *(Thunb.) Less.*, Syn. Comp. 277 (1832); DC., Prodr. 6: 176 (1838); Harv. in F.C. 3: 224 (1865). Type: Cape of Good Hope, *Thunberg* (sheet 19246, UPS, holo.!).

Gnaphalium rotundifolium Thunb., Prodr. 152 (1800), Fl. Cap. 660 (1823). *Spiralepis rotundifolia* (Thunb.) D. Don in Mem. Wern. nat. Hist. Soc. 552 (1826).

Helichrysum nummularium Moeser in Bot. Jb. 44: 307 (1910); Levyns in Adamson & Salter, Fl. Cape Penins. 782 (1950). Lectotype: Cape, Sir Lowry's Pass, *Schlechter* 7231 (Z!; BM; E; G; PRE; SAM, isolecto.!).

Perennial herb, much branched from a thick (up to 10 mm diam.) woody, creeping underground stock, aerial flowering branches up to c. 150 mm long, several radiating from each crown, mostly simple, decumbent-ascending, grey-woolly, leafy. *Radical leaves* rosetted, up to 30 × 18 mm, broadly elliptic to subrotund, very obtuse, densely greyish-white woolly; *cauline leaves* similar but smaller and tending towards oblong-spathulate, half-clasping. *Heads* homogamous, campanulate, 5−8 × 4−5 mm, rarely solitary, usually 2−7 in terminal clusters surrounded by leaves. *Involucral bracts* in c. 5 series, subequal, loosely imbricate, outer series pale brown, loosely webbed together with wool, inner slightly exceeding flowers, sometimes purplish above the stereome, tips opaque white, very obtuse, crisped, radiating. *Receptacle* nearly smooth. *Flowers* 24−54. *Achenes* not seen, ovaries with myxogenic duplex hairs. *Pap-*

pus bristles many, in more than one series, about equalling the corolla, tips scabrid, scabridulous below, fused at base in bundles. Fig. 38: 2.

On the mountains in the SW. and S. Cape, from Great Winterhoek near Tulbagh south to the mountains of the Cape Peninsula, Wellington Sneeukop and east across the Langeberg and Swartberg to the mountains near Uniondale, between c. 650 and 1 800 m. Grows in open places; flowering mainly between November and January. Map 127.

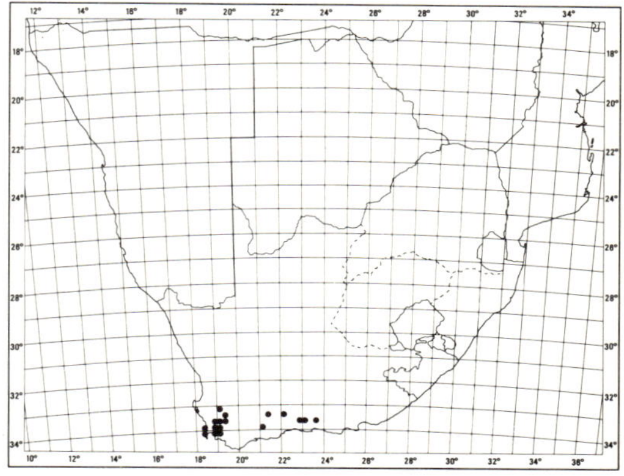

MAP 127.— **Helichrysum rotundifolium**

Vouchers: *Acocks* 19984 (PRE); *Compton* 19576 (NBG); *Esterhuysen* 33072 (BOL; E; PRE); *Pillans* 8482 (BOL; PRE); *Stokoe* SAM 68696 (SAM).

121. **Helichrysum altigenum** *Schltr. & Moeser* in Bot. Jb. 44: 322 (1910). Type: Cape, Koude Bokkeveld, Elandsfontein, 5 000 ft, 18 i 1897, *Schlechter* 10026 (BM; BOL; E; G; PRE; S; Z, iso.!).

Dwarf subshrub, rootstock creeping, branched, each branch producing a rosette of leaves with 2 'to' several aerial stems radiating from the crown, prostrate to ascending, 10−150 mm long, simple or once or twice forked near the tip, silky grey-woolly, closely leafy. *Leaves* imbricate, blade c. 5−13 × 2−7 mm, scarcely diminishing upwards, elliptic-ovate, base narrowed, c. 3−4 × 2 mm, apex obtuse or subacute, apiculate, slightly recurved, both surfaces enveloped in silvery grey 'tissue-paper' indumentum that webs the leaf bases to the stems. *Heads* homogamous or heterogamous, campanulate, c. $8-10 \times 16$ mm across the radiating bracts, solitary at the branch tips, closely invested by leaves. *Involucral bracts* in c. 5 series, loosely imbricate, graded, inner slightly exceeding flowers, all acute or subacute, outer membranous, palest brown, inner opaque, dull white, tipped palest brown or pure white. *Receptacle* shortly honeycombed. *Flowers* c. $44-58$, $0-5 \female$, $44-58 \male$, yellow, sometimes tipped pink. *Achenes* narrowly obovate in outline, with myxogenic duplex hairs. *Pappus* bristles many, tips scabrid, bases cohering by patent cilia. Fig. 38: 3.

Endemic to the SW. and S. Cape, from the Koude Bokkeveld in Ceres division across the mountains (Baviaansberg, Seven Weeks Poort, Swartberg, Kouga Mountains) to the Great Winterhoek Mountains near Uitenhage between c. 900 and 2 000 m. Grows on stony slopes; flowering between November and January, and seems to be conspicuous after burning. Map 128.

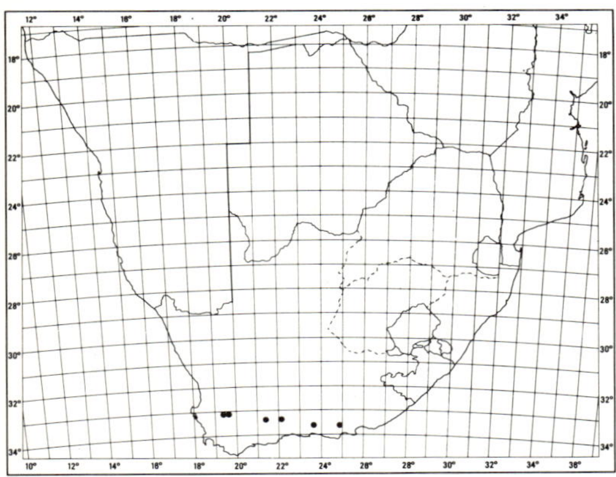

MAP 128.— **Helichrysum altigenum**

Vouchers: *Esterhuysen* 27983 (BOL; PRE); *Nordenstam* 1925 (E; S); *Stokoe* 4544 (BOL).

122. **Helichrysum crispum** *(L.) D. Don* in Loudon, Hort. Brit. 341 (1830); Less., Syn. Comp. 310 (1832); Hilliard & Burtt in Notes R. bot. Gdn Edinb. 32: 346 (1973). Lectotype: Plukenet, Phytogr. t. 298 fig. 3 (1694), Almagest. 171 (1696), herb. Sloane 100: folio 104, top left (BM!).

Gnaphalium crispum L., Sp. Pl. edn 2: 1197 (1762), Mantissa 91 (1767).

G. polyanthos Thunb., Prodr. 151 (1800), non p. 147, Fl. Cap. 657 (1823). *G. multiflorum Willd,*, Sp. Pl. 3: 1900 (1804); Thunb., Mus. Upsal. Append. 13: 3 (1806). Type: Cape of Good Hope, *Thunberg* (sheet 19205, UPS!).

Helichrysum leucophyllum DC., Prodr. 6: 175 (1838). *G. leucophyllum* (DC.) Sch. Bip. in Bot. Ztg 3: 169 (1845). Type: Cape, Worcester, *Ecklon* 1425 (G-DC!).

H. crassifolium sensu DC., Prodr. 6: 175 (1838); Harv. in F.C. 3: 224 (1865); Moeser in Bot. Jb. 44: 309 (1910), non (L.) D. Don.

H. rotundifolium sensu Moeser in Bot. Jb. 44: 309 (1910), non (Thunb.) Less.

Perennial herb with tufts of soft-wooded simple or branched stems up to 600 mm long, 4 mm diam., erect or spreading from the crown, rooting at nodes, tips ascending, white silky-woolly, closely leafy. *Leaves* often with dwarf axillary branchlets, primary leaves up to 35×12 mm, diminishing upwards, spathulate to oblong-obovate, apex obtuse to subacute, apiculate, base broad, half-clasping, soft-textured, both surfaces white silky-woolly. *Heads* homogamous or rarely heterogamous, campanulate, c. $3-6 \times 3,5-5$ mm, many in dense terminal corymbose clusters arranged in compact corymbose panicles. *Involucral bracts* in c. 7 series, outermost short, thinly woolly, others subequal, loosely imbricate, about equalling the flowers, not radiating, obovate, tips almost truncate, markedly crisped-undulate, opaque creamy white. *Receptacle* scarcely honeycombed. *Flowers* $12-50$, very rarely $3-4 \female$. *Achenes* 0,75 mm long, glabrous or with duplex hairs. *Pappus* bristles many, about equalling corolla, scabrid, bases with patent cilia, not cohering. Fig. 39: 3.

Mainly along the Cape coast (but also recorded from Tulbagh Kloof) from Blouberg strand to George, on coastal or inland dunes and other sandy places; flowering between October and December. Map 129.

There is some variation in the size of the heads and the degree of crisping of the involucral bracts, the larger heads often having the bracts less crisped than the smaller. Nevertheless, the species is not easily confused with any other.

Vouchers: *Acocks* 19075 (PRE); *Compton* 12782 (NBG); *Pillans* 9764 (BOL); *Rogers* 2150a (BOL; PRE); *Salter* 7079 (BOL; PRE).

123. **Helichrysum outeniquense** *Hilliard, species nova* H. felino *Less. affinis sed*

habitu a basi ramosiore, foliis plus lanatis, capitulis campanulatis c. 4 × 5 mm (nec sub-globosis 5−7 × 6−10 mm), bracteis involucralibus c. 5-seriatis (nec 8−12-seriatis), floribus 20−25 (nec c. 40−100) distinguenda.

Herba perennis, e basi ramosa; caules c. 250 mm alti, tenuiter cano-lanati, foliati, sursum pedunculoidei et prope capitula nudi. Folia plerumque 10−25 × 5−7 mm, elliptico-oblonga, sursum lanceolata et in bracteas paucas remotas transeuntia, apice obtuso in foliis summis acuto, basi lata semi-amplectente, utrinque cano-lanata et insuper glanduloso-setosa. Capitula homo-gama, campanulata, c. 4 × 3 mm, 5 mm diametro trans bracteas radiantes, multa in glomerulos parvos corymbosos paniculam corymbosam compactam formantes disposi-ta. Bracteae involucrales c. 5-seriatae, extimae breves pallide fuscae, ceterae sub-aequales flores vix superantes, apicibus obtusis opacis albis leviter crispatis radianti-bus. Receptaculum breviter fimbrilliferum. Flores c. 20−25. Achenia non visa; ovaria pilis duplicibus induta. Pappi setae multae, corollam aequantes, apicibus barbellatae, inferne scabridae, basibus ciliis patentibus valde cohaerentes.

Type: Cape, Uniondale div., N. face of Outeniquas near Joubertina, Die Hoek, peaty soil near spring in valley, 15 i 1947, *Esterhuysen* 13632 (BOL, holo.!).

Perennial herb or subshrub, branching from the base, stems c. 250 mm tall, thinly greyish-white woolly, leafy, becoming ped-unculoid and nude near the heads. *Leaves* mostly 10−25 × 5−7 mm, elliptic-oblong becoming lanceolate towards the heads and passing rapidly into a few distant bracts, apex obtuse becoming acute in the upper-most leaves, base broad, half-clasping, both surfaces greyish-white woolly, glandular-setose as well. *Heads* homogamous, campan-ulate, c. 4 × 3 mm, 5 mm across the radiating bracts, many in small corymbose clusters arranged in a compact corymbose-panicle. *Involucral bracts* in c. 5 series, outermost short, pale fuscous, others sub-equal, scarcely exceeding the flowers, tips obtuse, opaque white, slightly crisped. *Receptacle* with short fimbrils. *Flowers* c.

20−25. *Achenes* not seen, ovaries with duplex hairs. *Pappus* bristles many, equal-ling corolla, tips barbellate, shaft scabrid, bases with patent cilia cohering strongly.

Known only from the type collection. Allied to *H. felinum* (below) but distinguished by its different habit, woollier leaves, and smaller heads with bracts in fewer series, arranged in a spreading corymbose panicle. Map 129.

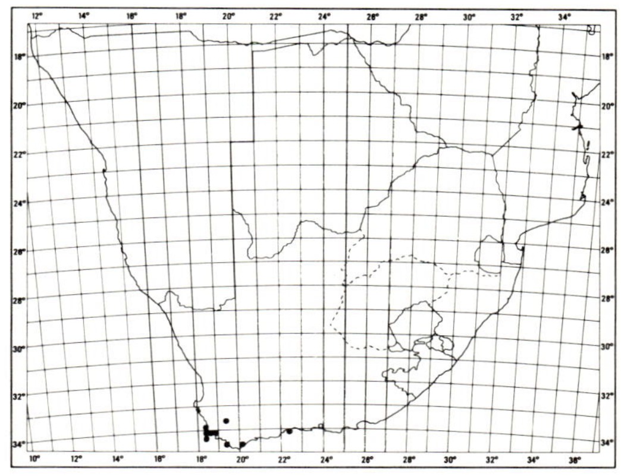

MAP 129.— ● **Helichrysum crispum**
○ **Helichrysum outeniquense**

124. **Helichrysum felinum** *Less.,* Syn. Comp. 287 (1832); DC., Prodr. 6: 176 (1838) incl. vars; Harv. in F.C. 3: 225 (1865); Moeser in Bot. Jb. 44: 309 (1910); Hilliard & Burtt in Notes R. bot. Gdn Edinb. 32: 348 (1973); Hilliard, Compositae in Natal 232 (1977). Type: Cape, *Thunberg* (sheet 19159, UPS, holo.!).

Gnaphalium frutescens, tomentosum, folio oblongo, floribus comosis Burm., Pl. Afr. Rar. 224, t. 79 fig. 4 (1739) (specimen in herb. Burm. G!).

G. elongatum Lam., Encycl. 2: 741 (1788); Ker-Gawl. in Bot. Reg. 3, t. 243 (1813); Curtis's Bot. Mag. t. 2328 (1822), non *Helichrysum elongatum* Moench (1794). Type: Cape of Good Hope (P-LAM!).

G. felinum Thunb., Prodr. 140 (1800), Fl. Cap. 648 (1823), (nom. illegit.) quoad spec, excl. syn. *(Thun-berg,* sheet 19159, UPS!).

G. fastigiatum Schrank in Denkschr. K. Akad. Wiss. Münch. 8: 148 (1824). Type: Cape of Good Hope, *Brehm* (M, holo.!).

G. serratum sensu Thunb., Prodr. 149 (1800), Fl. Cap. 659 (1823), non L.

Stout perennial herb up to c. 1 m tall, stem woody, up to 5 mm diam., simple below, subsimple or branched above, branches erect, rod-like, rough with leaf scars below, loosely greyish-white woolly and densely leafy above. *Leaves* spreading or deflexed, up to 35 (−50) × 8 (−12) mm, sometimes smaller and distant upwards, lanceolate to elliptic, apex more or less acute, mucronate, bases broad, clasping, upper surface cobwebby at first, later harshly pubescent, sometimes glabrescent, generally rugose, lower grey woolly-felted. *Heads* homogamous, subglobose, 5−7 mm long, 7−10 mm across the fully radiating bracts, many in dense, rounded corymbose clusters c. 20−40 mm across, sometimes more open or becoming so with age. *Involucral bracts* in c. 8−12 series, subequal, loosely imbricate, about equalling the flowers, milk-white, or tinged rose-pink, often crisped, radiating. *Receptacle* shallowly honeycombed or shortly toothed. *Flowers* 38−103, yellow. *Achenes* not seen, ovaries with duplex hairs. *Pappus* bristles many, about equalling corolla, tips barbellate, bases cohering strongly by patent cilia.

Ranges from the SW. Cape, including the Peninsula, through the SE. and E. districts and the Transkei to southern Natal. Found on rough scrubby slopes; flowering between September and December. Very variable, particularly in leaf size and colour of the involucral bracts, but easily recognized. Map 130.

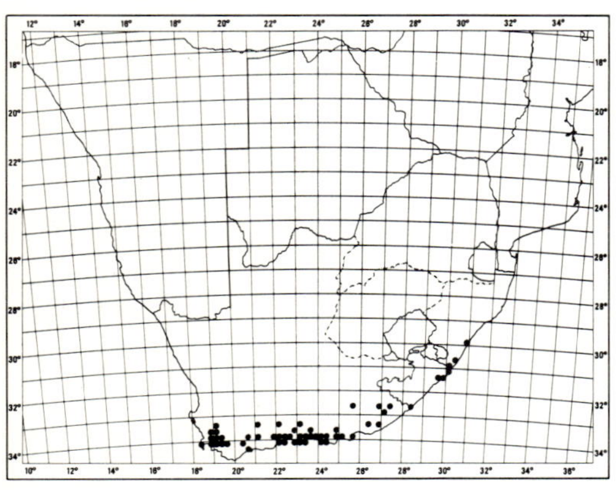

MAP 130.— **Helichrysum felinum**

Vouchers: *Codd* 3583 (NU; PRE); *Hilliard & Burtt* 10935 (E; K; MO; NU; PRE; S); *Nordenstam*

1914 (LD; NU); *Pillans* 6752 (BOL); *Tyson* 944 (SAM).

The epithet *'felinum'* derives from Hermann (Cat. Pl. Afr. 13, 1737) who wrote *'floribus umbellatis, facie Pedis Cati'*. He must have had a cat with white paws!

125. **Helichrysum grandiflorum** *(L.) D. Don* in Sweet, Hort. Brit. 222 (1826); Less., Syn. Comp. 289 (1832); DC., Prodr. 6: 175 (1838); Harv. in F.C. 3: 224 (1865); Moeser in Bot. Jb. 44: 309 (1910); Levyns in Adamson & Salter, Fl. Cape Penins. 782 (1950). Type: Cape of Good Hope (LINN 989.3!).

Gnaphalium grandiflorum L., Sp. Pl. 850 (1753).

G. verbascifolium Schrank in Denkschr. K. Akad. Wiss. Münch. 8: 150 (1824). Type: Cape of Good Hope, *Brehm* s.n. (M, holo.!).

Perennial, base woody, stems several from the crown, decumbent then erect to c. 450 mm, simple or subsimple, loosely woolly, densely leafy near the base, then distantly so, pedunculoid and bracteate under the compound inflorescences. *Radical leaves* up to 100 × 60 mm, obovate or broadly elliptic becoming oblong to oblong-lanceolate then linear-lanceolate upwards, apex obtuse to subacute, uppermost reduced leaves acute with a white papery appendage, base broad, clasping, both surfaces loosely greyish-white woolly. *Heads* homogamous, campanulate, c. 7 × 7 mm, many in dense terminal spherical clusters 20−50 mm across. *Involucral bracts*

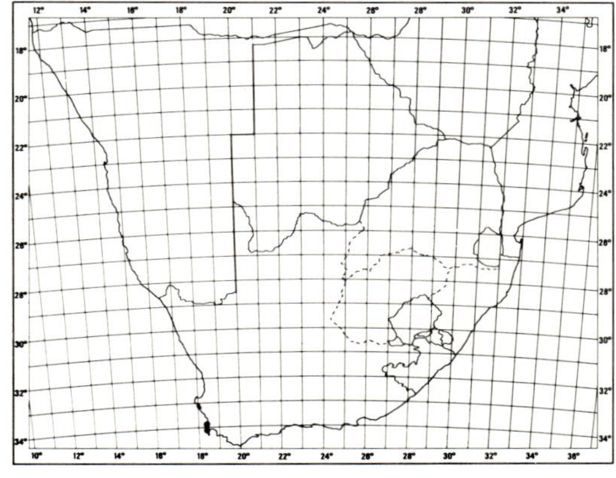

MAP 131.— **Helichrysum grandiflorum**

in c. 10 series, graded, imbricate, inner shortly exceeding flowers, minutely radiating, obtuse or subacute, dull, creamy white. *Receptacle* nearly smooth. *Flowers* 70—107. *Achenes* 0,75 mm long, barrel-shaped, ribbed, glabrous. *Pappus* bristles many, equalling corolla, tips subplumose, bases cohering strongly by patent cilia.

Endemic to the mountains of the Cape Peninsula; flowering between December and February. Sometimes confused with *H. fruticans:* see under that species, below. Map 131.

Vouchers: *Bolus* 3874 (BOL); *Compton* 8218 (NBG); *Esterhuysen* 26853 (BOL); *Gillett* 3498 (PRE).

126. **Helichrysum fruticans** *(L.) D. Don* in Sweet, Hort. Brit. 222 (1826); Less., Syn. Comp. 288 (1832); DC., Prodr. 6: 175 (1838); Harv. in F.C. 3: 225 (1865); Moeser in Bot. Jb. 44: 310 (1910); Levyns in Adamson & Salter, Fl. Cape Penins. 782 (1950). Type: Cape of Good Hope (LINN 989.5!).

Gnaphalium fruticans L, Mant. alt. 282 (1771); J. Sims in Curtis's Bot. Mag. t. 1802 (1816). *Astelma fruticans* (L.) Ker-Gawl., Bot. Reg. t. 726 (1821).

Stem robust, c. 1 m tall, nude below, densely leafy above, simple or subsimple, flowering stem white-woolly, closely leafy becoming pedunculoid below the heads. *Leaves* up to 60 (—100) × 30 mm, diminishing in size upwards and becoming distant, elliptic to oblong-elliptic, apex obtuse to subacute, mucronate, base broad, half-clasping, upper surface thinly white-woolly, lower densely so, 3-nerved. *Heads* homogamous or occasionally heterogamous, campanulate, c. 11 × 20 mm across the radiating bracts, many in a compact corymbose panicle up to 110 mm across. *Involucral bracts* in c. 10 series, graded, loosely imbricate, inner much exceeding flowers, radiating, acute to subobtuse, dull, snow-white. *Receptacle* nearly smooth. *Flowers* 50—78, occasionally 1 ♀. *Achenes* 1,5 mm long, barrel-shaped, ribbed, with duplex hairs, not myxogenic. *Pappus* bristles many, tips subplumose, bases cohering strongly by patent cilia.

Endemic to the mountains of the Cape Peninsula, flowering between September and December. Map 132.

Close to *H. grandiflorum* (above) but distinguished by its larger heads (involucre c. 11 mm long,

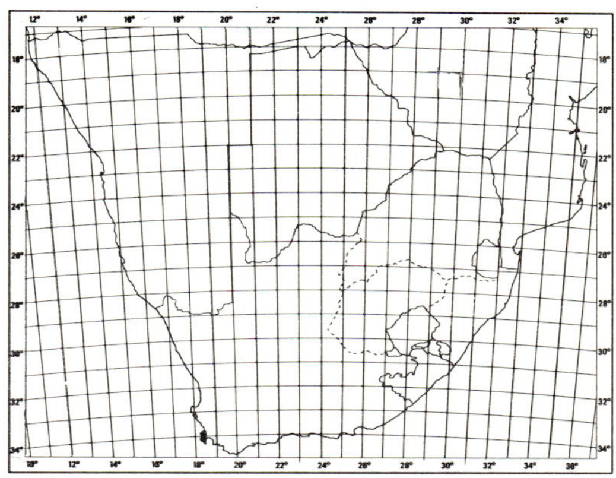

MAP 132.— **Helichrysum fruticans**

not 7 mm, with involucral bracts much exceeding the flowers) arranged in a spreading corymb, not in a tight cluster.

Vouchers: *Bolus* 4502 (BOL); *Compton* 14025 (NBG; PRE); *Galpin* 4140 (PRE); *Rodin* 3219 (BOL).

127. **Helichrysum acrophilum** *H. Bol.* in Trans. S. Afr. phil. Soc. 18,3: 389 (1907). Lectotype: Cape, Koude Bokkeveld, Gydouwberg, 1 800 m, 19 i 1897, *Schlechter* 10049 (BOL!; BM; E; G; PRE; S; Z, isolecto.!).

Gnaphalium deltoides Thunb. Mus. Upsal. auct. 1827: 16 (1827), nomen (*Thunberg,* sheet 19127, UPS!).

Helichrysum spathulatum Moeser in Bot. Jb. 44: 310 (1910), nom. illegit.

Small loosely branched shrub up to c. 300 mm tall, branches sometimes decumbent then rooting, thinly grey silky-woolly, young stems distantly leafy, old stems with leaves crowded on dwarf axillary shoots. *Leaves* up to 15 × 6 mm, smaller on the dwarf shoots, spathulate, tip almost truncate, mucronate, recurved, base much narrowed, sessile, both surfaces thinly grey woolly-felted. *Heads* homogamous, campanulate, c. 7 × 12 mm across the radiating bracts, solitary or 2—5 loosely corymbose at the branch tips. *Involucral bracts* in c. 5 series, loosely imbricate, graded, woolly on backs below, inner exceeding flowers, tips radiating, subacute to obtuse, dull white, outermost sometimes overlaid palest brown,

1a
2a
2
1
3
3a
H. Wouda-du Toit

sometimes crimson above stereome. *Receptacle* with fimbrils about equalling ovaries. *Flowers* 28—61, yellow. *Achenes* 1,25 mm long, cylindric, obscurely ribbed, with myxogenic duplex hairs. *Pappus* bristles many, about equalling corolla, scabrid, tips subplumose, bases lightly fused. Fig. 39: 2.

On the mountains of the SW. Cape (excluding the Peninsula) from Pakhuis Pass and the Cedarberg to the Swartberg, between 900 and 1 900 m above sea level, but most records above 1 500 m. A constituent of shrub communities in rocky places. Flowers mainly in December and January, but as early as November and as late as March. The small grey spathulate leaves with their recurved tips, allied to the white or somewhat rosy campanulate heads, make recognition easy. Map 133.

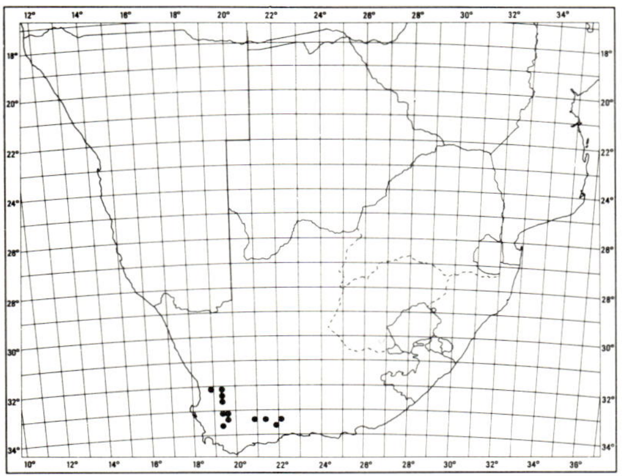

MAP 133.— ● **Helichrysum acrophilum**
○ **Helichrysum amplectens**

Vouchers: *Compton* 8431 (BOL; NBG); *Esterhuysen* 20007 (BOL; PRE); *Oliver* 5541 (PRE); *Stokoe* SAM 56574 (PRE; SAM).

128. **Helichrysum amplectens** *Hilliard* in Notes R. bot. Gdn Edinb. 34: 78 (1975), Compositae in Natal 231 (1977). Type: Natal, Bergville distr., Cathedral Peak area, Cleft Peak, c. 2 800 m, *Nordenstam* 2127 (NU; holo.!; LD, iso.!).

Straggly, loosely branched perennial herb, possibly up to 600 mm tall, base woody, branches erect or ascending, brittle, loosely woolly, flowering stems in the axils of dry withered leaves below new leafy growth, up to 400 mm long, slender, simple below, forking above into the lax inflorescence, loosely woolly, remotely leafy. *Leaves* on vegetative shoots crowded, those at the tips spreading, c. 30—80 × 10—30 mm, panduriform, apex subacute to subobtuse, base auricled, amplexicaul, upper surface thinly cobwebby, lower white-felted; leaves on flowering stems smaller, distant, oblong-elliptic becoming lanceolate upwards, base broad, cordate-clasping, indumentum as in primary leaves. *Heads* homogamous, broadly campanulate, c. 6—8 mm long, 10—13 mm across when fully radiating, 2—10 in a very open corymbose panicle. *Involucral bracts* in c. 7 series, graded, loosely imbricate, inner much exceeding flowers, bases woolly, tips obtuse or subacute, opaque, snow-white. *Receptacle* honeycombed. *Flowers* c. 100, yellow. *Achenes* 1,5 mm long, ellipsoid, glabrous. *Pappus* bristles 12—15, about equalling corolla, delicate, tips subplumose, bases not cohering. Fig. 39: 1.

Recorded only from Cleft Peak, Cathkin Peak and Monk's Cowl in the Natal Drakensberg (Bergville and Estcourt districts) between c. 2 400 and 2 800 m, growing on rock ledges and crannies at the foot of cliffs on steep rocky S. facing slopes; flowering in February. Map 133.

Vouchers: *Esterhuysen* 20257 and 26100 (BOL; PRE).

129. **Helichrysum diffusum** *DC.*, Prodr. 6: 175 (1838); Harv. in F.C. 3: 223 (1865); Moeser in Bot. Jb. 44:309 (1910); Hilliard, Compositae in Natal 230 (1977). Type: Cape, Drakensteensberg, Dutoitskloof, *Drège* 1792 (G-DC, holo.!; BM; E; PRE; SAM; TCD, iso.!).

Gnaphalium diffusum (DC.) Sch. Bip. in Bot. Ztg 3: 170 (1845).

Wiry perennial herb, roots fibrous, stems slender, soft-wooded, decumbent or ascending, up to 500 mm long, simple

---

FIG. 39.—1, **Helichrysum amplectens**, flowering branch, × 0,7; 1a, head, × 2,6 (*Nordenstam* 2127). 2, **H. acrophilum**, flowering branch, × 1,3; 2a, head, × 4 (*Stokoe* 68367). 3, **H. crispum**, part of plant, × 0,7; 3a, head, × 4,6 (*Esterhuysen* 26442).

below, sparingly branched above, greyish-white woolly, densely leafy. *Leaves* up to 40 × 25 mm, smaller and more distant upwards, oblong or elliptic, apex acute, mucronate, base broad, clasping, both surfaces greyish-white woolly. *Heads* homogamous, campanulate, c. 5—7 mm long, double that across the radiating bracts, up to c. 12 in a dense terminal corymbose cluster. *Involucral bracts* in c. 9 series, loosely imbricate, much exceeding flowers, acute, glossy, white, mostly tipped dark brown or purplish brown outside, rarely whole bract dull purple. *Receptacle* smooth or shortly honeycombed. *Flowers* 26—64, yellow. *Achenes* c. 1 mm, with duplex hairs. *Pappus* bristles many, scabrid, tips barbellate, white or purplish, bases cohering by patent cilia.

Recorded from the mountains in Paarl, Stellenbosch, Worcester and Caledon divisions of the SW. Cape, then a disjunction to southernmost Natal, on the Table Mountain Sandstone cliffs and outcrops above the Umtamvuna river and its tributaries in Port Shepstone district. Flowers between September and December. Map 134.

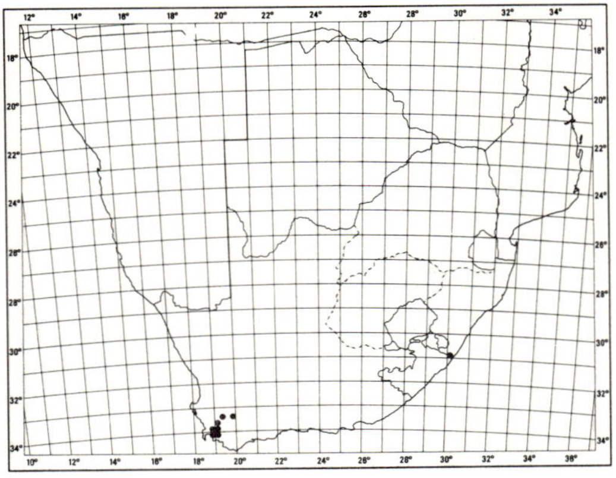

MAP 134.— **Helichrysum diffusum**

Natal specimens have slightly greyer foliage, more closely leafy flowering stems, and fewer flowers in the head than Cape specimens, but these are scarcely specific differences.

Vouchers: *Esterhuysen* 8846 (BOL); *Hilliard* 3996 (E; K; MO; NU; PRE; S); *Marsh* 635 (PRE); *Rodin* 3250 (BOL; PRE); *Stokoe* 8127 (BOL).

Group 18

More or less scandent, soft-wooded shrubs; *leaves* small or medium-sized, broadly elliptic or broadly ovate to rhomboid or orbicular, either petiolate or panduriform; *heads* homogamous, 4—8 × 3—10 mm, mainly in corymbose panicles; *involucral bracts* radiating, opaque white or rarely yellowish or tinged with rose; *receptacle* fimbrilliferous; *flowers* 12—68, corolla funnel-shaped; *achenes* glabrous; *pappus* bristles scabrid, bases cohering strongly by patent cilia, sometimes lightly fused as well.

Species 130—133, one species (*H. panduratum;* no. 133) widely distributed from Gabon to tropical E. Africa and south to the Transkei (with a subspecies confined to the E. highlands of Zimbabwe, the E. Transvaal, W. Swaziland and N. Natal), 2 species endemic to the Cape, one to the Cape and Transkei. Found on forest margins or, in the Cape, in shrub communities.

1a Involucral bracts very obtuse, concave:

    2a Leaves panduriform, margins crisped-undulate .... 131. *H. patulum*

    2b Leaves subrotund, broadly ovate or rhomboid-elliptic, distinctly petiolate, petiole sometimes winged,
        leaf margins flat ....132. *H. petiolare*

1b Involucral bracts subacute to acute:

    3a Leaves orbicular to broadly ovate, abruptly contracted to a broad petiole-like ear-clasping base.
        SW. Cape .... 130. *H. pandurifolium*

    3b Leaves panduriform, sessile, or ovate-rhomboid to elliptic and distinctly petiolate:

        4a Leaves panduriform, sessile .... 133. *H. panduratum*

        4b Leaves ovate-rhomboid to elliptic, petiolate .... 133. *H. panduratum* var. *transvaalense*

130. Helichrysum pandurifolium *Schrank* in Denkschr. K. Akad. Wiss. Münch. 8: 169 (1824). Type: Cape of Good Hope, *Brehm* s.n. (M, holo.!).

Gnaphalium auriculatum Thunb., Prodr. 151 (1800), Fl. Cap. 657 (1823), non Lam. 1788. *Helichrysum auriculatum* Less., Syn. Comp. 311 (1832); DC., Prodr. 6: 209 (1838); Harv. in F.C. 3: 253 (1865) excl. var.

panduratum Harv.; Moeser in Bot. Jb. 44: 311 (1910); Levyns in Adamson & Salter, Fl. Cape Penins. 785 (1950). Type: Cape, Table Mountain, *Thunberg* (sheet 19103, UPS, holo.!).

H. auriculatum var. *oblongifolium* DC., Prodr. 6: 209 (1838). Type: Cape, Table Mtn, *Drège* 295 (G-DC, holo.!; BM, iso.!).

A loosely branched, soft-wooded shrub, branches long, slender, thinly grey-woolly, leafy becoming pedunculoid and distantly bracteate below the inflorescence. *Leaves* mostly 8−16 × 8−10 mm, orbicular to broadly ovate abruptly narrowed to a broad petiole-like ear-clasping base, margins markedly crisped-undulate, both surfaces grey-woolly. *Heads* homogamous, broadly campanulate, c. 5−8 × 6−10 mm (−14 mm when fully radiating), few to many in loose terminal corymbose panicles. *Involucral bracts* in c. 6 series, graded, loosely imbricate, outer sometimes palest brown, inner exceeding the flowers, backs loosely woolly above stereome, radiating, tips acute or subacute, somewhat crisped, opaque milk-white, sometimes suffused with rose. *Receptacle* with fimbrils exceeding ovaries. *Flowers* 12−39, yellow. *Achenes* 0,75 mm, barrel-shaped, glabrous. *Pappus* bristless many, scabrid, about equalling corolla, bases cohering strongly by patent cilia. Fig. 40: 3.

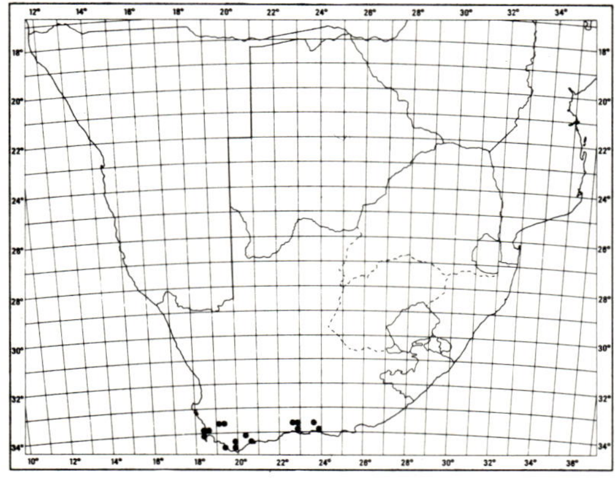

MAP 135.— **Helichrysum pandurifolium**

Ranges from the Cape Peninsula, Bainskloof and Waboom's River through the coastal districts to the Tsitsikama and Kouga Mountains, on hill and mountain slopes in sandy or rocky places, from near sea level to c. 1 500 m. Flowers between September and January, but mainly from October to December. Map 135.

Can be confused with both *H. patulum* (below) and *H. petiolare* (no. 132) but distinguished by its more pointed involucral bracts.

Vouchers: *Bolus* 3315 (BOL; SAM); *Esterhuysen* 387 (BOL; PRE); *Hafström & Acocks* 1608 (PRE); *Taylor* 4470 (PRE).

131. **Helichrysum patulum** *(L.) D. Don* in Loudon, Hort. Brit. 341 (1830); Hilliard & Burtt in Notes R. Bot. Gdn Edinb. 32: 355 (1973). Type: Cape of Good Hope, *Gnaphalium* no. 15 (Herb. Cliff. BM, holo.!).

Gnaphalium patulum L., Sp. Pl. 855 (1753).

G. divaricatum Berg., Descr. Pl. Cap. 250 (1767). Type: Cape, *Ekeberg* (STB, holo.!).

G. spatulatum Burm. f., Prodr. Fl. Cap. 25 (1768); Breyn., Prodr. t. 18 fig. 3 (1739). Type: Cape of Good Hope, *Breynius* (W!).

G. auriculatum Lam., Encycl. 2: 754 (1788), non *Helichrysum auriculatum* Less. Type: Cape of Good Hope (P-LAM!).

Helichrysum crispum var. *subrufescens* DC., Prodr. 6: 208 (1838). Types: ex herb. De la Roche (G-DC!); spec. labelled *Gnaphalium spatulatum* Breyn. (G-DC!). [These excluded from IDC microfiche].

H. crispum var. *citrinum* Harv. in F.C. 3: 253 (1865). Lectotype: Cape Flats, *Wallich* (BM!).

Helichrysum crispum sensu Less., Syn. Comp. 310 (1832); DC., Prodr. 6: 208 (1838); Harv. in F.C. 3: 253 (1865); Moeser in Bot. Jb. 44: 311 (1910); Levyns in Adamson & Salter, Fl. Cape Penins. 784 (1950); non (L.) D. Don.

Straggling well-branched subshrub up to 1 m tall, branches virgate, white-woolly, closely leafy becoming nude and pedunculoid below the inflorescence. *Leaves* mostly 6−20 × 2−12 mm, panduriform, upper part broadly elliptic to suborbicular, abruptly contracted about the middle, apex very obtuse, base broad, ear-clasping, margins crisped-undulate, both surfaces greyish-white woolly, upper less densely so. *Heads* homogamous, campanulate to subglobose, 4−5 × 3−5 mm, many in small corymbose clusters often corymbose-paniculately arranged. *Involucral bracts* in 5 series, graded, loosely imbricate, inner about equalling flowers, tips not radiating, concave, very obtuse, more or less undulate-crisped, opaque creamy white to pale primrose yellow, or rosy. *Receptacle* with fimbrils equalling ovaries. *Flowers* 14−30. *Achenes* 0,75 mm long, barrel-shaped, c. 8-ribbed, glabrous. *Pappus* bristles many, a little shorter than corolla, scabrid, bases cohering strongly by patent cilia. Fig. 40: 4.

Confined to the SW. and S. Cape, from the Peninsula, Cape Flats and Stellenbosch along the coast to Groot Brak River, Mossel Bay. Grows in shrub communities on coastal dunes and inland on S. facing mountain slopes up to c. 600 m above sea level. Flowers between September and February, mainly in December and January. Map 136.

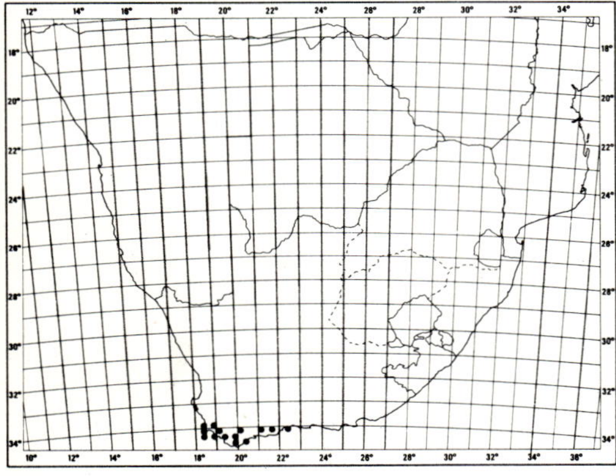

MAP 136.— **Helichrysum patulum**

Often confused with *H. pandurifolium* (no. 130), from which it is easily distinguished by its concave obtuse involucral bracts; distinguished from *H. petiolare* (below) by its crisped-undulate leaf margins.

Vouchers: *Acocks* 21553 (PRE); *Bolus* 4420 (BOL); *Compton* 8117 (NBG); *Schlechter* 9777 (E); *Taylor* 7024 (PRE).

132. **Helichrysum petiolare** *Hilliard & Burtt* in Notes R. bot. Gdn Edinb. 32: 357 (1973); Clapham in Tutin *et al.*, Fl. Europ. 4: 131 (1976). Type: Cape, Zuurberg, *Drège* 264 (G-DC, holo.!).

Gnaphalium tomentosum, foliis orbiculatis subtus incanis Burm., Rar. Afr. Pl. 214, t. 76, fig. 2 (1739).

H. petiolatum sensu DC., Prodr. 6: 208 (1838) quoad descr. et spec., excl. syn.; Harv. in F.C. 3: 252 (1865); Moeser in Bot. Jb. 44: 312 (1910); non (L.) DC.

A straggling, loosely branched, soft-wooded shrub forming tangled masses several metres across and as high when supported by other vegetation, branches long, slender, thinly grey-woolly, leafy becoming pedunculoid and distantly bracteate below the inflorescences. *Leaf* blades mostly 10−35 × 10−30 mm, subrotund to broadly ovate or elliptic-rhomboid, apex rounded, base truncate, subcordate or cuneate, both surfaces grey-woolly-felted, upper sometimes only cobwebby, petiole up to 10 mm, flat, sometimes winged, base auriculate. *Heads* homogamous, subglobose opening broadly campanulate, c. 5 × 5 mm, many in loose terminal corymbose panicles. *Involucral bracts* in c. 5 series, graded, loosely imbricate, about equalling the flowers, radiating, backs loosely grey-woolly above stereome, tips rounded, concave, opaque milk-white. *Receptacle* with fimbrils equalling or exceeding ovary. *Flowers* 18−30, yellow, sweetly scented. *Achenes* 1 mm long, barrel-shaped, 5-ribbed, glabrous. *Pappus* bristles many, scabrid, about equalling corolla, bases cohering strongly by patent cilia, lightly fused as well. Fig. 40: 5.

Ranges from the Cedarberg and the mountains near Stellenbosch to the S. Cape coast forests, thence NE. to Insizwa and Tabankulu Mountains in the Transkei, at about 900−1 050 m in the SW. Cape, but descending to lower altitude in the coast forests, then between c. 1 050 and 1 400 m from the Amatola Mountains and nearby to the Transkei. Naturalized in Portugal.

Grows in shrubby places in damp sheltered kloofs and on forest margins; flowering in December and January. In the NE. part of the range (roughly Katberg, Hogsback and King William's Town to the Transkei), leaves tend to be rhomboid-elliptic and only cobwebby above, not ovate to subrotund and grey-woolly-felted as they often are in the western part of the range. Specimens from Suurberg Pass, Storms

---

FIG. 40.−1, **Helichrysum panduratum** var. **panduratum**, part of flowering stem, × 0,7 (*Hilliard* 4071); 1a, head, × 6,6; 1b, hermaphrodite flower, × 10; 1c, pappus bristle, × 10 (*Gordon Gray* 6362). 2, **H. panduratum** var. **transvaalense**, leaf, × 1,3 (*Hilliard & Burtt* 8424). 3, **H. pandurifolium**, part of flowering stem, × 1; 3a, head, × 5,3 (*Sheasby* 28). 4, **H. patulum**, leaf, × 1,3; 4a, head, × 5,3 (*Acocks* 21533). 5, **H. petiolare**, head, × 6,6 (*Ross* 2411); 5a, leaf from western part of range, × 1,3 (*Hilliard & Burtt* 10802); 5b, leaf from eastern part of range, × 1,3 (*Hilliard & Burtt* 10974).

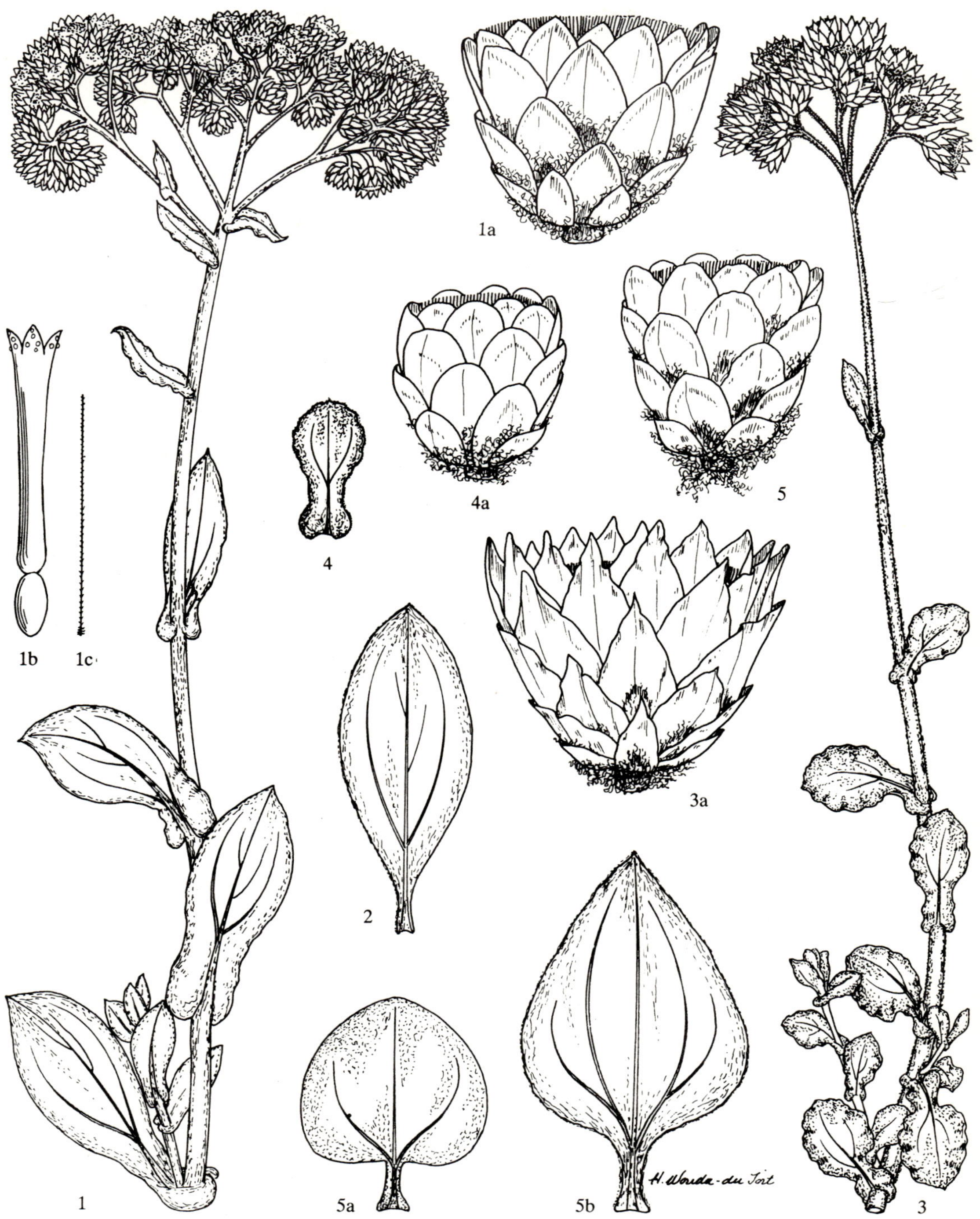
H. Wonda-du Toit

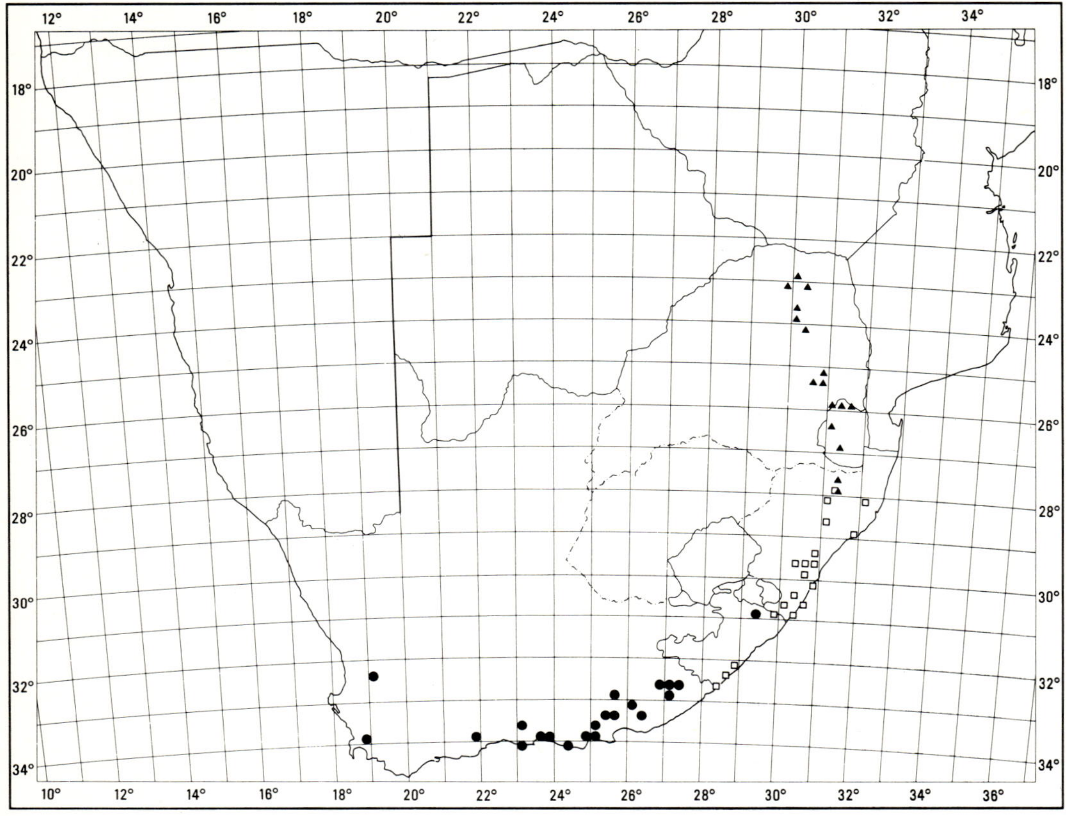

MAP 137.— ● **Helichrysum petiolare**
    □ **Helichrysum panduratum** var. **panduratum**
    ▲ **Helichrysum panduratum** var. **transvaalense**

River and Komadagga are more or less intermediate in shape and cobwebby above. Map 137.

H. petiolare is distinguished from its close allies, *H. patulum* (no. 131) and *H. pandurifolium* (no. 130), by its flat, not crisped-undulate, leaf margins, and from the latter also by its more obtuse, concave involucral bracts.

Vouchers: *Compton* 23115 (NBG); *Fourcade* 561 (BOL; PRE); *Galpin* 2909 (PRE); *Hilliard & Burtt* 6503 (E; K; MO; NBG; NU; PRE; S).

133. **Helichrysum panduratum** *O. Hoffm.* in Bull. Herb. Boissier 2 sér., 1: 827 (1901); Moeser in Bot. Jb. 44: 312 (1910); Brenan in Mem. N.Y. bot. Gdn 8,5: 468 (1954); Hilliard, Compositae in Natal 191 (1977). Lectotype: Congo, Kisantu, *Gillet* 1293 (BR!).

H. auriculatum Less. var. *panduratum* Harv. in F.C. 3: 253 (1865). Lectotype: Natal, near Durban and between Maritzburg and Ladysmith, *Gerrard* 256 (K!).

H. auriculatum sensu Oliv. & Hiern in F.T.A. 3: 347 (1877), non Less.

A diffuse, loosely branched, soft-wooded sub-shrub, stems loosely and thinly grey-woolly, leafy, becoming pendunculoid upwards with small distant leaves passing into inflorescence bracts. *Leaves* up to c. 70 × 35 mm, panduriform, sessile, base broadly auricled. half-clasping, (ovate-rhomboid, petiolate in var. *transvaalense*), margins somewhat crisped, thinly grey-woolly above, densely so below. *Heads* homogamous, subglobose, c. 5—6 mm long, double that across the radiating bracts, many in loose terminal corymbose panicles. *Involucral bracts* in c. 5 series, graded, loosely imbricate, inner about equalling flowers, backs woolly above the stereome, tips subacute or obtuse, often erose, opaque white. *Receptacle* with fimbrils about equalling the ovary. *Flowers* 27−48 (−68 in var. *transvaalense*), yellow, honey-scented. *Achenes* c. 1 mm long, barrel-shaped, obscurely ribbed, glabrous. *Pappus* bristles many, scabrid, about equalling corolla, bases cohering strongly by patent cilia, lightly fused as well.

Two varieties are recognized:

(a) var. **panduratum.**

Ranges from the Great Kei, the southern border of the Transkei, along the coast to Natal, where it has been recorded from sea level to c. 1 200 m. Also in Uganda, Congo, Kenya, Tanzania, Malawi, Zambia, Zimbabwe, Mozambique, Angola and Gabon, but apparently absent from the Transvaal, where it is entirely replaced by var. *transvaalense*. Grows in large tangled clumps in high-rainfall areas, particularly in mixed scrub-grassland near forest margins; flowering mainly in December and January. Fig. 40: 1; Map 137.

Vouchers: *Acocks* 13961 (PRE); *Flanagan* 1750 (BOL; PRE; SAM); *Hilliard* 4071 (E; K; NH; NU); *Pegler* 677 (BOL); *Tyson* 2822 (PRE; SAM).

(b) var. **transvaalense** *Moeser* in Bot. Jb. 44: 312 (1910); Hilliard, Compositae in Natal 637 (1977). Type: E. Transvaal, Shilouvane, Aug. 1899, *Junod* 566 (Z, holo.!; G; K; PRE, iso.!).

Similar in habit and general facies to var. *panduratum* but differing in its ovate-rhomboid to elliptic leaves tapering into a distinct petiole minutely eared at the base. Fig. 40: 2.

H. panduratum var. *transvaalense* ranges from the eastern highlands of Zimbabwe to the Soutpansberg, through the eastern highlands of the Transvaal and western Swaziland to northern Natal, where it has been recorded from Ngome and near Louwsburg; Ngome is the only place in South Africa where the two varieties are known to be sympatric. Wild (pers. comm.) also records subsp. *transvaalense* from Malawi and Gúruè in Mozambique, but I have seen no specimens. Map 137.

Grows on forest margins, but also favours rocky, bushy places; flowering between August and November, mainly August to October.

Vouchers: *Codd* 1682 (NU; PRE); *Compton* 27994 (NBG; PRE); *Moss* 15463 (BM; PRE); *Scheepers* 1172 (G; PRE).

Group 19

Small shrubs; leaves medium-sized, broadly elliptic, ovate, rhomboid or suborbicular, petiolate; *heads* homogamous, 4−5 × 4−6 mm, in terminal corymbose clusters; *involucral bracts* minutely radiating, white or straw-coloured; *receptacle* honeycombed or fimbrilliferous; *flowers* 10−32, corolla narrowly campanulate above; *achenes* glabrous or hairy; *pappus* bristles scabrid, bases usually cohering by patent cilia, sometimes free.

Species 134−139, one (*H. lepidissimum;* no. 134) widely distributed from Malawi and E. Zimbabwe to the Transkei, three endemic to the E. Transvaal, two endemic to Natal. Often on cliffs and rock outcrops.

1a Heads in small clusters often further arranged in compact corymbose panicles; flowering stems leafy:

  2a Ovaries glabrous:

    3a Upper leaf surface eventually free of wool to reveal long coarse shaggy hairs; leaf margins crisped ....134. *H. lepidissimum*

    3b Upper leaf surface persistently thinly grey-tomentose; leaf margins flat:

      4a Involucral bracts crisped, straw-coloured .... 135. *H. mimetes*

      4b Involucral bracts smooth, white .... 136. *H. rudolfii*

  2b Ovaries hairy:

    5a Involucral bracts glossy; leaves ovate or elliptic-ovate abruptly contracted to a broad flat petiolar part ....137. *H. homilochrysum*

    5b Involucral bracts dull; leaves elliptic to obovate gradually narrowed to the petiolar part ....138. *H. woodii*

1b Heads webbed together at the base and forming a tight flattish terminal cluster; flowering stem becoming pedunculoid and either nude or remotely bracteate below the compound head.....................
....139. *H. drakensbergense*

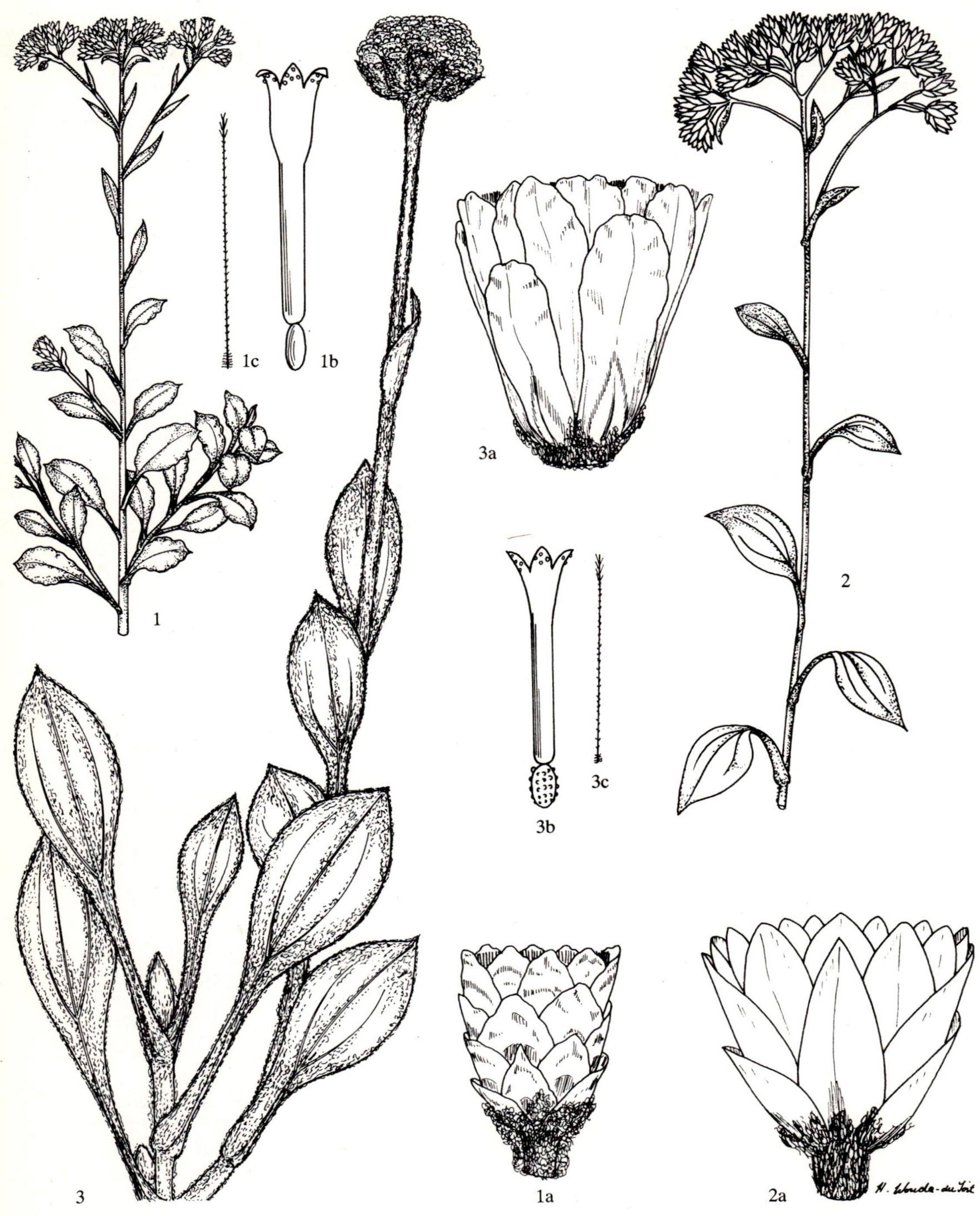

134. **Helichrysum lepidissimum** *S. Moore* in J. Bot., Lond. 31: 399 (1903); Moeser in Bot. Jb. 44: 311 (1910); Compton, Fl. Swaziland 630 (1976); Hilliard, Compositae in Natal 190 (1977). Type: Transvaal, Johannesburg, among sandstone rocks to northward, *Rand* 1294 (BM, holo.!).

H. lepidissimum var. *flavidum* Moeser in Bot. Jb. 44: 311 (1910). Lectotype: Transvaal, between Pietersburg and Haenertsburg, *Junod* 515 (G!).

H. gazense S. Moore in J. Linn. Soc., Bot. 40: 110 (1911). Type: Rhodesia, Melsetter, *Swynnerton* 1853 (BM, holo.!; K, iso.!).

Dense twiggy shrublet 100−600 mm high, lower branches often decumbent, rooting, all closely leafy, the younger parts loosely greyish-white woolly. *Leaves* mostly 7−30 × 3−15 mm, diminishing slightly upwards, usually broadly elliptic, rarely suborbicular, apex subacute to obtuse or rounded, mucronate, abruptly contracted to a flat petiolar part roughly ¼ to ½ the total leaf length, base auriculate in larger leaves, margins more or less crisped, lightly and loosely greyish-white woolly above, wool glabrescent to reveal long coarse shaggy hairs, greyish-white woolly-felted below. *Heads* homogamous, campanulate, 4−6 × 3−5 mm, double that across the fully radiating bracts, few to many in corymbose clusters terminating the branchlets. *Involucral bracts* in c. 5 series, graded, imbricate, inner about equalling flowers, outer acute, inner more or less obtuse, crisped, glossy, white, creamy or pale straw-coloured. *Receptacle* with fimbrils at least equalling ovaries. *Flowers* 10−30, yellow. *Achenes* 1 mm, barrel-shaped, obscurely ribbed, glabrous. *Pappus* bristles many, scabrid, bases cohering by patent cilia. Fig. 41: 1.

H. lepidissimum ranges widely, from Dedza district and the Shire Highlands in Malawi to Inyanga in the E. highlands of Zimbabwe and Mt Gorongoza in Mozambique south to the Soutpansberg and Blouberg, the mountains of the E. Transvaal including the Lulu Mountains and the Steenkampsberge, the Maga-liesberg, Witwatersrand, Transvaal Highveld, western Swaziland and the mountainous parts of N. Natal and Zululand; south of the Tugela confined to Table Mountain Sandstone formations and recorded as far south as the S.bank of the Msikaba River and Fraser's Falls in Pondoland, Transkei. Always in rocky places, particularly rocky mountain tops and cliff edges, flowering mainly from April to August, but flowers can be found in any month. Map 138.

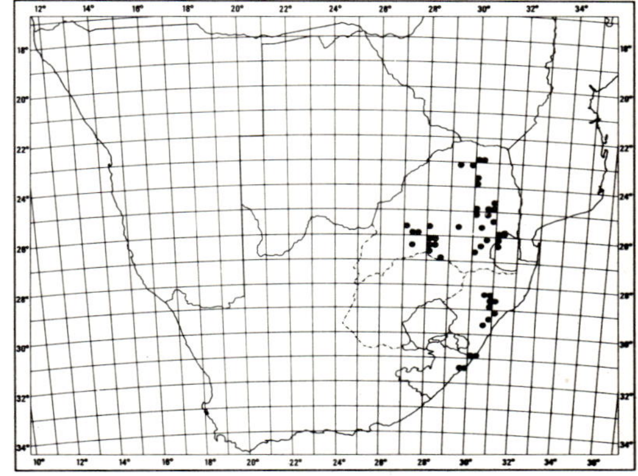

MAP 138.— **Helichrysum lepidissimum**

The chief distinguishing mark of *H. lepidissimum* is its leaves with crisped margins and upper surface eventually free of wool to reveal long shaggy hairs, green when fresh, brown when dried; in contrast, the leaves of both *H. mimetes* (below) and *H. rudolfii* (no. 136), its close allies, are persistently tomentose above and grey, with smooth margins.

Colour of the involucral bracts varies from white to cream to pale or deep straw-colour. Plants with white or straw-coloured bracts may grow together, though plants with white bracts are commoner on the Transvaal Highveld and in the Soutpansberg than plants with straw-coloured bracts, and the opposite is true in Natal. There is also some variation in size of heads, big-headed plants being commonest in the Soutpansberg and in the mountains of the NE. Transvaal.

Vouchers: *Acocks* 11592 (K; PRE); *Codd* 2872 (NU; PRE); *Galpin* 14889 (K; PRE); *Hilliard & Burtt* 10232 (E; K; MO; NU; S); *Rogers* 18854 (K; PRE).

135. **Helichrysum mimetes** *S. Moore* in J. Bot., Lond. 37: 322 (1899). Moeser in

---

FIG. 41.—1, **Helichrysum lepidissimum**, twig, × 0,7; 1a, head, × 6,6; 1b, hermaphrodite flower, × 13; 1c, pappus bristle, × 13 (*Codd* 2872). 2, **H. rudolfii**, twig, × 1; 2a, head, × 6,6 (*Hilliard & Burtt* 14363). 3, **H. drakensbergense,** part of stem, × 0,7; 3a, head, × 6,6; 3b, hermaphrodite flower, × 6,6; 3c, pappus bristle, × 6,6 (*Wright* 915).

Bot. Jb. 44: 313 (1910). Type: Transvaal, Lydenburg distr., Spitzkop, Devil's Knuckles, *Wilms* 727 (BM, holo.!).

Shrublet up to c. 600 mm high, branches thinly greyish-white tomentose, leafy. *Leaves* on coppice shoots and sappy growth ovate to subrotund, tapering to a petiole-like base, blade mostly 25—35 × 20—30 mm, apex very obtuse or rounded, mucronate, petiolar part c. 10—20 mm long, leaves on twigs often smaller, blade elliptic-ovate, ovate or subrotund, mostly 13—24 × 13—18 mm, becoming smaller and more distant upwards, petiolar part c. 6—10 mm long, base slightly expanded, half-clasping, upper surface of all leaves thinly grey-tomentose and glandular-puberulous beneath the wool, lower more densely tomentose. *Heads* homogamous, campanulate, c. 4—5 × 4—5 mm, c. 8 mm acros the radiating bracts, many in corymbose clusters arranged in corymbose panicles. *Involucral bracts* in c. 5 series, graded, imbricate, inner slightly exceeding flowers, acute to obtuse, crisped, straw-coloured. *Receptacle* with fimbrils about equalling the ovaries. *Flowers* 16—24. *Achenes* not seen, ovaries glabrous. *Pappus* bristles many, scabrid, bases cohering by patent cilia.

Endemic to the E. Transvaal and W. Swaziland, ranging from Mariepskop and the Blyde River area to the Barberton Mountains, just entering Swaziland around Havelock and Pigg's Peak. Favours rock outcrops and broken rocky cliffs, flowering mainly between May and September. Map 139.

H. mimetes is easily confused with *H. lepidissimum* (above), but can be recognized by its decidedly grey leaves, persistently woolly on both surfaces and with flat margins; in *H. lepidissimum* the leaves tend to be green above (drying brown) because the wool is soon shed from the upper surface to reveal the underlying coarse shaggy hairs, and the leaf margins are crisped. *H. mimetes* can be distinguished from *H. rudolfii* (no. 136) by its graded, crisped, straw-coloured involucral bracts (not subequal, smooth, white). There is a difference in leaf shape too, but this is difficult to express in words.

All three species are sympatric and may grow in close proximity. It seems likely that the original material of *H. mimetes* was of hybrid origin: the leaves are more coarsely hairy above than is usual and the wool tends to be evanescent, so that the leaves are not distinctly grey; also, the leaf margins are very slightly

crisped. *Codd* 1618 (NU; PRE) and *Acocks* 12865 (PRE), from the same site in the Barberton Mountains, somewhat resemble the type.

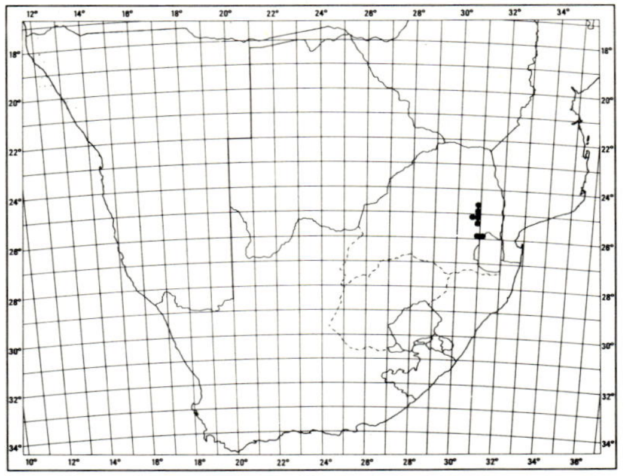

MAP 139.— **Helichrysum mimetes**

Vouchers: *Codd* 7827 (K; PRE); *Compton* 27825 (K; PRE); *Moss* 18861 (BM; J); *Nixon* s.n. (E; K; NU; PRE); *Rogers* 18714 (BM; J; PRE).

136. **Helichrysum rudolfii** *Hilliard* in Notes R. bot. Gdn Edinb. 40: 265 (1982). Type: Transvaal, 2530 BA, Sabie to Lydenburg, Long Tom Pass, 15 iii 1981, common on rocky banks and low cliffs, *Hilliard & Burtt* 14363 (NU, holo.!; E; K; M; PRE; S, iso.!).

Shrubby, up to 1 m tall, branches greyish-white felted, closely leafy, more distantly so on flowering twigs. *Leaves* 10—40 × 3—30 mm, diminishing slightly upwards, rhomboid-elliptic or ovate, gradually, or sometimes abruptly, tapering to a slender petiole roughly ¼ to nearly ½ the total leaf length, base expanded, half-clasping, apex acute to very acute in narrower leaves, mucronate, margins flat, upper surface thinly and persistently woolly, glandular-puberulous beneath the wool, lower surface white-tomentose. *Heads* homogamous, campanulate, 5 × 4—5 mm, double that across the radiating bracts, many in cymose-corymbose clusters terminating the branch-

---

FIG. 42.—1, **Helichrysum homilochrysum,** part of stem, × 0,7; 1a, head, × 6,6; 1b, hermaphrodite flower, × 6,6; 1c, pappus bristle, × 6,6 (*Rogers* 20306). 2, **H. woodii,** part of stem, × 1; 2a, head, × 6,6 (*Hilliard* 4832).

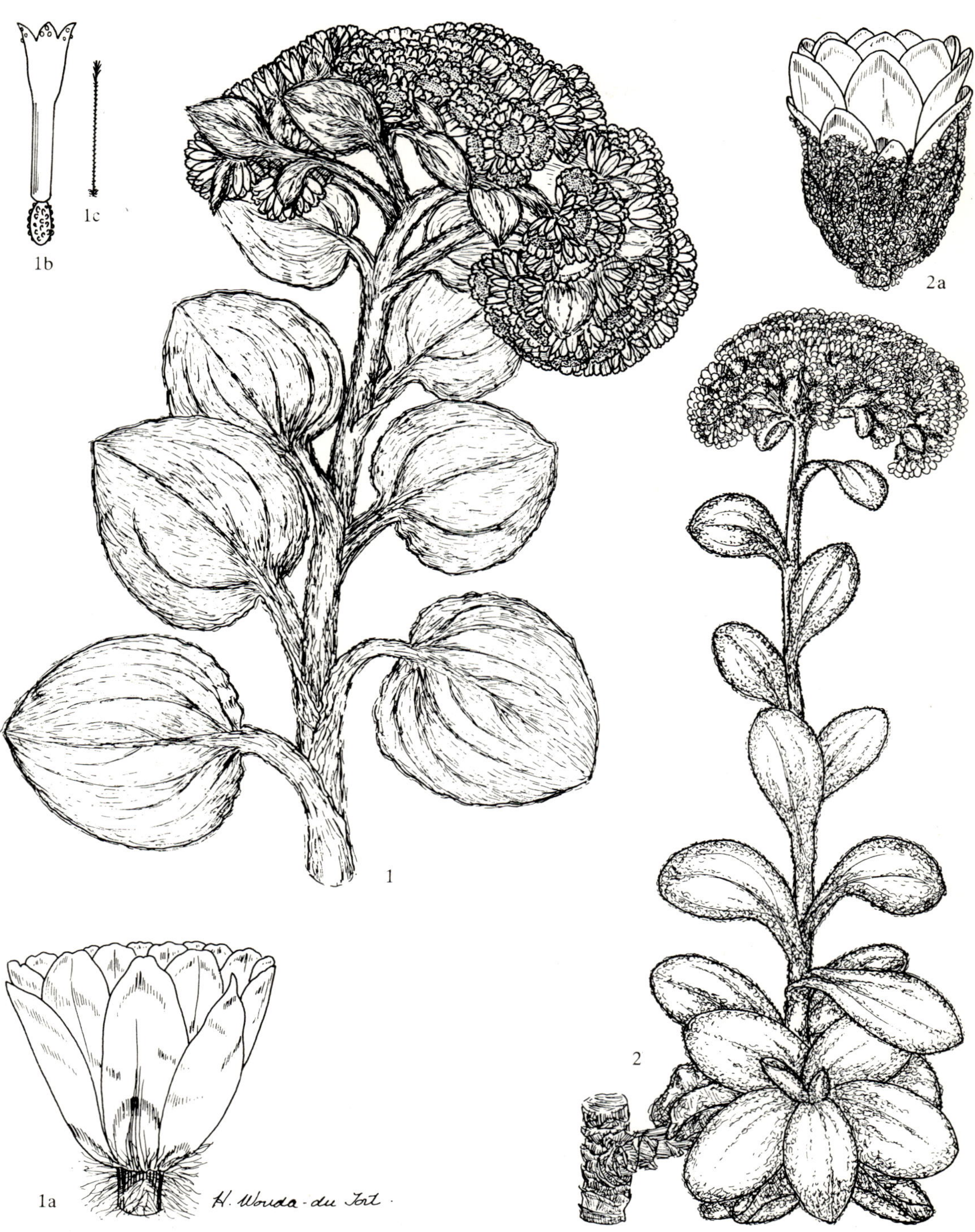
1c
1b
2a
1
1a
H. Wouda-du Tort
2

lets. *Involucral bracts* in c. 5 series, outermost short, inner subequal, very loosely imbricate, slightly exceeding flowers, subacute to obtuse, smooth or slightly crisped, glossy, tips opaque white. *Receptacle* with fimbrils exceeding the ovaries. *Flowers* 21—32, yellow. *Achenes* 1 mm long, ± cylindric, obscurely and irregularly ribbed, glabrous. *Pappus* bristles many, equalling corolla, scabrid, bases cohering strongly by patent cilia. Fig. 41: 2.

Endemic to the E. Transvaal mountains, and recorded from about Woodbush south to Elandshoogte near Machadodorp. Grows among rocks on the mountain tops; flowering mainly between April and June, but flowers can possibly be found in any month. Map 140.

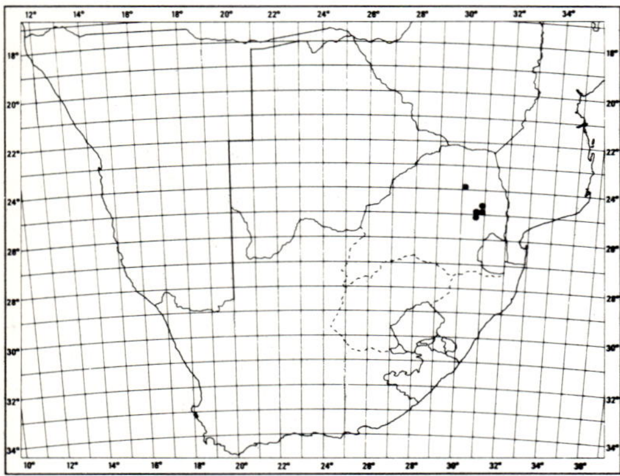

MAP 140.— **Helichrysum rudolfii**

Easily confused with *H. mimetes* (above), but distinguished by its smooth white involucral bracts (see also comments under *H. mimetes*).

Vouchers: *Kerfoot* 7894 (J; NU; PRE); *Meeuse* 9987 (M; PRE); *Stirton* 1820 (K; PRE).

137. **Helichrysum homilochrysum** *S. Moore* in J. Bot., Lond. 37: 371 (1899); Moeser in Bot. Jb. 44: 313 (1910). Type: Transvaal, Lydenburg distr., Devil's Knuckles, Aug. 1884, *Wilms* 716 (BM, holo.!; K, iso.!).

Shrubby, stems possibly up to 450 mm long, stout (8 mm diam. near base), woody, erect or decumbent, loosely branched, upper parts thickly white-woolly, closely

leafy. *Leaves* 25—50 × 10—35 mm, sometimes smaller and distant below the heads, ovate or elliptic-ovate, abruptly contracted to a broad flat petiolar part roughly ⅓ the total leaf length, base auriculate, clasping, apex subacute to obtuse, mucronate, upper surface thinly woolly, lower thickly so. *Heads* homogamous, campanulate, c. 5 × 4 mm, double that across the radiating bracts, many in compact clusters corymbosely arranged at the branch tips. *Involucral bracts* in c. 5 series, subequal, slightly exceeding flowers, very loosely imbricate, outer acute, inner obtuse, slightly crisped, semi-opaque, glossy, pale straw-yellow. *Receptacle* with fimbrils about equalling ovaries. *Flowers* 14—26, yellow. *Achenes* 1 mm long, cylindric, with myxogenic duplex hairs. *Pappus* bristles shorter than corolla, scabrid, bases cohering lightly by patent cilia. Fig. 42: 1.

Endemic to the E. Transvaal highlands, and recorded from Mariepskop, Magalieskop, MacMac near Sabie, and Mount Anderson, at c. 1 500 m. Grows on cliff faces and edges; flowering between July and September. Map 141.

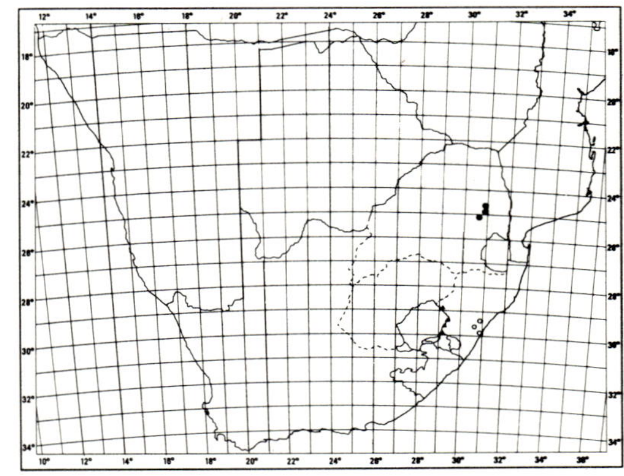

MAP 141.— ● **Helichrysum homilochrysum**
        ○ **Helichrysum woodii**
        ▲ **Helichrysum drakensbergense**

Vouchers: *MacLea* 3013 (BOL; K; SAM); *Rogers* in TM 20306 (PRE), prob. = *Rogers* 22998 (Z).

138. **Helichrysum woodii** *N.E. Br.* in Kew Bull. 1906: 21 (1906); Hilliard, Compositae in Natal 215 (1977). Type:

Natal, Pinetown district, on rocks near Emberton, *Wood* 5761 (K, holo.!; M, iso.!).

A small well-branched shrub up to c. 300 mm tall, branches greyish-white woolly, densely leafy. *Leaves* 15—50 × 5—15 mm, diminishing upwards, elliptic to obovate, obtuse, gradually narrowed to a broad petiole-like base, thinly woolly above, thickly white-woolly below, triplinerved, nerves faintly raised below. *Heads* homogamous, campanulate, 4—5 mm long, 5—6 mm across the radiating bracts, in compact rounded corymbose clusters 30—50 mm across terminating the branches. *Involucral bracts* in c. 5 series, graded, loosely imbricate, outer woolly, inner about equalling the flowers, tips obtuse, opaque, dull, pale straw-yellow. *Receptacle* with pit margins slightly produced. *Flowers* 11—17, yellow. *Achenes* 1 mm, cylindric, with myxogenic duplex hairs. *Pappus* bristles many, scabrid, about equalling the corolla, bases not cohering. Fig. 42: 2.

Endemic to Natal; recorded only from a very small area in the contiguous Pinetown, Camperdown and New Hanover districts, on the edges and faces (southerly aspect) of Table Mountain Sandstone cliffs 600—900 m above sea level. Flowers in mid-winter (July), and rarely collected. Map 141.

Allied to the Transvaal mountain endemic, *H. homilochrysum* (above) but easily distinguished by its differently shaped leaves and dull, opaque involucral bracts, the outer much shorter than the inner.

Voucher: *Hilliard* 4832 (E; K; NU; PRE; S).

139. **Helichrysum drakensbergense** *Killick* in Bothalia 7: 23 (1958); Hilliard, Compositae in Natal 162 (1977). Type: Natal, Bergville distr., Cathedral Peak Forest Research Station, below Organ Pipes Pass, *Killick* 1879 (PRE, holo.!; K, iso.!).

Bushy perennial herb or subshrub up to c. 400 mm tall, stems often decumbent at base and rooting, simple below, forking several times above, grey-woolly, branches erect, leafy in lower half becoming pedunculoid upwards. *Leaves* up to 90 × 30 mm, more or less spathulate, apex rounded or subacute, petiole-like base expanded below and half-clasping, both surfaces closely grey-woolly. *Heads* homogamous, campanulate, c. 6 × 4 mm, many felted together at the base in a very congested, flattish corymbose cluster 20—25 mm across. *Involucral bracts* in c. 3 series, loosely imbricate, about equalling flowers, not radiating, glossy, straw-coloured. *Receptacle* with pit margins thickened. *Flowers* c. 15—28, yellow. *Achenes* 1,5 mm long, with duplex hairs. *Pappus* bristles copious, delicate, scabrid above, bases lightly cohering by patent cilia. Fig. 41: 3.

Known only from a small area of the Natal Drakensberg, from Garden Castle Forest Reserve and Sani Pass, in Underberg district, north to Cleft Peak in Bergville district. It forms large but often highly localized colonies on grass slopes or in scrub, between c. 1 525 and 2 740 m above sea level. Flowers from November to January. Map 141.

Vouchers: *Hilliard* 4826 (E; K; NH; NU; PRE; S); *Hilliard & Burtt* 7045 (E; K; NU; S); *Schelpe* 497 (NU); *Wright* 432 (E; K; NH; NU).

Group 20

Perennial herbs with woody tubers; *leaves* medium-sized or large, elliptic to broadly elliptic, clasping, sometimes decurrent; *heads* homogamous, c. 5 × 5 mm, either webbed together and congested, or in a spreading corymbose panicle; *involucral bracts* not radiating, white or yellow; *receptacle* fimbrilliferous and paleate; *flowers* 33—55, corolla narrowly campanulate above; *achenes* hairy; *pappus* bristles with smooth shaft and barbellate tips, bases free or cohering by patent cilia or lightly fused in bundles.

Species 140—141, one species ranging from the E. Transvaal to the S. Cape mountains, the other narrowly endemic in the Natal Drakensberg, both in grassland.

1a Heads in a corymbose panicle, involucral bracts creamy white, leaf bases decurrent in long stem wings
.... 140. *H. platypterum*

1b Heads tightly congested in a terminal cluster webbed together with wool, involucral bracts yellow and brown; leaf bases not decurrent.. 141. *H. paleatum*

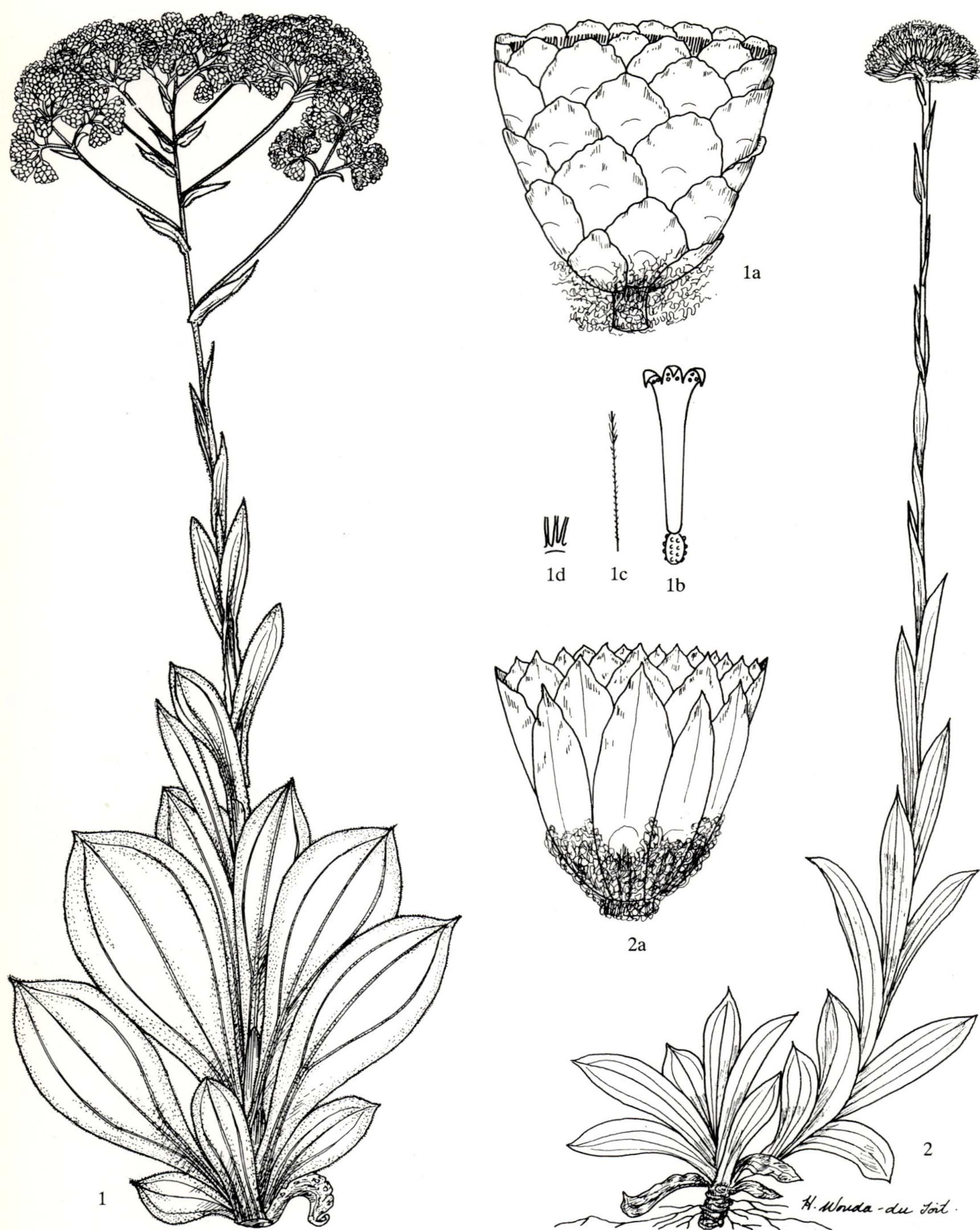
1a
1d
1c
1b
2a
1
2
H. Wouda-du Toit.

140. **Helichrysum platypterum** *DC.*, Prodr. 6: 201 (1838); Harv. in F.C. 3: 243 (1865); Hilliard, Compositae in Natal 192 (1977). Type: Cape, between the Umzimvubu and Umzimkulu Rivers, *Drège* 5001 (G-DC, holo.!).

Gnaphalium platypterum (DC.) Sch. Bip. in Bot. Ztg 3: 172 (1845).

G. amplum O. Kuntze, Rev. Gen. Pl. 3: 150 (1898). Type: Natal [Pinetown distr.] Krantz Kloof, *Kuntze* (NY, holo.; K, photo.!).

Cassinia alba O. Hoffm. in Bot. Jb. 24: 470 (1898). Type: Pondoland, June 1858, *Bachmann* 1431.

Perennial herb with a woody tuber up to 60 mm diam., stems solitary or several together, erect to c. 1 m, simple, thinly white cottony, leafy particularly below, bracteate above. *Leaves* up to 250 × 80 mm, lower broadly elliptic, contracted to the broadly winged base, upper narrower, elliptic then lanceolate upwards, passing into bracts, apex acute, base broad, clasping, decurrent in long conspicuous stem wings, upper surface thinly and harshly pubescent, lower more markedly so particularly on the nerves, triplinerved, or 5-nerved in the broadest leaves. *Heads* homogamous, campanulate, c. 5 × 5 mm, many in a large spreading corymbose panicle up to 200 mm across. *Involucral bracts* in c. 5—6 series, graded, closely imbricate, base woolly, inner about equalling the flowers, tips obtuse, opaque silvery white sometimes tinged palest brown, not radiating. *Receptacle* fimbrilliferous and paleate, paleae resembling the inner involucral bracts, conduplicate, enfolding each flower, caducous. *Flowers* 33—55. *Achenes* not seen, ovaries with myxogenic duplex hairs. *Pappus* bristles several (c. 18), tips barbellate, shaft nearly smooth, bases cohering lightly by patent cilia, some light fusion as well, or bases nude and free. Fig. 43: 1.

Ranges from the highlands of the eastern Transvaal and western Swaziland to the mountainous NE. Orange Free State and neighbouring Lesotho (Leribe plateau and Molimo Nthuse), Natal (where it is widespread), Transkei and Cape as far west as the Tsitsikama Mountains between Port Elizabeth and George. Found in the rank growth on forest margins or on damp grassy mountain slopes. Flowers mainly between February and April. Map 142.

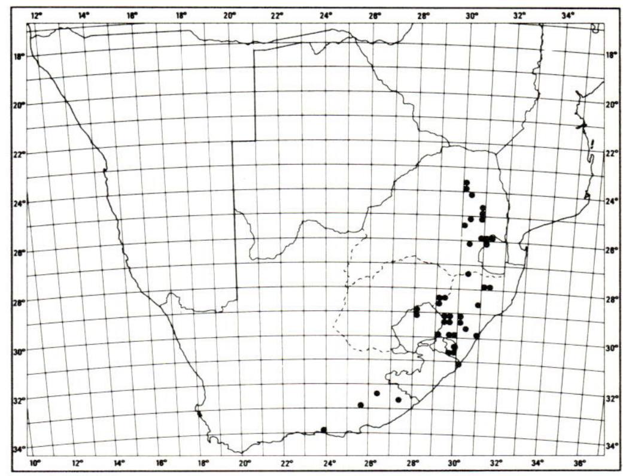

MAP 142.— **Helichrysum platypterum**

Vouchers: *Hilliard* 4679 (E; K; NH; NU); *Meeuse* 9860 (K; PRE); *Schlechter* 6972 (BOL); *Tyson* 1218 (BOL; SAM); *Wright* 453 (E; K; NH; NU).

141. **Helichrysum paleatum** *Hilliard* in Notes R. bot. Gdn Edinb. 32: 354 (1973), Compositae in Natal 193 (1977). Type: Natal, Estcourt distr., Giant's Castle Game Reserve, Bannerman Pass, c. 8 000 ft, *Hilliard* 4828 (NU, holo.!; E, iso.!).

Perennial herb with a small oval woody tuber c. 10 mm in diam. crowned by a leaf rosette, flowering stems 1—4, lateral, slightly decumbent then erect to about 300 mm, simple, loosely white-woolly, leafy. *Radical leaves* up to 70 × 1—5 mm, elliptic, with 5 parallel veins, firm-textured, apex acute, mucronate, base broad, clasping, margins thickened, glabrous or with sparse rough hairs, veins and margins glabrous or woolly; *cauline leaves* up to 75 × 10 mm, rapidly smaller upwards and passing into bracts, more or less distichous, clasping, linear-lanceolate, lowermost conduplicate, slightly falcate, upper erect, straight, otherwise as the radical leaves. *Heads* homogamous,

---

FIG. 43.—1, **Helichrysum platypterum,** whole plant, × 0,3 (*Hilliard* 4817); 1a, head, × 8; 1b, hermaphrodite flower, × 8; 1c, pappus bristle, × 8; 1d, base of pappus bristles much enlarged to show fusion (*Hilliard* 5466). 2, **H. paleatum,** whole plant, × 0,7 (*Hoener* 2188); 2a, head, × 6,6 (*Hilliard & Burtt* 12545).

cylindric, c. 5 × 5 mm, many in a terminal compound head 20−30 mm across, tightly congested, the bases webbed together with white wool. *Involucral bracts* in c. 4 series, subequal or the outer a little shorter, loosely imbricate, not radiating, equalling the flowers, translucent, golden-yellow overlaid light brown. *Receptacle* paleate. *Flowers* c. 45−55, yellow. *Achenes* 1 mm, elliptic, with myxogenic duplex hairs. *Pappus* bristles c. 4−8, delicate, tips barbellate, bases nude, not cohering. Fig. 43: 2.

Known only from a small area of the Natal Drakensberg, from Bushman's Nek, in Underberg district, to the Giant's Castle area, in Estcourt district, between c. 1 900 and 2 500 m. Grows socially in short grassland on stony mountain slopes; flowering from January to April. Easily recognized by its stiff, somewhat falcate leaves and flowers subtended by paleae. Map 143.

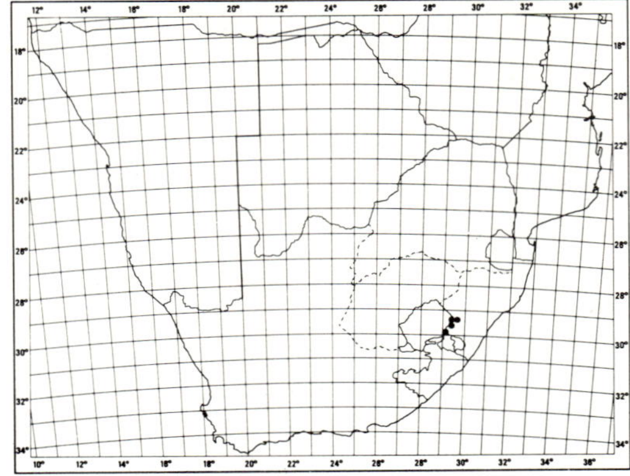

MAP 143.— **Helichrysum paleatum**

Vouchers: *Hilliard* 4810 (E; K; M; MO; NU; PRE; S); *Hilliard & Burtt* 9754 (E; K; MO; NU; S); *Wright* 459 (E; K; NU; S).

Group 21

Stoloniferous perennial herbs; *leaves* medium-sized or large, radical leaves tufted, elliptic, ovate-elliptic, oblanceolate to obovate, base often petiole-like; *heads* homogamous, 5−9 × 3−5 mm in small clusters, these often arranged in corymbose panicles; *involucral bracts* not radiating, yellow; *receptacle* shortly honeycombed; *flowers* 7−16, corolla narrowly campanulate above; *achenes* hairy; *pappus* bristles scabrid, bases cohering by patent cilia, some fusion as well.

Species 142−144, in the grasslands of E. Southern Africa, with close allies in tropical Africa.

1a Heads c. 5−7 mm long:

    2a Leaves glandular-pubescent or lightly cobwebby to thickly tomentose, involucral bracts closely imbricate .... 142. *H. acutatum*

    2b Leaves silky-tomentose, involucral bracts loosely imbricate .... 144. *H. oreophilum*

1b Heads c. 9 mm long .... 143. *H. dasymallum*

142. **Helichrysum acutatum** *DC.*, Prodr. 6: 186 (1838); Harv. in F.C. 3: 234 (1865); Moeser in Bot. Jb. 44: 275 (1910); Compton, Fl. Swaziland 627 (1976); Hilliard, Compositae in Natal 194 (1977). Type: Cape, between the Umzimkulu and Umtentu Rivers, *Drège* 4999 (G-DC, holo.!).

Gnaphalium acutatum (DC.) Sch. Bip. in Bot. Ztg 3: 171 (1845).

Helichrysum floccosum Klatt in Bull. Herb. Boissier 4: 836 (1896). Type: Natal, Durban distr., Northdene, *Wood* 5231 (Z, holo.!; BM; K; PRE, iso.!).

H. danaë S. Moore in J. Bot., Lond. 37: 371 (1899). Type: Natal, Zululand, *Gerrard* s.n. (BM, holo.!; K, iso.!).

H. schlechteri H. Bol. in Trans. S. Afr. phil. Soc. 18: 387 (1907); Compton, Fl. Swaziland 634 (1976). Lectotype: Transvaal, Elandsspruitberg, 2 070 m, *Schlechter* 3952 (BOL!; BM; G; K; PRE; S; Z, isolecto.!).

H. acutatum var. *rhombifolium* Moeser in Bot. Jb. 44: 275 (1910). Type: Natal, Inanda, *Rehmann* 8296 (Z, holo.!).

FIG. 44.−1, **Helichrysum acutatum,** whole plant, × 0,3; 1a, head, × 10; 1b, hermaphrodite flower, × 6,6; 1c, pappus bristle, × 6,6; 1d, base of pappus bristles much enlarged to show fusion as well as cohesion by patent cilia (*Hilliard* 2379). 2, **H. oreophilum,** whole plant, × 0,7; 2a, head, × 10 (*Hilliard* 2295).

1a
2a
1d
1c
1b
1
2
H. Wouda-du Toit

H. galpinii Schltr. & Moeser in Bot. Jb. 44: 274 (1910), non N.E. Br. (1896). *H. transvalense* Staner in Annls Soc. scient. Brux., sér. B, 56: 249 (1936). Type: Transvaal, Barberton, Saddleback Range, *Galpin* 607 (BOL; K; NH; PRE; Z, iso.!).

H. eriophorum Conrath in Kew Bull. 1914: 133 (1914). Type: Transvaal, Irene, *Conrath* 432 (K, holo.!).

Perennial herb, rootstock stout, woody, crown sometimes woolly, flowering stems solitary or several together, erect to 450 (−600) mm, simple, cobwebby to woolly-tomentose, leafy in the lower part, pedunculoid and bracteate upwards. *Radical leaves* up to 250 (−350) × 50 (−80) mm, c. a third to one half of the length petiolar, blade elliptic or ovate-elliptic tapering smoothly or more abruptly into the long, winged petiole, base expanded and clasping, apex acute, both surfaces glandular-punctate, often thinly or thickly glandular-pubescent, frequently lightly cobwebby to densely greyish-woolly as well, conspicuously triplinerved; *cauline leaves* similar but soon sessile, oblong-lanceolate, very acute to long acuminate passing into distant linear-lanceolate acuminate bracts. *Heads* homogamous, oblong-campanulate, c. 5−7 × 3 mm, many in dense corymbose clusters corymbosely arranged. *Involucral bracts* in c. 4 series, closely imbricate, outermost webbed together with wool, inner slightly shorter than or equalling the flowers, tips acute or very acute, pellucid, bright yellow, not radiating. *Receptacle* shortly honeycombed. *Flowers* 7−13. *Achenes* 1,25 mm long, with myxogenic duplex hairs. *Pappus* bristles many, slightly shorter than corolla, scabridulous, bases cohering strongly by patent cilia, some light fusion as well. Fig. 44: 1.

Ranges from Baziya (W. of Umtata) through Transkei, Natal, Swaziland, the Transvaal Highveld and eastern highlands to the Soutpansberg, with closely related plants in tropical Africa (Wild, pers. comm., records *H. acutatum* from Zimbabwe, Malawi and Tanzania). Grows in grassland, in Natal, where it is widespread, from near sea level to c. 1 950 m; flowering between September and January. Map 144.

Plants are commonly thickly tomentose, but all degrees of development of woolly hairs occur, and sometimes wool is either completely wanting or confined to the leaf margins.

Vouchers: *Bremer* 603 (NU); *Codd* 3306 (NU; PRE); *Hilliard & Burtt* 5979 (E; K; NU; PRE); *Moll* 1889 (NU); *Wood* 11855 (NU).

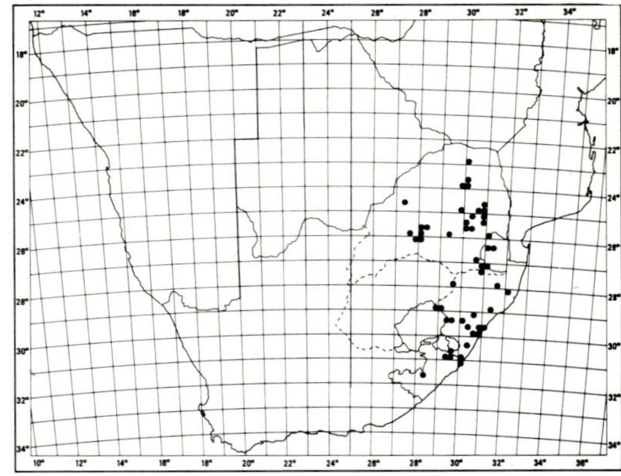

MAP 144.— **Helichrysum acutatum**

143. **Helichrysum dasymallum** *Hilliard* in Notes R. bot. Gdn Edinb. 32: 347 (1973), Compositae in Natal 195 (1977). Lectotype: Transvaal, Magaliesberg, Sand River, *Zeyher* 878 (BM!; E; K; P; S, isolecto.!).

H. lanatum Harv. in F.C. 3: 233 (1865), non Schrank (1824). Type as above.

Perennial herb, rootstock stout, woody, crown very woolly from the thick rather loose brownish wool on the young leaves, stems one or several from crown, simple, erect to 400 mm, greyish-white woolly, leafy particularly below. *Leaves* up to 130 × 50 mm, diminishing in size upwards and passing into bracts, lower obovate to elliptic, obtuse, apiculate, contracted into a broad, clasping, petiole-like base, upper obovate to oblanceolate, subacute to acute, all thickly clothed in greyish-white wool. *Heads* homogamous, turbinate-campanulate, c. 9 mm long, many in small clusters corymbose-paniculately arranged. *Involucral bracts* in 4−5 series, subequal, loosely imbricate, about equalling flowers, bases woolly, tips very acute, lemon-yellow, not radiating. *Receptacle* honeycombed. *Flowers* 10−16. *Achenes* not seen, ovaries with myxogenic duplex hairs. *Pappus* bristles many, about equalling corolla, scabrid, bases cohering by patent cilia, some light fusion as well.

Recorded from scattered localities in the Transvaal (from Woodbush in the NE. to Potgietersrus and

Nylstroom, the Magaliesberg and other hills near Pretoria), Nqutu and Utrecht districts in northern Natal, and one record from Swaziland, unlocalized but probably Hlatikulu. Grows in grassland; flowering between November and February. Closely allied to *H. oreophilum* (below) but easily distinguished by its larger heads and woolly indumentum. Map 145.

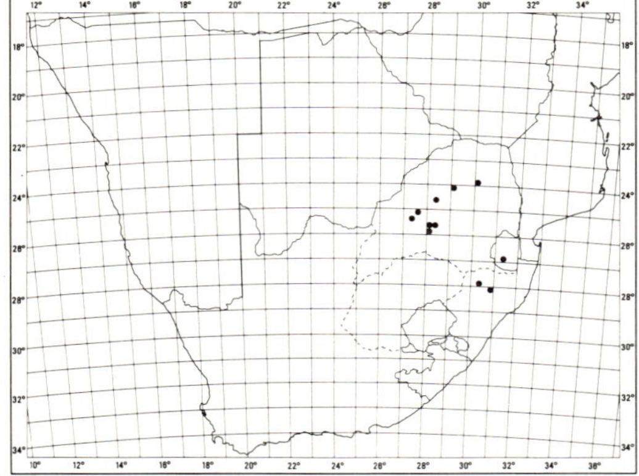

MAP 145.— **Helichrysum dasymallum**

Vouchers: *Codd* 6800 (K; PRE); *Hutchinson* 2356 (BOL; K; PRE); *Rehmann* 6097 (K; Z).

144. **Helichrysum oreophilum** *Klatt* in Bull. Herb. Boissier 4: 837 (1896); Moeser in Bot. Jb. 44: 276 (1910); Compton, Fl. Swaziland 632 (1976); Hilliard, Compositae in Natal 196 (1977). Lectotype: Transvaal, Klein Olifantsrivier, *Schlechter* 3805 (Z!; K; PRE, isolecto.!).

Perennial herb with a woody branching rhizome, flowering-stems erect to c. 300 mm, simple, thinly grey silky-woolly, leafy. *Leaves* up to 80 × 15 mm, progressively smaller upwards, oblanceolate becoming lanceolate upwards, apex acute, base half-clasping, both surfaces grey silky-woolly,

generally only the main nerve visible below. *Heads* homogamous, turbinate-campanulate, c. 7 × mm, several to many in a compact corymbose panicle. *Involucral bracts* in c. 5 series, subequal, loosely imbricate, cobwebby at base, tips acute, slightly exceeding the flowers, not radiating, lemon-yellow. *Receptacle* honeycombed. *Flowers* 9—12. *Achenes* c. 1,25 mm long, with myxogenic duplex hairs. *Pappus* bristles many, about equalling corolla, scabrid, bases cohering strongly by patent cilia, some light fusion as well. Fig. 44: 2.

Ranges from the E. highlands of the Transvaal, the Witwatersrand and Suikerbosrand to the mountains in Swaziland, the low Drakensberg on the Transvaal-Natal border, the mountainous NE. corner of the Orange Free State, and the Natal Drakensberg and its foothills and outliers. Grows in grassland and readily invades overgrazed and eroded areas. Flowers mainly between October and December. Map 146.

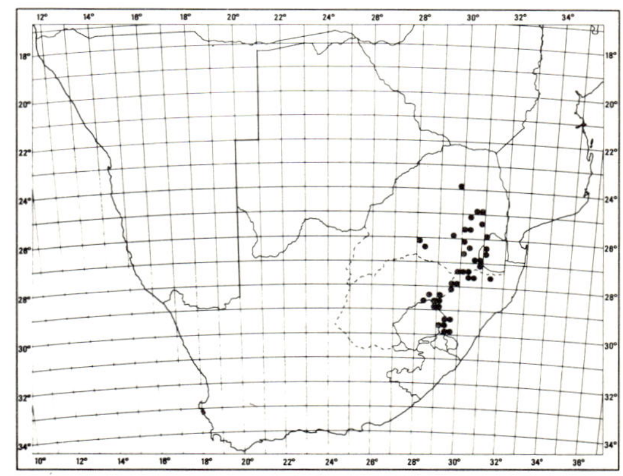

MAP 146.— **Helichrysum oreophilum**

Vouchers: *Bremer* 605 (NU); *Codd* 3131 (NU; PRE); *Hilliard* 1020 (NU); *Wright* 286 (E; NU).

Group 22

Shrubs or tufted perennial herbs or rarely (one species) annual; *leaves* small, linear or oblong to spathulate, obovate or ovate-lanceolate; *heads* usually heterogamous, rarely homogamous and then sometimes within a single species, 3—7 × 2—7 mm, in lax or congested corymbose panicles; *involucral bracts* minutely radiating, yellow or some golden-brown; *receptacle* more or less honeycombed, rarely fimbrilliferous; *flowers* 7—156, 6—60 ♀, corolla of ☿ flowers narrowly campanulate above, of ♀ flowers tubular; *achenes* glabrous or hairy, sometimes within a single species; *pappus* bristles scabrid, tips sometimes barbellate, bases cohering by patent cilia, rarely some fusion as well.

Species 145−155, *H. splendidum* on the mountains from Ethiopia to the SW. Cape, nos. 146−149 endemic to the Drakensberg Centre, nos. 152−155 endemic to the Cape, no. 150 endemic to the E. Transvaal and N. Natal, no. 151 in the Cape, Lesotho, S. Transvaal and N. Natal. Mostly on the mountains.

1a Shrubs, sometimes mat-forming:

  2a Heads hemispherical at maturity, all the bracts bright yellow, loosely imbricate:

    3a Virgate shrub, usually erect, occasionally sprawling, leaves linear, linear-oblong or linear-lanceolate, acute or subacute . . . . 145. *H. splendidum*

    3b Mat-forming dwarf shrub with short closely congested branches, leaves obovate, lingulate or subspathulate, very obtuse . . . . 146. *H. montanum*

  2b Heads campanulate at maturity, outer bracts often brownish or straw-coloured, remaining closely imbricate, inner bright yellow (sometimes with reddish tints), more loosely imbricate:

    4a Leaves linear or oblong, tips recurved:

      5a Leaves either woolly on both surfaces or glabrous above and then minutely gland-dotted . . . . 147. *H. trilineatum*

      5b Leaves glabrous above, woolly below, upper surface covered, at least when young, with minute elongated appressed trichomes, not gland-dotted . . . . 149. *H. tenuifolium*

    4b Leaves ovate-lanceolate, tips not or scarcely recurved . . . . 148. *H. witbergense*

1b Either bushy or tufted perennial herbs, or annuals:

  6a Leaf tips long-acuminate . . . . 150. *H. chrysargyrum*

  6b Leaf tips obtuse to acute:

    7a Both leaf surfaces greyish-white woolly:

      8a Heads c. 5 mm long . . . . 151. *H. psilolepis*

      8b Heads 3−4 mm long:

        9a Stereome of outermost but one series of involucral bracts c. 2 mm long (outermost bracts often reduced), ♀ flowers mostly 3−4 times as many as ♀ . . . . 152. *H. versicolor*

        9b Stereome of outermost but one series of involucral bracts c. 1 mm long, ♀ flowers at least 5 times as many as ♀ . . . . 153. *H. moeserianum*

    7b Upper leaf surface either without wool at maturity or wool confined to margins and midline:

      10a Heads c. 4 mm long, upper leaf surface coarsely hairy, lower white-tomentose . . . . 154. *H. intricatum*

      10b Heads c. 7 mm long, leaves with wool confined to margins and midline . . . . 155. *H. isolepis*

145. Helichrysum splendidum *(Thunb.) Less.*, Syn. Comp. 286 (1832); DC., Prodr. 6: 185 (1838); Harv. in F.C. 3: 234 (1865) excl. var. *montanum* (DC.) Harv.; Moeser in Bot. Jb. 44: 284 (1910) excl. *var. montanum* (DC.) Harv.; Hedberg, Afroalpine Vascular Plants, Symb. Bot. Ups. 15: 204 and 342 (1957); Compton, Fl. Swaziland 635 (1976); Hilliard, Compos-itae in Natal 200 (1977). Type: Cape of Good Hope, *Thunberg* (sheet 19257, UPS, holo.!).

Gnaphalium splendidum Thunb., Prodr. 149 (1800), Fl. Cap. 648 (1823).

G. strictum Lam., Encycl. 2: 747 (1788). *Helichrysum strictum* (Lam.) Druce in Rep. bot. Soc. Exch. Club Br. Isl. 1916: 626 (1917), non *H. strictum* Moench (1794). Type: Cape of Good Hope (P-LAM!).

FIG. 45.−1, **Helichrysum splendidum**, part of plant, × 1; 1a, head, × 6; 1b, hermaphrodite flower, × 8; 1c, female flower, × 8; 1d, pappus bristle, × 8 (*Hilliard & Burtt* 12248). 2, **H. trilineatum**, part of plant, × 1; 2a, narrow glabrous leaf, tip hooked, × 3,3 (*Hilliard & Burtt* 6618); 2b, broad woolly leaf, tip hooked, × 5,3 (*Hilliard & Burtt* 8650); 2c, head, × 6 (*Hilliard & Burtt* 6618). 3, **H. witbergense**, leaf, to show differences in shape and leaf tip from that of *H. trilineatum*, × 4 (*Hilliard & Burtt* 6668). 4. **H. montanum**, part of plant, × 1 (*Hilliard* 5350).

Helichrysum xanthinum DC., Prodr. 6: 185 (1838). *H. splendidum var. xanthinum* (DC.) Harv. in F.C. 3: 234 (1865). *Gnaphalium xanthinum* (DC.) Sch. Bip. in Bot. Ztg 3: 171 (1845). Type: Cape, between the Storm- and Witteberg, *Drège* 5745 (G-DC, holo.!).

G. abyssinicum Sch. Bip. in Bot. Ztg 3: 174 (1845). *Helichrysum abyssinicum* (Sch. Bip.) A. Rich., Tent. Fl. Abyss. 423 (1848); Oliv. & Hiern in F.T.A. 3: 351 (1877). Type: Ethiopia, Mt Kubbi, 2 450 m, 26 June 1837, *Schimper* I: 127 (P, holo.!; BM; G; K; S, iso.!).

H. hendersonae S. Moore in J. Linn. Soc., Bot. 37: 316 (1906). Type: Nyasaland [Malawi], Nyika Plateau, 2 150−2 400 m, *Henderson* (BM, holo.!).

H. acrobates Bullock in Kew Bull. 1933: 149 (1933). Type: Tanganyika [Tanzania], Hanang, summit, 3 400 m, *B. D. Burtt* 2279 (K, holo.!).

Virgate shrub up to 1,5 m tall, or rarely sprawling, branches thinly grey-woolly, closely leafy, older parts glabrescent, rough with leaf scars. *Leaves* 10−30 (−40) × 1−2 (−6) mm, linear, linear-oblong or linear-lanceolate, apex acute or subacute, mucronate, base broad, half-clasping, margins commonly revolute, upper surface thickly or thinly greyish-white woolly or glabrous or nearly so, lower surface greyish-white woolly. *Heads* heterogamous, hemispherical, 4−5 mm long, 5−8 mm across the radiating bracts, bases woolly, many in congested or open corymbose panicles terminating long leafy branches. *Involucral bracts* in c. 5 series, graded, loosely imbricate, inner about equalling or slightly exceeding flowers, tips minutely radiating, bright canary-yellow, rarely orange, glossy. *Receptacle* very shortly honeycombed. *Flowers* 47−105, 8−14 ♀, 39−84 ☿, yellow. *Achenes* 0,75 mm long, with duplex hairs. *Pappus* bristles many, equalling corolla, scabrid, bases cohering by patent cilia. Fig. 45:1.

Widely distributed on the mountains from Ethiopia and the high mountains of East Africa and Malawi to Zimbabwe, the highlands of the Transvaal and Swaziland, the Midlands and Uplands of Natal, the mountainous NE. Orange Free State, Lesotho and Transkei, and the Cape mountains from the Witteberg and Cape Drakensberg, the Stormberg and Sneeuberg, Great Winterberg and Amatola Mountains to the Swartberg near Oudtshoorn and George. Favours rocky places, on forest margins, in stream gullies, and on mountain tops; flowering between October and January. Map 147.

Vouchers: *Codd* 1965 (NU; PRE); *Compton* 26097 (NBG); *Hilliard & Burtt* 7067 (E; K; MO; NU; PRE; S); *Hoener* 1994 (E; NU); *Wright* 782 (E; K; NH; NU; S).

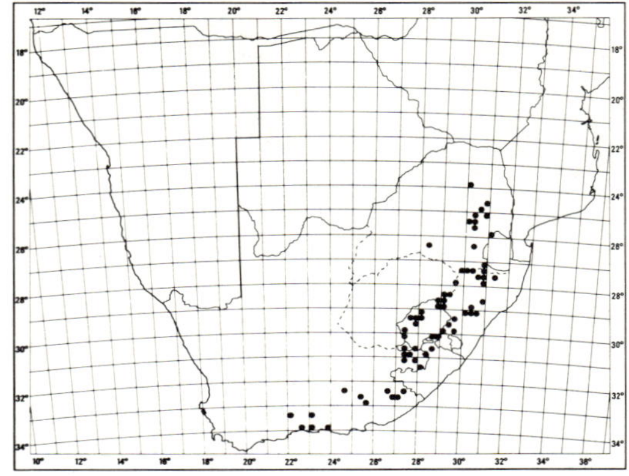

MAP 147.— **Helichrysum splendidum**

146. **Helichrysum montanum** *DC.*, Prodr. 6: 186 (1838); Hilliard, Compositae in Natal 202 (1977). Type: Cape, Wittebergen, *Drège* (G-DC, holo.!; B; BM; E; K; S, iso.!).

Gnaphalium montanum (DC.) Sch. Bip in Bot. Ztg 3: 171 (1845), non Willd. (1804). *H. splendidum* var. *montanum* (DC.) Harv. in F.C. 3: 234 (1865); Moeser in Bot. Jb. 44: 285 (1910).

Mat-forming dwarf shrub c. 100−450 mm high, often 1 m or more across, branches very short, congested, densely leafy, the leaves appearing rosetted from above, flowering stems terminal, erect, up to 200 mm long, greyish-white woolly, leafy throughout. *Leaves* up to 25 (−50) ×8 (−10) mm, smaller and more distant below the inflorescences, obovate, lingulate or subspathulate, apex very obtuse, base broad, half-clasping, both surfaces thickly greyish-white woolly, distinctly striped from the 3−5 parallel nerves. *Heads* heterogamous, hemispherical, 4−5 mm long, 5−8 mm across the radiating bracts, bases woolly, many in a compact terminal corymbose panicle. *Involucral bracts* in 6−7 series, graded, loosely imbricate, inner slightly exceeding flowers, tips minutely radiating, glossy, bright canary-yellow. *Receptacle* honeycombed. *Flowers* 53−133, 12−26 ♀, 34−112 ☿. *Achenes* 0,75 mm, with myxogenic duplex hairs. *Pappus* bristles many, tips barbellate or subplumose, bases with patent cilia, cohering or not. Fig. 45:4.

On the high mountains in Natal, Lesotho and E. central Cape, from The Sentinel area in Natal and Ficksburg in the Orange Free State south to Sani Pass in Natal, the Witteberg at Lady Grey, the Stormberg, Great Winterberg, Sneeuberg and mountains near Murraysburg (c. 32°S 24°E). Forms thick mats on cliff faces, in rocky gullies, and on rock outcrops on steep mountain slopes; flowering between January and April. Map 148.

MAP 148.— **Helichrysum montanum**

Distinguished from *H. splendidum* (above) by its compact habit and by its leaves, always broadest near the tip. It generally grows at higher altitudes than *H. splendidum* and flowers later.

Vouchers: *Esterhuysen* 17350 (BOL); *Hilliard & Burtt* 8651 (E; K; MO; NU; PRE; S); *Schelpe* 541 (NU); *Wright* 145 (E; NH; NU).

147. **Helichrysum trilineatum** *DC.*, Prodr. 6: 192 (1838); Harv. in F.C. 3: 245 (1865); Moeser in Bot. Jb. 44: 286 (1910); Hilliard, Compositae in Natal 203 (1977). Type: Cape, Wittebergen, *Drège* 3760 (G-DC, holo.!).

Gnaphalium trilineatum (DC.) Sch. Bip. in Bot. Ztg 3: 172 (1845).

Helichrysum alveolatum DC., Prodr. 6: 192 (1838). *G. alveolatum* (DC.) Sch. Bip. in Bot. Ztg 3: 171 (1845). *H. trilineatum* var. *tomentosum* Harv. in F.C. 3: 245 (1865). Type: Cape, Wittebergen, *Drège* 3759 (G-DC, holo.!; BM, iso.!).

H. trilineatum var. *glabriusculum* Harv. in F.C. 3: 245 (1865). Type as for *H. trilineatum*.

H. trilineatum var. *brevifolium* Harv. in F.C. 3: 245 (1865). Type: Natal, 100 miles inland at 6 500 ft, *Sutherland* (K, holo.!).

Twiggy rounded shrub c. 150 mm — 1 m high, old branches bare, rough with leaf scars, branchlets greyish-white woolly, closely leafy. *Leaves* 3—25 × 1—5 mm, linear or oblong, apex mucronate, acute, subacute or obtuse, but generally appearing very blunt because of the recurved tip, base broad, half-clasping, margins revolute, upper surfaces minutely gland-dotted, glabrous, cobwebby or greyish-white woolly, lower surface white-woolly, ribbed by the parallel veins. *Heads* heterogamous or rarely homogamous, campanulate, 4—6 mm long, 4—8 mm across the radiating bracts, often woolly or cobwebby at the base, many in terminal compact corymbose clusters. *Involucral bracts* in c. 5 series, graded, closely imbricate, pellucid, outer pale golden-brown, inner more or less equalling flowers, tips obtuse. bright canary-yellow, minutely radiating. *Receptacle* honeycombed. *Flowers* 25—60, (0—) 4—13 ♀, 19—53 ☿. *Achenes* c. 1 mm, barrel-shaped, with duplex hairs. *Pappus* bristles many, about equalling corolla, scabrid, bases cohering by patent cilia. Fig. 45:2.

Ranges from the Koudeveld Mountains and Sneeuberg, NW. and N. of Graaff-Reinet, across the mountains to the Stormberg, Great Winterberg, Katberg, Witteberg and Cape Drakensberg, thence along the Drakensberg and its outliers on the Cape-Lesotho-Natal border as far north as Mont aux Sources; also on the Lesotho mountains. Often forms large colonies on steep mountain slopes or on the summit plateau of the Drakensberg with grasses or with other shrubby species (fynbos) or among rock outcrops or in rocky gullies. Flowers mainly from August to February, but can be found in flower in almost any month. Map 149.

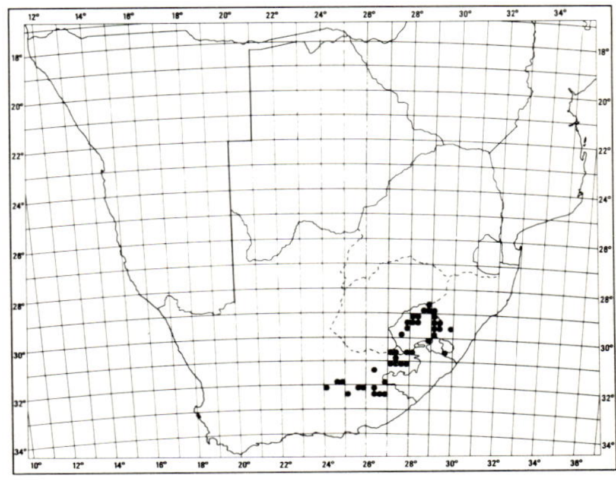

MAP 149.— **Helichrysum trilineatum**

Vouchers: *Esterhuysen* 19715 (BOL; NBG; NU); *Hilliard* 5442 (E; K; NU; PRE; S); *Hilliard & Burtt* 6609 (E; K; MO; NU; PRE; S); *Tyson* 494 (SAM); *Wright* 451 (E; K; NH; NU; S).

Very variable in leaf size and indumentum and has the curious habit, shared with related and other species, of sometimes producing broad grey-woolly leaves at the base of the branches, narrow glabrous bright yellow-green ones at the tips. But more usually plants are either grey-woolly or green, growing in mixed communities or in pure stands of grey-leaved or green-leaved plants. Green-leaved plants are sometimes confused with *H. tenuifolium* (no. 149) and *H. witbergense* (no. 148), but *H. tenuifolium* has minute appressed trichomes, not gland dots, on the upper leaf surface, while *H. witbergense* has differently shaped leaves.

Not eaten by stock, but used as fuel, the branches burning readily when green.

148. **Helichrysum witbergense** H. Bol. in Trans. S. Afr. phil. Soc. 18: 387 (1907); Hilliard, Compositae in Natal 205 (1977). Type: Cape, Barkly East distr., Witbergen, Ben Macdhui, c. 2 925 m, *Galpin* 6692 (BOL, holo.!; K; NH; SAM, iso.!).

H. splendidum var. *basuticum* Phill. in Ann. S. Afr. Mus. 9: 344 (1917). Type: Lesotho, Maseru distr., slopes of Machache Mtn, *Jacottet* in herb. *Dieterlen* 1068 (SAM, holo.!; PRE, iso.!).

Spindly shrub up to 900 mm high, old stems bare, rough with leaf bases, branchlets closely leafy. *Leaves* commonly up to 20 × 3,5 mm, or 35 × 10 mm on coppice growth, ovate-lanceolate, apex acute, mucronate, base broad, half-clasping, margins revolute, both surfaces sometimes grey-woolly, or commonly upper surface very lightly cobwebby, hairs soon deciduous or confined to the midrib, often minutely gland-dotted and glutinous, sometimes glandular-hairy as well when young, lower surface white-felted, striate. *Heads* heterogamous, campanulate, 5—6 mm long, c. 8 mm across the radiating involucral bracts, bases white-woolly, many in compact rounded corymbose clusters 15—20 mm across at the branch tips. *Involucral bracts* in 5—6 series, closely imbricate, graded, glossy, outer light brown, inner slightly exceeding flowers, tips obtuse, bright canary-yellow, minutely radiating. *Receptacle* honeycombed. *Flowers* 34—56, 6—12 ♀, 26—50 ☿. *Achenes* not seen, ovaries 1 mm long with duplex hairs. *Pappus* bristles many, equalling corolla, tips subplumose, bases cohering by patent cilia. Fig. 45:3.

On the high Lesotho mountains, the Witteberg and the Cape Drakensberg, in large but localized colonies on moist grassy mountain slopes or valley bottoms. Flowers between November and February. Can be confused with *H. trilineatum* (no. 147) but has differently shaped leaves, broadest at the base. Map 150.

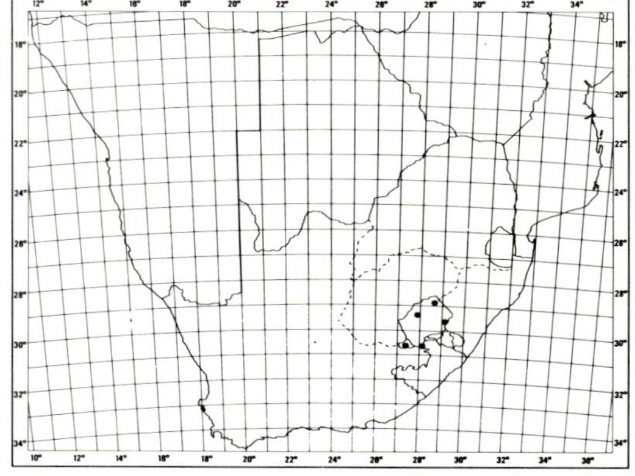

MAP 150.— **Helichrysum witbergense**

Vouchers: *Esterhuysen* 13192 (BOL); *Hilliard & Burtt* 7109 (E; K; MO; NU; PRE; S); *Hilliard & Burtt* 9685 (E; K; MO; NU; S); *Nordenstam* 2084 (S).

149. **Helichrysum tenuifolium** Killick in Bothalia 6: 424 (1954); Hilliard, Compositae in Natal 204 (1977). Lectotype: Natal, Bergville distr., Cathedral Peak Forest Research Station, *Killick* 1591 (PRE!; BM; K; NU; S, isolecto.!).

Twiggy, rounded shrub up to 2 m high, old branches bare, rough with leaf bases, branchlets white-woolly, closely leafy. *Leaves* 7—18 × 1—1,5 mm, linear, apex mucronate, acute, hooked, base broad, half-clasping, margins revolute, upper surface with minute, elongated appressed trichomes, later glabrous or nearly so, lower surface white-woolly. *Heads* heterogamous, campanulate, 4—5 mm long and as much across the radiating bracts, bases woolly, many in compact terminal corymbose clusters 10—20 mm across. *Involucral bracts* in c. 4—5 series, graded, closely imbricate, glossy, outer palest brown often tipped reddish or orange when young, inner about equalling the flowers, tips obtuse, bright canary-yellow, minutely radiating. *Receptacle* shortly honeycombed. *Flowers* 27—45, 7—15 ♀, 19—30 ☿. *Achenes* 0,75 mm long, with duplex hairs. *Pappus* bristles many, equalling corolla, scabrid, bases cohering by patent cilia.

Recorded from only the Natal Drakensberg between Cathedral Peak and the upper Polela, west of Himeville. Grows in colonies in rocky gullies, along streamsides, in boulder beds, among rock outcrops, or as a constituent of fynbos, above 1 650 m. Flowers in November and December; very showy. Can be confused with *H. trilineatum* (above), but has appressed trichomes, not gland dots, on the upper leaf surface. Map 151.

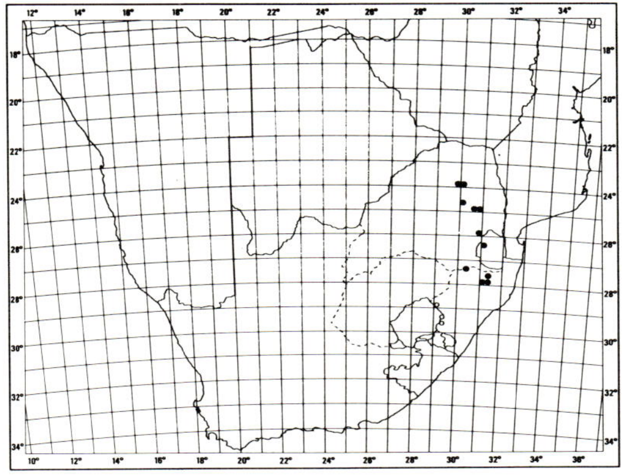

MAP 151.— ○ **Helichrysum tenuifolium**
  ● **Helichrysum chrysargyrum**

Vouchers: *Hilliard & Burtt* 9290 (E; K; NU; S); *Schelpe* 960 (NU); *Wright* (E; NH; NU).

150. **Helichrysum chrysargyrum**

Moeser in Bot. Jb. 44: 286 (1910); Compton, Fl. Swaziland 629 (1976); Hilliard, Compositae in Natal 206 (1977). Lectotype: Transvaal, Mpome, *Schlechter* 4728 (Z!; BM; K; PRE; S, isolecto.!).

Bushy perennial herb up to 400 mm high, old stems bare and woody below, otherwise leafy throughout, stems, leaves and base of each head enveloped in silvery, silky, skin-like indumentum. *Leaves* up to c. 75 (—90) ×5 (—6) mm, narrowly lanceolate, apex long-acuminate, base slightly narrowed, half-clasping. *Heads* heterogamous, broadly campanulate, c. 5 mm long, 8 mm

across the fully radiating bracts, up to c. 15 in rather congested terminal corymbose panicles. *Involucral bracts* in c. 5 series, graded, closely imbricate, glossy, outer palest golden-brown or straw-coloured, inner about equalling the flowers, tips obtuse, lemon-yellow, radiating. *Receptacle* honeycombed. *Flowers* 38—75, 4—13 ♀, 30—62 ☿. *Achenes* c. 1 mm long, broadly cylindric, with myxogenic duplex hairs. *Pappus* bristles many, about equalling corolla, scabridulous, bases cohering lightly by patent cilia. Fig. 46: 1.

Recorded from the E. Transvaal, from Woodbush and Haenertsburg in the north, south to the Barberton Mountains and the low Drakensberg on the Transvaal-Natal border, western Swaziland, and the mountainous parts of northern Natal. Flowers between February and April. Map 151.

Vouchers: *Compton* 25077 (NBG; PRE); *Devenish* 1694 (E; K; MO; NU); *Hilliard & Burtt* 10001 (E; K; M; MO; NU; PRE; S); *Scheepers* 930 (PRE).

151. **Helichrysum psilolepis**

Harv. in F.C. 3: 235 (1865); Moeser in Bot. Jb. 44: 285 (1910); Hilliard, Compositae in Natal 207 (1977). Lectotype: Cape, between Zwart Kei and Stormberg, *Drège* 3755 (G-DC!).

H. adscendens sensu DC., Prodr. 6: 185 (1838), non (Thunb.) Less.

Tufted perennial herb, stems many from a rhizomatous stock, erect, or ascending to 400 mm, simple or branched from the base, greyish-white woolly, closely leafy throughout. *Leaves* up to 30 × 4 mm, but often only c. 15×2 mm, linear or linear-lanceolate, apex acute, mucronate, base broad, half-clasping, both surfaces greyish-white woolly. *Heads* heterogamous, campanulate, c. 5 mm long and as broad, many in congested terminal rounded corymbose clusters 15—20 mm across. *Involucral bracts* in c. 7 series, 4 inner subequal, loosely imbricate, tips slightly exceeding flowers, obtuse, pellucid, canary-yellow, crisped, slightly radiating, outer tipped pale golden-brown, bases woolly. *Receptacle* minutely tuberculate. *Flowers* 112—156, 54—81 ♀, 62—102 ☿, ♀ very rarely outnumbering ☿. *Achenes* not seen, ovaries with myxogenic duplex hairs. *Pappus* bristles many, equalling corolla, scabrid, bases cohering by patent cilia.

Ranges from Vereeniging, Standerton and Wakkerstroom in the southernmost Transvaal through Dundee, Newcastle and Utrecht districts in northern Natal to the mountainous NE. corner of the Orange Free State, western Lesotho, and the E. and E. central Cape as far west as the mountains about Graaff-Reinet and the Suurberg north of Port Elizabeth, between c. 600 and 2 100 m. Grows in grassland; flowering between November and January. Map 152.

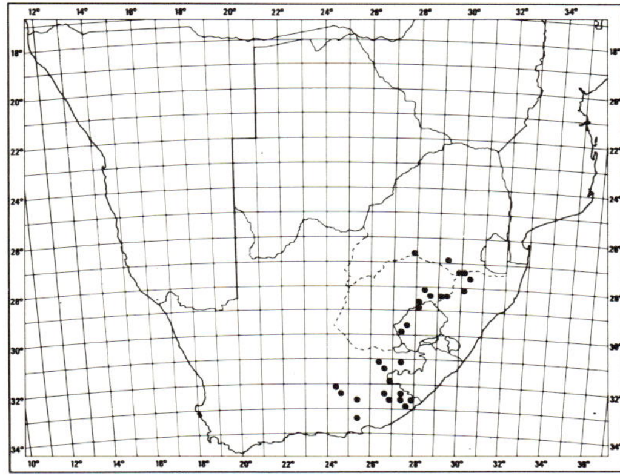

MAP 152.— **Helichrysum psilolepis**

Vouchers: *Acocks* 18675 (PRE); *Dieterlen* 143 (NBG); *Flanagan* 363 (PRE, SAM); *Galpin* 9811 (PRE); *Hilliard & Burtt* 10975 (E; K; MO; NU; S).

152. **Helichrysum versicolor** *O. Hoffm. & Muschl.* in Annln naturh. Mus. Wien 24: 317 (1910). Type: Albany distr., Grahamstown, *Penther* 1266 (W, holo.!).

Annual herb, stems several from the base, up to 300 mm long, erect or decumbent, simple or sparingly branched, white-woolly, leafy. *Leaves* mostly 15−40 × 4−12 mm, diminishing upwards, oblong-lanceolate or oblanceolate becoming lanceolate upwards, apex subobtuse to acute, mucronate, base cordate-clasping, margins flat or crisped-undulate, both surfaces thinly greyish-white woolly. *Heads* heterogamous, campanulate, 3 × 2 mm, many in congested woolly clusters in compact corymbose panicles at the branch tips. *Involucral bracts*

in c. 6 series, loosely imbricate, backs white woolly, outermost short, others subequal, about equalling flowers, tips semi-pellucid or opaque, very obtuse, crisped, bright yellow, minutely radiating. *Receptacle* scarcely honeycombed. *Flowers* 12−30, 3−7 ♀, 9−23 ☿. *Achenes* c. 0,5 mm long, with myxogenic duplex hairs. *Pappus* bristles many, equalling corolla, scabrid, bases not cohering. Fig. 46:5.

The type reputedly came from Grahamstown, but all subsequent collections have been from the coast, often on the seashore or in dune scrub, from Knysna in the west to Port Alfred and the mouth of the Fish River in the east, flowering mostly in October and November, but as late as February. Map 153.

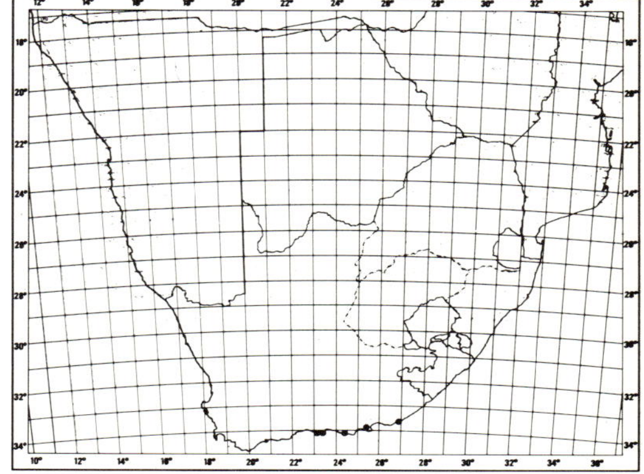

MAP 153.— **Helichrysum versicolor**

Although this species was first collected by Burchell in 1814 at Krakakamma, Port Elizabeth district, it has either passed as 'H. rutilans' or remained unnamed. It is distinguished from *H. moeserianum* (=*H. rutilans* auctt.) (below) by its perennial habit, leaves less narrowed to the base, and subtly different heads, the differences difficult to express, but clearly seen; measurable parameters are: stereome ± 2mm long, not ± 1mm, in the outermost but one series of bracts (the outermost are reduced) and hermaphrodite flowers mostly 3−4 times as many as female, not mostly at least 5 times as many. Also, the two species are allopatric.

Vouchers: *Burchell* 4562 (K); *MacOwan* 1442 (BOL; BM; K; S); *Taylor* 4425 (PRE; STE); *Wall* 116 (S).

FIG. 46.−1, **Helichrysum chrysargyrum,** part of plant, × 1 (*Hilliard & Burtt* 5913 and 9844). 2, **H. isolepis,** part of plant, × 1 (*Hilliard & Burtt* 12387). 3, **H. intricatum,** part of plant, × 1 (*Hilliard & Burtt* 12373). 4, **H. moeserianum,** head, × 9,3; 4a, involucral bract from outer series but one, × 13 (*Acocks* 17218). 5, **H. versicolor,** head, × 9,3; 5a, involucral bract from outer series but one, × 13 (*Smart* 27981).

153. **Helichrysum moeserianum** *Thell.*

in Vierteljahrsschr. naturf. Ges. Zürich 66: 239 (1921); Hilliard & Burtt in Bot. J. Linn. Soc. 82: 369 (1981). Type: Cape, Touws River, Nov. 1898, *Schimper* (Z, holo.!).

Gnaphalium bracteatum Lam., Encycl. 2: 751 (1788), non *Helichrysum bracteatum* Andr. (1805). Type: Cape of Good Hope (P-LAM!).

Helichrysum rutilans sensu Less., Syn. Comp. 275 (1832); DC., Prodr. 6: 185 (1838); Harv. in F.C. 3: 234 (1865); Moeser in Bot. Jb. 44: 287 (1910); Levyns in Adamson & Salter, Fl. Cape Penins. 781 (1950), non (L.) D. Don.

Annual herb mostly 100—300 mm tall, stems erect, solitary or branched from the base, simple or sparingly branched above, greyish-white woolly, distantly leafy. *Leaves* mostly 15—30 (—60) × 3—5 (—13) mm, diminishing slightly upwards, oblong-spathulate becoming linear upwards, apex obtuse to acute, mucronate, base half-clasping, both surfaces greyish-white woolly. *Heads* homogamous or heterogamous, campanulate, 3—4 × 2—3 mm, many in terminal corymbose clusters, these often in corymbose panicles. *Involucral bracts* in c. 5 series, graded, loosely imbricate, outer woolly on the backs, tips subacute or obtuse, inner about equalling or exceeding the flowers, tips opaque canary-yellow, minutely radiating, obtuse, well crisped or only slightly so. *Receptacle* scarcely honeycombed. *Flowers* 12—30, (0—) 3—6 ♀, 12—27 ☿. *Achenes* 0,75 mm long, obscurely ribbed, with myxogenic duplex hairs or very rarely glabrous. *Pappus* bristles many, about equalling corolla, tips barbellate, bases not cohering. Fig. 46: 4.

Recorded mainly from the W. and SW. Cape, from Vanrhynsdorp south to the Peninsula, Hex River and Touws River and east to Mossel Bay, on the mountains and on the flats. There is one record from Ookiep in Namaqualand, but the locality 'Windhoek' for *Schlechter* 8362 is deceptive; it is near Vanrhynsdorp. Map 154.

Vouchers: *Acocks* 17562 (PRE); *Compton* 21740 (NBG); *Nordenstam & Lundgren* 1508 (E, S); *Pillans* 7980 (BOL); *Schlechter* 8362 (BM; E; G; PRE); *Tyson* 706 (E; SAM).

Some specimens have more or less acute rather than obtuse outer bracts and there is some variation in the degree of crisping of the inner, obtuse to more or less truncate, bracts; however, *H. moeserianum* is easily recognized by its annual habit allied to bright yellow crisped bracts. It can be confused with *H. versicolor* (above; see under that species for discriminating characters).

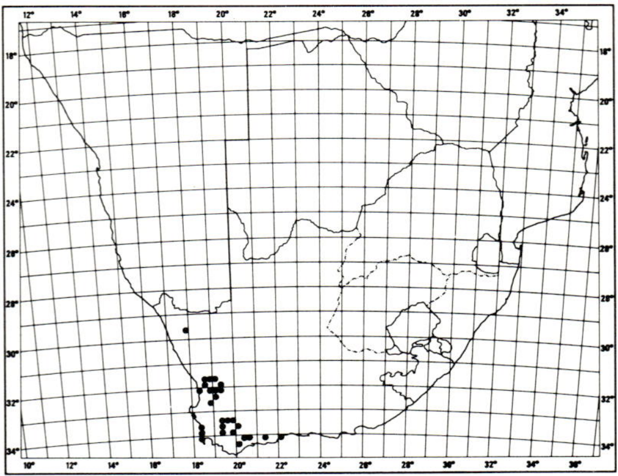

MAP 154.— **Helichrysum moeserianum**

154. **Helichrysum intricatum** *DC.*,

Prodr. 6: 204 (1838); Harv. in F.C.3: 248 (1865). Type: Cape, between Swellendam and George, *Ecklon* 299 (G-DC, holo.!; S, iso.!).

Gnaphalium intricatum (DC.) Sch. Bip. in Bot. Ztg 3: 172 (1845).

Helichrysum anomalum var. *lanatum* DC., Prodr. 6: 204 (1838), nomen. Van Staadensberg, 1 000—2 000 Fuss, *Drège* (distributed as *H. anomalum* var. *lanata* DC. a, BM!; G!).

Perennial herb with tufts of delicate wiry stems decumbent then erect, up to 600 mm long, virgate, glabrous below, white-woolly-felted above, leafy, becoming pedunculoid upwards. *Leaves* mostly 12—30 × 2,5—8 mm, diminishing upwards and passing into small distant bracts, oblong, oblong-spathulate or lanceolate, apex subacute or acute, mucronate, base cordate-clasping in larger leaves, margins slightly revolute, upper surface with coarse spreading hairs, cobwebby as well initially, lower surface white woolly-felted. *Heads* homogamous or heterogamous, campanulate, 4 × 2,5 mm, many felted at the base into terminal flat-topped glomerules 10—20 mm across, becoming somewhat looser with age. *Involucral bracts* in 4 series, graded, loosely imbricate, outermost pale brown or yellow, webbed together with wool, inner lemon-yellow, sub-pellucid, tips obtuse, about equalling flowers, minutely radiating. *Receptacle* with fimbrils about equalling or exceeding ovaries. *Flowers* 8—15, 0—3 ♀,

6—15 ♀. *Achenes* 1 mm long, broadly cylindric, glabrous or with myxogenic duplex hairs. *Pappus* bristles many, equalling corolla, scabrid, bases cohering by patent cilia, some light fusion as well. Fig. 46: 3.

Recorded from 'between Swellendam and George' (the type locality), Van Staadensberg (Drège), 'Groote Hoogte and Assagai Bosch' (Pappe), then Hogsback, Katberg, Thomas Mountain and Evelyn Valley near Stutterheim. Grows in marshy places, forming large tangled masses at the muddy margins of streams or in the sponges at their sources; flowering in December and January. Map 155.

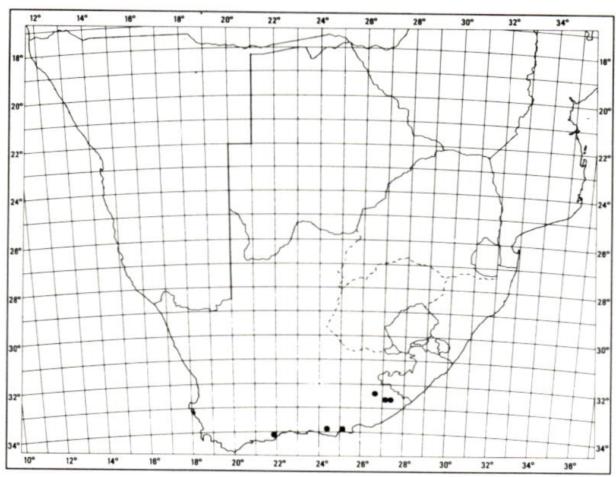

MAP 155.— **Helichrysum intricatum**

In the type collection and that made by Pappe the heads contain 1—3 ♀ flowers and the achenes are hairy; the specimens from Hogsback, Katberg, and Evelyn Valley have homogamous heads and glabrous ovaries, but I can see no other feature to distinguish them. The species is rarely collected, and should be sought between Swellendam and the Amatola Mountains so that its variability can be properly assessed.

Vouchers: *Compton* 19268, 19142 (NBG); *Hilliard & Burtt* 10987 (E; K; M; MO; NU; PRE; S); *Leighton* 2717 (BOL; K; NBG); *Pappe* s.n. (SAM; TCD).

155. **Helichrysum isolepis** *H. Bol.* in Trans. S. Afr. phil. Soc. 18: 385 (1907). Type: Cape, mountainside Great Winterberg, 8 iii 1900, c. 2 250 m, *Galpin* 2657 (K!; PRE!).

Much-branched, straggling perennial herb, stems branching at ground level,

decumbent, rooting, then erect to c. 300 mm, simple or subsimple, loosely white-woolly, closely leafy, more distantly so under the heads. *Leaves* c. 10—25 × 3—5 mm, diminishing upwards, oblong-obovate to oblong, apex subacute or obtuse, mucronate, uppermost with a yellow scarious appendage, base narrowed, half-clasping, both surfaces coarsely glandular-pubescent, cobwebby initially, margins thickened, they and the midvein below white-woolly. *Heads* homogamous, campanulate, c. 7 × 6—7 mm, several in a spherical cluster c. 20 mm across at the branch tips. *Involucral bracts* in 5 series, subequal, loosely imbricate, bases webbed together with wool, oblong-elliptic, tips very obtuse, somewhat crisped, about equalling flowers, semiglossy, bright canary-yellow, radiating. *Receptacle* shortly honeycombed. *Flowers* c. 24—35. *Achenes* 1,5 mm long, glabrous. *Pappus* bristles many, equalling corolla, tips barbellate, yellow, bases cohering by patent cilia. Fig. 46: 2.

Known only from the Great Winterberg and Katberg Pass in the E. Cape. Grows in stony mountain grassland, forming loose tangled clumps; in flower in December and January. Rarely collected. Map 156.

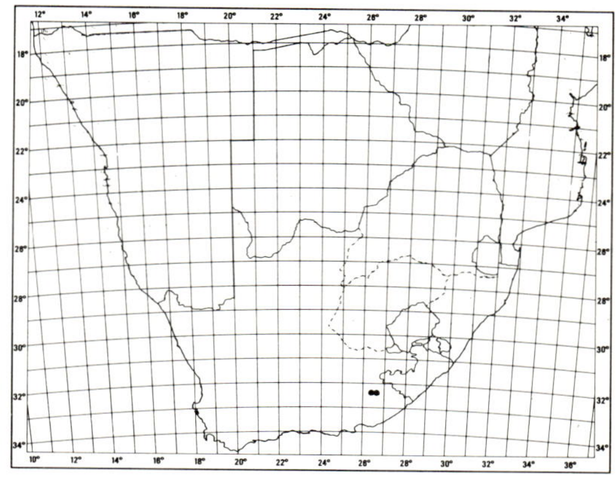

MAP 156.— **Helichrysum isolepis**

Voucher: *Hilliard & Burtt* 10957 (E; GRA; K; M; MO; NU; PRE; S).

Group 23

Perennial herbs, often stoloniferous, often with a turf of radical leaves; *leaves* medium-sized or large, mostly elliptic to ovate, obovate or suborbicular, sometimes linear-lanceolate, base often petiole-like; *heads* homogamous, 4−8 × 3−8 mm, in lax or compact corymbose-panicles; *involucral bracts* minutely radiating or not, or sometimes squarrose, creamy, straw-coloured, tawny, yellow, or sometimes rosy; *receptacle* fimbrilliferous; *flowers* 8−69, corolla campanulate above; *achenes* glabrous; *pappus* bristles scabrid, bases cohering strongly by patent cilia, sometimes lightly fused as well.

Species 156−175, six widely distributed in the grasslands of Africa, the rest endemic to Southern Africa, some narrowly so. Mostly in grassland, but *H. mundtii* (no. 175) favours marshy places.

1a Flowering stem leafy mainly in lower half, nude or with a few reduced leaves or bracts in upper part:

  2a Crown clothed in brown silky wool, the wool sometimes hidden in the leaf axils:

    3a Involucral bracts either minutely radiating or not, but not squarrose:

      4a Upper leaf surface variously pubescent or woolly at least on the main veins and margins:

        5a Upper leaf surface clad in long coarse hairs.. 157. *H. pilosellum*

        5b Upper leaf surface with minute soft pubescence or woolly as well or wool confined to margins and main veins:

          6a Both leaf surfaces closely silky-woolly, or upper surface with only minute soft pubescence or with traces of wool as well .... 159. *H. gerberifolium*

          6b Both leaf surfaces loosely woolly or wool confined to margins and main veins......... 161. *H. thapsus*

      4b Upper leaf surface glabrous and smooth, though initially it may be covered with papery indumentum or a little wool may cling to the midline:

        7a Involucral bracts brown .... 158. *H. pedunculatum*

        7b Involucral bracts straw-coloured... 163. *H. coriaceum*

    3b Involucral bracts very acute to acuminate, squarrose ....160. *H. griseum*

  2b Crown not clothed in brown silky wool:

    8a Radical leaves c. 4−6, obovate to subrotund, spreading flat on ground .... 156. *H. longinquum*

    8b Leaves not spreading flat on ground:

      9a Upper leaf surface glabrous at maturity except for traces of wool sometimes clinging to the veins:

        10a Involucral bracts reddish brown; leaves 5−7-nerved................................. 158. *H. pedunculatum*

        10b Inner involucral bracts pale yellow or whitish, outer often purplish or reddish; leaves triplinerved ....170. *H. allioides*

      9b Upper leaf surface variously hairy:

        11a Heads subglobose, bracts white or pink..170. *H. calocephalum*

        11b Heads campanulate or cylindric-campanulate, bracts mostly various shades of cream, yellow or brown, or rarely whitish, the outer then purplish or reddish:

          12a At least the radical leaves 5−7-nerved, or if 3-nerved, heads 6−7 mm long and inner bracts whitish or creamy:

            13a Involucral bracts creamy, yellow, straw-coloured, golden-brown or combinations of yellow and brown; leaves 5−7-nerved:

              14a Leaves rigid, involucral bracts mostly pale or lemon-yellow, only the outermost sometimes pale brown... 164. *H. nudifolium*

              14b Leaves soft, involucral bracts straw-coloured, golden-brown, creamy, yellow or yellow and brown:

                15a Involucral bracts straw-coloured or golden-brown:

                  16a Leaf blade elliptic, tapering above and below .... 159. *H. gerberifolium*

                16b Leaf blade broadly elliptic, base suddenly contracted then gradually tapering into the petiolar part... 174. *H. qathlambanum*

              15b Involucral bracts creamy, yellow, or yellow and brown:

17a Tertiary veins in leaves not conspicuous ....167. *H. mollifolium*

17b Tertiary veins on lower leaf surface scalariform, raised or not, but clearly visible .... 172. *H. pallidum*

13b Outer involucral bracts purplish or reddish, inner white or creamy; leaves generally 3-nerved, rarely 5-nerved.. 171. *H. oxyphyllum*

12b Radical leaves 3-nerved, heads 4−5 mm long, bracts light golden-brown or yellow or a combination of the two:

18a Flowers 16−32, involucral bracts in c. 6 series....................................... 164. *H. nudifolium*

18b Flowers 9−13, involucral bracts in c. 4 series .... 173. *H. inornatum*

1b Flowering stem either leafy throughout or with more or less crowded reduced leaves or bracts below the compound inflorescence:

19a Cauline leaves not decurrent in long narrow wings:

20a Heads campanulate, bracts not convex:

21a Heads c. 6 × 3 mm, bracts closely imbricate, lemon-yellow................................ 162. *H. pannosum*

21b Heads c. 4−5 × 3−5 mm, bracts loosely imbricate, either tawny, yellow or yellow and brown:

22a Involucral bracts tawny, mostly obtuse...165. *H. harveyanum*

22b Involucral bracts yellow often washed light golden-brown, mostly acute to acuminate ....166. *H. miconiifolium*

20b Heads subglobose, bracts deeply convex ....168. *H. krebsianum*

19b Cauline leaves decurrent in long narrow wings....................................... 175. *H. mundtii*

---

156. **Helichrysum longinquum** *Hilliard* in Notes R. bot. Gdn Edinb. 40:258 (1982). Type: Natal, Underberg distr., Cobham Forest Station, Upper Polela Cave area, c. 2 285 m, 16 ii 1979, *Hilliard & Burtt* 12613 (NU holo.!; E; K; PRE, iso.!).

Perennial herb, roots fusiform, tuberous, stock small, producing slender stolons, flowering stem terminal, solitary, simple, 200−300 mm tall, white-woolly-felted, remotely bracteate. *Leaves* c. 4−6, spreading flat on ground, 25−30 × 18−25 mm, obovate to subrotund, slightly narrowed to a broad clasping base, apex rounded or very obtuse, upper surface with long coarse multicellular hairs, lower thinly white-felted, soon glabrescent, prominently 5-nerved, side nerves invisible; bracts few, distant, small, oblong to lanceolate, indumentum as in leaves, but wool persistent. *Heads* homogamous, campanulate, c. 4 × 4 mm, c. 15−25 in a compact rounded terminal cluster, bases webbed together with white wool. *Involucral bracts* in c. 5 series, imbricate, subequal, about equalling flowers, straw-coloured or tawny, tips slightly darker, opaque, very obtuse, erose, crisped, scarcely radiating. *Receptacle* with fimbrils much exceeding ovaries. *Flowers* 19−26, yellow. *Achenes* not seen, ovaries 0,75 mm long, glabrous. *Pappus* bristles many, equalling corolla, scabrid, bases cohering strongly by patent cilia, sometimes lightly fused as well. Fig. 47: 1.

Known only from a small area of the southern Natal Drakensberg in Mpendhle and Underberg districts, but should be looked for particularly further south. Grows on moist grass slopes at c. 2 100 to 2 300 m; flowering in January and February. Readily recognized by its basal rosette of small spreading rounded leaves prominently 5-nerved below but lacking reticulate venation, remotely bracteate flowering stem and terminal cluster of tawny heads. Map 157.

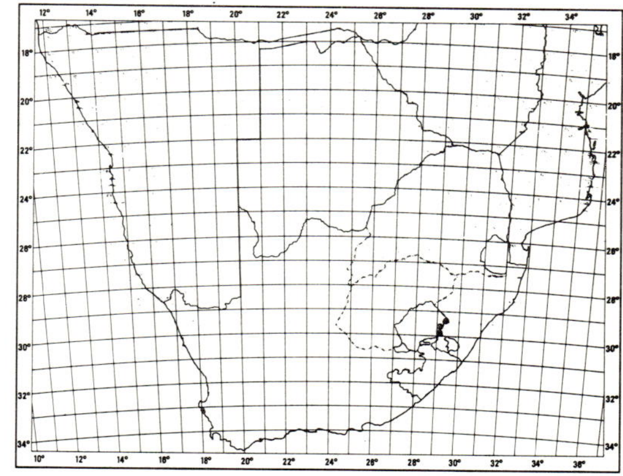

MAP 157.— **Helichrysum longinquum**

Voucher: *Stewart* 2082 (E; K; NU).

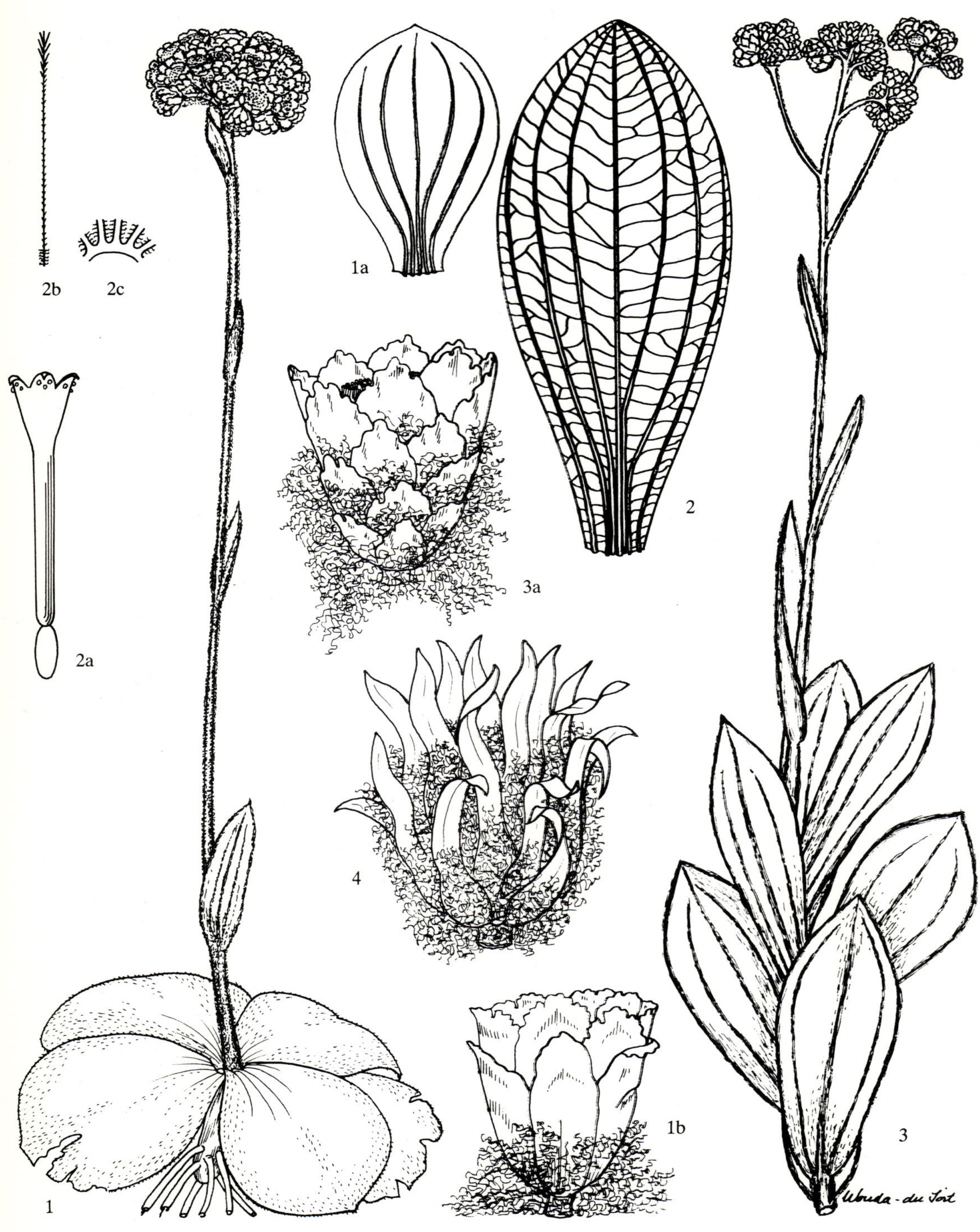

157. **Helichrysum pilosellum** *(L.f.) Less.*, Syn. Comp. 297 (1832) quoad syn. excl. descr.; Hilliard & Burtt in Notes R. bot. Gdn Edinb. 32: 358 (1973); Compton, Fl. Swaziland 633 (1976); Hilliard, Compositae in Natal 172 (1977). Type: Cape of Good Hope, *Bäck* [i.e. Forster] (LINN 989.79!).

Gnaphalium pilosellum L. f., Suppl. 364 (1781).

G. latifolium Thunb., Prodr. 152 (1800), Fl. Cap. 660 (1823). *Helichrysum latifolium* (Thunb.) Less., Syn. Comp. 297 (1832); DC., Prodr. 6: 198 (1838); Harv. in F.C. 3: 237 (1865); Moeser in Bot. Jb. 44: 262 (1910). Type: Cape, Galgebosch, *Thunberg* (sheet 19183, UPS!).

H. latifolium var. *reticulatum* Harv. in F.C. 3: 237 (1865). Type: Natal, between Maritzburg and Ladysmith, *Gerrard & M'Ken* 281 (TCD!).

H. pedunculare (L.) DC. var. *pilosellum* (L. f.) Harv. in F.C. 3: 238 (1865) quoad syn., excl. spec.

Perennial herb, roots fusiform, tuberous, rootstock stout, woody, crown brown silky-woolly, flowering stem terminal, simple, up to 450 mm tall, often leafy at the base only, more rarely distantly leafy upwards, woolly-felted. *Leaves* often only 3 or 4, spreading or ascending, up to 150 (−230) × 40 (−60) mm, elliptic, ovate or subrotund, narrowed to a broad clasping base, stem leaves, if present, cordate-clasping, shortly decurrent, apex obtuse or subacute, upper surface harshly pubescent with long coarse multicellular hairs, lower white-felted, strongly 5−9 nerved, reticulate. *Heads* homogamous, campanulate, c. 5 × 5 mm, many in compact corymbose clusters in a rounded or flat-topped cyme. *Involucral bracts* in c. 3 series, imbricate, subequal, about equalling the flowers, base woolly, tips more or less obtuse, crisped, pellucid, pale brown, minutely radiating. *Receptacle* with fimbrils longer than ovaries. *Flowers* 23−58, bright yellow. *Achenes* c. 1 mm long, glabrous. *Pappus* bristles many, equalling corolla, scabrid, bases lightly cohering by patent cilia, lightly fused as well. Fig. 47: 2.

Ranges from Zambia, Malawi and Zimbabwe to the Transvaal, Swaziland, Orange Free State, Lesotho, Natal and the Cape about as far west as Uitenhage and Galgebosch (roughly 25°E). Widespread in grassland, down to sea level in Natal, Transkei and Cape; flowering mainly between August and December. Map 158.

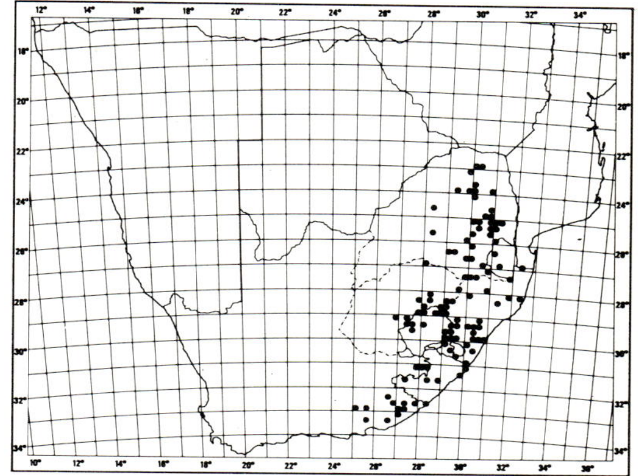

MAP 158.— **Helichrysum pilosellum**

Vouchers: *Compton* 27079 (NBG); *Devenish* 1606 (E; K; MO; NU; S); *Strey* 7648 (NU); *Tyson* 1101 (BOL).

158. **Helichrysum pedunculatum** *Hilliard & Burtt* in Notes R. bot. Gdn Edinb. 32: 356 (1973). Type: Cape, Beaufort distr., Kat River valley, 1860, *Cooper* 415 (E, holo.!; BM; K; NH; Z, iso.!).

H. pilosellum sensu Less., Syn. Comp. 297 (1832) quoad descr. excl. syn., non (L.f.) Less.

H. pedunculare sensu DC., Prodr. 6: 198 (1838) quoad descr. excl. syn.; Harv. in F.C. 3: 238 (1865) excl. syn. et var.; Moeser in Bot. Jb. 44: 262 (1910); Batten & Bokelmann, Wild Flow. E. Cape Prov. 154 plate 123 (1966), non (L.) DC.

Perennial herb, rootstock woody, some silky-brown wool hidden in the axils of the leaf bases, stem solitary, simple, erect to c. 500 mm, white-woolly, leafy below, becoming bracteate upwards. *Leaves* mostly radical, mostly 80−130 × 20−40 mm, elliptic, apex acute, tapering below to a broad, flat, clasping petiole-like base, upper

---

FIG. 47.−1, **Helichrysum longinquum**, whole plant, × 1; 1a, lower leaf surface to show venation, × 1,3; 1b, head, × 6,6 (*Hilliard & Burtt* 12613). 2, **H. pilosellum**, undersurface of leaf to show venation, × 1,3; 2a, hermaphrodite flower, × 11; 2b, pappus bristle, × 11; 2c, base of pappus bristles much enlarged to show fusion as well as cohesion by patent cilia (*Hilliard* 887). 3, **H. thapsus**, whole plant, × 0,7; 3a, head, × 6,6 (*Wilms* 733). 4, **H. griseum**, head, × 6,6 (*Hilliard* 2872).

surface glabrous, lower with a white silky-woolly-felted skin-like indumentum, stem leaves similar but soon sessile, passing rapidly into distant linear-lanceolate acuminate bracts. *Heads* homogamous, campanulate, c. 7−8 × 6−8 mm, many in a compact or loose terminal corymbose panicle. *Involucral bracts* in c. 5 series, graded, inner about equalling flowers, loosely imbricate, oblong, loosely woolly below, tips pullucid, subacute, reddish brown, crisped, not radiating. *Receptacle* with fimbrils exceeding ovaries. *Flowers* 30−65. *Achenes* not seen, ovaries glabrous. *Pappus* bristles many, about equalling corolla, scabrid, bases lightly fused.

Ranges from southern Lesotho to the southern Transkei and eastern Cape with isolated records from Graaff-Reinet (Cave Mountain, *Bolus* 245, BOL) Great Swartberg, Prince Albert div. (*Stokoe* 6768, BOL), Uitenhage (coll. Burchell) and Spiegel River, Riversdale (coll. Burchell). Grows in grassland, flowering between November and February. Distinguished from *H. pilosellum* (above) by its larger heads and leaves, glabrous above, silky below. Map 159.

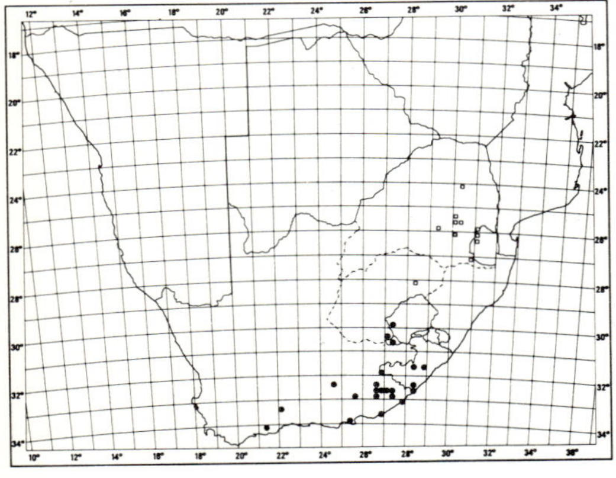

MAP 159.—● **Helichrysum pedunculatum**
         □ **Helichrysum gerberifolium**

Vouchers: *Barker* 1159 (NBG); *Burchell* 7201 (K); *Dieterlen* 1078 (K; PRE); *Galpin* 1696 (PRE); *Pegler* 1322 (BOL; PRE).

159. **Helichrysum gerberifolium** [*Sch. Bip. ex*] *A. Rich.*, Tent. Fl. Abyss. 1: 425 (1848); Moeser in Bot. Jb. 44: 263 (1910).

Type: Abyssinia [Ethiopia], near Adua, Scholada, 7 June 1837, *Schimper* 203 (P, holo.!; G; K; S, iso.!).

H. davyi S. Moore in J. Bot., Lond. 43: 169 (1905); Compton, Fl. Swaziland 630 (1976). Type: Transvaal, Carolina distr., 1 mile N. of Robinson's, 10 Jan. 1905, *Burtt Davy* 2972 (BM, holo.!; K; BOL and PRE under 2973, iso.!).

H. brunneum Burtt Davy in Jl S. Afr. Bot. 1: 107 (1935). Type: Transvaal, Pietersburg distr., Shilouvane, *Junod* 1632 (K, holo.!).

Perennial herb, roots narrowly fusiform, rootstock woody, crowned with old fibrous leaf bases with a little brown wool hidden in the axils, flowering stem solitary, erect, 100−450 (−750) mm tall, white-woolly-felted, leafy below, pedunculoid above with 1 or 2 much reduced leaves. *Leaves* mostly radical, c. 60−200 × 13−33 mm, roughly half the length petiolar, blade elliptic, tapering above and below, apex acute, apiculate, upper surface with minute soft pubescence, usually with some traces of wool as well or persistently thinly greyish silky-woolly, lower surface greyish silky-woolly-felted, strongly 5−7-nerved, more or less reticulate as well, *stem leaves* similar but soon sessile, broad-based, clasping, shortly decurrent, reduced leaves and bracts lanceolate-acuminate. *Heads* homogamous, campanulate, c. 4−5 × 3−5 mm, many in a congested globose terminal cluster c. 15−30 mm in diam., or sometimes the compound inflorescence loose with clusters in a corymbose panicle. *Involucral bracts* in c. 4 series, outer 2 short, inner 2 subequal, equalling flowers, not radiating, all woolly on backs, tips translucent, crisped, obtuse, often erose, pale golden-brown. *Receptacle* with fimbrils exceeding ovaries. *Flowers* 18−45. *Achenes* 1 mm long, cylindric, faintly ribbed, glabrous. *Pappus* bristles many, equalling corolla, scabrid, bases lightly fused.

Ranges from the highlands of Ethiopia south through the elevated parts of E. Africa and Zimbabwe to the highlands of the E. and SE. Transvaal and neighbouring Swaziland with one record from Bethlehem in the Orange Free State. Favours rocky grassland between 1 200 and 2 000 m; flowering between October and January. Map 159.

Closely allied to the widespread *H. pilosellum* (above), but in that species the crown is conspicuously brown silky-woolly, the leaves are usually scarcely petiolate, harshly pubescent above with long coarse multicellular hairs, and mostly lack wool on the upper surface. Sometimes confused with *H. coriaceum* (no. 163), which is immediately distinguished by its different leaves and straw-coloured involucral bracts.

The type of *H. gerberifolium* has leaves initially silky grey-woolly on both surfaces, later with only traces of wool; the type of *H. davyi* also has leaves persistently grey silky-woolly above, but plants with leaves scarcely woolly above seem just as common. The type of *H. brunneum* has only traces of wool clinging to the upper leaf surface.

Vouchers: *Codd & De Winter* 3223 (K; NU; PRE); *Compton* 27189 (NBG; PRE).

160. **Helichrysum griseum** *Sond.* in Linnaea 23: 65 (1850); Harv. in F.C. 3: 237 (1865); Moeser in Bot. Jb. 44: 262 (1910); Wood, Natal Plants 6,2: t. 549 (1910); Hilliard, Compositae in Natal 173 (1977). Lectotype: Port Natal, *Gueinzius* 322 (S!; G (herb. Boiss.); P; W, isolecto.!).

H. agrostophilum Klatt in Bull. Herb. Boissier 4: 833 (1896) p.p. quoad spec. *Schlechter* 6216 (Z!).

Perennial herb, rootstock stout, woody, crown brown silky-woolly, flowering stems up to c. 700 mm high, simple, one or several tufted together, woolly-felted, remotely leafy in lower part, remotely leafy or bracteate upwards. *Leaves* often only 3 or 4, ascending, up to 200 × 80 mm, broadly elliptic or ovate, apex more or less acute, base somewhat narrowed in lower leaves, broad, cordate-clasping in upper, upper surface thinly cobwebby at first, later harshly pubescent, lower greyish-white felted, faintly 5−9-nerved, reticulate. *Heads* homogamous, campanulate, c. 6 × 5 mm, many in a congested or lax corymbose cyme. *Involucral bracts* in c. 4 series, subequal, closely imbricate, base woolly, tips glossy, subopaque, bright deep or pale pink or buff, very acute to acuminate, squarrose. *Receptacle* with fimbrils longer than ovaries. *Flowers* 25−31. *Achenes* c. 1 mm long, glabrous. *Pappus* bristles many, equalling corolla, scabrid, bases lightly fused. Fig. 47: 4.

From Umvoti district in Natal south to East London, eastern Cape. Grows in grassland, up to c. 1200 m; flowering between August and October, after the grass has been burnt off. Can be confused with *H. pilosellum* (no. 157) but easily distinguished by its very acute squarrose bracts. Map 160.

Vouchers: *Batten* s.n. (E; K; NU); *Flanagan* 349 (BOL; K; SAM); *Hilliard* 1624 (E; NU); *Pegler* 86 (BOL); *Rudatis* 1712 (PRE).

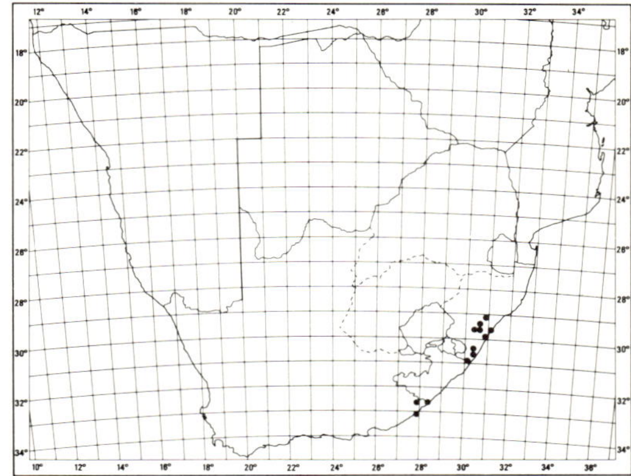

MAP 160.— **Helichrysum griseum**

161. **Helichrysum thapsus** *(O. Kuntze) Moeser* in Bot. Jb. 44: 263 (1910); Compton, Fl. Swaziland 635 (1976); Hilliard, Compositae in Natal 173 (1977). Type: Natal, Lions River distr., Highlands station, *Kuntze* (NY, holo.; K, iso.!; PRE photo.!).

Gnaphalium thapsus O. Kuntze, Rev. Gen. Pl. 3,2: 154 (1898).

Helichrysum lanatomarginatum Markötter in Annale Univ. Stellenbosch 8 sect. A, 1: 47 (1930). Type: Bergville distr., Oliviershoek Pass, c. 5 500 ft, *Thode* s.n., STE 2796 (STE, holo.!).

Perennial herb, crown brown silky-woolly, stems one or several, simple, erect to c. 600 mm, thinly white-felted, leafy in lower half, bracteate above. *Leaves* up to c. 120 × 30 mm, elliptic, or elliptic-spathulate, apex acute, base narrowed, half-clasping, both surfaces thinly to thickly greyish-white woolly, sometimes glabrescent, the wool then clinging to the margins and the three prominent nerves. *Heads* homogamous, campanulate or broadly cylindric, c. 5 × 3 mm, up to c. 12 in tight clusters, these in turn arranged in an open corymbose panicle. *Involucral bracts* in c. 6 series, graded, closely imbricate, woolly, inner semipellucid, straw-coloured, or pale brown, plicate, equalling the flowers, not radiating. *Receptacle* with fimbrils about equalling ovary. *Flowers* c. 15−30. *Achenes* not seen, ovaries glabrous. *Pappus* bristles many, about equalling corolla, scabrid, bases cohering lightly by patent cilia. Fig. 47: 3.

Ranges from the highlands of the eastern Transvaal and Swaziland to the Natal Midlands and Umzimkulu district, Transkei. Grows in grassland; flowering between December and March. Rarely collected. Map 161.

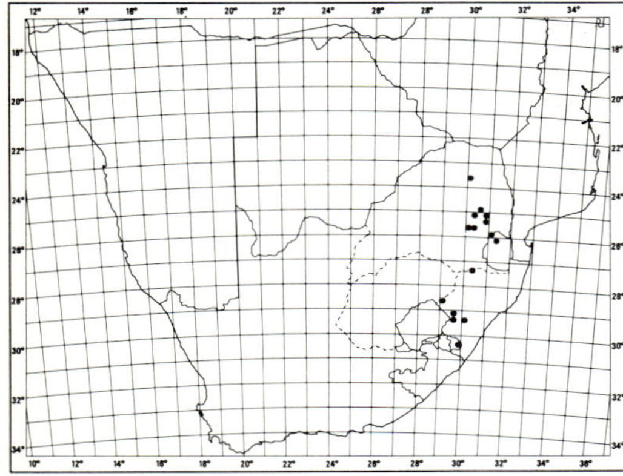

MAP 161.— **Helichrysum thapsus**

Can be confused with *H. acutatum* (no. 142) but that species has nearly smooth receptacles and hairy achenes, and with *H. pannosum* (below), which has similar receptacle and ovaries, but smooth, lemon-yellow, acute involucral bracts.

Vouchers: *Devenish* 1925 (E; K; NU); *Hilliard & Burtt* 14286 (E; K; NU; PRE; S); *Rogers* 18642 (K; Z); *Tyson* 1473 (G; K; Z); *Wilms* 733 (BM; E; G; K; NU; Z).

162 Helichrysum pannosum *DC.*,

Prodr. 6: 204 (1838); Harv. in F.C. 3: 248 (1865); Moeser in Bot. Jb. 44: 320 (1910); Hilliard, Compositae in Natal 174 (1977). Type: Cape, between the Umsikaba and Umzimvubu Rivers, *Drège* 5003 (G-DC, holo.!; BM; K; P; S; TCD, iso.!).

Gnaphalium pannosum (DC.) Sch. Bip. in Bot. Ztg 3: 172 (1845).

Perennial herb up to c. 1 m high, stem from a creeping woody stock, crown brown silky-woolly, sparingly branched low down, slender, woody, brittle, often zigzag, thinly greyish white-woolly, leafy throughout. *Leaves* up to 100 × 20 (−30) mm, smaller upwards, oblanceolate or oblong-elliptic, lanceolate upwards, apex more or less acute, base broad, cordate-clasping, both

surfaces greyish-white woolly, often drying fulvous, sometimes glabrescent, wool then clinging to the three prominent nerves. *Heads* homogamous, narrowly campanulate, c. 6 × 3 mm, many in tight corymbose clusters corymbosely arranged. *Involucral bracts* in c. 6 series, graded, closely imbricate, woolly, inner acute, glossy, lemon-yellow, about equalling flowers, not radiating. *Receptacle* with fimbrils about equalling ovaries. *Flowers* 8−15. *Achenes* 1 mm, faintly ribbed, glabrous. *Pappus* bristles many, about equalling corolla, scabrid, bases cohering by patent cilia.

Ranges from Stanger, Camperdown, Pinetown and Inanda districts in Natal south to Port St Johns in the Transkei, in grassland up to c. 650 m, often on hillslopes near forest patches. Flowers mainly between April and June. Map 162.

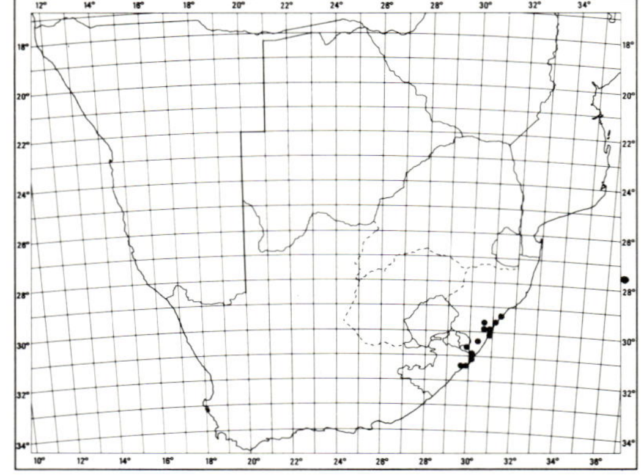

MAP 162.— **Helichrysum pannosum**

Vouchers: *Hilliard* 1337 (E; NU); *Hilliard & Burtt* 10250 (E; K; NU; S); *Rudatis* 957 (K; PRE); *Strey* 8544 (NU; PRE).

163. Helichrysum coriaceum *Harv.* in

F.C. 3: 239 (1865), non (DC.) Harv., l.c. 230; Moeser in Bot. Jb. 44: 264 (1910); Hilliard, Compositae in Natal 175 (1977). Lectotype: Transvaal, Magaliesberg, *Zeyher* 880 (K!; BM; S, isolecto.!).

Gnaphalium plantaginifoliatum O. Kuntze, Rev. Gen. Pl. 3: 153 (1898). Type: Cape, Cathcart, *O. Kuntze* (Z!).

Helichrysum plantaginifolium [O. Hoffm. ex] Moeser in Bot. Jb. 44: 264 (1910), non C.H. Wr. (1901).

Perennial herb up to c. 750 mm tall, crown woody, brown silky-woolly, stems often several tufted together, simple, thinly white-felted, leafy particularly in lower part. *Leaves* up to 270 × 25 mm, half the length petiolar in the radical leaves, diminishing upwards and becoming sessile, blade coriaceous, elliptic, radical leaves tapering at both ends, petiolar region relatively broad, base expanded, clasping, cauline leaves broad-based, clasping, sometimes shortly decurrent, upper surface at first with a thin skin-like indumentum, soon glabrous, lower white-felted or silky, 3–5 prominent main veins, often reticulate. *Heads* homogamous, campanulate, c. 5–7 × 4–6 mm, many in small corymbose clusters corymbosely arranged. *Involucral bracts* in c. 6 series, graded, closely imbricate, woolly, inner about equalling flowers, pellucid, straw-coloured, not radiating. *Receptacle* with fimbrils as least equalling ovaries. *Flowers* 37–69. *Achenes* not seen, ovaries glabrous. *Pappus* bristles many, about equalling corolla, scabrid, bases cohering by patent cilia.

Widely distributed from the highlands of Kenya to the highlands of Zimbabwe, the Transvaal Highveld and E. highlands, Swaziland, NE. Orange Free State, Lesotho, northern Natal and Midlands (but uncommon there), Transkei and E. Cape to 27°E. Grows in grassland; flowering from November to January. Sometimes confused with forms of *H. nudifolium* (below) but distinguished by its brown silky-woolly crown and larger heads. Map 163.

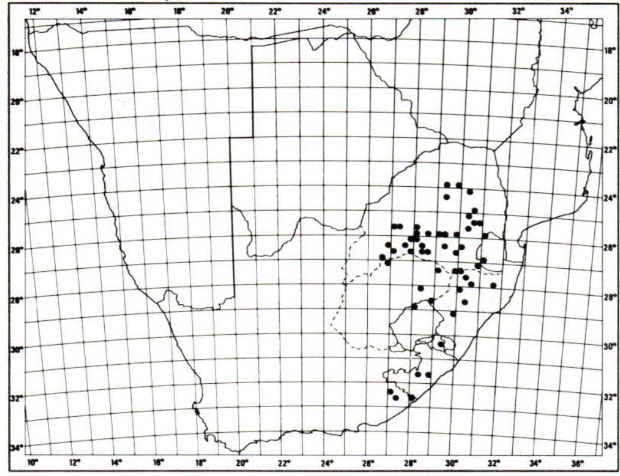

MAP 163.— **Helichrysum coriaceum**

Vouchers: *Codd* 2133 (NU; PRE); *Dieterlen* 573 (K; PRE; Z); *Hilliard* 2207 (E; NU); *Schlechter* 6322 (BOL; G; K; PRE; Z); *Wright* 800 (E; K; NU).

164. Helichrysum nudifolium *(L.) Less.*, Syn. Comp. 299 (1832); DC., Prodr. 6: 200 (1838); Harv. in F.C. 3: 240 (1865); Levyns in Adamson & Salter, Fl. Cape Penins. 783 (1950); Adams in F.W.T.A. 2, 2: 264 (1963); Compton, Fl. Swaziland 632 (1976); Hilliard, Compositae in Natal 176 (1977). Lectotype: Breynius, Cent. t. 71 (1674-1678).

Gnaphalium nudifolium L., Pl. Rar. Afr. 19 (1760), Sp. Pl. edn 2: 1196 (1763). *Anaxeton nudifolium* (L.) Gaertn., Fruct. 2: 407 (1791). *Lepiscline nudifolia* (L.) Cass., Dict. 26: 50 (1823). *Euchloris nudifolia* (L.) D. Don in Mem. Wern. nat. Hist. Soc. 5: 548 (1826). *Helichrysum nudifolium* var. *scorzoneraefolium* DC., Prodr. 6: 200 (1838).

G. plantagineum Burm. f., Prodr. 25 (1768), non L. (1767). *Helichrysum nudifolium* var. *plantagineum* DC., Prodr. 6: 200 (1838). Type: Specimen in herb. Burm. (G!).

G. quinquenerve Thunb., Prodr. 152 (1800), Fl. Cap. 658 (1823). *Helichrysum quinquenerve* (Thunb.) Less., Syn. Comp. 300 (1832); Harv. in F.C. 3: 240 (1865); *H. nudifolium* var. *quinquenerve* (Thunb.) Moeser in Bot. Jb. 44: 265 (1910). Lectotype: Cape of Good Hope, *Thunberg* (sheet 19241, right hand specimen, UPS!).

Helichrysum multinerve DC., Prodr. 6: 199 (1838), incl. vars. *Gnaphalium multinerve* (DC.) Sch. Bip. in Bot. Ztg 3: 172 (1845). Type: Cape, Van Staadens-riviersberg, *Drège* 4996 (G-DC, holo.!; var. *pallens* DC.).

H. nudifolium var. *medium* DC., Prodr. 6: 200 (1838). Lectotype: Cape, Camdeboo Berg, *Drège* 852 (G-DC!).

H. leiopodium DC., Prodr. 6: 200 (1838); Harv. in F.C. 3: 239 (1865). *Gnaphalium leiopodium* (DC.) Sch. Bip. in Bot. Ztg 3: 172 (1845). *Helichrysum nudifolium* var. *leiopodium* (DC.) Moeser in Bot. Jb. 44: 266 (1910); Compton, Fl. Swaziland 632 (1976); Hilliard, Compositae in Natal 178 (1977). Lectotype: Cape, between the Kei and Bashee Rivers, *Drège* 5742 (G-DC!).

H. nudifolium var. *obovatum* Harv. in F.C. 3: 240 (1865). Type: without precise locality, *Ecklon* s.n. (S!).

H. leiopodium var. *denudatum* Harv. in F.C. 3: 239 (1865). Lectotype: O.F.S., Thaba Nchu, *Zeyher* 881 (K!).

H. plantaginifolium C.H. Wr. in Kew Bull. 1901: 123 (1901); Moeser in Bot. Jb. 44: 267 (1910). Type: Malawi, Namasi, *Cameron* 6.

H. asperifolium Moeser in Bot. Jb. 48: 340 (1912). Type: Cape, Caledon div., Houw Hoek, 470 m, 30 iv 1896, *Schlechter* 7766 (BOL, iso.!; *Schlechter* sub TM 1228, PRE, is possibly another isotype).

1b
2a
3
1
1a
4
2

Perennial herb spreading by stout underground runners, flowering stems solitary, simple, up to c. 1, 5 m tall, leafy below, pedunculoid upwards, thinly woolly. *Radical leaves* up to 600 × 130 mm, but often only half that or less, linear-lanceolate, elliptic to ovate, more or less abruptly, or sometimes gradually, contracted to a narrow or broad and then winged, petiole-like base, apex acute, margins sometimes undulate, upper surface scabrid, lower scabrid, cobwebby or thinly to thickly white-woolly, (3−) 5 (−7)-nerved, nerves strongly raised below, linked by netted veins; *cauline leaves* similar, sessile, stem-clasping, lower ones particularly sometimes strongly decurrent, becoming smaller upwards and rapidly passing into distant brachts. *Heads* homogamous, narrowly campanulate, c. 4−5 × 2,5− 3 mm, very many in a large corymbose panicle. *Involucral bracts* in c. 6 series, closely imbricate, woolly at base, obtuse, pellucid or subopaque, pale or lemon-yellow, outer sometimes palest brown, equalling the flowers, not radiating. *Receptacle* with fimbrils at least equalling ovaries. *Flowers* 16−32. *Achenes* 1 mm long, glabrous. *Pappus* bristles many, equalling corolla, scabrid, bases lightly fused and also cohering by patent cilia. Fig. 48: 1.

Widely distributed from W. Africa to Yemen and Ethiopia, thence south through Kenya, Uganda, Tanzania, Malawi, Zimbabwe, Mozambique and the eastern half of S. Africa to the Cape Peninsula. Also in the Angolan highlands. Common in grassland, flowering mainly between November and March. Map 164.

Very variable in leaf shape, width and indumentum, which shows no clear geographical partitioning. In the Natal Drakensberg for instance, specimens with narrow glabrous leaves and wide-spreading inflorescences are common in grassland and stabilized boulder beds, while on moist rocky grass slopes, plants may have leaves white-tomentose below and the radical ones more distinctly petiolate than those of the glabrous-leaved plants and the inflorescence may be more contracted. These are possibly local ecological races: a thorough field study of the whole complex is needed, a formidable task.

Narrow-leaved specimens with only 3 main veins in the radical leaves are easily confused with *H. miconiifolium* (no. 166), but can usually be distinguished by the stems being only distantly bracteate

above. However, some specimens may be difficult to place, and it is not improbable that crossing takes place.

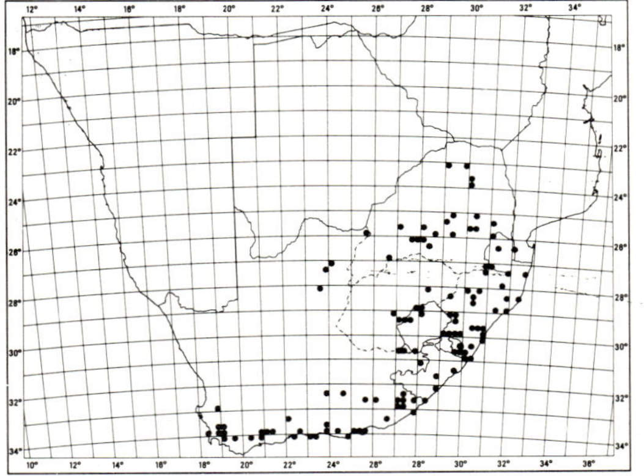

MAP 164.— **Helichrysum nudifolium**

Vouchers: *Bolus* 1179 (BOL; K; SAM); *Codd* 2406 (NU; PRE); *Hilliard & Burtt* 6658 (E; K; MO; NU); *Strey* 9197 (NU); *Wright* 345 (E; K; M; NH; NU).

165. **Helichrysum harveyanum** *Wild* in Notes R. bot. Gdn Edinb. 40:255 (1982). Lectotype: Transvaal, Mooye River, *Burke* 411 (K; TCD, isolecto.!).

H. subulifolium Harv. in F.C. 3: 241 (1865), non F. Muell. (1863); Moeser in Bot. Jb. 44: 267 (1910). Type as for *H. harveyanum*.

Perennial herb, roots with slender tubers, rootstock slender, woody, flowering stem solitary, simple, erect to 300−750 mm, glandular, loosely cobwebby woolly, closely leafy but more distantly so above. *Radical leaves* often wanting, when present c. 100−300 × 5−10 mm, roughly ⅓ wiry petiole, blade linear-lanceolate, triplinerved, margins weakly revolute, both surfaces scabrid, hairs often only near margins and above veins, often cobwebby-woolly below; *cauline leaves* mostly 40−170 × 1−5 mm, smaller and more distant upwards and passing into bracts, linear-lanceolate or linear, acuminate, base broad,

---

FIG. 48.−1, **Helichrysum nudifolium**, whole plant, × 0,7; 1a, leaf to show 5 main veins, × 1; 1b, head, × 6,6 (*Gordon Gray* 1309). 2, **H. harveyanum**, whole plant, × 0,7; 2a, head, × 6,6 (*Huntley* 1106). 3, **H. krebsianum**, head, × 6,6 (*Hilliard* 2509). 4, **H. miconiifolium**, head, × 6,6 (*Hilliard & Burtt* 12233).

clasping, shortly decurrent, margins strongly revolute, upper surface and midrib below scabrid, often some woolly hairs as well. *Heads* homogamous, campanulate, 4−5 × 3−4 mm, many in congested corymbose clusters arranged in a spreading corymbose panicle. *Involucral bracts* in c. 5 series, graded, loosely imbricate, about equalling flowers, not radiating, tawny, pellucid or subopaque, tips obtuse, or rarely subacute, somewhat crisped. *Receptacle* with fimbrils much exceeding ovaries. *Flowers* 17−30. *Achenes* 1,5 mm long, glabrous. *Pappus* bristles many, about equalling corolla, scabrid, bases cohering strongly by patent cilia, lightly fused as well. Fig. 48: 2.

Recorded from the Bushveld in the Transvaal and Botswana, from about Zebediela and the Loskop Dam area in the east to Sterkfontein and Lobatsi in the west. Also in Zambia fide Wild (pers. comm.). Flowers mainly between October and January. Map 165.

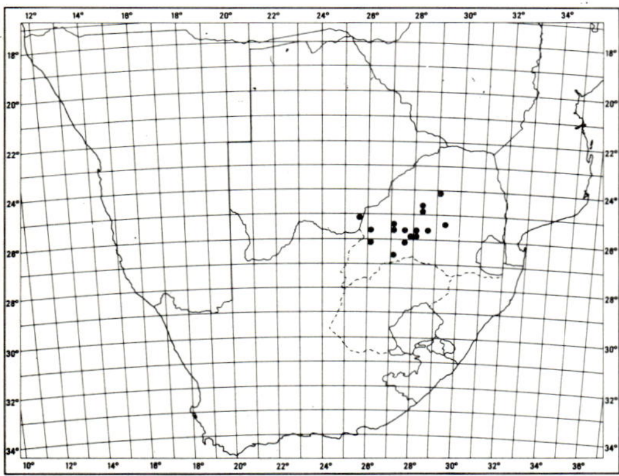

MAP 165.— **Helichrysum harveyanum**

Closely allied to *H. miconiifolium* (below), from which it is most easily distinguished by its tawny and generally obtuse involucral bracts. The two species occupy different areas.

Vouchers: *Acocks* 12406 (K; PRE); *Codd* 2269 (K; NU; PRE); *Repton* 2779 (PRE); *Rogers* 18969 (PRE).

166. **Helichrysum miconiifolium** *DC.*, Prodr. 6: 200 (1838); Harv. in F.C. 3: 240 (1865); Moeser in Bot. Jb. 44: 267 (1910); Compton, Fl. Swaziland 631 (1976); Hilliard, Compositae in Natal 178 (1977). Type: Cape, Wittebergen, *Drège* 3747 (G-DC, holo.!; BM, iso.!).

Gnaphalium miconiifolium (DC.) Sch. Bip. in Bot. Ztg 3: 172 (1845).

H. miconiifolium var. *minus* DC., Prodr. 6: 200 (1838). Type: Albany and Grahamstown, *Ecklon* 1582 (G-DC, holo.!).

Perennial herb, rootstock stout, woody, crowned with fibrous leaf bases, roots producing narrow tubers, flowering stem usually solitary, up to 600 mm tall, thinly woolly, leafy but more distantly so above. *Radical leaves* up to 300 × 45 mm, blade narrowly to broadly elliptic, triplinerved, abruptly contracted to a thin wiry petiole half to one third the total leaf length, expanded and clasping below, apex more or less acute, upper surface scabridulous or occasionally smooth, lower white-felted, sometimes glabrescent, margins weakly revolute; *cauline leaves* similar to radical but soon sessile, linear-lanceolate, acuminate, margins strongly revolute, passing upwards into bracts. *Heads* homogamous, campanulate, c. 5 × 5 mm, many in a loose or compact, flat-topped, corymbose panicle. *Involucral bracts* in c. 6 series, graded, loosely imbricate, about equalling flowers, base woolly, limb glossy, ovate, acute to acuminate, opaque or subopaque, lemon-yellow often washed light golden-brown, not radiating. *Receptacle* with fimbrils at least equalling ovaries. *Flowers* 12−23. *Achenes* 1 mm long, glabrous. *Pappus* bristles many, scabrid, bases cohering strongly by patent cilia, lightly fused as well. Fig. 48: 4.

Ranges from the E. highlands of the Transvaal, the Magaliesberg, Witwatersrand and Suikerbosrand, and Swaziland to the mountainous NE. part of the Orange Free State, Lesotho, Natal, Transkei, Cape Drakensberg and Witteberg, and E. Cape about as far south as Grahamstown and west to the mountains about Graaff-Reinet. Grows in grassland from near sea level to c. 2 000 m; flowering mainly between November and February.

Can be confused with some forms of *H. nudifolium* (no. 164) but is distinguished by its stems more closely bracteate near the summit, its triplinerved leaves, and loosely imbricate acute involucral bracts. See also *H. harveyanum* (above). Map 166.

Vouchers: *Codd* 2582 (NU; PRE); *Compton* 26471 (NBG); *Hilliard & Burtt* 7790 (E; K; MO; NU; S); *Tyson* 1497 (BOL; SAM); *Wright* 346 (E; K; NH; NU; S).

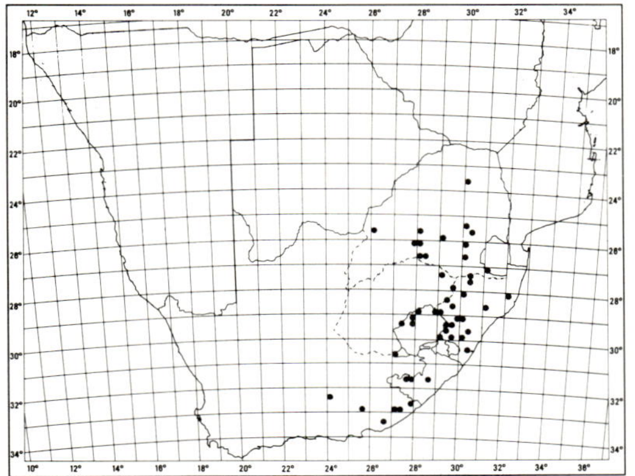

MAP 166.— **Helichrysum miconiifolium**

167 **Helichrysum mollifolium** *Hilliard* in Notes R. bot. Gdn Edinb. 40: 261 (1982). Type: Natal, Underberg, distr., Cobham Forest Station, Upper Polela Cave area, c. 2 350 m, 16 ii 1979, *Hilliard & Burtt* 12611 (NU, holo.!; E; K; PRE; S, iso.!).

Perennial herb, roots with narrow fusiform tubers, stock small, producing slender rhizomes, flowering stem terminal, solitary, simple, 350—600 mm tall, with long multicellular hairs and thinly white-woolly as well, reduced leaves near base, distantly bracteate upwards. *Leaves* mostly radical, c. 4—6, petiole thin, wiry, c. 20—120 mm long, blade c. 30—120 × 25—50 mm, broadly elliptic to ovate-elliptic, soft-textured, apex obtuse to subacute, apiculate, base rounded or broadly cuneate, upper surface rough with long multicellular hairs, lower conspicuously 5-nerved, hairs best developed on nerves, elsewhere with shining yellow sessile glands, margins weakly revolute; *cauline leaves* similar but soon sessile and narrower, passing rapidly into few, distant acuminate bracts. *Heads* homogamous, campanulate, c. 5—6 × 4—5 mm, many in a loose or compact corymbose panicle. *Involucral bracts* in c. 4 series, slightly graded, loosely imbricate, about equalling flowers, bases loosely woolly, limb subopaque, slightly glossy, outer golden-brown, inner yellow, tips obtuse, slightly crisped, not radiating. *Receptacle* with fimbrils twice as long as ovaries. *Flowers* 19—30. *Achenes* not

seen, ovaries glabrous. *Pappus* bristles many, scabrid, bases cohering strongly by patent cilia, lightly fused as well.

Known from Royal Natal National Park, Bergville district, then the southern Natal Drakensberg, from the great ridge running SE. from Giant's Castle south to Garden Castle Forest Reserve, at c. 2 100— 2 450 m. Grows in moist grassy depressions running down the mountain slopes, along flats near streams, or on the damp margins of forest patches, flowering from November to January. Distinguished from *H. miconiifolium* (no. 166) by its flowering stem only remotely bracteate above, soft, 5-nerved leaves and blunt involucral bracts, and from *H. nudifolium* (no. 164) by its soft leaves crowded at the base of the stem and more golden and brown involucral bracts. Map 167.

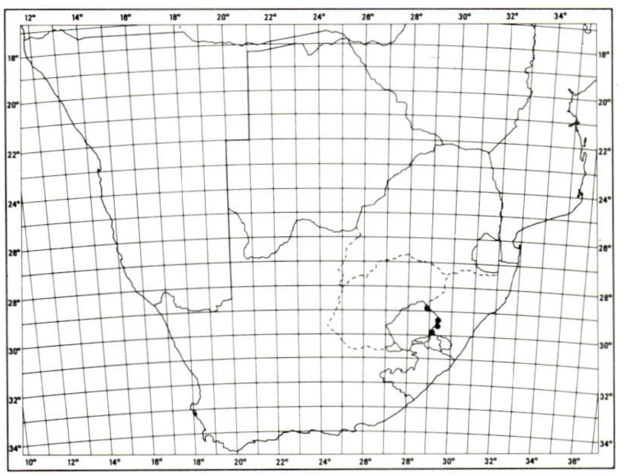

MAP 167.— **Helichrysum mollifolium**

Vouchers: *Hilliard & Burtt* 12555 (E; K; NU); *Wright* 407 (E; NU).

168. **Helichrysum krebsianum** *Less.*, Syn. Comp. 308 (1832); DC., Prodr. 6: 204 (1838); Harv. in F.C. 3: 241 (1865); Moeser in Bot. Jb. 44: 267 (1910); Hilliard, Compositae in Natal 179 (1977). Type: Cape, *Krebs* 154 (B†; fragment S!).

Gnaphalium krebsianum (Less.) Sch. Bip. in Bot. Ztg 3: 172 (1845).

Helichrysum crassinerve DC., Prodr. 6: 199 (1838). *G. crassinerve* (DC.) Sch. Bip. in Bot. Ztg 3: 171 (1845). Type: Kaffraria, Katberg, *Ecklon* 1891 (G-DC, holo.!).

Perennial herb up to 600 mm tall, roots narrowly fusiform, stock stout, woody, stem solitary, simple, thinly white cottony, leafy at the base becoming bracteate upwards. *Radical leaves* up to 200 × 50 mm, including the c. 70−100 mm petiole, blade elliptic, apex and base tapering, strongly triplinerved, petiole thin, wiry, base expanded, clasping, upper leaf surface harshly pubescent, lower thinly woolly; *cauline leaves* crowded, soon linear-lanceolate, acuminate, base broad, clasping, margins strongly revolute, harshly pubescent above, white-cobwebby below. *Heads* homogamous, subglobose, c. 5 × 5 mm, many in a spreading flat-topped corymbose panicle. *Involucral bracts* in 5−6 series, graded, loosely imbricate, convex, tips rounded, usually opaque creamy white, sometimes pale yellow, equalling the flowers, not radiating. *Receptacle* with fimbrils at least equalling the ovaries. *Flowers* 19−29. *Achenes* 0,75 mm long, glabrous. *Pappus* bristles many, about equalling the corolla, scabrid, bases lightly fused, also cohering by patent cilia. Fig. 48: 3.

Ranges from the Midlands of Natal and Zululand south to the Amatola Mountains and Katberg in the E. Cape, in grassland from near sea level to c. 2 000 m; flowering mainly between October and February. Easily recognized by its globose heads with smooth convex involucral bracts. Map 168.

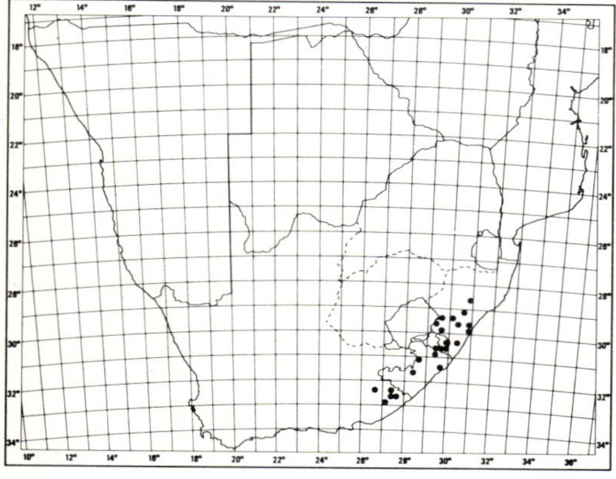

MAP 168.— ● **Helichrysum krebsianum**
○ **Helichrysum calocephalum**

Vouchers: *Beverly & Hoener* 683 (NU); *Hilliard* 4678 (E; NH; NU); *Schlechter* 6527 (K); *Tyson* 741 (BOL; K; PRE); *Wright* 370 (E; NH; NU).

169. **Helichrysum calocephalum** *Klatt* in Bull. Herb. Boissier 4: 834 (1896); Moeser in Bot. Jb. 44: 267 (1910). Type: E. Transvaal, near Barberton, Saddleback Mtn, *Galpin* 608 (Z, holo.!; BOL; K; PRE, iso.!).

Perennial herb c. 450 mm tall, stock crowned with old fibrous leaf bases, flowering stem solitary, simple, thinly woolly, leafy near base, distantly bracteate upwards. *Radical leaves* up to 100 × 10 mm, a third to half the length petiolar, blade elliptic, tapering above and below, strongly triplinerved, petiole thin, wiry, base expanded, clasping, both leaf surfaces harshly pubescent; *cauline leaves* linear-lanceolate, apex acute to acuminate, base broad, clasping, margins strongly revolute, harshly pubescent, *Heads* homogamous, subglobose, c. 5 × 5 mm, many in a spreading flat-topped corymbose panicle. *Involucral bracts* in c. 5 series, graded, loosely imbricate, somewhat convex, tips rounded, crisped, erose, opaque white or whole head suffused pink, equalling flowers, not radiating. *Receptacle* with fimbrils exceeding ovaries. *Flowers* c. 22−29, yellow or pink. *Achenes* not seen, ovaries glabrous. *Pappus* bristles many, equalling corolla, scabrid, bases lightly fused, also cohering by patent cilia.

Recorded only from the mountains around Barberton, Transvaal; flowering in September and October. Very closely allied to *H. krebsianum* (above) but distinguished by its flowering stems only distantly bracteate upwards and by its crisped involucral bracts. Map 168.

Vouchers: *Bolus* 7808 (BOL; PRE); *Galpin* 445 (PRE).

170. **Helichrysum allioides** *Less.*, Syn. Comp. 299 (1832); DC., Prodr. 6: 198 (1838); Harv. in F.C. 3: 239 (1865); Hilliard, Compositae in Natal 175 (1977). Type: Cape, *Krebs* 145 (G, iso.!).

Gnaphalium allioides (Less.) Sch. Bip. in Bot. Ztg 3: 172 (1845).

Helichrysum allioides var. *latiusculum* DC., Prodr. 6: 198 (1838). Type: Cape, Berg am Zwart Key, *Drège* 5743 (G-DC, holo.!).

Perennial herb, stock woody, crowned with thin wiry leaf bases, roots producing thin tubers, flowering stems usually solitary, up to 650 mm tall, leafy below becoming pedunculoid upwards, thinly woolly. *Leaves* mostly radical, up to 600 × 20 mm, linear-lanceolate, apex acuminate, base gradually narrowed to a thin wiry petiole accounting for up to half the leaf length, margins often undulate, upper surface thinly woolly, glabrescent, then smooth and coriaceous, lower surface woolly-felted, triplinerved, tertiary veins invisible, *cauline leaves* similar but soon sessile, cordate-clasping, rapidly passing into distant bracts. *Heads* homogamous, campanulate, c. 6 × 4—6 mm, many in a compact, flat-topped corymb. *Involucral bracts* in c. 4 series, closely imbricate, bases woolly, tips ovate, subobtuse, outer often purplish or reddish, inner pale yellow or sometimes whitish, equalling the disc, scarcely radiating. *Receptacle* with fimbrils equalling the ovary. *Flowers* 20—30 (—43). *Achenes* c. 1 mm long, glabrous. *Pappus* bristles many, scabrid, bases cohering strongly by patent cilia.

Ranges from the Barberton mountains in the SE. Transvaal south through Natal and the Transkei to the mountains about Queenstown and Molteno and the Amatola Mountains. Grows socially in open grassland from near sea level to c. 1 250 m; flowering from August to December. Map. 169.

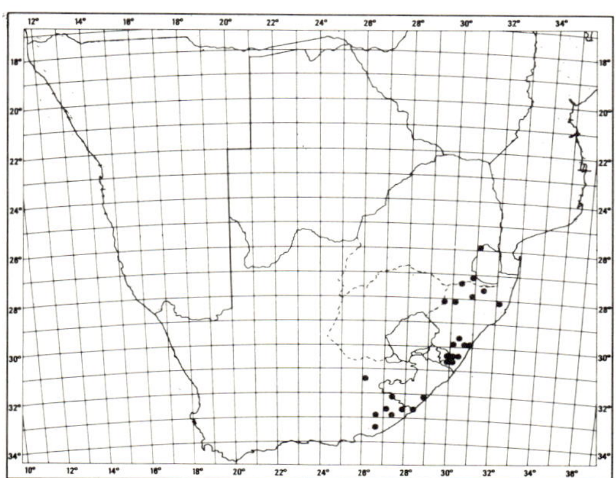

MAP 169.— **Helichrysum allioides**

H. allioides is characterized by its long narrow leaves, smooth above with traces of white wool often clinging to the veins, white-felted below. The tips of the inner involucral bracts are nearly always yellow, rarely creamy white. The rare specimens with broader leaves (but retaining the characteristic indumentum) have

been recorded from Hluhluwe Game Reserve and Nfabanana in Zululand, and approach *H. oxyphyllum* (below) both in this character and in the creamier-coloured involucral bracts. Some crossing may have taken place.

Vouchers: *Bolus* 7811 (BOL); *Devenish* 1213 (PRE); *Esterhuysen* 29156 (BOL); *Pegler* 63 (PRE); *Strey* 7665 (K; NU; PRE); *Tyson* 1456 (BOL; SAM).

171. **Helichrysum oxyphyllum** *DC.,*

Prodr. 6: 199 (1838); Harv. in F.C. 3: 241 (1865); Hilliard & Burtt in Notes R. bot. Gdn Edinb. 32,3: 353 (1973); Hilliard, Compositae in Natal 179 (1977). Type: Cape, Gouritz River, *Burchell* 4733 (G-DC, holo.!; K, iso.!).

Gnaphalium crispum L.f., Suppl. 363 (1781), non L. (1767). *G. elatum* Lam., Encycl. 2: 742 (1788), non *Helichrysum elatum* DC. (1838). *G. undatum* J. F. Gmel., Syst. Veg. 2: 1213 (1792), nom. illegit. *Helichrysum undatum* Less., Syn. Comp. 298 (1832), nom. illegit. Type: Cape of Good Hope, *Thunberg* s.n.

Helichrysum undatum var. *elongatum* DC., Prodr. 6: 199 (1838). Type: Cape, Zuurberg, *Drège* 5740 (G-DC, holo.!).

H. allioides var. *flavo-rubens* DC., Prodr. 6: 198 (1838). Type: Cape, Uitenhage, Zuurberg, *Ecklon* 1878 (G-DC, holo.!).

H. amoenum Moeser in Bot. Jb. 44: 268 (1910). Type: Cape, Langeberg near Zuurbraak, 1 500 ft, 18 i 1893, *Schlechter* 2116 (Z, holo.!; BM; J, iso.!).

Perennial herb, stock woody, flowering stem usually solitary, simple, up to c. 600 mm tall, leafy below, pedunculoid upwards, thinly woolly. *Radical leaves* up to 350 × 40 mm, about half the length petiolar, blade elliptic to ovate, abruptly or gradually narrowed to the slender petiole, apex acute, margins often undulate, upper surface scabridulous, lower white-felted, 3-, rarely 5-nerved, tertiary nerves not or faintly visible; *cauline leaves* similar to radical, but soon sessile and cordate-clasping, passing into distant acuminate bracts. *Heads* homogamous, campanulate, c. 6—7 mm long and as broad, many in a compact, flat-topped corymbose panicle. *Involucral bracts* in c. 4 series, graded, closely imbricate, woolly at the base, tips ovate, subobtuse, more or less opaque, outer purplish or reddish, inner tipped white or creamy, equalling the flowers, scarcely radiating. *Receptacle* with fimbrils at least equalling the ovaries. *Flowers* 18—29. *Achenes* not seen, ovaries glabrous. *Pappus* bristles scabrid, bases lightly fused, also cohering by patent cilia.

Ranges from Zimbabwe, Malawi and Mozambique to the E. Transvaal and Swaziland Lowveld and Zululand, through the coastal and Midlands districts of Natal, Transkei and eastern Cape to Caledon in the W. Cape. Grows in grassland or, in the W. Cape, among restiads; flowering between August and January. Map 170.

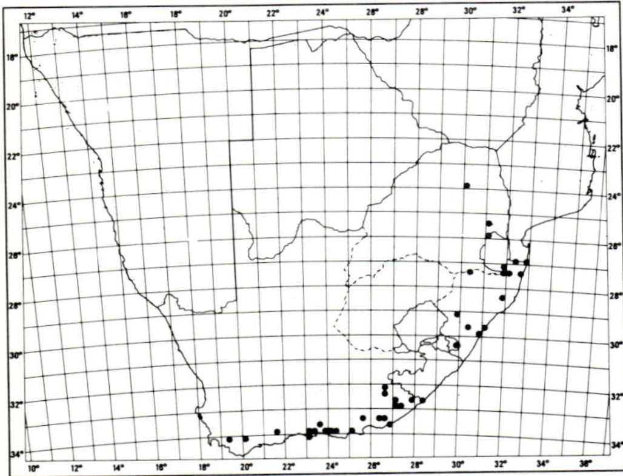

MAP 170.— **Helichrysum oxyphyllum**

Very closely allied to *H. pallidum* (below), but distinguished by its triplinerved (rarely 5-nerved) leaves with tertiary veins not or scarcely visible, and heads with graded involucral bracts, the outer ones tinged pink to purplish; *H. pallidum* has leaves with 5—7 main veins, tertiary veins clearly visible, and involucral bracts subequal and pale-coloured. Intermediate specimens occur; see note under *H. pallidum* (below). See also under *H. allioides* (above).

Vouchers: *Esterhuysen* 16923 (BOL; NBG); *Flanagan* 868 (PRE; SAM); *Galpin* 8363 (PRE); *Hutchinson* 1545 (BOL); *Pott* 5405 (E; NU); *Strey* 8126 (NU).

172. **Helichrysum pallidum** *DC.*,
Prodr. 6: 199 (1838); Hilliard & Burtt in Notes R. bot. Gdn Edinb. 32: 355 (1973); Compton, Fl. Swaziland 633 (1976); Hilliard, Compositae in Natal 180 (1977). Type: Cape, between the Umzimkulu and Umtentu Rivers, *Drège* 5008 (G-DC, holo.!; BM, iso.!; *H. pallidum* var. *ellipticum* DC.).

Gnaphalium pallens Sch. Bip. in Bot. Ztg 3: 171 (1845). *H. undatum* var. *pallidum* (DC.) Harv. in F.C. 3: 238 (1865).

H. pallidum DC. var. *intermedium* DC., Prodr. 6: 199 (1838). Lectotype: Cape, between the Kei and Buffalo Rivers, *Drège* 3746 (G-DC!; BM, isolecto.!).

H. pallidum DC. var. *longifolium* DC., Prodr. 6: 199 (1838). Lectotype: Cape, Kat River Mts, *Drège* 3746 (G-DC!).

H. agrostophilum Klatt in Bull. Herb. Boissier 4: 833 (1896) p.p. *H. undatum* var. *agrostophilum* (Klatt) Moeser in Bot. Jb. 44: 270 (1910). Lectotype: Transvaal, Barberton, Saddleback Mountain, *Galpin* 703 (Z!; BOL; K; NH, isolecto.!).

H. agrostophilum var. *nemorosum* H. Bol. in Trans. S. Afr. phil. Soc. 18: 393 (1907). Type: Transvaal, Pietersburg distr., Houtbosch, c. 1 500 m, Feb., *Bolus* 10991 (BOL, holo.!; K, iso.!).

Perennial herb, rootstock stout, woody, roots producing narrow tubers, flowering stem usually solitary, up to 650 mm tall, often much shorter, leafy in lower half becoming pedunculoid upwards, thinly woolly. *Radical leaves* up to 400 × 120 mm, up to ⅔ of the length petiolar, blade broadly to narrowly elliptic, gradually contracted to the long, flat petiole-like base, expanded and clasping below, apex acute to acuminate, margins often undulate, upper surface harshly pubescent, sometimes lightly cobwebby at first, lower thinly greyish-white felted, 5—7-nerved, tertiary nerves scalariform, raised or not, but clearly visible; *cauline leaves* similar, becoming sessile and cordate-clasping upwards and passing into distant bracts. *Heads* homogamous, campanulate, c. 6—7 mm long and as broad, many in a compact, flat-topped corymb. *Involucral bracts* in c. 4 series, subequal, closely imbricate, equalling the flowers, backs woolly, tips ovate, subobtuse, creamy to pale yellow, sometimes washed palest brown, more or less opaque, scarcely radiating. *Receptacle* with fimbrils at least equalling the ovaries. *Flowers* 18—34. *Achenes* not seen, ovaries glabrous. *Pappus* bristles many, scabrid, bases lightly fused. Fig. 49: 2.

Ranges from the eastern highlands of the Transvaal and neighbouring Swaziland to Natal (sea level to c. 2 700 m), the NE. part of the Orange Free State and nearby parts of Lesotho, Transkei and E. Cape as far as East London and the Amatola Mountains, Katberg, Boschberg and the Suurberg N. of Port Elizabeth. Grows in grassland; flowering between August and February. Map 171.

At higher altitudes, the radical leaves of *H. pallidum* tend to be lanceolate-elliptic; at lower altitudes they tend to be ovate-elliptic, more like those of *H. oxyphyllum,* and some specimens are difficult to place. It may be significant that difficulty arises when the two species are sympatric; see note under *H. oxyphyllum* (no. 171).

Vouchers: *Hilliard* 1748 (NU); *Hilliard* 5015 (E; K; NH; NU); *MacOwan* 1681 (SAM); *Wright* 402 (E; NH; NU).

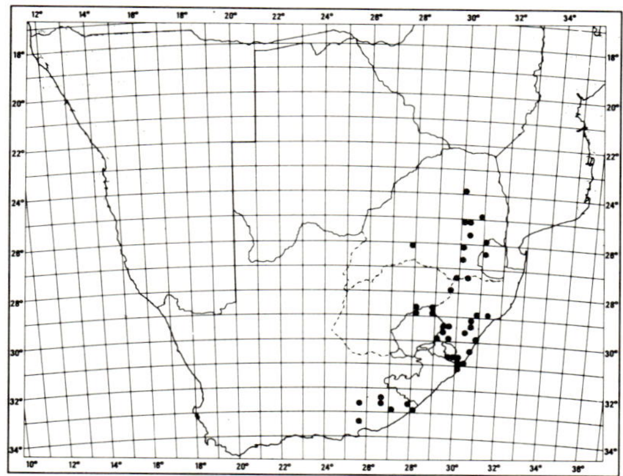

MAP 171.— **Helichrysum pallidum**

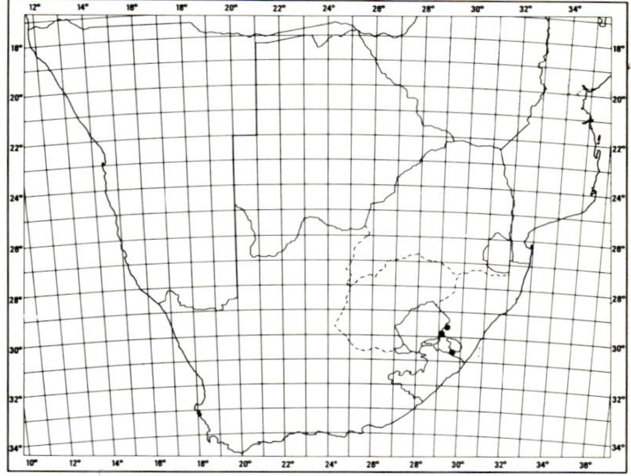

MAP 172.— **Helichrysum inornatum**

173. **Helichrysum inornatum** *Hilliard & Burtt* in Notes R. bot. Gdn Edinb. 34: 81 (1975); Hilliard, Compositae in Natal 181 (1977), Type: Natal, Alfred distr., Ngeli Mtn, SE. end, c. 1 800 m, *Hilliard & Burtt* 5766 (NU, holo.!; E; K; S, iso.!).

Perennial herb, stock slender, woody, stem solitary, erect to c. 500 mm, simple, slender, white-felted. *Leaves* mostly radical, up to 150 × 10 mm, linear-lanceolate, thin, apex acuminate, base tapered, flat, petiole-like, upper surface scabridulous, often lightly cobwebbed, lower thinly greyish-white felted, triplinerved, side veins invisible, *cauline leaves* similar but sessile, broad-based, clasping, rapidly passing into distant bracts. *Heads* homogamous, narrowly campanulate, c. 4 × 2,5 mm, many in compact cymose clusters arranged in a dense flat-topped corymbose panicle 20—30 mm across. *Involucral bracts* in c. 4 series, graded, loosely imbricate, woolly at base, inner equalling flowers, obtuse, pellucid, light golden-brown, not radiating. *Receptacle* with fimbrils a little longer than the ovaries. *Flowers* 9—13. *Achenes* 0,75 mm long, glabrous. *Pappus* bristles many, scabrid, bases cohering by patent cilia, lightly fused as well. Fig. 49: 1.

Known only from Mpendhle, Underberg and Alfred districts in Natal, where it grows in damp or marshy grassland, often on steep slopes, 1 800 to 2 100 m; flowering in December and January. Map 172.

Vouchers: *Hilliard & Burtt* 7888 (E; K; NU; PRE; S); *Wright* 378 (E; K; NU).

174. **Helichrysum qathlambanum** *Hilliard* in Notes R. bot. Gdn Edinb. 40: 264 (1982). Type: Natal, Underberg distr., Garden Castle Forest Reserve, valley beyond Forester's house, in scrub on steep stony slope, 1 950 m, 4 xi 1980, *Hilliard & Burtt* 13783 (NU, holo.!; E, iso.!).

Perennial herb, stock slender (c. 4 mm diam.), probably rhizomatous, flowering stem erect to c. 750 mm, solitary or paired, simple, glandular-hairy, thinly woolly as well, leafy in lower part, becoming pedunculoid upwards. *Radical leaves* few, 130—250 × 25—50 mm, roughly half the length petiolar, blade broadly elliptic, apex acute, base suddenly then gradually narrowed into the petiolar part, expanded and clasping below, both surfaces glandular-pilose, lower thinly white-woolly as well, 5-nerved; *cauline leaves* similar, rapidly becoming smaller, narrower and sessile, and passing into a few distant, lanceolate-acuminate bracts. *Heads* homogamous, cylindric-campanulate, 4 × 3 mm, many in crowded clusters webbed together with wool, forming a very compact corymbose compound head. *Involucral bracts* in c. 4 series, subequal, closely imbricate, equalling the flowers, not radiating, subpellucid, straw-coloured or pale golden-brown. *Receptacle* with large fimbrils exceeding the ovaries. *Flowers* c. 9—13. *Achenes* not seen, ovaries glabrous. *Pappus* bristles many, scabrid, bases cohering by patent cilia, sometimes light fusion as well. Fig. 49: 3.

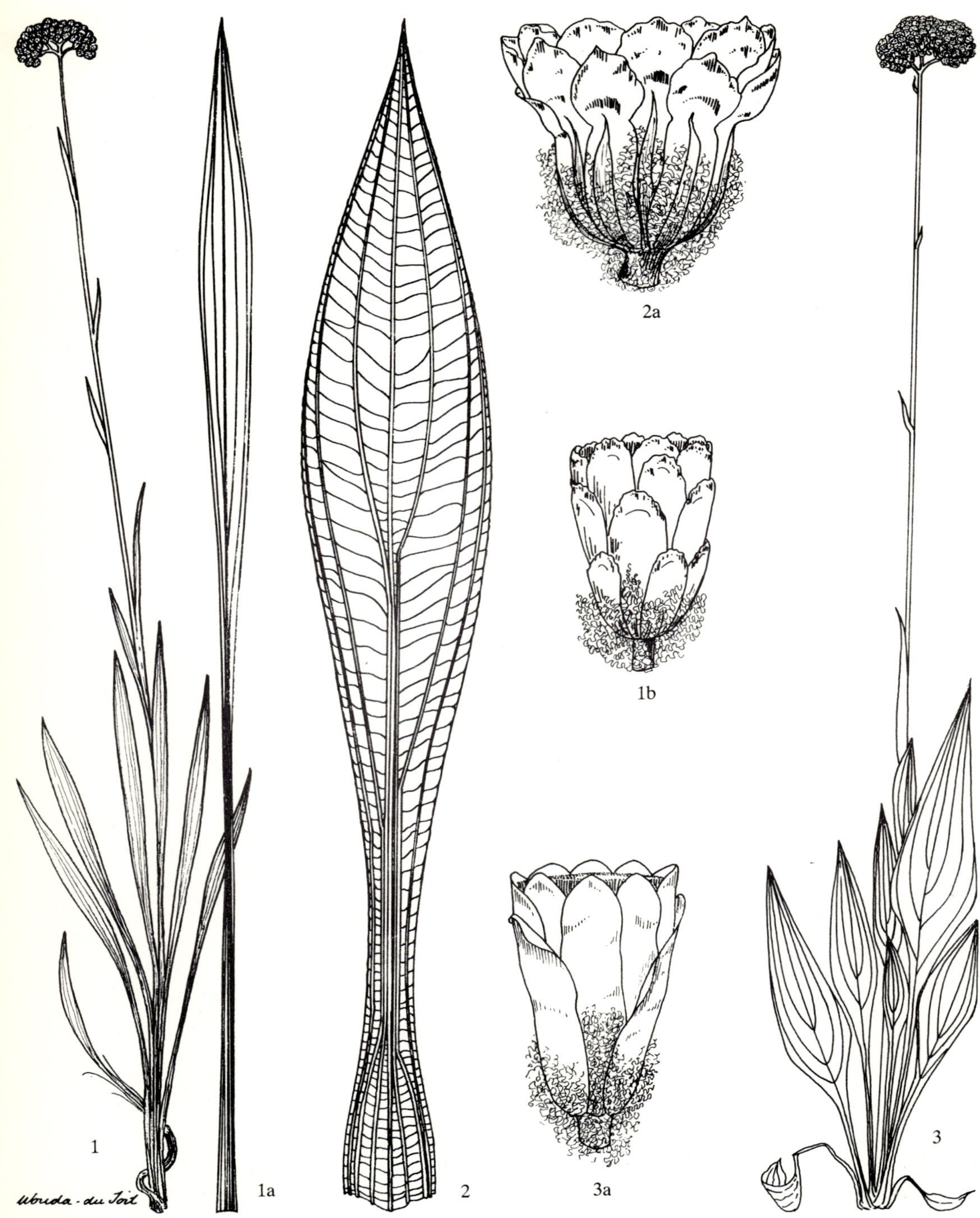

2a
1b
1
1a
2
3a
3
ubreda. du Toit

Recorded only from Butha Buthe district in Lesotho and Mpendhle and Underberg districts in Natal, between c. 1 800 and 3 000 m above sea level, on moist rocky grassy or scrubby slopes; flowering between November and February. Easily overlooked, despite its size, because it somewhat resembles the common *H. nudifolium* (no. 164); readily distinguished, however, by its soft-textured leaves and small heads with brownish involucral bracts. Map 173.

rugose, lower white-felted, prominently triplinerved, reticulate; *cauline leaves* oblong-lanceolate, much narrowed to the base, soon linear-lanceolate upwards, acute, base decurrent in long narrow stem wings; indumentum as in radical leaves. *Heads* homogamous, campanulate, c. 4 × 3,5 mm, many in cymose clusters arranged in a large, much branched corymbose panicle. *Involucral bracts* in 5−6 series, graded, closely imbricate, inner about equalling flowers, not radiating, opaque creamy-white. *Receptacle* with fimbrils equalling ovaries. *Flowers* 18−24. *Achenes* not seen, ovaries glabrous. *Pappus* bristles many, equalling corolla, scabrid, bases cohering strongly by patent cilia, some light fusion as well.

Widespread, from the highlands of Angola, Zambia, Tanzania and Zimbabwe to the Transvaal, Swaziland, Lesotho, NE. Orange Free State, Natal Midlands, W. Transkei, and the Cape as far west as Swellendam and Toise River. Grows in marshy places, often along streamsides or in seepage, forming dense stands because of its creeping rootstock. Flowers between February and April. Easily recognized by its winged stems, discolorous rugose leaves and large panicles of small creamy heads. Map 174.

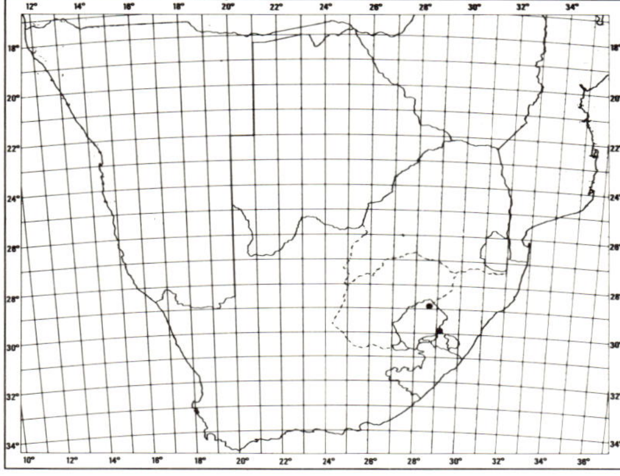

MAP 173.— **Helichrysum qathlambanum**

Vouchers: *Coetzee* 430 (NBG); *Lubke* 246 (PRE).

175. **Helichrysum mundtii** *Harv*. in F.C. 3: 243 (1865); Moeser in Bot. Jb. 44: 271 (1910); Compton, Fl. Swaziland 631 (1976); Hilliard, Compositae in Natal 182 (1977). Type: Cape, Swellendam, Wagenmaker's Bosch, *Mund* (TCD, holo.!; K, iso.!).

Robust perennial herb up to 1, 5 m tall, stock stout, woody, creeping, slender leafy runners as well, flowering stem simple, woody, up to 8 mm diam. at base, herbaceous, upper part thinly white-felted, leafy throughout. *Radical leaves* up to 600 mm long, half of this petiolar, blade elliptic, up to 60 mm broad, apex acute or subacute, base tapering into the long narrowly winged petiole, expanded below, clasping, upper surface glabrous, finely reticulate, often

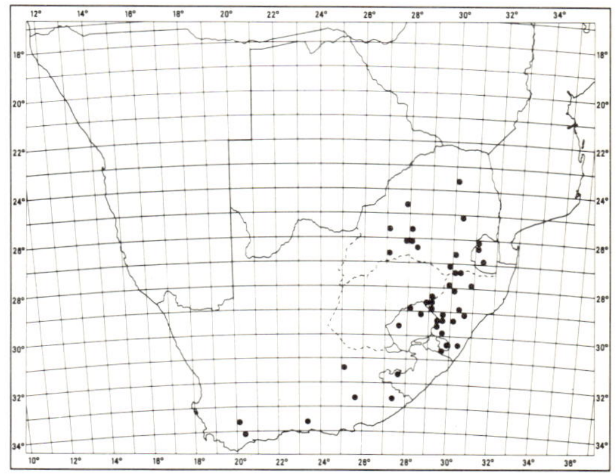

MAP 174.— **Helichrysum mundtii**

Vouchers: *Compton* 25887 (NBG); *Hilliard* 5459 (E; K; MO; NU; S); *MacOwan* 1472 (BOL); *Wilms* 728 (E; NU); *Tyson* 2761 (SAM).

FIG. 49.−1, **Helichrysum inornatum**, whole plant, × 0,5; 1a, leaf, × 1; 1b, head, × 10 (*Hilliard & Burtt* 5766). 2, **H. pallidum**, undersurface of leaf to show venation, × 1; 2a, head, × 5,3 (*Wright* 144b). 3, **H. qathlambanum**, whole plant, × 0,3; 3a, head, × 10 (*Lubke* 246).

Group 24

Perennial herbs with basal leaf rosettes; *leaves* medium-sized, oblong, elliptic, lanceolate, ovate or obovate; *heads* homogamous, c. 5—15 mm long, in lax or congested corymbose panicles; *involucral bracts* radiating, white, creamy or yellow, often with some light or dark brown, crimson or purple; *receptacle* fimbrilliferous; *flowers* 23—335, corolla narrowly funnel-shaped; *achenes* glabrous; *pappus* bristles scabrid, tips sometimes barbellate, bases cohering strongly by patent cilia.

Species 176—185, mostly confined to the grasslands of E. South Africa, two species reaching the E. highlands of Zimbabwe.

1a Involucral bracts pure white, or white tipped light or dark brown:

  2a Involucral bracts white, leaves markedly discolorous, green often drying black, white-felted below...177. *H. opacum*

  2b Involucral bracts white, tipped brown, leaves more or less concolorous, grey-woolly ....178. *H. albo-brunneum*

1b Involucral bracts creamy to pale or bright yellow, often with brown as well, or crimson or purple:

  3a Flowering stems leafy or bracteate throughout:

    4a Heads (6—) 7—14 mm long, involucral bracts dull, creamy to pale yellow, often overlaid brown, purplish or crimson, acute to acuminate, often more or less squarrose .... 176. *H. appendiculatum*

    4b Involucral bracts generally glossy, if dull or dullish, heads either smaller or bract colour and shape not as above (the species following are closely allied and not all specimens can be placed; see notes after the species descriptions):

      5a Heads in congested compound inflorescences, the peduncles often, though not always, webbed together with wool; involucral bracts subequal, all acute and then heads very tightly congested and webbed together at base with wool; or outer acute, inner obtuse or subobtuse; or slightly graded and then outer bracts acute, inner obtuse, webbed together with wool or not:

        6a Radical leaves acute to obtuse, persistently woolly on both surfaces, or sometimes lacking wool above but then leaf tips broad:

          7a Heads 4—6 mm long:

            8a Plants growing in well-drained grassland, flowering mainly from December onwards, radical leaves often woolly above, outer involucral bracts acute, inner broadly obtuse, apiculate...179. *H. cephaloideum*

            8b Plants growing in marshy grassland, flowering mainly in October and November, radical leaves lacking wool above, either all involucral bracts acute or inner subacute or narrowly subobtuse ....181. *H. subluteum*

          7b Heads 6—8 mm long, very tightly congested, all the bracts acute or the inner subacute or subobtuse ....180. *H. auriceps*

        6b Radical (and cauline) leaves linear-lanceolate, very acute, becoming discolorous through loss of wool on the upper surface...182. *H. ingomense*

      5b Heads in relatively loose inflorescences with all the individual peduncles visible, involucral bracts graded, all more or less acute or the inner sometimes narrowly obtuse:

        9a All the leaves woolly above and below...183. *H. mixtum*

        9b Radical leaves and usually the cauline as well discolorous at maturity, without wool and green above, drying brown, woolly below...184. *H. longifolium*

  3b Flowering stem pedunculoid in upper half...185. *H. xerochrysum*

176. Helichrysum appendiculatum (*L. f.*) *Less.*, Syn. Comp. 308 (1832); DC., Prodr. 6: 208 (1838); Harv. in F.C. 3: 242 (1865); Moeser in Bot. Jb. 44: 318 (1910); Compton, Fl. Swaziland 628 (1976); Hilliard, Compositae in Natal 183 (1977). Lectotype: Cape of Good Hope, *Thunberg* (sheet 19095, UPS!).

Gnaphalium appendiculatum L. f., Suppl. 363 (1781).

G. lupulaceum Lam., Encycl. 2: 752 (1788). Type: Cape of Good Hope (P-LAM!).

G. humile Thunb., Prodr. 151 (1800), Fl. Cap. 655 (1823). *Helichrysum appendiculatum* var. *humile* (Thunb.) DC., Prodr. 6: 209 (1838). Type: Cape of Good Hope, *Thunberg* (sheet 19177, UPS, holo.!).

Helichrysum folliculatum DC., Prodr. 6: 197 (1838). *Gnaphalium folliculatum* (DC.) Sch. Bip. in Bot. Ztg 3: 171 (1845). Lectotype: Cape, not far from the Kowie River, *Burchell* 4145 (G-DC!; K, isolecto.!).

H. folliculatum var. *purpurascens* DC., Prodr. 6: 197 (1838). Type: Cape, Winterberg, highest mountain between Tarka and Katberg, 5 000−6 000 ft, *Ecklon* 1320 (G-DC, holo.!).

H. discolor DC., Prodr. 6: 197 (1838). *Gnaphalium discolor* (DC.) Sch. Bip. in Bot. Ztg 3: 172 (1845). *H. appendiculatum* var. *discolor* (DC.) Harv. in F.C. 3: 242 (1865). Type: Natal, between the Umzimkulu River and Port Natal, *Drège* 5016 (G-DC, holo.!).

Perennial herb, stock thick, woody, flowering stems 1 or several from the crown, up to c. 550 mm tall, simple or occasionally forked, thinly greyish woolly, closely leafy throughout. *Radical leaves* up to 80 × 20 mm, spreading, elliptic-oblong, apex acute, base broad, clasping, both surfaces woolly; *cauline leaves* smaller, diminishing in size upwards, lanceolate-oblong to lanceolate, apex acute, uppermost tipped with a small scarious bract, base broad, half-clasping, very shortly decurrent, upper surface grey-woolly, at least initially, often glandular-hispid, or nearly glabrous with age, lower surface grey-woolly. *Heads* homogamous, campanulate, (6−) 7−14 mm long, many in corymbose clusters arranged in a compact or open corymbose panicle. *Involucral bracts* in c. 6 series, graded, loosely imbricate, inner equalling or exceeding flowers, tips radiating, acute to acuminate, often squarrose, dull, creamy white or yellowish often washed brown, purplish or crimson. *Receptacle* with fimbrils at least equalling the ovaries. *Flowers* 28−58, yellow. *Achenes* 1 mm long, barrel-shaped, glabrous. *Pappus* bristles many, about equalling corolla, scabrid, bases cohering strongly by patent cilia.

Ranges from the E. highlands of the Transvaal and western Swaziland to the NE. mountainous corner of the Orange Free State and nearby Leribe in Lesotho, Natal from near sea level to c. 2 100 m, Transkei, and the Cape mountains as far west as Swellendam. Grows in grassland, or in short fynbos in the southern Cape; flowering mainly between December and February. Map 175.

There is much variation in the degree of woolliness of the upper leaf surface and in the degree to which the tips of involucral bracts are produced and recurved. The typical plant has leaves woolly on both surfaces and acute involucral bracts that were probably pale yellow, the outermost reddish (the bracts have faded);

the types of *H. lupulaceum* and *H. folliculatum* are similar and the bracts are clearly pale yellow with some red outside, while that of *G. humile* has concolorous woolly leaves and acuminate crimson bracts. The type of *H. discolor* has discolorous leaves (wool lacking above) and yellow acuminate bracts. Variation is continuous and cannot be partitioned satisfactorily. However, *H. appendiculatum* is easily distinguished from its allies by its dull, not glossy, involucral bracts.

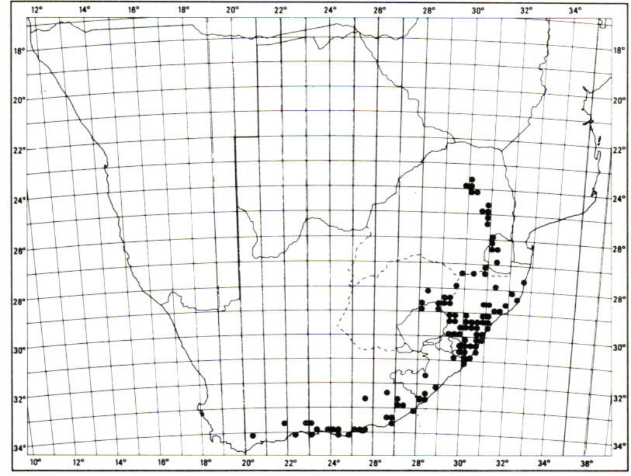

MAP 175.— **Helichrysum appendiculatum**

Vouchers: *Codd* 2794 (NU; PRE); *Compton* 24917 (NBG); *Devenish* 1629 (E; K; NU; S); *Hilliard & Burtt* 10982 (E; K; M; MO; NU; PRE; S); *MacOwan* 739 (BOL; SAM); *Tyson* 1501 (SAM); *Wright* 449 (E; K; M; NH; NU).

177. **Helichrysum opacum** *Klatt* in Bull. Herb. Boissier 4: 836 (1896); Moeser in Bot. Jb. 44: 319 (1910); Hilliard, Compositae in Natal 182 (1977). Type: Swaziland, Pigg's Peak, *Galpin* 1267 (Z, holo.!; BOL; K; NH; PRE, iso.!).

Perennial herb, roots fusiform, crown clad in loose brown wool, flowering stem up to 400 mm high, nearly always solitary, thinly greyish-white felted, remotely leafy. *Leaves* mostly radical, up to c. 150 × 12 mm, lanceolate, only slightly narrowed to the broad clasping base, markedly discolorous, green, often drying blackish, above with a few scattered hairs, white-felted below, striate from the strongly raised parallel veins, cauline leaves few, distant, much smaller than the radical, acuminate. *Heads* homogamous, campanulate, c. 5 mm long, double that across the fully radiating

1a
1c
1b
1
Woeda-du Toit
2a
3a
2
3

involucral bracts, many in a corymbose panicle. *Involucral bracts* in c. 8 series, inner subequal, loosely imbricate, much exceeding flowers, very acute, opaque, snow-white. *Receptacle* with fimbrils much exceeding ovaries. *Flowers* 23—28, yellow. *Achenes* 1 mm long, glabrous. *Pappus* bristles many, equalling corolla, scabrid, bases cohering strongly by patent cilia. Fig. 50: 1.

Ranges from Pilgrim's Rest and Dullstroom in the SE. Transvaal to Havelock, Pigg's Peak and Mbabane in Swaziland, thence to the low Drakensberg on the Transvaal-Natal-Orange Free State border, apparently with a disjunction to Mawahqua and Ngeli mountains in southernmost Natal. Grows on grassy mountain slopes; flowering in December and January. Map 176.

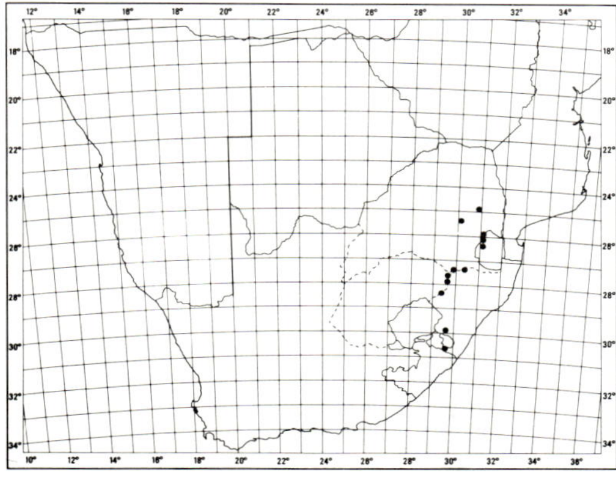

MAP 176.— **Helichrysum opacum**

Vouchers: *Galpin* 13042 (PRE); *Devenish* 1609 (E; K; NU; S); *Hilliard & Burtt* 9441 (E; K; MO; NU; PRE; S).

178. **Helichrysum albo-brunneum** S. *Moore* in Bot. J. Linn. Soc. 35: 334 (1903); Moeser in Bot. Jb. 44: 309 (1910); Hilliard, Compositae in Natal 229 (1977). Type: Cape, Koudeveld Mountains near Murraysburg, 6 500 ft, Jan. 1879, *Tyson* 98 (BM, holo.!; BOL; NH; SAM, iso.!).

H. setigerum H. Bol. in Trans. S. Afr. phil. Soc. 18: 390 (1907). Lectotype: Cape, Barkly East distr., summit Ben Mcdhui and Doodman's Krans Mtn, c. 2 900—3 000 m, *Galpin* 6676 (BOL!; NH, isolecto.!).

H. setigerum var. *minor* H. Bol. l.c. Type: Barkly East distr., summit Doodman's Krans Mtn, 2 940 m, *Galpin* 6686 (PRE!).

Perennial herb with several leaf rosettes crowning a slender woody rhizome, flowering stems simple, somewhat decumbent then erect to c. 150—300 mm, loosely greyish-white woolly, closely leafy. *Radical leaves* c. 30—80 × 10—35 mm, obovate or oblong-obovate, apex subacute, base slightly narrowed, clasping, both surfaces densely grey-woolly, *cauline leaves* similar, becoming oblong-lanceolate to linear-lanceolate and smaller upwards, appressed. *Heads* homogamous or occasionally heterogamous, broadly campanulate, c. 7—15 mm long, solitary or up to c. 14 crowded at the stem tip. *Involucral bracts* in c. 8—12 series, graded, loosely imbricate, much exceeding flowers, radiating, outer glossy white, tipped dark or light brown, becoming pure white inwards. *Receptacle* with fimbrils equalling or exceeding ovaries. *Flowers* 50—335, rarely 2—7 ♀, yellow sometimes tipped purplish. *Achenes* c. 1 mm long, glabrous. *Pappus* bristles many, equalling corolla, barbellate above, bases cohering by patent cilia. Fig. 50: 2.

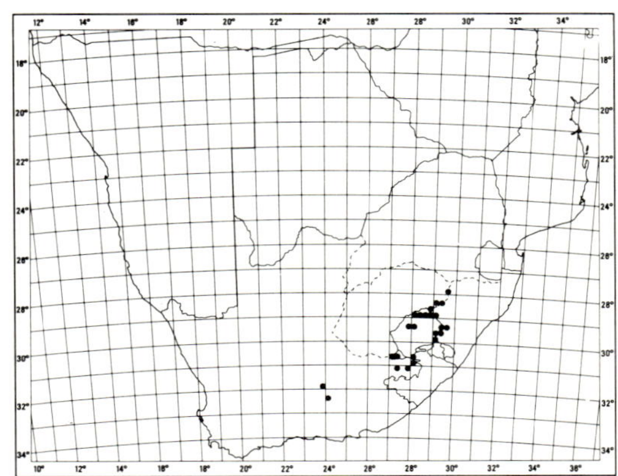

MAP 177.— **Helichrysum albo-brunneum**

FIG. 50.—1, **Helichrysum opacum**, whole plant, × 0,7; 1a, head, × 6,6; 1b, hermaphrodite flower, × 13; 1c, pappus bristle, × 13 (*Hilliard & Burtt* 9441). 2, **H. albo-brunneum**, whole plant, × 1 (*Hilliard* 2353); 2a, part of flowering stem with solitary head, × 1 (*Hilliard* 5389). 3, **H. ingomense**, whole plant, × 0,7; 3a, head, × 4,6 (*Hilliard & Burtt* 5925).

Ranges from Normandien Pass (across the low Drakensberg in Natal west of Newcastle) to the mountainous NE. corner of the Orange Free State, the mountainous parts of Lesotho, the Natal and Cape Drakensberg and the Witteberg and as far west as the Koudeveldberge near Graaff-Reinet and Murraysburg. Grows in colonies in short turf, often on stony or rocky slopes; flowering mainly between December and February. Map 177.

H. albo-brunneum displays considerable variation in size of the heads; the larger heads tend to be solitary or few together, while the smaller ones are clustered in groups of about 5—15.

Vouchers: *Acocks* 21779 (PRE); *Compton* 21270 (NBG; PRE); *Hilliard* 5389 (E; K; MO; NU; S); *Wright* 441 (E; K; NH; NU; S).

179. **Helichrysum cephaloideum** *DC.*, Prodr. 6: 197 (1838) p.p.; Harv. in F. C. 3: 242 (1865); Hilliard & Burtt in Notes R. bot. Gdn Edinb. 32: 345 (1973); Hilliard, Compositae in Natal 186 (1977). Lectotype: Cape, Albany, Zwartehoogde, Zuurberg range, *Ecklon* 1887 (G-DC!).

H. adscendens var. *cephaloideum* (DC.) Moeser in Bot. Jb. 44: 319 (1910). *Gnaphalium ecklonianum* Sch. Bip. in Bot. Ztg 3: 171 (1845). *G. hoffmannii* O. Kuntze, Rev. Gen. 3,2: 151 (1898).

H. cephaloideum var. *polycephalum* DC., Prodr. 6: 197 (1838). Type: Albany, Zwartehoogde, Zuurberg range, *Ecklon* 1862 (G-DC, holo.!).

H. campaneum S. Moore in J. Bot., Lond. 41: 399 (1903). Type: Transvaal, Roodepoort, *Rand* 1301 (BM, holo.!).

H. polyphyllum Conrath in Kew Bull. 1908: 225 (1908). Type: Transvaal, Modderfontein, *Conrath* 444 (K, holo.!).

H. adscendens sensu Moeser in Bot. Jb. 44: 319 (1910), non (Thunb.) Less.

H. infuscum Burtt Davy in Jl S. Afr. Bot. 1: 108 (1935). Type: Transvaal, Pietersburg distr., The Downs, *Rogers* 21959 (BOL and K, holo.!; PRE, iso.!).

Perennial herb, stock stout, woody, flowering stems several from the crown, simple or sometimes forked near base, erect to c. 400 mm, loosely grey-woolly, leafy throughout. *Radical leaves* rosetted, often withered at flowering, c. 20—50 × 10—20 mm, lanceolate-oblong or oblanceolate-oblong, apex acute, base broad, clasping, both surfaces loosely grey-woolly, upper also with coarse patent hairs, wool sometimes wanting; *cauline leaves* erect, imbricate, smaller than radical, decreasing in size upwards, oblong-lanceolate to lanceolate, acute to acuminate, uppermost tipped with a small scarious bract, base half-clasping, very shortly decurrent, both surfaces grey-woolly, upper sometimes glabrescent. *Heads* homogamous, campanulate, 4—5 (—6) mm long, several in a compact subglobose cluster up to c. 30 mm across, often webbed together at the base with wool. *Involucral bracts* in c. 6 series, subequal, loosely imbricate, exceeding flowers, radiating, glossy, bright yellow or deep straw-colour, outer often overlaid or tipped golden-brown or rarely purple-brown, outer acute, inner obtuse, apiculate. *Receptacle* with fimbrils exceeding ovaries. *Flowers* 30—70 (—103). *Achenes* 0,75 mm, barrel-shaped, glabrous. *Pappus* bristles many, equalling corolla, scabrid, tips barbellate, bases cohering by patent cilia. Fig. 51: 1.

Ranges from the E. highlands of Zimbabwe to the high parts of the Transvaal (Soutpansberg, E. highlands, Highveld, Magaliesberg), western Swaziland, Natal, NE. Orange Free State, Lesotho, Transkei and E. Cape as far west as the Amatola Mountains and Great Winterberg, and the Suurberg near Port Elizabeth. Common in grassland; flowering between November and May, but principally from December to March. Map 178.

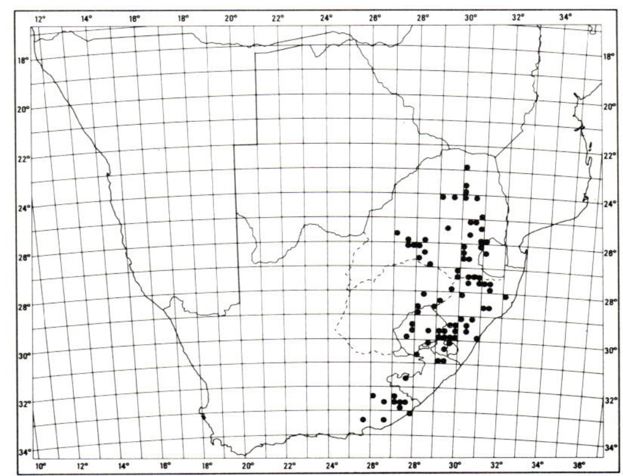

MAP 178.— **Helichrysum cephaloideum**

FIG. 51.—1, **Helichrysum cephaloideum**, whole plant, × 1; 1a, head, × 4 (*Devenish* 1604). 2, **H. subluteum**, whole plant, × 1; 2a, head, × 4 (*Hilliard & Burtt* 7249). 3, **H. auriceps**, whole plant, × 1; 3a, head, × 4 (*Meidner* 89).

Vouchers: *Devenish* 1604 (E; K; NU); *Flanagan* 2278 (PRE; SAM); *Hilliard & Burtt* 7755 (E; K; MO; NU; S); *Moss* 6881 (J; NU); *Story* 3467 (PRE).

Typical *H. cephaloideum* from the E. Cape is characterized by its congested compound heads, the individual heads 4–5,5 mm long with subequal involucral bracts, the outer acute, the inner very obtuse and apiculate, and containing c. 30–60 flowers. Similar plants occur over the rest of the range, but in Natal and the Transvaal heads are often larger (5–6 mm long) and may contain up to 100 flowers. In both provinces, the area of *H. cephaloideum* overlaps with that of its close ally *H. mixtum* (no. 183), and in Natal with *H. auriceps* (below) as well, and it is possible that some crossing takes place. However, the species seems also to be inherently variable, particularly in leaf size and woolliness and in the colour and imbrication of the involucral bracts. The bracts are often bright yellow overlaid or tipped golden-brown. Plants from the Transvaal with tawny or straw-coloured, closely imbricate, bracts were described as *H. polyphyllum* and *H. campaneum*; apart from the colour of the bracts, they closely resemble *H. cephaloideum* var. *polycephalum*, described from Grahamstown. The type of *H. infuscum* is very young, but it is perhaps no more than a specimen of *H. cephaloideum* in which the bracts are heavily overlaid with chestnut or purplish brown; plants with similarly coloured bracts occur sporadically in both Natal and the Transvaal, often together with ordinary yellow-headed plants. Two other Transvaal specimens need special mention: *Venter* 1163 (PRE), from the top of Mt Lebojane near The Downs, whence came the type of *H. infuscum*, and *Van der Merwe* 1315 (PRE) from Kliprivier north of Dullstroom, have straw-coloured involucral bracts tipped with purplish brown; the heads resemble those of *H. infuscum*, but are larger, 8 mm long, not 5 mm.

Specimens with remarkably large radical leaves (up to 200 × 40 mm) have been collected in the grassland adjoining Ngome and Nkandla forests in Natal (*Hilliard & Burtt* 5945, E; K; NH; NU; and *Hilliard* 2652, NU) and are possibly a local race.

H. cephaloideum is part of a very closely allied group that includes *H. subluteum* (no. 181) and *H. ingomense* (no. 182), both of which may have been derived from *H. cephaloideum*, as well as *H. auriceps* (below), *H. longifolium* (no. 184) and *H. mixtum* (no. 183). Further study, including experimental work, is desirable.

180. **Helichrysum auriceps** *Hilliard* in Notes R. bot. Gdn Edinb. 34: 258 (1976), Compositae in Natal 188 (1977). Lectotype: Natal, Richmond, *Schlechter* 6722 (BOL; PRE; Z!).

H. araneosum Klatt in Bull. Herb. Boissier 4: 834 (1896), non Bak. (1887). Types: Natal, between Howick and Pinetown, *Junod* 186 (Z!); Richmond, *Schlechter* 6722 (BOL!; PRE!; Z!).

Perennial herb, stock stout, woody, flowering stems several from the crown, simple or rarely forked, erect to c. 600 mm, loosely grey-woolly, leafy throughout.

Radical leaves rosetted, c. 65–150 × 8–16 (–27) mm, oblong-lanceolate, apex acute, base broad, clasping, both surfaces loosely grey-woolly, upper surface also with long coarse hairs, wool ocasionally wanting with age; *cauline leaves* erect, imbricate, smaller than the radical, decreasing in size upwards, oblong-lanceolate to lanceolate, acute to acuminate, base half-clasping, very shortly decurrent, both surfaces loosely grey-woolly. *Heads* homogamous, campanulate, (6–) 7–8 mm long, several to many in a tight subglobose cluster up to 45 mm across, base of compound head webbed with wool. *Involucral bracts* in c. 6 series, subequal, loosely imbricate, exceeding the flowers, radiating, glossy, bright yellow, outer sometimes tinged golden brown, all acute or inner sometimes subacute or subobtuse and this often associated with smaller (i.e. shorter) heads. *Receptacle* with fimbrils exceeding ovaries. *Flowers* 36–59. *Achenes* 1 mm long, barrel-shaped, glabrous. *Pappus* bristles many, about equalling the corolla, scabrid, bases cohering by patent cilia. Fig. 51: 3.

Recorded principally from Natal, where it ranges from Dundee southwards to the Transkei border and Bizana district in the Transkei. Grows in grassland, from sea level to c. 2 000 m; flowering between January and July but chiefly in February, March and April. Map 179.

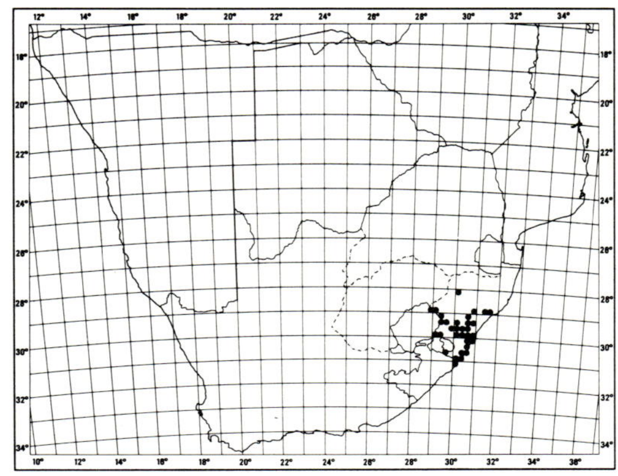

MAP 179.— **Helichrysum auriceps**

Closely allied to *H. cephaloideum* (above) from which it is distinguished by its larger heads, (6–) 7–8 mm long with, typically, all the involucral bracts acute, and longer, narrower and more acute radical leaves, a character difficult to define precisely. It is possible that some crossing takes place.

Vouchers: *Codd* 2796 (NU; PRE); *Hilliard & Burtt* 8022 (E; K; NU); *Ross* 2099 (NU; PRE); *Wright* 472 (E; K; NH; NU).

181. **Helichrysum subluteum** *Burtt Davy* in Jl S. Afr. Bot. 1: 110 (1935); Hilliard, Compositae in Natal 189 (1977). Lectotype: Transvaal, Woodbush Hill, Nov. 1913, *Leendertz-Pott* 4532 (K!); (*Pott* s.n., T.M. 1335 (PRE!), probably isolecto.!).

Perennial herb, stock woody, c. 5—8 mm diam., flowering stems several from the crown, erect to c. 200 (−350) mm, loosely greyish-white woolly, leafy throughout. *Radical leaves* rosetted, spreading, 28—55 × 17—22 mm, elliptic or ovate, apex subacute to obtuse, base broad, clasping, upper surface with long coarse patent hairs, sometimes very lightly cobwebby as well, lower greyish-white felted; *cauline leaves* erect, imbricate, c. 20—34 × 2—4 (−4,5) mm at the middle of the stem, diminishing in size upwards, lanceolate-oblong to lanceolate, acute to acuminate, uppermost with a small scarious appendage, base half-clasping, very shortly decurrent, both surfaces loosely greyish-white woolly, upper sometimes glabrescent to reveal long coarse patent hairs. *Heads* homogamous, campanulate, 5—6 mm long, several in a sub-globose compound head up to 25 mm across, base webbed with wool. *Involucral bracts* in c. 6 series, subequal or slightly graded, about equalling the flowers, loosely imbricate, radiating, glossy, bright yellow, outer acute, tipped golden-brown, inner often acute, sometimes subacute or narrowly obtuse. *Receptacle* with fimbrils at least equalling the ovaries. *Flowers* 37—66. *Achenes* 0,75 mm long, barrel-shaped, obscurely ribbed, glabrous. *Pappus* bristles many, equalling corolla, scabrid, tips barbellate, bases cohering by patent cilia. Fig. 51: 2.

Recorded from Woodbush in the NE. Transvaal southwards on the high ground to western Swaziland, Natal and East Griqualand (c. 1 500 to 2 400 m), Sehlabathebe in Lesotho and Naude's Nek on the Cape Drakensberg. Grows in grassy marshy places, often along streams or seepage lines; flowering between September and December, mainly in November, rarely as late as January. Map 180.

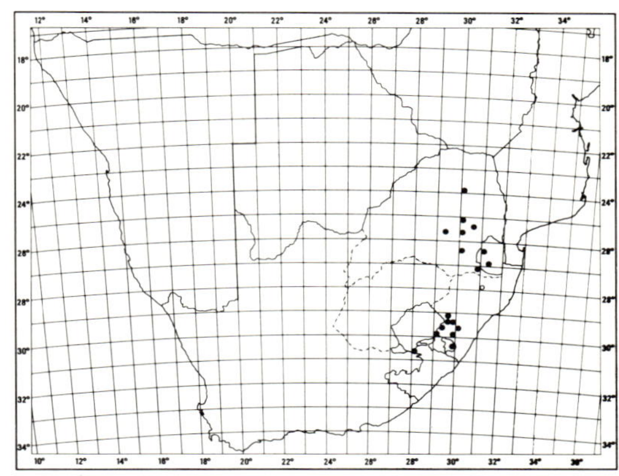

MAP 180.— ● **Helichrysum subluteum**
○ **Helichrysum ingomense**

Very closely allied to *H. cephaloideum* (no. 179) from which it is most easily distinguished by its short broad radical leaves lacking wool on the upper surface. Also, *H. subluteum* favours marshy habitats rather than the stony grassland where *H. cephaloideum* generally grows, and flowers earlier, in spring rather than late summer.

Vouchers: *Codd* 3210 (NU; PRE); *Compton* 26133 (PRE); *Hilliard & Burtt* 7190 (E; K; NU; PRE); *Moss* 15462 (PRE); *Wright* 1202 (E; K; MO; NU).

182. **Helichrysum ingomense** *Hilliard* in Notes R. bot. Gdn Edinb. 40: 255 (1982). Type: Natal, Vryheid distr., Ngome Forest Reserve, c. 4 000 ft, 12 i 1969, *Hilliard & Burtt* 5925 (NU, holo.!; E; K; NH, iso.!).

Helichrysum sp.; Hilliard, Compositae in Natal 189 (1977).

Perennial herb, stock slender, woody, flowering stems several from the crown, simple, erect to c. 300 mm, loosely greyish-white woolly, leafy throughout. *Radical leaves* rosetted, up to c. 90 × 7 mm, linear-lanceolate, apex very acute, base broad, clasping, upper surface thinly pilose with coarse jointed hairs, lower greyish-white woolly; *cauline leaves* similar but smaller and thinly woolly above at first, passing upwards into bracts, the uppermost with a small scarious appendage, erect, imbricate, very shortly decurrent. *Heads* homogamous, campanulate, c. 5 mm long, several in a tight subglobose cluster up to 20 mm across webbed at base with wool. *Involucral bracts* in c. 6 series, slightly graded, loosely imbricate, exceeding the flowers, radiating, glossy, inner tawny yellow, obtuse, apiculate, outer acute, tipped brown. *Receptacle* with fimbrils at least equalling the ovaries. *Flowers* c. 33. *Achenes* not seen, ovaries glabrous. *Pappus* bristles many, about equalling the corolla, tips barbellate, bases not cohering (but material young). Fig. 50: 3.

Known only from the type collection made at Ngome in northern Natal. It was in young flower in mid-January. The plant is common around the edges of rock sheets in grassland where water oozes to the surface, while its close ally, *H. cephaloideum* (no. 179) is found nearby in drier grassland. *H. ingomense* is easily recognized even in the vegetative state by its linear-lanceolate, very acute leaves that soon lose the wool from the upper surface. It should be looked for elsewhere in northern Natal and in the SE. Transvaal. Map 180.

183. **Helichrysum mixtum** (*O. Kuntze*) *Moeser* in Bot. Jb. 44: 320 (1910); Hilliard, Compositae in Natal 185 (1977). Type: Cape, Cathcart, *O. Kuntze* (NY, holo.; K, photo.!; Z, iso.!).

Gnaphalium mixtum O. Kuntze, Rev. Gen. Pl. 3,2: 152 (1898).

Perennial herb, stock stout, woody, flowering stems several from the crown, simple, erect to c. 450 mm, thinly greyish-felted, leafy throughout. *Radical leaves* rosetted, c. 40−100 (−190) × 10−20 (−40) mm, elliptic, elliptic-lanceolate or oblanceolate, apex acute or subacute, slightly narrowed to a broad clasping base, both

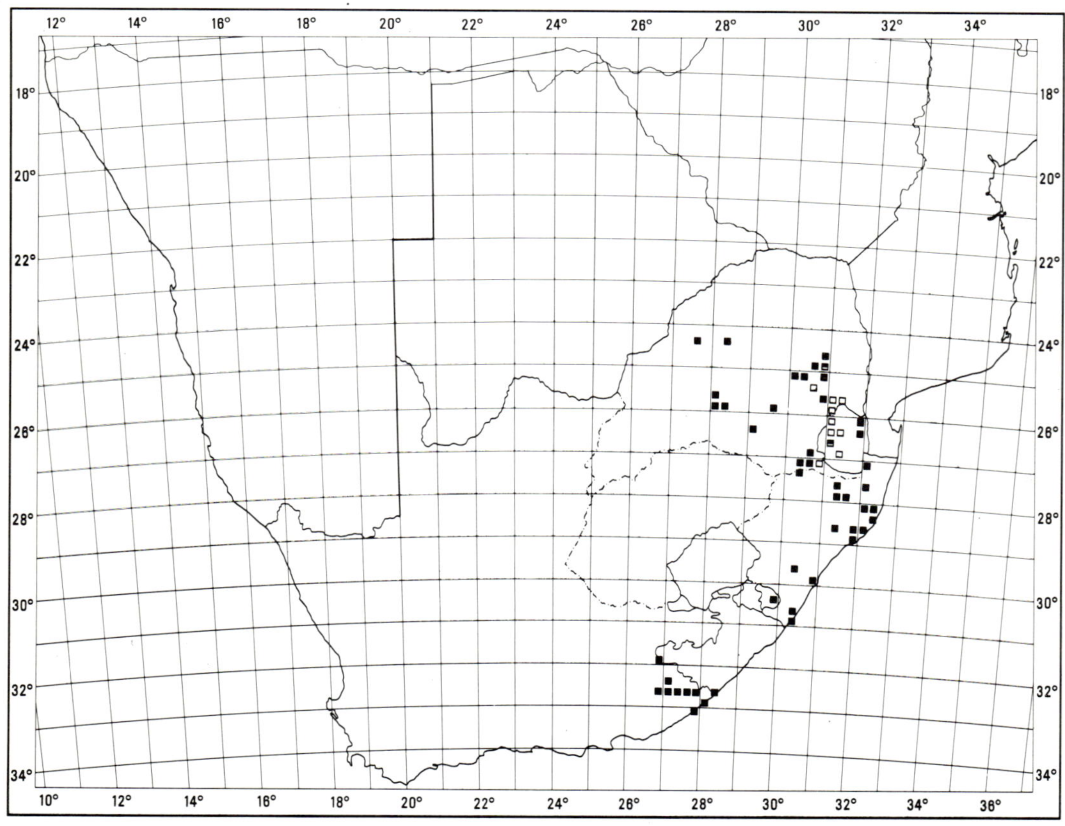

MAP 181.— ■ **Helichrysum mixtum** var. **mixtum**
    □ **Helichrysum mixtum** var. **grandiceps**
    ▱ sympatric

surfaces thinly grey-woolly; *cauline leaves* much smaller, diminishing in size upwards, lanceolate to linear-lanceolate, acute to acuminate, base broad, clasping, shortly decurrent, both surfaces thinly grey-woolly. *Heads* homogamous, campanulate, 6,5—9 mm long, several in a congested or open corymbose panicle. *Involucral bracts* in c. 6—8 series, subequal or slightly graded, loosely imbricate, inner slightly exceeding flowers, radiating, glossy, bright yellow, all acute or subacute or sometimes inner obtuse. *Receptacle* with fimbrils exceeding ovaries. *Flowers* 48—132, yellow. *Achenes* 1 mm long, barrel-shaped, glabrous. *Pappus* bristles many, scabrid, tips barbellate, bases cohering strongly by patent cilia.

Two varieties are recognized:

(a) var. **mixtum.**

Widely distributed from the E. highlands of Zimbabwe to the high parts of the Transvaal and Swaziland, through the eastern half of Natal to the E. Cape as far west as East London, the Amatola Mountains and the mountains about Queenstown. Grows in grassland; flowering mainly between November and January in the Cape and Natal, but mostly later in the Transvaal, principally March and April. Map 181. Fig. 52: 2.

Vouchers: *Compton* 25453 (PRE); *Devenish* 1624 (E; K; NU); *Galpin* 6295 (PRE); *Hilliard & Burtt* 10980 (E; K; MO; NU; PRE; S); *Story* 3323 (PRE).

H. mixtum is aptly named, for it combines the foliage of *H. cephaloideum* (no. 179) with the loose inflorescence and acute involucral bracts of *H. longifolium* (below), as Kuntze himself pointed out, and it is not always possible to place specimens with certainty. In the Cape, despite the inner involucral bracts often being as obtuse as in *H. cephaloideum*, there is no difficulty in distinguishing *H. mixtum* with heads consistently larger than those of *H. cephaloideum* and loosely arranged. In Natal and the Transvaal, however, the situation is not so clearcut. In Natal, it is often difficult to decide whether a specimen is better placed as *H. mixtum* or *H. cephaloideum* or even *H. longifolium*. As suggested before (Hilliard *l.c.*), this may be due to crossing, or speciation in this whole difficult group may be incomplete. In the Transvaal, the situation is complicated by the presence of a plant that differs from *H. mixtum* in its larger heads with less glossy involucral bracts, and in its earlier flowering time. For convenience, it is accorded varietal rank, but experimental work on the whole group is highly desirable.

(b) var. **grandiceps** *Hilliard var. nov. a typo capitulis magnis 12—15 mm longis (nec 7—9 mm) differt.*

Type: Transvaal, Barberton, Dec. 1916, *Pott* 5408 (PRE, holo.!).

Distinguished from typical *H. mixtum* by its larger heads (12—15 mm long, containing 95—155 flowers) with involucral bracts generally less glossy. Specimens with heads not fully expanded may be difficult to place.

Recorded from Graskop and MacMac south through western Swaziland to Piet Retief; growing on grassy mountain slopes. Flowers in December, January and February, mostly earlier than var. *mixtum* in the Transvaal. Map 181.

Transvaal. — Barberton, Rimer's Creek, 1 500 m, January, *Galpin* 1293 (PRE); Mountain tops between Louw's Creek and Maid of the Mist Mountain, 5 i 1929, *Hutchinson* 2417 (PRE); from top of mountain behind Barberton, c. 1 600 m and over, 18 ii 1931, *Liebenberg* 2430 (PRE); Ida Doyer Nature Reserve 38 km SE. of Barberton, 1 140 m, 7 xii 1971, *Muller* 2018 (PRE); Endahwin, Zeist farm, 1 500 m, 11 i 1974, *De Villiers* 95 (PRE). Pilgrim's Rest distr., above MacMac, c. 1 500 m, 21 xii 1932, *Smuts & Gillett* 2260 (PRE); between Sabie and Graskop, 1 100—1 400 m, 1 ii 1959, *Werdermann & Oberdieck* 2115 (PRE); Graskop, 1 600 m, 18 i 1921, *Irvine* 13 (PRE). Piet Retief distr., ii 1959, *Sidey* 3348 (PRE).

Swaziland. — Mbabane distr., Usutu Forests, c. 1 000 m, 19 xii 1957, *Compton* 27357 (PRE); Malandela, c. 1 100 m, 17 xii 1965, *Compton* 32487 (PRE); Mbabane, 1 300 m, Jan. 1914, *Rogers* 11460 (PRE); near Komati Bridge, 16 xii 1955, *Prosser* 1953 (PRE); Bremersdorp [Manzini], 860 m, 5 i 1905, *Burtt Davy* 3008 (PRE); Hlatikulu distr., Ngudwane River, c. 660 m, 9 i 1957, *Compton* 26424 (PRE).

184. **Helichrysum longifolium** *DC.,* Prodr. 6: 198 (1838); Harv. in F.C. 3: 242 (1865); Moeser in Bot. Jb. 44: 320 (1910); Hilliard, Compositae in Natal 185 (1977). Type: Natal, between the Umzimkulu River and Port Natal, *Drège* 5024 (G-DC, holo.!; B; BM; P; TCD, iso.!).

Gnaphalium longifolium (DC.) Sch. Bip. in Bot. Ztg 3: 172 (1845), non Blume (1826). *G. caffrum* O. Kuntze, Rev. Gen. 3,2: 151 (1898).

Perennial herb, stock stout, woody, flowering stems several from the crown, simple, erect to c. 450 mm, thinly greyish-white felted, leafy throughout. *Radical leaves* rosetted, c. (70—) 100—250 × 7—20 (—40) mm, linear-lanceolate to oblong-lanceolate, apex more or less acute, often apiculate, base broad, clasping, upper surface cobwebby at first, generally glabrous or occasionally with scattered coarse hairs later, lower greyish-white felted, often silky; *cauline leaves* much smaller, diminishing in size upwards, lanceolate to linear-

lanceolate, acute to acuminate, uppermost with a small scarious appendage, base half-clasping, very shortly decurrent, both surfaces cobwebby or thinly felted, upper at least usually glabrescent. *Heads* homogamous, campanulate, 7−9 (−11) mm long, several in a congested or open corymbose panicle. *Involucral bracts* in c. 6−8 series, more or less graded, loosely imbricate, inner exceeding flowers, radiating, glossy, bright yellow, all acute or subacute. *Receptacle* with fimbrils much exceeding the ovaries. *Flowers* 53−154, yellow. *Achenes* 1 mm long, barrel-shaped, obscurely 5-ribbed, glabrous. *Pappus* bristles many, about equalling the corolla, scabrid, tips barbellate, bases cohering strongly by patent cilia. Fig. 52: 1.

Ranges from Ricatla (near Delagoa Bay) in Mozambique through Natal and the Transkei to about East London in the E. Cape, in sandy grassland from near sea level to c. 900 m. Flowers between July and February, but mainly in November, December and January. Map 182.

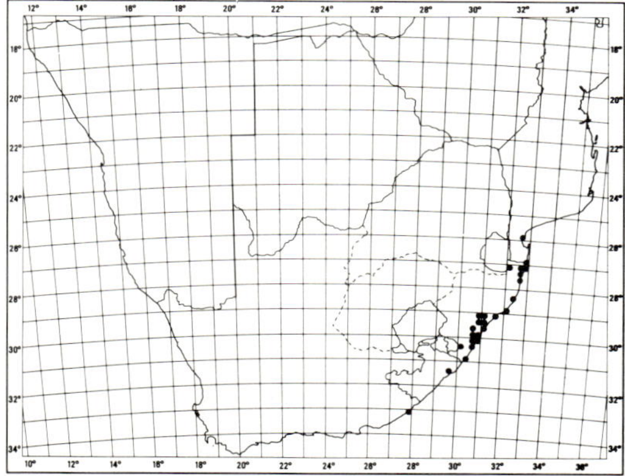

MAP 182.— **Helichrysum longifolium**

H. longifolium is distinguished by its long and often relatively narrow leaves, glabrous above at maturity, woolly below, heads in an open panicle (though this is often congested initially), and glossy yellow acute involucral bracts. It is very closely allied to *H. mixtum* (above); see under that species.

Vouchers: *Hilliard & Burtt* 5642 (E; K; NH; NU); *Pooley* 1937 (E; K; MO; NU; S); *Strey* 6836 (NU; PRE); *Wood* 11571 (PRE).

185. **Helichrysum xerochrysum** *DC.*,

Prodr. 6: 201 (1838); Harv. in F. C. 3: 243 (1865); Moeser in Bot. Jb. 44: 321 (1910). Type: Cape, between the Kei and Bashee, *Drège* 5020 (G-DC, holo.!; BM; P, iso.!).

Gnaphalium xerochrysum (DC.) Sch. Bip. in Bot. Ztg 3: 171 (1845).

Perennial herb, rootstock woody, crowned with one or a few leaf rosettes, flowering stems one or a few lateral to the rosette, erect, simple, leafy below, becoming pedunculoid and distantly bracteate upwards, tawny- or grey-woolly. *Radical leaves* c. 60−100 × 25−50 mm, oblong-elliptic, apex subacute, base slightly narrowed, clasping, both surfaces grey-woolly; *cauline leaves* similar but becoming lanceolate upwards, broad-based, decurrent in long narrow wings. *Heads* homogamous, subglobose, c. 10 × 20 mm across when fully radiating, few on nude peduncles in a lax corymbose panicle. *Involucral bracts* in c. 10 series, graded, loosely imbricate, inner exceeding flowers, radiating, acute, golden-yellow overlaid reddish brown. *Receptacle* with fimbrils exceeding ovaries. *Flowers* 122−240. *Achenes* not seen, ovaries glabrous. *Pappus* bristles many, equalling corolla, yellow, barbellate in upper half, bases cohering strongly by patent cilia. Fig. 52:3.

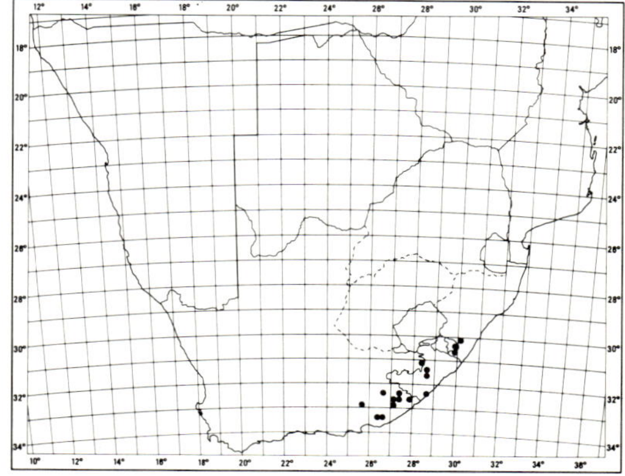

MAP 183.— **Helichrysum xerochrysum**

FIG. 52.−1, **Helichrysum longifolium,** whole plant, × 0,7; 1a, head, × 2,6 (*Hart* 13). 2, **H. mixtum** var. **mixtum,** whole plant, × 0,7; 2a, head, × 2 (*Hilliard & Burtt* 12366). 3, **H. xerochrysum,** whole plant, × 0,3; 3a, head, × 2,6 (*Hilliard & Burtt* 11206).

Ranges from southernmost Natal (Alfred and Ixopo districts) through the Transkei to the Amatola Mountains, Katberg, Boschberg at Somerset East, and Grahamstown. Grows on moist grassy mountain and hill slopes; flowering in December and January.

Distinctive and easily recognized. Map 183.

Vouchers: *Acocks* 9573 (PRE); *Flanagan* 2650 (PRE; SAM); *Galpin* 6296 (PRE); *Hilliard & Burtt* 11206 (E; K; MO; NU; PRE; S); *Pegler* 1571 (BM).

Group 25

Mat-forming perennial herbs or subshrubs; *leaves* small to medium-sized, linear, linear-lanceolate or lanceolate-elliptic, margins sometimes weakly involute; *heads* homogamous, c. 7—18 mm long, solitary or few clustered at the branch tips; *involucral bracts* radiating, white or whitish; *receptacle* more or less honeycombed; *flowers* 50—200, corolla narrowly campanulate above; *achenes* hairy; *pappus* bristles scabrid, bases cohering strongly by patent cilia.

Species 186—188, two ranging from the E. highlands of Zimbabwe through the E. part of S. Africa, one (*H. grandibracteatum*) endemic. Rocky places in grassland.

1a Involucral bracts glossy, heads either solitary or few at the branch tips on distinct peduncles:

    2a Heads 7—12 (—15) mm long, leaves linear or linear-lanceolate, 1,5—5 mm broad with 3 parallel nerves .... 186. *H. chionosphaerum*

    2b Heads 13—18 mm long, leaves lanceolate-elliptic, c. 10 mm broad, nerves scarcely visible .... 188. *H. swynnertonii*

1b Involucral bracts dull, heads several crowded at the stem tip, peduncles very short .... 187. *H. grandibracteatum*

186. **Helichrysum chionosphaerum** *DC.*, Prodr. 6: 174 (1838); Harv. in F.C. 3: 221 (1865); Moeser in Bot. Jb. 44: 317 (1910); Hilliard, Compositae in Natal 218 (1977). Lectotype: Cape, Stormberg, *Drège* 3743 quoted as *Ecklon* (G-DC!; TCD, isolecto.!).

Gnaphalium chionosphaerum (DC.) Sch. Bip. in Bot. Ztg 3: 170 (1845).

Helichrysum randii S. Moore in J. Bot., Lond., 41: 132 (1903); Moeser in Bot. Jb. 44: 317 (1910). Type: near Johannesburg, *Rand* 889 (BM, holo.!).

H. pondoense Schltr. in Bot. Jb. 40: 95 (1908). Types: Natal, Alexandra County, 'Fairfield', *Rudatis* 98; Cape, Pondoland, Dorkin, *Bachmann* 1567.

Mat-forming perennial herb or subshrub, stock stout, woody, up to 10 mm diam., stems many, well-branched, branches up to 5 mm diam., prostrate, older parts rooting, young parts leafy, leaves more or less rosetted when branchlets very short. *Leaves* 10—100 × 1,5—3 (—5) mm, linear or linear-lanceolate, upper surface sometimes greyish-white or white-felted, but usually only glandular-pubescent, lower surface sometimes silky-woolly or silky-felted, but usually at maturity wool confined to the margins and 3 parallel veins, margins slightly involute. *Heads* homogamous, campanulate, 7—12 (—15) mm long, 12—25 mm across the fully radiating bracts, solitary or up to 4 clustered at the tips of erect leafy stems up to 150 mm tall. *Involucral bracts* in c. 6 series, graded, loosely imbricate, about equalling flowers, glossy white, occasionally washed palest yellowish brown outside. *Receptacle* shortly honeycombed. *Flowers* 71—154, yellow. *Achenes* c. 1,25 mm, barrel-shaped, with myxogenic duplex hairs. *Pappus* bristles many, equalling corolla, scabrid, bases cohering strongly by patent cilia. Fig. 53:2.

Recorded from Inyangani in the E. highlands of Zimbabwe, the E. highlands of the Transvaal, the Witwatersrand and the Magaliesberg, W. Swaziland, Natal from c. 500—2 250 m, but commoner above c. 1 200 m, the mountainous NE. corner of the Orange

---

FIG. 53.—1, **Helichrysum grandibracteatum**, whole plant, × 1; 1a, head, × 2,6 (*Hilliard & Burtt* 7429). 2, **H. chionosphaerum**, part of plant with long grassy leaves, × 1 (*Hilliard & Burtt* 8613); 2a, part of plant with short broad leaves, × 1; 2b, head, × 2; 2c, hermaphrodite flower, × 8; 2d, pappus bristle, × 5,3 (*Hilliard & Burtt* 12129). 3, **H. swynnertonii**, part of plant, × 1 (*Hilliard & Burtt* 7448).

2c
2d
2a
2b
2
1
1a
3
Wouda . du Toit

Free State, Lesotho, Transkei, and NE. Cape on the Witteberg, Drakensberg, Stormberg and Katberg. Forms large, flowery mats up to a metre or more across, in stony grassland, over rock sheets, or draped over low rock or earth cliffs and big boulders, sometimes in tufts in the cracks of rock pavements. Flowers from July to December. Map 184.

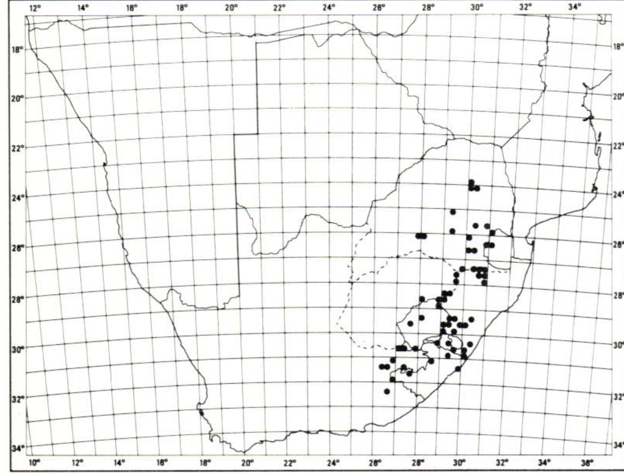

MAP 184.— **Helichrysum chionosphaerum**

Displays considerable variation in leaf length and degree of woolliness. Some variation may be induced by growing conditions: plants on rock sheets for instance often have much shorter leaves than nearby plants in grassland.

Vouchers: *Codd* 3366 (NU; PRE); *Flanagan* 1596 (NU); *Hilliard & Burtt* 5873 (E; K; NH; NU; S); *Hoener* 1686 (E; NU); *Wright* 190 (E; K; M; NH; NU).

187. **Helichrysum grandibracteatum** *M. D. Henderson* in Bothalia 6: 422 (1954); Hilliard, Compositae in Natal 219 (1977). Type: Natal, Bergville distr., Cathedral Peak Forest Research Stn., *Killick* 1149 (PRE, holo.! K; NH, iso.!).

Tufted perennial herb, stock stout (up to 10 mm diam.), woody, crowned with several leaf rosettes, flowering stems terminal, solitary, up to 200 mm high, grey-woolly, leafy. *Radical leaves* up to 120 × 4 mm, cauline leaves much smaller, linear-lanceolate, apex acute, uppermost tipped with a white scale, base broad, clasping, glabrous above, silvery silky-woolly-felted below, striate from the strongly raised parallel veins, margins slightly involute. *Heads* homogamous, campanulate, 8—10 mm long, 18—20 mm across

the fully radiating bracts, several clustered at the stem apex. *Involucral bracts* in c. 12 series, loosely imbricate, inner subequal, much exceeding the flowers, acute to acuminate, snow-white. *Receptacle* shortly honeycombed. *Flowers* 51—89, yellow. *Achenes* 1,5 mm, elliptic, with myxogenic duplex hairs. *Pappus* bristles many, equalling corolla, barbellate, bases cohering strongly by patent cilia. Fig. 53: 1.

Ranges from the Mont aux Sources area in NW. Natal along the face of the Drakensberg and in the more elevated parts of Natal and the Transkei to the Amatola Mountains, Katberg and Boschberg in the E. Cape. Relatively common in the Natal Drakensberg between c. 1 800 and 2 600 m, but also recorded from the top of Swartkop (c. 1 350 m) overlooking Pietermaritzburg. Grows scattered in poor stony grassland; flowering between September and November, and then sometimes making conspicuous white patches. Map 185.

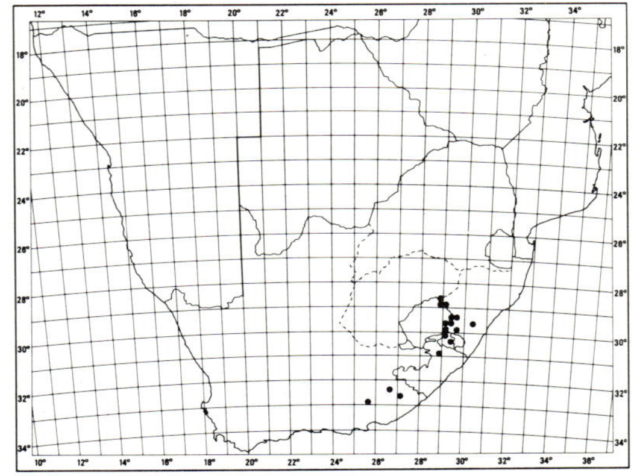

MAP 185.— **Helichrysum grandibracteatum**

Vouchers: *Hilliard* 5177 (E; K; MO; NU; S); *Hilliard & Burtt* 10932 (E; K; NU); *Wright* 1048 (E; MO; NU).

188. **Helichrysum swynnertonii** *S. Moore* in Bot. J. Linn. Soc., 40: 109 (1911); Hilliard, Compositae in Natal 219 (1977). Type: Rhodesia, Melsetter, 6 000 ft, *Swynnerton* 6110 (BM, holo.!).

Mat-forming perennial herb or subshrub, stock creeping, becoming very stout and woody, stems many, branching only below, ascending, c. 100—150 mm long, closely leafy, flowering stems solitary, terminal, one-headed, leafy below, bracteate upwards, grey-felted. *Leaves* up to 80 × 10 mm, more or less erect, lanceolate-elliptic, apex acute, base slightly narrowed, half-clasping, webbed to the stems by loosely interwoven stringy-woolly hairs, both surfaces and margins enveloped in similar loosely-woven grey stringy-woolly felt underlain by patent glandular hairs, margins slightly involute. *Heads* homogamous, broadly campanulate, c. 13—18 mm long, solitary at the branch tips. *Involucral bracts* in c. 10 series, graded, loosely imbricate, about equalling the flowers, acute, radiating, glossy white with a faint yellowish or brownish overcast. *Receptacle* with flattened tubercles. *Flowers* 129—201, yellow. *Achenes* c. 1,5 mm, barrel-shaped, with myxogenic duplex hairs. *Pappus* bristles many, equalling corolla, scabrid, bases cohering strongly by patent cilia. Fig. 53: 3.

Recorded from Mt Kilimanjaro, the eastern highlands of Zimbabwe (Melsetter district), the eastern highlands of the Transvaal (Mariepskop, Mt Anderson, Ohrigstad, Kaapse Hoop, Witklip Forest Reserve), a few localities in the southern Drakensberg of Natal and the Suurberg near Weza, and Fort Cunynghame, Stutterheim district, E. Cape. Forms spreading clumps or mats on stony mountain slopes; flowering between October and December. Allied to *H. chionosphaerum* (no. 186) but distinguished by its broader and differently shaped leaves with nerves scarcely visible. Map 186.

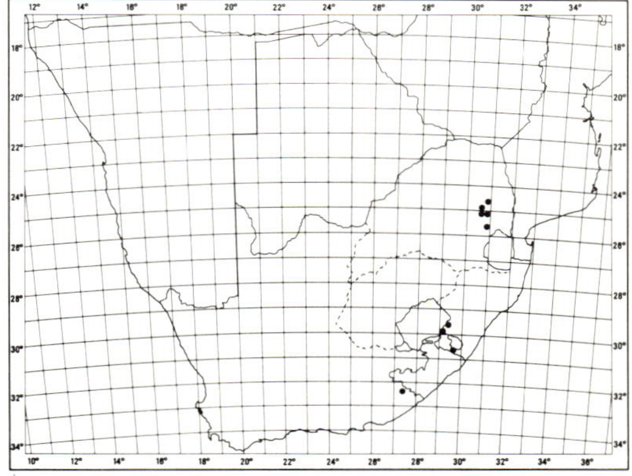

MAP 186.— **Helichrysum swynnertonii**

Vouchers: *Hilliard & Burtt* 7448 (E; K; MO; NBG; NU; PRE; S); *Nordenstam 3196* (LD; NU); *Rogers 24024* (PRE).

Group 26

Half-shrubs or mat-forming perennial herbs with straggling and sometimes rooting branches; *leaves* small or medium-sized, linear-lanceolate to oblong or obovate; *heads* heterogamous or rarely homogamous in the same species, c. 10—40 mm long, solitary; *involucral bracts* radiating, white often overlaid brown or crimson; *receptacle* smooth or shortly honeycombed; *flowers* 18—282, (0—) 1—44 ♀, corolla of ♀ flowers narrowly funnel-shaped, of ♀ flowers narrowly tubular; *achenes* glabrous or hairy, sometimes in the same species; *pappus* bristles with scabrid shaft, tips barbellate or subplumose, bases either shortly fused into a ring or partly fused and cohering by patent cilia as well.

Species 189—194, three endemic to the S. and SW. Cape, 3 to the extended Drakensberg Centre. Most species on the mountains, one on the seashore.

1a Heads c. 15—40 mm long:

  2a Heads c. 25—40 mm long:

    3a Flowering stem leafy to the summit...189. *H. retortum*

    3b Heads terminating scaly peduncles sharply distinguishable from the leafy sterile branches
    .... 191. *H. lancifolium*

  2b Heads c. 15—20 mm long:

    4a Leaves wholly enveloped in silvery 'tissue-paper' indumentum.............................190. *H. stoloniferum*

    4b Upper leaf surface clad in 'tissue-paper' indumentum, lower closely white-felted ....
    .... 194. *H. retortoides*

1b Heads c. 8—10 mm long:

  5a Densely tufted mat-forming perennial herb; pappus bristles fused at the base in a smooth ring
....192. *H. praecurrens*

  5b Diffuse undershrub with long slender straggling branches; pappus bristles cohering at the base by
    patent cilia, some light fusion as well .... 193. *H. scitulum*

189. **Helichrysum retortum** *(L.)*
Willd., Sp. Pl. 3: 1907 (1804); DC., Prodr. 6: 177 (1838), excl. var. *minus* DC.; Harv. in F.C. 3: 228 (1865) excl. var. *minus* DC.; Moeser in Bot. Jb. 44: 321 (1910), excl. *Schlechter* 10095 and 10041; Levyns in Adamson & Salter, Fl. Cape Penins. 782 (1950). Type: Cape of Good Hope, Herb. Cliff. (BM!).

Xeranthemum retortum L., Sp. Pl. 858 (1753). *Argyrocome retorta* (L.) Gaertn., Fruct. 2: 410 t. 167, fig. 3 (1791). *Astelma retorta* (L.) [D. Don ex] Steud., Nom. 2,1: 153 (1840). *Gnaphalium retortum* (L.) Sch. Bip. in Bot. Ztg 3: 169 (1845).

X. polyfolium Crantz, Inst. 1: 290 (1766). Lectotype: *Xeranthemoides procumbens* Dill., Hort. Elth. 423 t. 322, fig. 415 (1732).

X. radicans Thunb., Prodr. 153 (1800). *Helichrysum radicans* (Thunb.) Thunb., Fl. Cap. 663 (1823); Less., Syn. Comp. 292 (1832). Type: Cape, seashore near Lion's Head, *Thunberg* (sheet 19312, UPS, holo.!).

Suffrutex, main stems up to 450 mm long or straggling to greater lengths, branching, prostrate, rooting at intervals, flowering branchlets erect, all closely leafy throughout. *Leaves* imbricate, 11—30 × 2—6 mm, scarcely diminishing in length upwards but often narrower, oblong, oblanceolate or obovate, lower half appressed, webbed to stem, upper half flat, spreading, often somewhat recurved, tip more or less acute, mucronate, hooked, both surfaces enveloped in silky silvery 'tissue-paper' indumentum. *Heads* heterogamous, turbinate, 25—40 mm long, nearly double that across the radiating bracts, solitary at the tips of the branchlets, closely surrounded by leaves. *Involucral bracts* in c. 7 series, loosely imbricate, graded, inner much exceeding flowers, linear-lanceolate, acute, glossy white, outer ovate-acuminate, golden-brown sometimes with a rosy-crimson overcast. *Receptacle* very shortly honeycombed. *Flowers* 94—246, 17—44 ♀, 77—204 ☿. *Achenes* 1 mm long, cylindric, with myxogenic duplex hairs. *Pappus* bristles many, equalling corolla, tips barbellate, shaft scabrid, bases lightly fused. Fig. 54: 1.

Along the SW. Cape coast from Blouberg, N. of Table Bay, to Rietvlei, Riversdale district, on sea cliffs, coastal dunes, sandy slopes or flats never far from the sea, from sea level to c. 45 m. A sprawling sandbinder, sometimes straggling up through neighbouring bushes; flowering between August and December. Map 187.

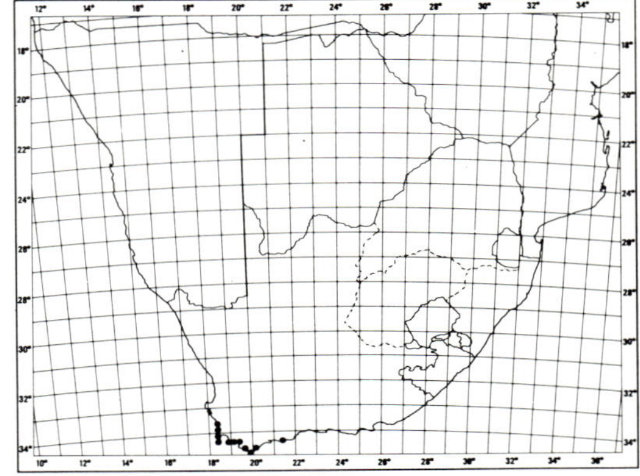

MAP 187.— **Helichrysum retortum**

Vouchers: *Barker* 2727 (NBG); *MacOwan* 2410 (SAM); *Rogers* 26452 (PRE); *Schlechter* 9486 (BM, PRE); *Taylor* 5267 (PRE).

190. **Helichrysum stoloniferum** *(L.f.)*
Willd., Sp. Pl. 3: 1907 (1804); Thunb., Fl. Cap. 662 (1823); Less., Syn. Comp. 291

FIG. 54.—1, **Helichrysum retortum**, part of plant, × 1 (*Leighton* 2102). 2, **H. retortoides**, part of plant, × 1 (*Wright* 650). 3, **H. praecurrens**, part of plant, × 1 (*Wright* 339). 4, **H. lancifolium**, part of plant, × 1; 4a, hermaphrodite flower, × 6,6; 4b, female flower, × 6,6; 4c, pappus bristle, × 6,6; 4d, base of pappus bristles much enlarged to show fusion as well as cohesion by patent cilia (*Acocks* 19928). 5, **H. stoloniferum**, part of a diffuse plant showing the difference between leaves on a vegetative twig and on a flowering twig, × 1 (*Esterhuysen* 3939); 5b, part of a compact plant, leaves on all twigs similar, × 1 (*Stokoe* SAM 68368). 6, **H. scitulum**, part of plant, × 1 (*Hilliard & Burtt* 10629).

1
2
3
4
4a
4b
4c
4d
5
5a
6
H. Wouda-du Toit

(1832); Harv. in F.C. 3: 228 (1865), excl. *Drège* spec. Type: Cape of Good Hope, *Thunberg* (990.24 LINN!).

Xeranthemum stoloniferum L.f., Suppl. 366 (1781). *Gnaphalium stoloniferum* (L.f.) Sch. Bip. in Bot. Ztg 3: 169 (1845).

Argyrocome retorta sensu Cass. in Dict. Sci. nat. 34: 39 (1825), non (L.) Gaertn.

Helichrysum retortum var. *minus* DC., Prodr. 6: 178 (1838); Harv. in F.C. 3: 228 (1865). Lectotype: Cape, Kamiesberge, *Drège* 2834 (G-DC!).

Subshrub, main stems up to c. 300 mm long, branched, prostrate or diffuse, flowering branchlets ascending, all closely leafy throughout. *Leaves* imbricate, those on sterile shoots mostly 7–15 × 2–6 mm, subequal, narrowly obovate, lower part appressed, webbed to stem, upper part ascending or spreading, often somewhat recurved, tip acute, apiculate, hooked, leaves on flowering twigs often much narrower, c. 1,5–3 mm broad, somewhat spreading or strongly appressed, oblong-lanceolate or oblong, acute or subacute, uppermost often tipped with papery brown scales, all leaves wholly enveloped in silvery, silky 'tissue-paper' indumentum. *Heads* heterogamous, turbinate-campanulate, mostly 15–20 mm long, nearly double that across the radiating bracts, solitary at tips of branchlets in upper leaf axils, these branches penduncle-like when leaves much reduced and appressed. *Involucral bracts* in

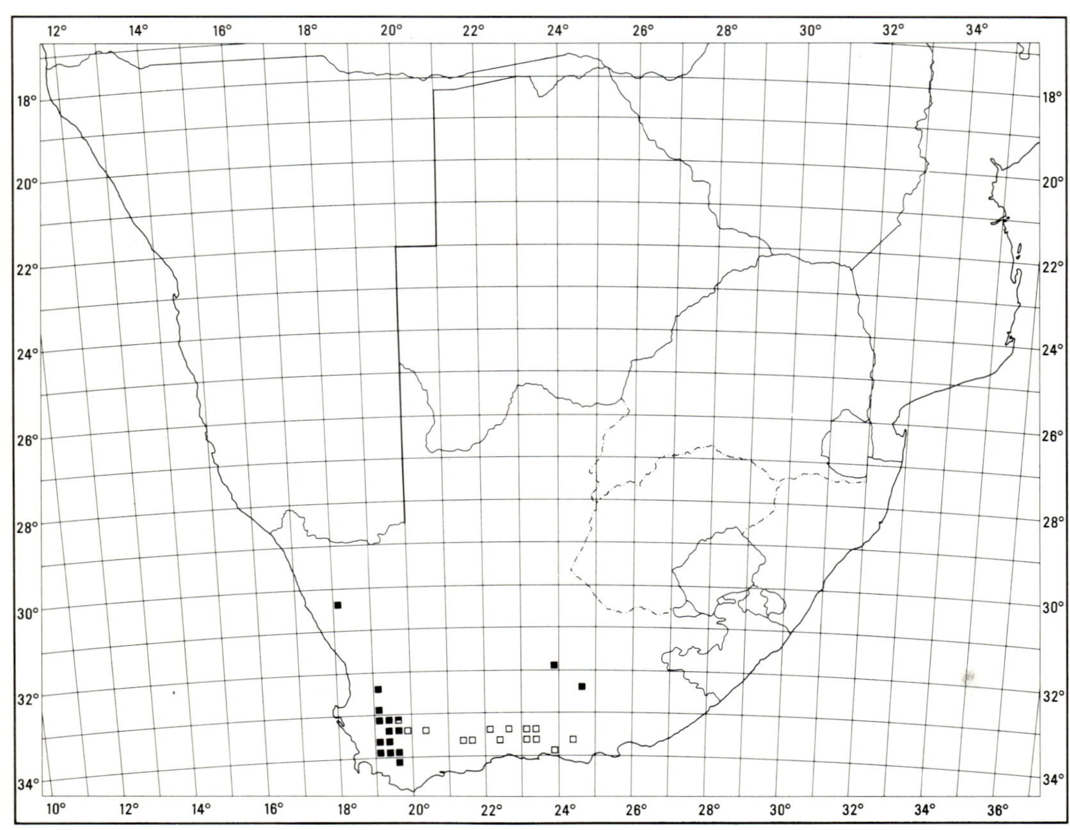

MAP 188.— ■ **Helichrysum stoloniferum**
      □ **Helichrysum lancifolium**
      ◧ sympatric

c. 7 series, loosely imbricate, graded, inner much exceeding flowers, linear-lanceolate, mostly subacute, sometimes acute, glossy white or pale to deep rose, or white tipped rose, outer ovate-lanceolate, subobtuse to acuminate, golden-brown sometimes tinged rose. *Receptacle* very shortly honeycombed. *Flowers* 58—282, 6—24 ♀, 52—243 ☿. *Achenes* 1 mm long, cylindric, with myxogenic duplex hairs. *Pappus* bristles many, equalling corolla, tips subplumose, shaft scabrid, bases lightly fused. Fig. 54: 5.

On the mountains of the W., SW. and central Cape, from the Kamiesberg to the Cedarberg, Cold Bokkeveld Mountains, Winterberge, and the mountains around Worcester, Ceres, French Hoek and Genadendal with one record from the mountains near Murraysburg and another from Cave Mountain, Graaff-Reinet, always above 1 200 m. Grows in rocky places, often rooted in rock crevices; flowering between November and February. Map 188.

The salient features of *H. stoloniferum* are the relatively small heads with generally bluntish inner involucral bracts. Plants may be compact (probably in response to hard growing conditions) and then the flowering twigs are short and there is little or no difference in size and form of leaves on sterile and flowering shoots. If the plants are well-grown and diffuse, there tends to be a marked difference between leaves on sterile shoots and those on flowering ones; the latter become pedunculoid with reduced, appressed, bract-like leaves. *Schlechter* 10095 (PRE), and *Esterhuysen* 3939 (PRE), both from the Koude Bokkeveld Tafelberg at 1 800 m, demonstrate this very well. This phenomenon is paralleled in other species e.g. *Edmondia pinifolia*.

Closely allied to and much confused with *H. retortum* (no. 189) but that is a coarser plant confined to the coast. See also *H. lancifolium* (below).

Vouchers: Compton 12855 (NBG); *Esterhuysen* 1421 (BOL); *Galpin* 12585 (PRE); *Schlechter* 10041 (BM; G; PRE); *Stokoe* in SAM 68368 (SAM; PRE).

191. **Helichrysum lancifolium** *(Thunb.) Thunb.*, Fl. Cap. 662 (1823); Harv. in F.C. 3: 227 (1865); Moeser in Bot. Jb. 44: 322 (1910). Type: Cape of Good Hope, *Thunberg* (sheet 19304, UPS, holo.!; S, iso.!).

Xeranthemum lancifolium Thunb., Prodr. 152 (1800). *Gnaphalium lancifolium* (Thunb.) Sch. Bip. in Bot. Ztg 3: 169 (1845).

Helichrysum xeranthemoides DC., Prodr. 6: 178 (1838). *Gnaphalium xeranthemoides* (DC.,) Sch. Bip. in Bot. Ztg 3: 169 (1845). Lectotype: Cape, Langekloof, *Drège* 2167 (G-DC!; K; S; SAM, isolecto.!).

Subshrub, branches weak, straggling, up to 600 mm long, closely leafy. *Leaves* imbricate, 6—18 × 1,5—4 mm, scarcely diminishing upwards on sterile twigs, abruptly passing into smaller, appressed bracts on flowering stems, linear-lanceolate, lower half appressed, webbed to stem, base broad, half-clasping, upper half flat or conduplicate, recurved, apex acute to acuminate, apiculate, both surfaces enveloped in closely woven silvery, silky indumentum. *Heads* heterogamous, turbinate, mostly c. 25—35 mm long, 40—50 mm across the radiating bracts, solitary at branchlet tips in upper leaf axils, branchlets peduncle-like, 50—200 mm long, closely bracteate, bracts appressed, uppermost tipped with papery brown scales passing into involucral bracts. *Involucral bracts* in c. 7 series, loosely imbricate, graded, inner much exceeding flowers, linear-lanceolate, acuminate, white usually overlaid pale to deep rose or crimson, outer bracts ovate to ovate-lanceolate, acute to acuminate, golden-brown. *Receptacle* very shortly honeycombed. *Flowers* 73—130, 20—35 ♀, 45—110 ☿. *Achenes* 1,25 mm long, cylindric, with myxogenic duplex hairs. *Pappus* bristles many, about equalling corolla, tips subplumose, shaft barbellate to scabrid, bases lightly fused. Fig. 54: 4.

On the mountains flanking the Little Karoo, from the Baviaansberg and Bonteberg N. and E. of Ceres, and the Witteberg at Laingsburg, across the Klein and Groot Swartberg to the Baviaanskloof Mountains, Kouga Mountains, Outeniqua and Langekloof Mountains, above c. 700 m. Found among shrubs and restiads; flowering between September and January. Map 188.

H. stoloniferum (no. 190) *and H. lancifolium* are closely allied, but they are nearly allopatric; their areas appear to overlap only in the Baviaansberg NE. of Ceres.

H. lancifolium is distinguished by its leaves, generally narrower and more pointed than those of *H. stoloniferum*, and by its heads, generally larger with more pointed inner involucral bracts. Peduncle-like flowering twigs seem to be a constant feature of *H. lancifolium*.

Vouchers: Acocks 19928 (PRE); *Compton* 21142 (NBG); *Esterhuysen* 6587 (BOL; PRE); *Oliver* 5406 (PRE); *Stokoe* (SAM 68674).

192. **Helichrysum praecurrens** *Hilliard* in Notes R. bot. Gdn Edinb. 32: 359 (1973), Compositae in Natal 222 (1977). Type: Natal, Estcourt distr., summit of Drakens-

berg in vicinity of Giant's Castle Pass, c. 3170 m, *Wright* 1044 (NU, holo.!; E; K; PRE, iso.!).

Densely tufted mat-forming perennial herb, main stems diffuse, prostrate, nude, rooting, producing innumerable erect closely leafy branchlets up to c. 20 mm high. *Leaves* up to c. 7 × 2 mm, closely imbricate, appearing rosulate from above, linear to narrowly lanceolate, with smooth sericeous 'tissue-paper' indumentum, green with a silvery sheen, drying silvery grey. *Heads* homogamous or rarely heterogamous, campanulate, c. 10 mm long, nearly double that across the fully radiating bracts, sessile, solitary at the tips of the branches. *Involucral bracts* in c. 7 series, graded, closely imbricate, outer pellucid, palest brown, inner much exceeding the flowers, white, or tinged or tipped rose-pink, opaque. *Receptacle* smooth. *Flowers* 18−24, rarely 1−4 ♀. *Achenes* 1,5 mm long, ellipsoid, obscurely ribbed, glabrous or occasionally with a few duplex hairs. *Pappus* bristles many, equalling corolla, scabrid, bases fused in a smooth ring. Fig. 54: 3.

On the high Lesotho mountains and on the summit of the high Drakensberg along the Natal-Lesotho border, always more than 3 000 m above sea level. Forms dense mats colonizing bare or sparsely grassed areas; flowering from November to early December. Map 189.

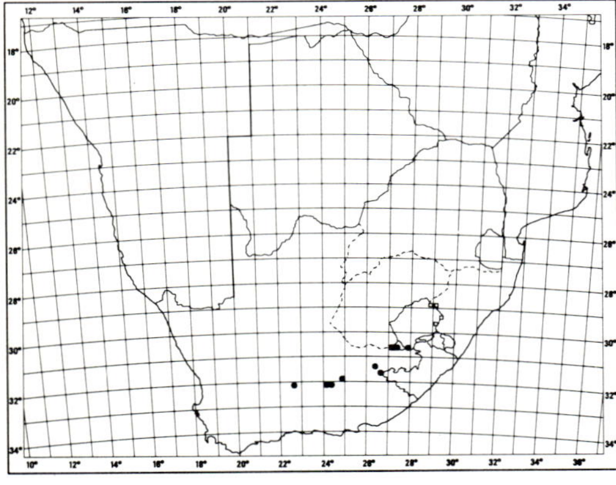

MAP 189.— □ **Helichrysum praecurrens**
        ● **Helichrysum scitulum**

Can be confused with *H. sessilioides* (no. 197) but is easily distinguished by its different habit, leaves silky below, not woolly, and different pappus.

Vouchers: *Edwards* 349 (NU); *Hilliard & Burtt* 7098 (E; K; MO; NBG; NU; PRE; S); *Wright* 339 (E; K; NU; S).

193. **Helichrysum scitulum** *Hilliard & Burtt* in Notes R. bot. Gdn Edinb. 40: 266 (1982). Type: Cape, Graaff-Reinet distr., S. extreme of Renosterberg above Lootsberg railway halt, farm Blaauwater, c. 1 800 m, 24 xi 1977, *Hilliard & Burtt* 10629 (NU holo.!; E; K; MO; PRE; S, iso.!).

H. stoloniferum sensu DC., Prodr. 6: 175 (1838) quoad spec., non (L.f.) Willd.

Diffuse undershrub, branches long, slender, straggling, closely leafy throughout. *Leaves* 3−7 × 0,5−1 mm, scarcely diminishing upwards, imbricate, spreading, somewhat recurved, linear, base broad, half-clasping, webbed to stem, tip acute, upper surface green covered in very thin 'tissue-paper' indumentum, lower closely greyish-white silky-felted. *Heads* heterogamous, turbinate-campanulate, c. 8−10 mm long, double that across the radiating bracts, solitary at the tips of short (10−30 mm) densely leafy erect branchlets often closely set along the main stems. *Involucral bracts* in c. 7 series, imbricate, graded, outer lanceolate, acute, rose, inner about equalling flowers, lanceolate or oblong-lanceolate, acute or obtuse, white often tipped rose. *Receptacle* honeycombed. *Flowers* 26−45, 3−6 ♀, 21−40 ☿, yellow, ♀ tipped pink. *Achenes* 1 mm long, cylindric, with myxogenic duplex hairs. *Pappus* bristles many, equalling corolla, scabrid above, smooth below, bases cohering by patent cilia, some fusion in bundles as well. Fig. 54: 6.

On the Cape mountains from the Nieuweveld Mountains, N. of Beaufort West, to the Renosterberg (E. end of the Sneeuberg), Bamboesberg, Stormberg and Witteberg near Lady Grey. Grows on rocky mountain tops, forming loose mats in rock crevices, or cascading down steep slopes, intertwined with grasses and other vegetation. Flowers in November and December. Map 189.

Much confused with *H. stoloniferum* (no. 190) and may be found in herbaria under that name, but easily distinguished by its different foliage, smaller heads, and different pappus.

Vouchers: *Drège*, Witteberg (BM; E; G-DC; K; PRE; S; SAM; TCD); *Flanagan* 1593 (NU; PRE; SAM); *Galpin* 6275 (PRE); *Oliver* 5219 (K; PRE; S).

Hilliard & Burtt 12167 (E; NU) and *Hilliard & Burtt* 6741 (E; NU) from the Witteberg, at Joubert's Pass and Lundean's Nek respectively, appear to be particularly well-grown specimens of *H. scitulum* with larger leaves (up to 12 × 3−4 mm) and larger heads (c. 14 mm long).

194. **Helichrysum retortoides** *N.E. Br.* in Kew Bull. 1906: 164 (1906); Hilliard, Compositae in Natal 223 (1977). Type: Natal, on the slopes of the Drakensberg, *Wilson* in herb. *Wood* 8265 (K, holo.!; NH, iso.!).

Diffusely branched perennial herb or subshrub, main stems prostrate, rooting, becoming thick and woody with age, flowering stems erect to c. 200 mm, uniformly and closely leafy. *Leaves* up to c. 10 × 3 mm, elliptic-oblong or subspathulate, upper surface with smooth sericeous 'tissue-paper' indumentum, green drying silvery grey, lower closely white-felted. *Heads* heterogamous, campanulate, c. 20 mm long, 30 mm across the fully radiating bracts, sessile, solitary at the tips of the branches. *Involucral bracts* in c. 9 series, closely imbricate, outer much shorter and broader than inner, glossy brown with a crimson tinge, inner much exceeding the flowers, white or tinged rose-pink to crimson, opaque. *Receptacle* shallowly honeycombed. *Flowers* 50−68, 3−8 ♀, 47−56 ☿. *Achenes* 1,75 mm long, ellipsoid, glabrous or with a few duplex hairs. *Pappus* bristles many, equalling corolla, tips subplumose, shaft scabrid, bases cohering by patent cilia with some fusion as well. Fig. 54: 2.

On the high Drakensberg and the high Lesotho mountains between c. 2 200 and 3 350 m. Forms diffuse mats on steep broken turf slopes or among boulders in stony grassland, or in the crevices of rock sheets; dominant in places on the summit plateau. Flowers from September to December. Map 190.

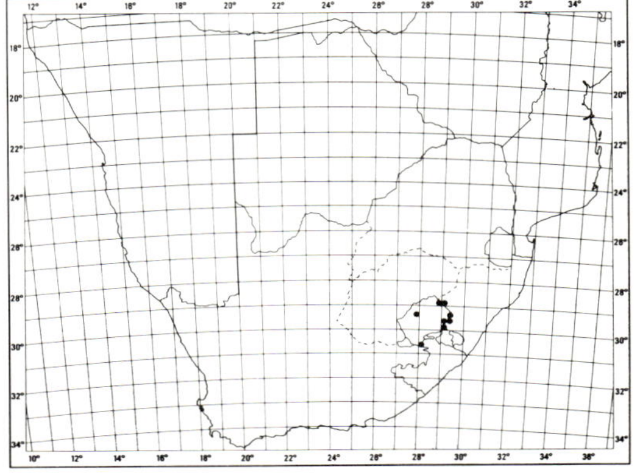

MAP 190.— **Helichrysum retortoides**

Vouchers: *Hilliard & Burtt* 7128 (E; K; MO; NBG; NU; PRE; S); *Trauseld* 901 (NU); *Wright* 639 (E; K; NH; NU).

Group 27

Shrubs or herbaceous perennials; *leaves* small or medium-sized, linear, elliptic, obovate or suborbicular; *heads* homogamous, 5−15 mm long, solitary or few tightly clustered; *involucral bracts* radiating, white, pink, or pink and white, often with some brown as well, pale yellow in one species; *receptacle* smooth or honeycombed; *flowers* 17−277, corolla campanulate above; *achenes* hairy, or rarely glabrous within the same species; *pappus* bristles scabrid, tips barbellate, bases mostly with patent cilia, cohering or not.

Species 195−200, all Drakensberg endemics, mainly on cliffs, rock platforms or stepped crags, one in moist turf.

1a Perennial herbs:

2a Plants with rosettes of radical leaves, heads solitary, c. 15 mm long ....199. *H. bellidiastrum*

2b Plants with long leafy prostrate rooting branches, heads in tight clusters, c. 5 mm long ....196. *H. hyphocephalum*

1b Dwarf shrubs or mat-forming sub-shrubs:

3a Heads c. 10−15 mm long, bracts white, pink, or pink and white, often with some brown as well:

4a Leaf blades suborbicular, obovate or broadly spathulate:

    5a Erect dwarf shrublet, leaves woolly, heads 10—13 mm long, bracts loosely imbricate
.... 200. *H. glaciale*

    5b Prostrate mat-forming subshrub with long stoloniferous branches, leaves silvery sericeous, heads
       c. 15 mm long, bracts closely imbricate .... 198. *H. haygarthii*

4b Leaf blades linear to lanceolate-elliptic.. 197. *H. sessilioides*

3b Heads c. 5—7 mm long, bracts pale lemon-yellow or whitish .... 195. *H. pagophilum*

195. **Helichrysum pagophilum** *M.D. Henderson* in Kirkia 1: 113 (1961); Hilliard, Compositae in Natal 224 (1977). Type: Basutoland [Lesotho], summit Drakensberg, Cleft Peak area, 9 800 ft., *Killick & Marais* 2177 (PRE, holo.!; K, iso.!).

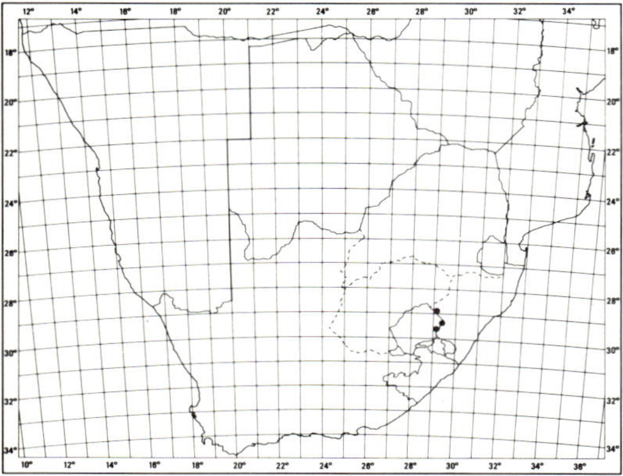

MAP 191. — **Helichrysum pagophilum**

Much-branched cushion-forming dwarf shrub, very compact, rounded in outline, old main stem woody, bare, gnarled, main branches prostrate, rooting, producing innumerable tightly congested, densely leafy erect branchlets. *Leaves* closely imbricate, rosulate when viewed from above, only the thick, broadly ovate or suborbicular blade visible, c. 4—6 × 4—6 mm, appressed grey silky-woolly, the oblong basal part c. 6 × 3 mm, membranous, glabrous above, thinly villous below, stem-clasping. *Heads* homogamous, campanulate, 5—7 mm long, double that across the fully radiating bracts, sessile, solitary at the tips of the branchlets. *Involucral bracts* in c. 6 series, graded, loosely imbricate, exceeding the flowers, obtuse to acute, glossy, lemon-yellow, pale lemon or whitish. *Receptacle* nearly smooth.

Flowers 41—64, yellow. *Achenes* 1,5 mm, ellipsoid, with sparse rather long duplex hairs, sometimes few or wanting, not mucilaginous when wet. *Pappus* bristles many, equalling corolla, scabridulous, bases not cohering by patent cilia, but some fusion in bundles. Fig. 55: 4.

Recorded from a small part of the high Drakensberg in Natal and Lesotho, from Cleft Peak (Bergville district) to Hodgson's Peaks and The Rhino (Underberg district), and the Black Mountains beyond Sani Pass, between c. 2 745 and 3 500 m. Forms dense hard cushions scarcely 100 mm high but as much as a metre across, pressed against rock surfaces. Flowers between November and January. Map 191.

Vouchers: *Hilliard* 5425 (E; K; M; MO; NU; PRE; S); *Hilliard & Burtt* 10483 (E; K; MO; NU; S); *Wright* 336 (E; K; NH; NU).

196. **Helichrysum hyphocephalum** *Hilliard* in Notes R. bot. Gdn Edinb. 32: 352 (1973), Compositae in Natal 226 (1977). Type: Natal, Estcourt distr., Kamberg Nature Reserve, 'Gladstone's Nose', stony ridges descending from summit cliffs of S.-facing slope, SW. aspect, c. 2 060 m, 29 xii 1967, *Wright* 363 (NU, holo.!; E; K; PRE, iso.!).

Well-branched mat-forming perennial herb, branches slender, prostrate, older parts nude or leafy, rooting, new shoots closely silvery sericeous, leafy, often much abbreviated and then leaves more or less rosetted. *Leaves* up to 20 × 4 mm, lanceolate-elliptic, slightly falcate, apex acute, base slightly narrowed, clasping, both surfaces with silvery grey silky 'tissue-paper' indumentum. *Heads* homogamous, turbinate-campanulate, c. 5 mm long, up to c. 10 in a tight cluster on remotely leafy peduncles up to c. 60 mm long terminating the branchlets, peduncle, leaves, base of compound inflorescence and base of individual heads all enveloped in silvery 'tissue-paper' indumentum. *Involucral bracts* in c. 4 series, scarcely imbricate, outer larger than inner, spathulate, up to 6 × 3 mm, inner shorter and narrower, all petaloid, limb milk-white. *Receptacle* smooth. *Flowers* c. 17—20, yellow. *Achenes* 1 mm long, with elongated duplex hairs. *Pappus* bristles 5 or 6, equalling the corolla, tips subplumose, bases not cohering. Fig 55: 5.

Known only from the southern part of the Natal Drakensberg, between Kamberg Nature Reserve (Estcourt district) and Thamatu Pass (Underberg district), and the Drakensberg outlier, Bamboo Mountain, also in Underberg district, between c. 1 830 and 2 350 m. Often grows rooted in the turf immediately above Cave Sandstone cliffs, down which it then hangs in big mats, or insinuates itself into the turf seams of stony ridges; flowering between December and February. A most distinctive species, with unusual involucral bracts. Map 192.

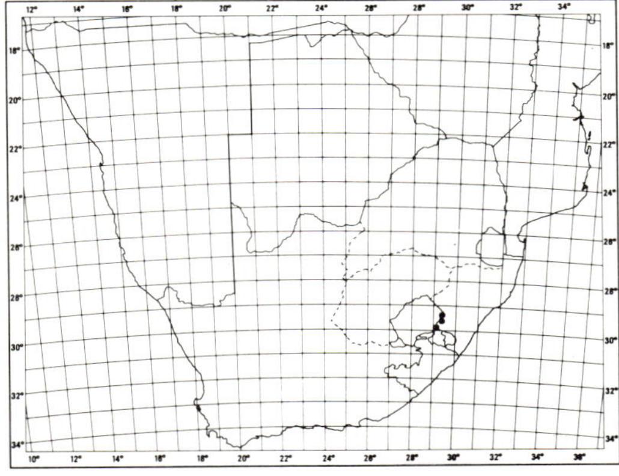

MAP. 192 — **Helichrysum hyphocephalum**

Voucher: *Hilliard & Burtt* 7833 (E; K; M; MO; NU; PRE; S).

197. **Helichrysum sessilioides** *Hilliard* in Notes R. bot. Gdn Edinb. 32: 360 (1973), Compositae in Natal 220 (1977). Type: Lesotho, Qacha's Nek distr., Mt Sauer, *Jacottett & Jacottet* (Z, holo.!).

H. aretioides Thell. in Vierteljahrsschrift naturf. Ges. Zürich 66: 241 (1921), non Turcz. (1851). Type as for *H. sessilioides*.

H. sessile sensu Killick, Mem. bot. Surv. S. Afr. 34: 140 (1963); Trauseld, Wild Flowers of Natal Drakensberg 200 and 201 (1969), non DC.

Much-branched cushion-forming dwarf shrub, very compact, smoothly rounded in outline, old main stems woody, gnarled, up to 10 mm diam., branches densely tufted, leafy. *Leaves* 4—15 × 1—5 mm, closely imbricate, rosulate when viewed from above, linear to lanceolate-elliptic, upper surface with smooth sericeous 'tissue-paper' indumentum, green with a silvery sheen, drying silvery grey, white-felted below. *Heads* homogamous, campanulate, 10—15 mm long, nearly double that when radiant, sessile, solitary at the tips of the branchlets. *Involucral bracts* in 8—12 series, graded, closely imbricate, much exceeding the flowers, outermost palest brown or sometimes rose-pink or crimson, inner white or tinged rose, opaque. *Receptacle* very shortly honeycombed. *Flowers* 37—118, yellow. *Achenes* 2 mm long, elliptic, with myxogenic duplex hairs. *Pappus* bristles many, equalling corolla, scabrid, bases cohering lightly by patent cilia. Fig. 55: 2.

On the high Lesotho mountains, the Drakensberg from the Cathedral Peak area in Natal to the Barkly East area in the E. Cape, the Witteberg near Lady Grey, and the Drakensberg outliers, Mount Currie near Kokstad and Ingeli Mountain on the Natal-Transkei border, between c. 2 000 and 3 200 m. Forms very dense, hard and compact cushions from 50 mm to c. 1 m across, tightly pressed against cliff faces or precipitous rock 'steps' down mountainsides, commonly on dolerite or basalt. Flowers from July to December. Map 193.

Much confused with *H. sessile* (no. 117), which is readily distinguished by its narrowly oblong leaves enveloped in thick white silky wool.

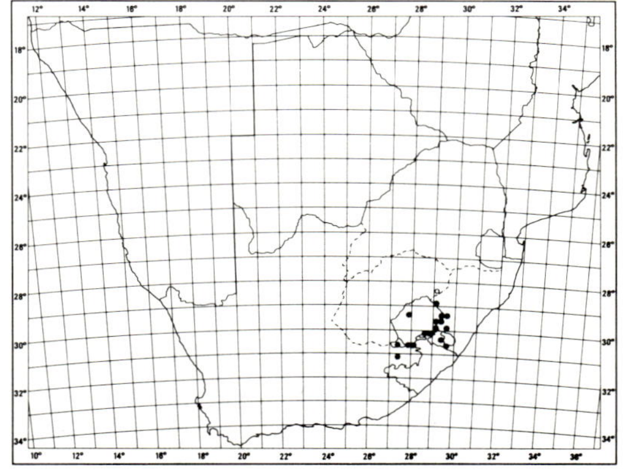

MAP 193.— ● **Helichrysum sessilioides**
○ **Helichrysum haygarthii**

Vouchers: *Beverly & Hoener* 717 (E; NU); *Hilliard* 5193 (E; K; MO; NU; PRE; S); *Wright* 187 (E; K; NU; S).

198. **Helichrysum haygarthii** *H. Bol.* in Trans. S. Afr. phil. Soc. 18: 382 (1907); Hilliard, Compositae in Natal 221 (1977). Type: Orange Free State, Harrismith distr., Rensburg's Kop, 6 000—7 000 ft., *Haygarth* in herb. *Wood* 9727 (BOL, holo.!; NH; P; PRE, iso.!).

H. multirosulatum O. Hoffm. & Muschl. in Annln naturh. Mus. Wien 24: 318 (1910). Type: Natal, Van Reenen's Pass, *Krook*, Pl. Penther. no. 1440 (W, holo.!).

Diffuse, prostrate, mat-forming subshrub, main stems bare, gnarled, up to c. 5 mm diam., ultimate branches very slender (c. 1 mm diam.), stoloniferous, leafy, producing dwarf side shoots with closely rosetted leaves. *Rosette leaves* up to 10 × 5 mm, obovate; *cauline leaves* c. 6 × 3 mm, imbricate, subspathulate, conduplicate, tips slightly recurved, all silvery sericeous. *Heads* homogamous, campanulate, c. 15 mm long, 20 mm across the fully radiating bracts, sessile, solitary at the tips of the dwarf shoots. *Involucral bracts* in c. 8 series, graded, closely imbricate, outer glossy crimson overlaid brown, much shorter and broader than inner, inner much exceeding flowers, white streaked crimson particularly below, opaque. *Receptacle* smooth. *Flowers* c. 50—91, yellow. *Achenes* 1,75 mm long, ellipsoid, with myxogenic duplex hairs. *Pappus* bristles many, equalling corolla, scabrid, bases with patent cilia scarcely cohering. Fig. 55: 3.

Recorded only from the two type localities, but not found again on Van Reenen's pass, only c. 10 km in a straight line from Rensburg's Kop, between Van Reenen and Harrismith in the Orange Free State. Grows on very broken weathered basalt cliffs at c. 2 225 m on Rensburg's Kop, flowering from about December to February. Map 193.

Voucher: *Hilliard* 4986 (E; K; NH; NU; PRE; S).

199. **Helichrysum bellidiastrum** *Moeser* in Bot. Jb. 48: 338 (1912); Hilliard,

Compositae in Natal 226 (1977). Type: Basutoland [Lesotho] Mont aux Sources, *Flanagan* 1966 (K; PRE, iso.!).

Rhizomatous perennial herb with one to several leaf rosettes at the crown, flowering stems usually solitary, c. 120 (—250) mm tall, loosely greyish-white woolly, leafy. *Radical leaves* prostrate, up to c. 70 × 30 mm, elliptic to obovate, loosely white- or greyish-woolly above at first, later shortly pubescent, white-felted below, the older leaves in particular commonly discolorous, drying greyish or blackish above, whitish below; *cauline leaves* up to c. 15 × 5 mm, becoming smaller upwards, distant, narrowly elliptic, greyish-white woolly. *Heads* homogamous, campanulate, c. 15 mm long, 30 mm across the fully radiating bracts, solitary. *Involucral bracts* in c. 15 series, graded, loosely imbricate, much exceeding the flowers, acute, outer palest golden-brown, sometimes pinkish or crimson, inner opaque white, pinkish at the base inside. *Receptacle* nearly smooth. *Flowers* 217—277. *Achenes* 1 mm, ellipsoid, with duplex hairs. *Pappus* bristles many, tips subplumose, bases with patent cilia, sometimes cohering. Fig. 55: 1.

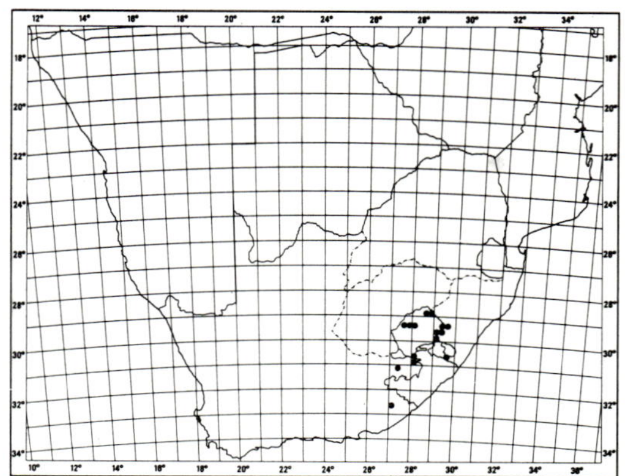

MAP 194.— **Helichrysum bellidiastrum**

FIG. 55.—1, **Helichrysum bellidiastrum,** whole plant, × 1; 1a, hermaphrodite flower, × 8; 1b, pappus bristle, × 11 (*Hilliard* 5416). 2, **H. sessilioides,** part of plant, × 1 (*Wright* 637). 3, **H. haygarthii,** part of plant, × 1 (*Hilliard* 4986). 4, **H. pagophilum,** part of plant, × 1 (*Hilliard* 5425). 5, **H. hyphocephalum,** part of plant, × 1; 5a, outer involucral bract, × 4,6 (*Hilliard & Burtt* 7833). 6, **H. glaciale,** part of plant, × 1 (*Hilliard* 5186).

Recorded on the high Lesotho mountains and the high Drakensberg from the Oxbow area and The Sentinel south to Naude's Nek (across the Cape Drakensberg near Maclear) and Saalboom Nek (near Barkly East) between 2 000 and 3 300 m; also on Ngeli Mtn on the Natal-Transkei border, then a disjunction (real ?) to the Amatola Mountains in the E. Cape. Favours damp turf slopes, particularly seepage areas and the banks of streams. Flowers between November and January. Map 194.

Vouchers: *Hilliard & Burtt* 7039 (E; K; MO; NU); *Hilliard & Burtt* 5826 (E; K; NH; NU; S; Z); *Trauseld* 891 (NU); *Wright* 1087 (E; K; NU).

200. **Helichrysum glaciale** *Hilliard* in Notes R. bot. Gdn Edinb. 32: 348 (1973), Compositae in Natal 225 (1977). Type: Cape, Maclear-Barkly East district boundary, Naude's Nek, c. 2 500 m, 27 xi 1971, *Hilliard* 5186 (NU, holo.!; BOL; E; GRA; K; M; MO; NBG; NH; NU; PRE; S, iso.!).

Well-branched rounded dwarf shrublet up to 200 mm high and 300 mm across, main stem very short, gnarled, woody, up to 5 mm diam., branches slender (c. 1 mm diam.), nude below and rooting, closely leafy above. *Leaves* closely imbricate, visible part suborbicular, c. 5 × 5 mm, thick, greyish-white woolly, base oblong, slightly shorter and narrower than blade, membranous, nearly glabrous, stem-clasping. *Heads* homogamous, campanulate, 10−13 mm long, c. 20 mm across the fully radiating bracts, solitary on terminal leafy peduncles, these leaves spathulate to elliptic, the uppermost sometimes with a scarious appendage. *Involucral bracts* in c. 11 series, descending on the peduncle, graded, loosely imbricate, outer palest brown, inner exceeding the flowers, obtuse, opaque, white, all tinged crimson at the base, giving a rosy cast to the whole head. *Receptacle* nearly smooth. *Flowers* 45−81, yellow. *Achenes* 1,25 mm long, with long (up to 0,5 mm) duplex hairs. *Pappus* bristles c. 12−14, about equalling corolla, tips shortly plumose, bases not cohering. Fig. 55: 6.

Recorded only from the high Drakensberg, at Naude's Nek north of Maclear in the Cape, nearby Satsanna's Peak in Lesotho, Sehlabathebe (Devil's Knuckles), Sani Pass and Mashai Pass in Natal, and Mota's Pass in Butha Buthe district of Lesotho, between 2 500 and 2 900 m. Grows mainly from cracks and ledges on the faces of basalt and dolerite cliffs; flowering between November and January. Very floriferous and pretty. Map 195.

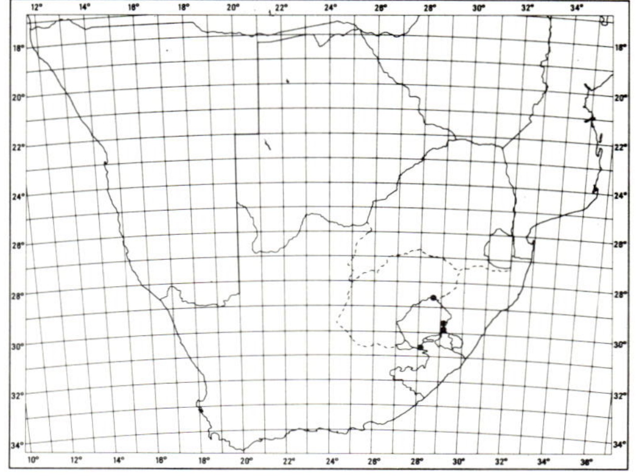

MAP 195.— **Helichrysum glaciale**

Vouchers: *Hilliard* 5417 (E; K; M; MO; NU; PRE; S); *Jacot Guillarmod* 2003 (PRE); *Nordenstam* 2053 (S).

Group 28

Perennial herbs with rosettes of radical leaves, sometimes forming small mats; *leaves* linear, oblong, elliptic, lingulate or obovate; *heads* heterogamous or rarely homogamous in the same species, 10−30 mm long, solitary or in few-headed corymbose panicles; *involucral bracts* radiating, white, pink or crimson, sometimes parti-coloured; *receptacle* shortly honeycombed; *flowers* c. 100−550, 0−90 ♀, corolla of ☿ flowers narrowly funnel-shaped, of ♀ flowers narrowly tubular; *achenes* glabrous or hairy; *pappus* bristles scabrid, tips barbellate to shortly subplumose, bases either free or cohering by patent cilia, occasionally some fusion as well.

Species 201−211, several endemic to the Drakensberg, one reaching Zimbabwe.

1a Ovaries hairy:

  2a Involucral bracts glossy:

    3a Heads c. 25–30 mm long:

      4a One or a few leaf rosettes tufted together, radical leaves membranous when dry; loosely woolly-cobwebby, wool sometimes stripping to leave only woolly margins, then upper surface clearly 3- to 5-nerved .... 201. *H. ecklonis*

      4b Mat-forming, with crowded leaf rosettes, radical leaves leathery, at maturity glabrous except for woolly margins, commonly only the main vein visible .... 202. *H. vernum*

    3b Heads c. 10–20 mm long:

      5a Leaves variously pubescent or woolly, but indumentum not silky:

        6a Leaves lingulate .... 203. *H. lingulatum*

        6b Leaves linear, lanceolate, elliptic-oblong or suborbicular:

          7a Flowering stems lateral to the leaf rosette; involucral bracts white, rose, crimson or parti-coloured:

            8a Heads 15–20 mm long; radical leaves elliptic to suborbicular .... 209. *H. adenocarpum*

            8b Heads 10 mm long, radical leaves linear .... 210. *H. petraeum*

          7b Flowering stem terminal; involucral bracts white .... 211. *H. monticola*

      5b Leaves silvery grey silky-woolly, linear or linear-lanceolate; involucral bracts white .... 204. *H. argentissimum*

  2b Involucral bracts dull white, wool on leaves confined to margins, or wanting .... 206. *H. marginatum*

1b Ovaries glabrous:

  9a Heads c. 20 mm long .... 205. *H. album*

  9b Heads c. 10–15 mm long:

    10a Heads c. 15 mm long, leaves glandular-pilose, margins white-woolly or not .... 207. *H. bellum*

    10b Heads c. 10 mm long, leaves grey silky-woolly .... 208. *H. palustre*

---

201. Helichrysum ecklonis *Sond.* in F.C. 3: 254 (1865). Type: Cape, Tambukiland, an der rechten Seite des Key-rivier, 3 000–4 000 Fuss, Nov., *Ecklon* 1410 (S, holo.!; B; G-DC; E, prob. iso.!).

H. calocephalum Schltr. in Bot. Jb. 40: 95 (1908), non Klatt (1896); Wood, Natal Plants 6,4: t. 589 (1912); Batten & Bokelmann, Wild Flow. E. Cape Prov. 149 plate 119,2 (1966). *H. lamprocephalum* H. Bol. in Trans. R. Soc. S. Afr. 1: 163 (1909); Hilliard, Compositae in Natal 235 (1977). *H. scapiforme* Moeser in Bot. Jb. 44: 334 (1910). Type: Natal, Umzinto distr., Dumisa, Moyeni, not far from Fairfield, c. 700 m, *Rudatis* 437 (errore 1317, BM; E; G; K; M; S; Z, iso.!).

Perennial herb, stock stout, woody, crowned with 1 or a few rosettes, flowering stem terminal, solitary, up to 500 m high, height depending on age and growing conditions, loosely woolly, leafy. *Radical leaves* up to c. 200 × 20 mm, but commonly less than 100 × 20 mm, oblong-lanceolate to elliptic-obovate, membranous, loosely woolly-cobwebby, upper surface sometimes glabrescent, then clearly 3- or 5-nerved, lower surface usually persistently woolly, margins always so; *cauline leaves* similar but narrower and decreasing rapidly in size upwards, uppermost often tipped with a scarious bract. *Heads* heterogamous, or rarely homogamous, campanulate, 25–30 mm long, nearly double that across the radiating bracts, solitary. *Involucral bracts* in c. 10 series, graded, loosely imbricate, much exceeding the flowers, glossy, white to deep rose pink. *Receptacle* shortly honey-combed. *Flowers* 231–407, (0–) 7–13 ♀, 224–403 ☿. *Achenes* 1,75 mm long, barrel-shaped, with myxogenic duplex hairs. *Pappus* bristles many, equalling corolla, scabrid, tips barbellate, bases cohering by patent cilia, some fusion in bundles as well. Fig. 56: 1.

Ranges from the low Drakensberg on the Transvaal-Natal border and the NE. corner of the Orange Free State through Lesotho and Natal (sea level to c. 2 750 m) to the Transkei and the King William's Town, Amatola Mountains and Katberg area of the E. Cape. Grows on grassy slopes, flowering between September and December. Very easily confused with *H. vernum* (below). Map 196.

Natural hybrids are suspected between *H. ecklonis* and *H. argentissimum* (no. 204); see Hilliard, l.c.

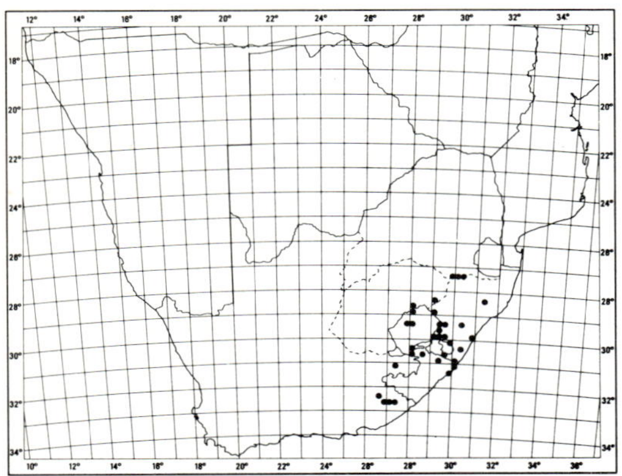

MAP 196.— **Helichrysum ecklonis**

Vouchers: *Devenish* 1564 (NU); *Hilliard & Burtt* 7320 (E; K; MO; NU; S); *Wright* 1008 (E; K; NU; S).

202. **Helichrysum vernum** *Hilliard* in Notes R. bot. Gdn Edinb. 32: 363 (1973), Compositae in Natal 236 (1977). Type: Natal, Mpendhle distr., Storm Heights, c. 2 100 m, *Wright* 229 (NU, holo.!; E; PRE, iso.!).

Mat-forming perennial herb with crowded leaf rosettes arising from a thick woody branching underground stem, flowering stems terminal, solitary, up to 250 mm long, height depending upon age and growing conditions, loosely woolly, closely leafy. *Basal leaves* up to c. 50 × 15 mm, elliptic-obovate, rarely lanceolate-oblong, apex obtuse to subacute, apiculate, base broad, clasping, loosely cobwebby at first, soon glabrous but margins persistently woolly, thick, leathery, commonly only the midvein visible; *cauline leaves* mostly oblong, up to 25 × 3 mm, decreasing in size upwards, uppermost tipped with a pink scarious bract. *Heads* heterogamous, campanulate, 25−30 mm long, double that across the radiating bracts, solitary. *Involucral bracts* in c. 10 series, graded, loosely imbricate, much exceeding the flowers, acute, rich glossy rose-pink to almost crimson. *Receptacle* shortly honeycombed. *Flowers* 172−304, 8−24 ♀, 164−280 ♀. *Achenes* 1,75 mm long, barrel-shaped, with myxogenic duplex hairs. *Pappus* bristles many, equalling corolla, scabrid, tips barbellate, bases cohering strongly by patent cilia.

On the Natal Drakensberg and its foothills and outliers from Oliviershoek Pass to Sehlabathebe in Lesotho, also the Suurberg at Weza, between 1 675 and 2 560 m. Forms compact mats in stony turf on mountain slopes; one of the first plants to flower in spring, between September and November. Map 197.

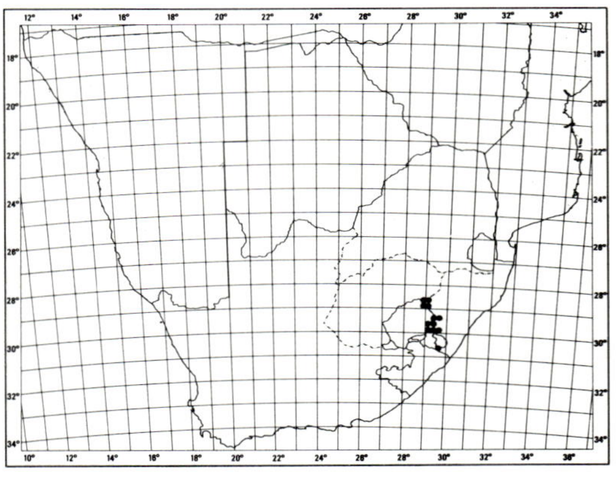

MAP 197.— **Helichrysum vernum**

Much confused with *H. ecklonis* (above) whose tufted rosettes contrast in the field with the tight mats of *H. vernum*, but in the herbarium this difference disappears. However, the leathery leaves of *H. vernum*, glabrous above and with side veins scarcely discernible, will generally serve to distinguish it from glabrous-leaved specimens of *H. ecklonis*, where the side veins are clearly visible in the membranous blade.

H. vernum probably hybridizes with *H. argentissimum* (no. 204); see Hilliard, l.c.

Vouchers: *Edwards* 340 (NU); *Hilliard* 4003 (E; K; NU); *Trauseld* 445 (NU); *Wright* 310 (E; MO; NU).

203. **Helichrysum lingulatum** *Hilliard* in Notes R. bot. Gdn Edinb. 32: 352 (1973),

---

FIG. 56.−1, **Helichrysum ecklonis**, whole plant, × 1; 1a, hermaphrodite flower, × 5,3; 1b, female flower, × 5,3; 1c, pappus bristle, × 5,3 (*Hilliard* 5172). 2, **H. bellum**, whole plant, × 1 (*Wright* 1107). 3, **H. lingulatum**, leaf, × 1 (*Hilliard* 5173). 4, **H. argentissimum**, leaf, × 1 (*Hilliard & Burtt* 7319). 5, **H. monticola**, whole plant, × 1 (*Hilliard* 2294).

Compositae in Natal 238 (1977). Type: Cape, Maclear distr., summit Pot River Pass, c. 1 950 m, *Hilliard* 5173 (NU, holo.!; E; K; M; MO; NBG; PRE; S, iso.!).

Mat-forming perennial herb with many crowded leaf rosettes from woody well-branched underground stems up to 10 mm diam., flowering stems terminal to each rosette, solitary, up to 180 mm tall, grey-woolly, leafy. *Radical leaves* up to 35 × 8 mm, lingulate, apex rounded, both surfaces closely grey-woolly; *cauline leaves* mostly narrowly elliptic, tipped with a white scarious bract. *Heads* homogamous or heterogamous, campanulate, c. 20 mm long, nearly double that across the radiating bracts, solitary. *Involucral bracts* in c. 7 series, graded, loosely imbricate, much exceeding the flowers, glossy, tinged delicate pink initially, fading to white, crimson blotch inside near base. *Receptacle* shortly honeycombed. *Flowers* 152−180, 0−12 ♀, 152−170 ♀. *Achenes* 1,5 mm, barrel-shaped, with myxogenic duplex hairs. *Pappus* bristles many, tips barbellate, bases cohering by patent cilia, some fusion in bundles as well. Fig. 56: 3.

Scanty records only from Underberg district in Natal, Mt Fletcher and Engcobo districts in Transkei, Maclear district in the E. Cape, and Qacha's Nek and Ramatseliso's Gate in Lesotho, between c. 1 500 and 1 950 m. Forms large mats on hard bare earth or in stony turf; flowering in September and October. Can be confused with *H. argentissimum* (below) but that species has linear or linear-lanceolate leaves with silky indumentum. Map 198.

Vouchers: *Coleman* 640 (NU); *Esterhuysen* 29166 (BOL); *Hilliard & Burtt* 13103 (E; K; NU; PRE; S); *Jacottet* 35 (G); *Werdermann & Oberdieck* 1116 (K).

204. **Helichrysum argentissimum** *J. M. Wood* in Kew Bull. 1907: 364 (1907); Hilliard, Compositae in Natal 237 (1977). Type: Natal, Lion's River distr., summit Mt Gilboa, *Wylie* in herb. Wood 10025 (NH, holo.!; BOL; E; K; PRE; S; SAM, iso.!).

Mat-forming perennial herb with numerous crowded leaf rosettes arising from a thick, woody, well-branched underground stem system, flowering stems terminal, solitary, up to 400 mm long, but often less than 150 mm, grey-woolly, leafy. *Radical leaves* up to 120 × 6 mm, but often only 30−50 × 3−4 mm, linear or linear-lanceolate, apex acute to subacute, silvery grey silky-woolly; *cauline leaves* similar but shorter, uppermost with a white scarious tip. *Heads* homogamous or heterogamous, campanulate, c. 20 mm long, double that across the fully radiating bracts, solitary. *Involucral bracts* in c. 10 series, graded, loosely imbricate, much exceeding the flowers, acute, glossy white, crimson or pink blotch at base inside. *Receptacle* shortly honeycombed. *Flowers* 111−201, 0−10 ♀, 103−191 ♀. *Achenes* 1,5 mm long, barrel-shaped, with myxogenic duplex hairs. *Pappus* bristles many, tips barbellate, bases cohering by patent cilia, some fusion in bundles as well. Fig. 56: 4.

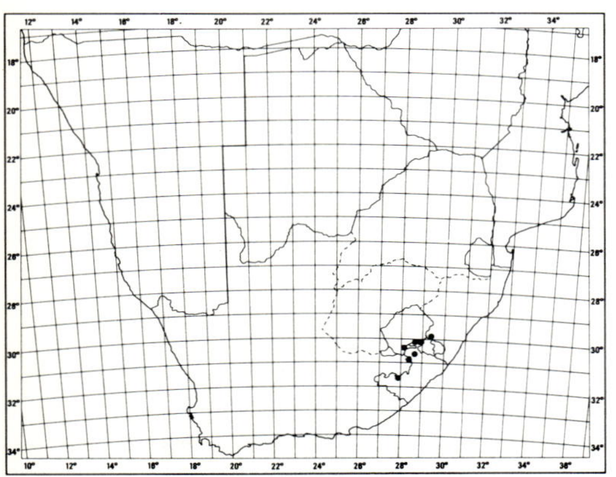

MAP 198.— **Helichrysum lingulatum**

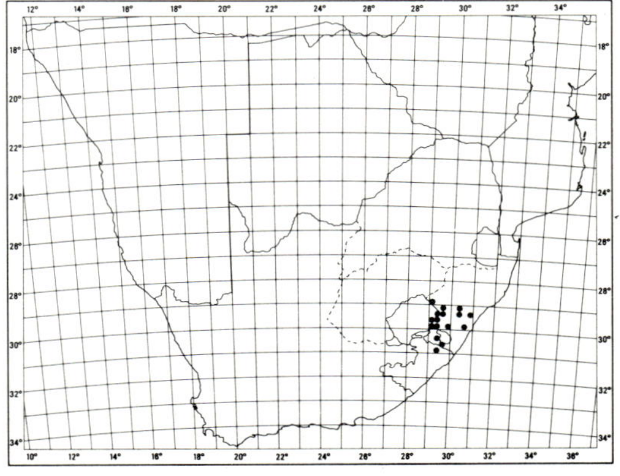

MAP 199.— **Helichrysum argentissimum**

Ranges from Bergville, Estcourt, Lion's River, Umvoti and New Hanover districts south to Mt Currie in East Griqualand and Insizwa and Tabankulu Mountains in the Transkei, between c. 900 and 2 300 m above sea level. Grows on stony grass slopes or on broken sandstone cliffs or platforms, forming thick mats as much as a metre across where protected from fire. Flowers between September and March, but mainly in spring. Map 199.

Closely resembles *H. lingulatum* (above), but is distinguished by its linear, not lingulate, leaves and silky-woolly, not merely woolly, indumentum. Possibly hybridizes with other species; see under *H. ecklonis* (no. 201).

Vouchers: *Hilliard & Burtt* 7319 (E; K; NU; S); *Strey* 10533 (NU); *Wright* 208 (E; K; NH; NU).

205. **Helichrysum album** *N.E. Br.* in Kew Bull. 1895: 24 (1895); Hilliard, Compositae in Natal 235 (1977). Type: Natal, on the Drakensberg at Bushman's River Pass, near the snowline, 7 000—8 000 ft., *Evans* 48 (K, holo!; NH, iso.!).

Perennial herb with 1 or a few leaf rosettes clustered on the crown of a woody stock, flowering stems 1 or occasionally 2 lateral to the new rosette, up to 150 mm tall, loosely greyish-white woolly, closely leafy. *Radical leaves* up to 60 × 25 mm, but commonly c. 35 × 25 mm, obovate, thickly and loosely greyish-white woolly; *cauline leaves* oblanceolate, up to 30 × 7 mm, decreasing in size upwards, woolly. *Heads* heterogamous, campanulate, c. 20 mm long, double that across the radiating bracts, solitary. *Involucral bracts* in c. 11 series, graded, loosely imbricate, much exceeding flowers, acute, glossy white, crimson at base inside. *Receptacle* shortly honeycombed. *Flowers* 172—224, 19—26 ♀, 153—199 ☿. *Achenes* c. 1,5 mm long, glabrous. *Pappus* bristles many, equalling corolla, tips barbellate, bases with patent cilia scarcely cohering.

Recorded from only a limited area of the high Drakensberg on the Natal-Lesotho border, from Cathedral Peak (Bergville distr.) to Garden Castle Nature Reserve (Underberg distr.) between c. 2 000 and 3 300 m above sea level. Grows in small colonies on steep, broken turf slopes and low basal cliffs; flowering between January and March. Rarely collected. Resembles solitary-headed specimens of *H. adenocarpum* (no. 209), but is readily distinguished by its glabrous ovaries. Map 200.

Vouchers: *Trauseld* 925 (NU); *Wright* 153 (E; NU).

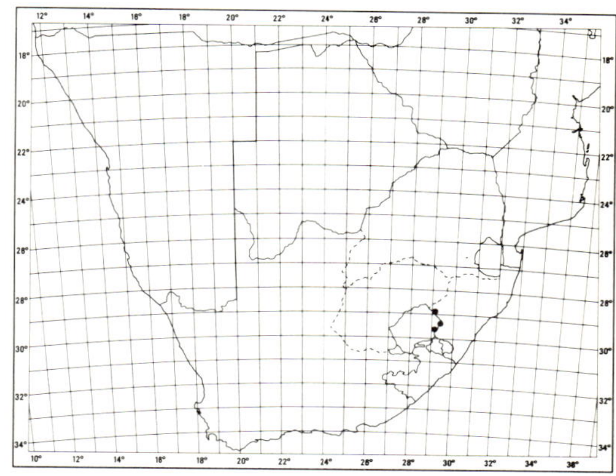

MAP 200.— **Helichrysum album**

206. **Helichrysum marginatum** *DC.,* Prodr. 6: 180 (1838); Harv. in F. C. 3: 229 (1865); Moeser in Bot. Jb. 44: 335 (1910); Hilliard, Compositae in Natal 234 (1977). Type: Cape, Wittebergen, *Drège* 3741 (G-DC, holo.!; BM; E; K; P; S; SAM; TCD, iso.!).

Gnaphalium marginatum (DC.) Sch. Bip. in Bot. Ztg 3: 170 (1845).

Mat-forming dwarf shrub, main stems prostrate, branching, thick (10 mm diam.) and woody with age, producing many crowded leaf rosettes, flowering stems terminal, often solitary, erect, up to c. 150 mm tall, loosely greyish-white woolly, leafy. *Radical leaves* lingulate, up to c. 40 × 10 mm, leathery, glandular punctate, glabrous at maturity or occasionally glandular-pubescent, margins white-woolly; *cauline leaves* similar but smaller and decreasing in size upwards. *Heads* heterogamous, campanulate, c. 15—20 mm long, double that across the radiating bracts, solitary. *Involucral bracts* in c. 10 series, graded, loosely imbricate, much exceeding the flowers, acute, snow-white, crimson blotch near base inside. *Receptacle* scarcely honeycombed. *Flowers* 195—259, 18—29 ♀, 175—234 ☿. *Achenes* 1,25 mm long, cylindric, with duplex hairs. *Pappus* bristles many, scabridulous, bases not cohering.

On the high Lesotho mountains and the high Drakensberg from Oxbow and Mont aux Sources south to the Cape Drakensberg and nearby Witteberg, between 2 440 and 3 300 m. Forms mats up to a metre across on steep stony mountain slopes and rock sheets; flowering between December and February. Map 201.

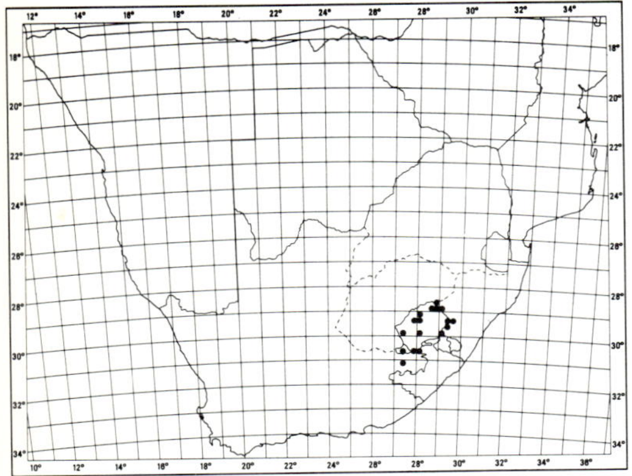

MAP 201.— **Helichrysum marginatum**

Can be confused with *H. bellum* (below) but distinguished by its different habit, leathery glandular-punctate leaves and hairy achenes.

Vouchers: *Hilliard* 4993 (E; K; NH; NU); *Nordenstam* 2089 (LD; NU); *Wright* 397 (E; NH; NU).

207. **Helichrysum bellum** *Hilliard* in Notes R. bot. Gdn Edinb. 31: 15 (1971), Compositae in Natal 233 (1977). Type: Natal-Lesotho border, top of Bushman's River Pass, c. 3 050 m, *Wright* 436 (NU, holo.!; E; K; NH; PRE; S, iso.!).

Perennial herb with 1 or a few leaf rosettes crowning a woody rhizome, flowering stems terminal, but lateral to the new rosettes, 1 or 2 from a rosette, decumbent then erect to c. 150 (−300) mm, loosely greyish-white woolly, closely leafy. *Radical leaves* up to 150 × 20 mm, but commonly not more than 50 × 15 mm, lanceolate-spathulate or lanceolate-elliptic, membran-ous, glandular-pilose, margins white-woolly particularly when young; *cauline leaves* similar but smaller and diminishing in size upwards. *Heads* heterogamous, campanu-late, c. 15 mm long, double that across the radiating bracts, solitary, very rarely paired. *Involucral bracts* in c. 10 series, graded loosely imbricate, much exceeding the

flowers, acute, snow-white. *Receptacle* nearly smooth. *Flowers* 287−544, 31−67 ♀, 256−478 ☿. *Achenes* 1,5 mm, ellipsoid, strongly 5-ribbed, glabrous. *Pappus* bristles many, about equalling corolla, scabridu-lous, bases sometimes lightly cohering by patent cilia. Fig. 56: 2.

On the high Lesotho mountains and the high Drakensberg from Mont aux Sources south to Barkly Pass in the Cape Drakensberg and the nearby Witteberg, between 2 400 and 3 300 m. Grows scattered in stony turf on steep slopes and on the summit plateau; flowering between January and March. See also *H. marginatum* (above) and *H. palustre* (below). Map 202.

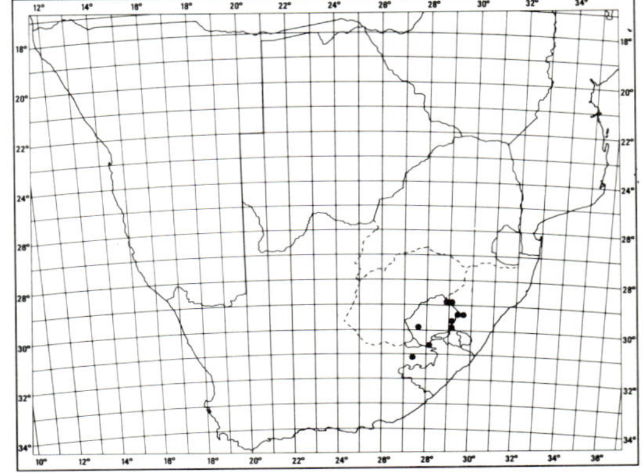

MAP 202.— **Helichrysum bellum**

Vouchers: *Hilliard* 5291 (E; K; NU; S); *Norden-stam* 2020 (LD; NU); *Wright* 1107 (E; K; NU).

208. **Helichrysum palustre** *Hilliard* in Notes R. bot. Gdn Edinb. 31: 17 (1971), Compositae in Natal 233 (1977). Type: Lesotho, plateau at headwaters of Loteni River, c. 400 yards from exit of Bushman's River Pass, c. 3 050 m, *Wright* 753 (NU, holo.!; E; K; NH, iso.!).

Perennial herb with tufts of leaf rosettes from a woody rhizome, flowering stems lateral, 1 or 2 from each tuft, erect to about 150 mm, loosely grey-woolly, closely leafy. *Radical leaves* up to 50 × 7 mm, lanceolate-lingulate, persistently silvery-sericeous; *cauline leaves* lanceolate, dimin-ishing in size upwards. *Heads* heterogamous, campanulate, c. 10 mm long, double that across the radiating bracts, solitary, or very rarely paired. *Involucral bracts* in 8—10 series, graded, loosely imbricate, much exceeding the flowers, acute, snow-white, dull crimson patch near base inside. *Receptacle* very shortly honeycombed. *Flowers* 145—224, 15—32 ♀, 130—192 ☿. *Achenes* 1,5 mm long, strongly 5-ribbed, glabrous. *Pappus* bristles many, equalling corolla, scabridulous, bases lightly fused in bundles.

Recorded only from the high Lesotho mountains and the summit plateau of the Drakensberg between c. 2 300 and 3 400 m, and rarely collected. Grows in marshes, either in the water or on raised hummocks, or along marshy streamsides; flowering in December and January. Map 203.

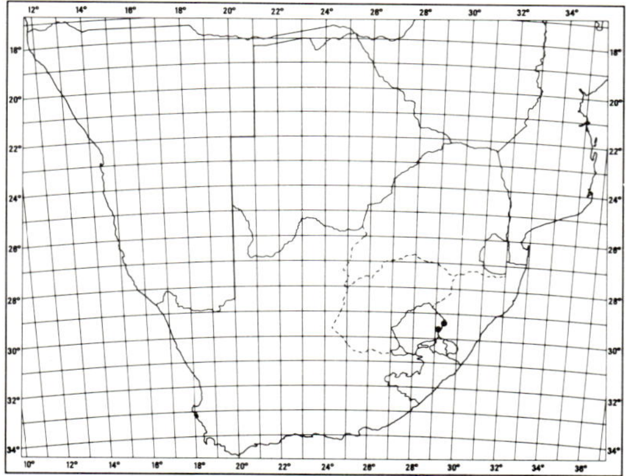

MAP 203.— **Helichrysum palustre**

Distinguished from *H. bellum* (above) by its smaller heads and silky leaves.

Vouchers: *Hilliard* 5387 (E; K; M; MO; NU; PRE; S); *Wright* 332 (E; K; NH; NU).

209. **Helichrysum adenocarpum** *DC.*, Prodr. 6: 180 (1838); Harv. in F.C. 3: 229 (1865); Moeser in Bot. Jb. 44: 335 (1910); Batten & Bokelmann, Wild. Flow. E. Cape Prov. 156, plate 125, 2 (1966); Compton, Fl. Swaziland 627 (1976); Hilliard, Compositae in Natal 239 (1977). Type: Natal, Port Natal [Durban], *Drège* 5019 (G-DC, holo.!; E; K; SAM, iso.!).

Gnaphalium adenocarpum (DC.) Sch. Bip. in Bot. Ztg 3: 170 (1845).

Perennial herb, stock woody, crowned with 1—several leaf rosettes, flowering stems lateral, decumbent or erect, 40—450 mm long, simple or forking above into a few- to many-headed very open corymb, glandular-pubescent or woolly, closely leafy. *Radical leaves* suborbicular to elliptic-oblong, prostrate, 20—40 (—140) × 15—25 (—40) mm, mostly loosely grey-woolly or cobwebby, sometimes only glandular-pubescent; *cauline leaves* oblong to lanceolate, smaller than the radical and passing into inflorescence bracts, woolly, cobwebby or glandular-pubescent, often without wool when the radical leaves are woolly. *Heads* heterogamous, campanulate, 15—20 mm long, c. 25—35 mm across the radiating bracts. *Involucral bracts* in 9—11 series, graded, loosely imbricate, much exceeding the flowers, glossy, acute, white, or white tipped rose, crimson or scarlet, or wholly rose or crimson. *Flowers* c. 165—520, 20—90 ♀, 145—500 ☿, yellow often tipped red. *Achenes* 0,75 mm long, barrel-shaped, with myxogenic duplex hairs. *Pappus* bristles several, barbellate to subplumose in upper part, bases nude, not cohering.

Two subspecies are recognized:

1a Radical leaves cobwebby or woolly, involucral bracts ranging from crimson to scarlet or pink, or these colours on white, or rarely pure white and then growing with plants with parti-coloured bracts ....
.... (a) subsp. *adenocarpum*

1b Radical leaves glandular, very rarely cobwebby or woolly, involucral bracts always pure white.....................(b) subsp. *ammophilum*

(a) subsp. **adenocarpum.**

The heads are always c. 15—20 mm long, but there is much variation in stature, leaf size and indumentum, number of heads on the flowering stem, and colour of the involucral bracts. Plants with 1-headed stems are common above c. 1 200 m, those with many-headed stems from sea-level to c. 1 800 m, but there are many exceptions. Plants with pure white bracts have been recorded only below c. 600 m and grow mixed with plants with parti-coloured bracts.

Widespread, from the Soutpansberg south through the highlands of the E. Transvaal and W. Swaziland to the mountainous NE. corner of the Orange Free State, Lesotho, Natal, Transkei, and E. Cape as far as the Amatola Mountains. Also in the E. highlands of Zimbabwe and neighbouring Mozambique. Found in grassland, often on moist slopes or in moist depressions from sea level to c. 3 000 m; flowering mainly between January and April. Map 204.

Vouchers: *Codd* 2777 (NU; PRE); *Compton* 25808 (PRE); *Hilliard* 8088 (E; K; M; MO; NU; S); *Hilliard & Burtt* 6584 (E; K; MO; NU; PRE).

H. adenocarpum is frequently confused with *H. monticola* (no. 211) but is easily distinguished by its lateral, not terminal, flowering stems.

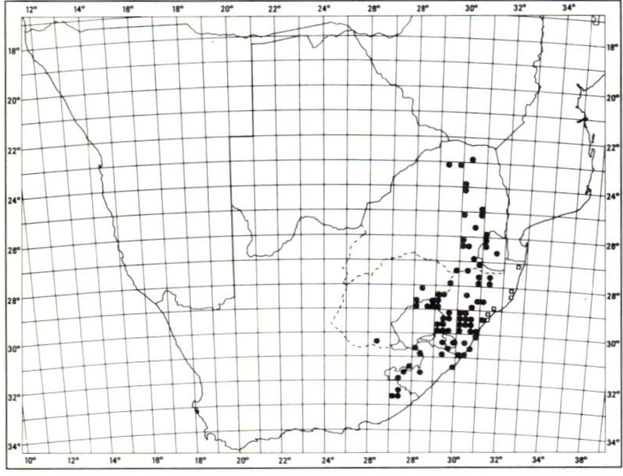

MAP 204.— ● **Helichrysum adenocarpum**
           subsp. **adenocarpum**
        □ **Helichrysum adenocarpum**
           subsp. **ammophilum**

(b) subsp. **ammophilum** *Hilliard* in Notes R. bot. Gdn Edinb. 32: 343 (1973), Compositae in Natal 240 (1977). Type: Natal, Ingwavuma distr., Ilala Flats past Mosi swamp on road to Maputa, 20 xii 1968, *Pooley* 268 (NU, holo.!; E; NH, iso.!).

Generally a much more robust plant than subsp. *adenocarpum* with stems sometimes attaining a height of 1 m, many-headed corymbs, pure white involucral bracts, and at least the radical leaves without wool at maturity.

Recorded on the Tongaland plain, from Delagoa Bay and Inhaca Island in Mozambique to Lower Tugela district in Natal, in sandy poorly drained grassland, often around the edges of marshy depressions; flowering between August and April. Map 204.

Vouchers: *Moll* 4716 (NU; PRE); *Pooley* 1712 (E; K; MO; NU; PRE; S); *Strey* 6803 (NU; PRE).

210. **Helichrysum petraeum** *Hilliard* in Notes R. bot. Gdn Edinb. 40: 263 (1982). Type: Natal, Richmond distr., ridge leading to Peak of Byrne, c. 1 525 m, 29 iv 1976, *Hilliard* 8092 (NU, holo.!; E; K; M; MO; PRE; S, iso.!).

Perennial herb, stock woody, crowned with several leaf rosettes, flowering stems lateral, decumbent, up to 450 mm long, loosely branched, loosely woolly, leafy. *Radical leaves* c. 30−50 × 2−4 mm, linear or linear-lanceolate, apex acute or subacute, mucronate, base broad, papery, clasping, purplish when young, both surfaces grey-woolly; *cauline leaves* c. 10−20 × 2−3 mm, oblong-lanceolate, acute, mucronate, base slightly narrowed, half-clasping, both surfaces loosely woolly, glandular-setose, wool glabrescent. *Heads* heterogamous, campanulate, 10 mm long, about twice that across the radiating bracts, solitary at the branch tips, the whole forming a very open corymbose panicle. *Involucral bracts* in c. 10 series, loosely imbricate, much exceeding the flowers, glossy, acute, white, tipped crimson. *Flowers* c. 120−230, 18−32 ♀, 118−195 ☿, yellow, sometimes tipped crimson. *Achenes* not seen, ovaries with myxogenic duplex hairs. *Pappus* bristles several, barbellate above, nude below, bases not cohering.

Recorded from Mbabane district in Swaziland, Nkandla, Mpendhle and Richmond districts in Natal. Grows around the edges and in the crevices of rock sheets in grassland; flowering in March and April. Map 205.

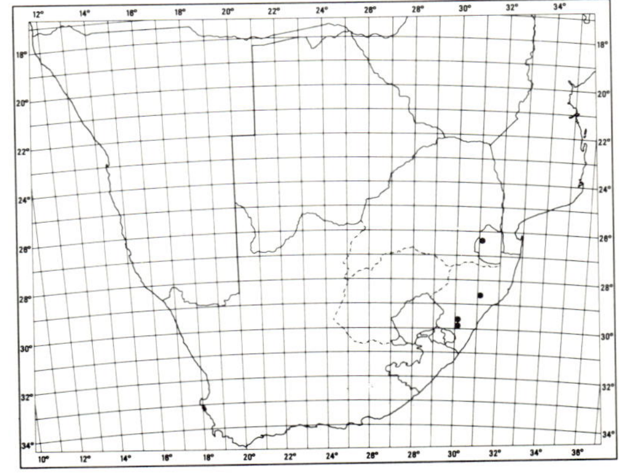

MAP 205.— **Helichrysum petraeum**

Easily confused with *H. adenocarpum* (above), but readily distinguished by its linear radical leaves and small heads.

Vouchers: *Compton* 25862 (M; PRE); *Galpin* 11939 (BOL; PRE); *Stewart* 2056 (E; K; MO; NU); *Wood* 9923 (BOL).

211. **Helichrysum monticola** *Hilliard* in Notes R. bot. Gdn Edinb. 31: 16 (1971), Compositae in Natal 240 (1977); Compton, Fl. Swaziland 631 (1976). Type· Natal, Vryheid distr., Zungwini Mt., 8 miles NE. Vryheid, c. 1 350 m, *Hilliard & Burtt* 5856 (NU, holo.!; E; K; NH, iso.!).

Perennial herb with 1 or several leaf rosettes crowning a thick woody stock, flowering stems solitary, terminal, up to c. 300 mm high, closely leafy. *Radical leaves* up to 90 × 15 mm, narrowly to broadly lanceolate, upper surface glandular-pubescent and cobwebby, wool usually deciduous, lower surface commonly greyish-white woolly, occasionally only cobwebby or glandular-pubescent; *cauline leaves* similar but smaller and passing into inflorescence bracts, more or less imbricate. *Heads* heterogamous, campanulate, 15—20 mm long, 20—30 mm across the fully radiating bracts, solitary or more usually in a few- to several-headed corymb. *Involucral bracts* in c. 10 series, graded, loosely imbricate, much exceeding the flowers, acute, glossy, white, inner tinged pink or crimson at base inside, rarely outer bracts with a brownish pink overcast (Barberton mountains). *Receptacle* shortly honeycomb-ed. *Flowers* 102—329, 10—26 ♀, 92—312 ♀. *Achenes* 1,5 mm long, with myxogenic duplex hairs. *Pappus* bristles c. 5, tips barbellate, shaft scabrid, bases not cohering. Fig. 56: 5.

Ranges from the environs of Lydenburg through the eastern highlands of the Transvaal to Mbabane district in Swaziland, the low Drakensberg on the Transvaal-Natal border, the NE. mountainous corner of the Orange Free State, the Natal Drakensberg and its outliers, Qudeni and Karkloof ranges, East Griqualand, the mountainous parts of the Transkei and the high parts of the E. Cape as far south as Pirie near King William's Town, mostly between 1 500 and 2 700 m. Grows on stony grass slopes; flowering between December and February. Map 206.

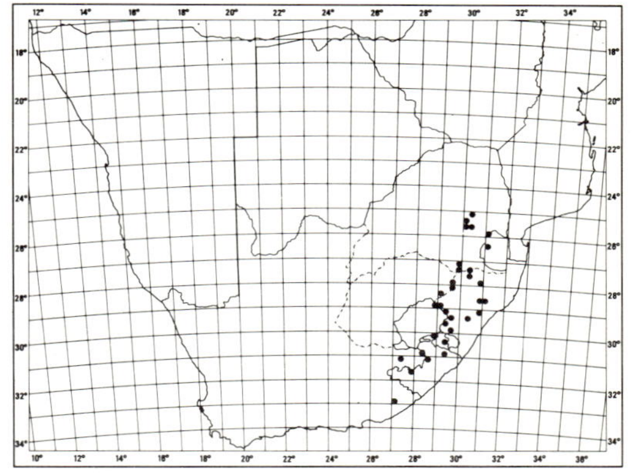

MAP 206.— **Helichrysum monticola**

Much confused with *H. adenocarpum* (no. 209), but most easily distinguished by its flowering stems, terminal in *H. monticola*, lateral in *H. adenocarpum*.

Vouchers: *Devenish* 1613 (E; K; MO; NU; S); *Hilliard* 5179 (E; K; MO; NU; S); *Hilliard & Burtt* 6548 (E; K; NU; PRE).

Group 29

Shrubs, subshrubs or mat-forming perennials becoming woody with age; *leaves* small or medium sized, linear, linear-lanceolate, oblong, elliptic or obovate; *heads* heterogamous or rarely homogamous in the same species, c. 10—25 mm long, solitary or few in lax corymbs; *involucral bracts* radiating, generally white with some brown, rose or crimson, sometimes yellow and brown; *receptacle* shortly honeycombed, paleate in one species; *flowers* 102—483, 0—50 ♀, corolla of ♀ flowers funnel-shaped, ♀ flowers tubular with a conspicuous limb; *achenes* glabrous; *pappus* bristles scabridulous below, barbellate above, bases nude, not cohering.

Species 212—221, one *(H. herbaceum)* widely distributed from Katanga and Tanzania to the S. Cape, the others mostly confined to Natal, the Transvaal and E. Cape, some narrowly endemic. Mainly in rocky grassland, rock outcrops and cliffs on mountain tops and slopes.

1a Involucral bracts yellow or yellow and brown:
  2a Heads c. 10 mm long; leaves enveloped in silvery white somewhat glossy indumentum
  ....212. *H. argyrophyllum*

  2b Heads c. 18—20 mm long; leaves loosely woolly ....213. *H. herbaceum*
1b Involucral bracts white, usually with some brown or crimson:
    3a Heads c. 25 mm long; stout shrub with large leaves 20—50 × 10—25 mm .... 214. *H. summo-montanum*

    3b Heads 10—20 mm long; leaves up to 9 mm broad, often much narrower:
      4a Shrubs or subshrubs, erect or forming tangled clumps:
        5a Plants closely branched:
          6a Heads c. 15 mm long ....215. *H. junodii*

          6b Heads c. 10—12 mm long:
            7a Leaves linear or linear-lanceolate, very acute, soon reflexed .... 218. *H. reflexum*

            7b Leaves oblong or lanceolate-oblong, apex obtuse to acute, not reflexed
            ....219. *H. mariepscopicum*

        5b Plants loosely branched:
          8a Leaves mostly 3—9 mm broad, narrowly panduriform to oblong becoming lanceolate-
            acuminate towards the heads ....216. *H. wilmsii*

          8b Leaves mostly 1—2 (—3) mm broad, linear or linear-lanceolate .... 217. *H. argyrolepis*
      4b Mat-forming perennial herbs or subshrubs:
        9a Radical leaves linear-lanceolate .... 220. *H. confertifolium*

        9b Radical leaves obovate to subspathulate .... 221. *H. milfordiae*

212. Helichrysum argyrophyllum *DC.,* Prodr. 6: 186 (1838); Harv. in F.C. 3: 233 (1865); Moeser in Bot. Jb. 44: 238 (1910); Batten & Bokelmann, Wild Flow. E. Cape Prov. 151, plate 121, 1 (1966). Lectotype: Cape, Kaffraria, *Ecklon* 1376 (G-DC!).

Gnaphalium argyrophyllum (DC.) Sch. Bip. in Bot. Ztg 3: 171 (1845).

Prostrate, well-branched, mat-forming dwarf shrub often forming extensive carpets, old stems bare, rooting, flowering branchlets decumbent or erect to c. 100 mm, simple, closely leafy, stems, leaves and base of each head closely enveloped in silvery white, closely-felted somewhat glossy indumentum. *Leaves* at first rosetted at the branch tips, obovate or obovate-spathulate, up to 30 × 10 mm, distant, smaller and narrower upwards on the flowering twigs, apex obtuse, hooked, recurved, base attenuated, half-clasping. *Heads* homogamous, broadly campanulate, c. 10 × 20 mm across the fully radiating bracts; solitary or few in lax corymbs. *Involucral bracts* in c. 8 series, graded, inner much exceeding the flowers, radiating, loosely imbricate, subacute, lemon-yellow, outer overlaid golden-brown. *Receptacle* paleate, paleae about twice the length of the ovaries, lanceolate, more or less conduplicate, yellow, scarious.

Flowers 149—202. *Achenes* not seen, ovaries glabrous. *Pappus* bristles many, about equalling corolla, yellow, scabridulous below, barbellate above, bases not cohering.

Recorded from a small area of the E. Cape and W. Transkei, from about Ugie and Engcobo to Katberg, the Amatola Mountains, and the mountains about Pirie. Favours dry places such as the edges of rock sheets in grassland and hard bare ground. Soon dominates overgrazed or otherwise mismanaged grassland and has become a problem particularly in the Amatola Mountains. Flowers between December and March. Map 207.

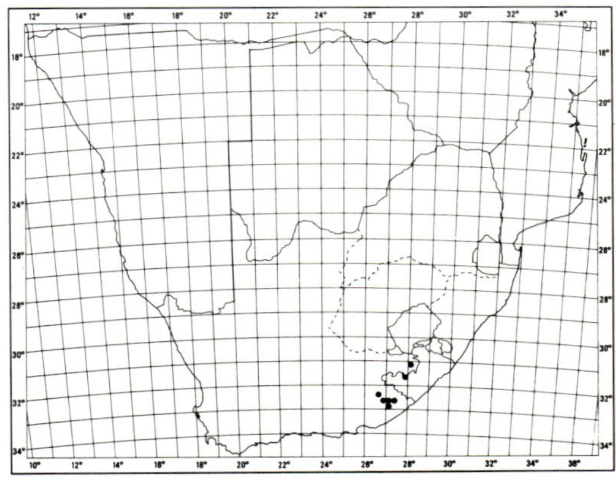

MAP 207.— **Helichrysum argyrophyllum**

Vouchers: *Dahlstrand* 2768 (PRE); *Pegler* 1608 (PRE); *Tyson* 2919 (E; PRE; SAM).

213. **Helichrysum herbaceum** *(Andr.) Sweet,* Hort. Brit. 1: 223 (1827); Hilliard & Burtt in Notes R. bot. Gdn Edinb. 32: 351 (1973); Hilliard, Compositae in Natal 223 (1978). Type: From the Cape, cult. George Hibbert, no specimen preserved.

Xeranthemum herbaceum Andr., Bot. Rep. 7, t. 487 (1807).

Helichrysum splendens Sims in Curtis's Bot. Mag. t. 1773 (1816). Type: From the Cape, cult. George Hibbert.

H. squamosum sensu Thunb., Fl. Cap. 661 (1823), excl. syn.; Less., Syn. Comp. 286 (1832) p.p.; DC., Prodr. 6: 185 (1838) p.p.; Harv. in F.C. 3: 233 (1865); Moeser in Bot. Jb. 44: 332 (1910), non (Jacq.) Thunb.

H. monocephalum Bak. in Kew Bull. 1898: 149 (1898). Types: Malawi, S. Nyika Mtns, *Whyte* s.n. (K!); between Kondowe and Karonga, *Whyte* s.n. (K!).

Stoloniferous perennial herb, flowering stems up to 400 mm tall, solitary or few together, simple or forking above, closely leafy throughout. *Radical leaves* more or less rosulate, spreading, elliptic or obovate, up to 50 × 20 mm, loosely cobwebby above at first, later glabrous except for a few coarse hairs, white-felted below; *cauline leaves* 10−20 (−25) × 1−2 (−5) mm, erect, more or less imbricate, commonly linear to linear-lanceolate, sometimes broader, acute to acuminate, base broad, clasping, margins often revolute, upper surface greyish-white cobwebby, lower loosely felted. *Heads* heterogamous, or rarely homogamous, campanulate, 18−20 mm long, 26−28 mm across the fully radiating bracts, solitary at the tips of the branches, or sometimes in a few-headed corymb. *Involucral bracts* in c. 8 series, often descending on the peduncle, closely imbricate, outer shorter and broader than inner, glossy, golden brown, inner lemon-yellow sometimes washed brown outside, much exceeding the flowers, acute or obtuse. *Receptacle* shortly honeycombed. *Flowers* 102−209, (0−2−) 14−38 ♀, 80−183 ☿. *Achenes* 1,25 mm, narrowly cylindric, glabrous. *Pappus* bristles many, equalling corolla, tips barbellate, bases not cohering.

Ranges from Katanga (Zaire), Tanzania, Malawi and the E. highlands of Zimbabwe to the E. highlands of the Transvaal, NE. Orange Free State, Lesotho, Natal, Transkei and E. and S. Cape as far west as Garcia's Pass between Riversdale and Ladismith. Map 208.

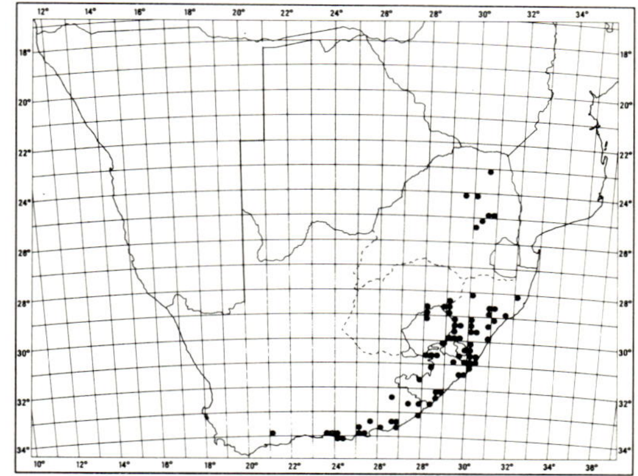

MAP 208.— **Helichrysum herbaceum**

In the Cape, Transkei and Natal, found from near sea level to c. 2 500 m; in the Transvaal and northwards, confined to the mountains. Grows in grassland or sometimes in shrub communities in the Cape, and will invade bare or disturbed areas. Flowers between October and April.

Vouchers: *Barker* 557 (NBG); *Codd* 2490 (NU; PRE); *Hilliard & Burtt* 7841 (E; K; MO; NU; S); *Wright* 390 (E; K; M; NH; NU).

214. **H. summo-montanum** *Verdoorn* in Flower. Pl. S. Afr. t. 483 (1933). Type: Transvaal, Pilgrim's Rest distr., near Sabie, summit Mt Anderson, 7 300 ft, June 1932, *Smuts* 44 (PRE, holo.!; K, iso.!).

H. andersoniense Burtt Davy in Jl S. Afr. Bot. 1: 107 (1935). Type: Lydenburg distr., summit Mt Anderson, Sept. 1919, *Rogers* 22990 (K, holo.!; BOL, fragment!).

Stoutly branched shrub attaining 1 m, old branches often decumbent, nude below, above clad in old withered leaves, branchlets short, densely white-woolly, very closely leafy. *Leaves* spreading, mostly 20−50 × 10−25 mm, becoming bracteate under the heads, elliptic, apex obtuse, mucronate, base broad, clasping, margins flat, upper surface glandular-pubescent, woolly over the 3 main nerves, greyish-white woolly below. *Heads* heterogamous, campanulate, c. 25 mm long, double that across the radiating bracts, solitary at tips of branchlets in axils of upper leaves. *Involucral bracts* in

c. 10 series, graded, imbricate, much exceeding flowers, glossy, outer pale brown, inner rosy fading to white, acute. *Receptacle* shortly honeycombed. *Flowers* c. 1 000, c. 50 ♀, 950 ☿. *Achenes* not seen, ovaries glabrous. *Pappus* bristles many, equalling corolla, barbellate above, bases not cohering. Fig. 57: 2.

Recorded only from the summit of Mount Anderson in the E. Transvaal, above c. 2 100 m, growing among rocks and flowering between March and June. Map 209.

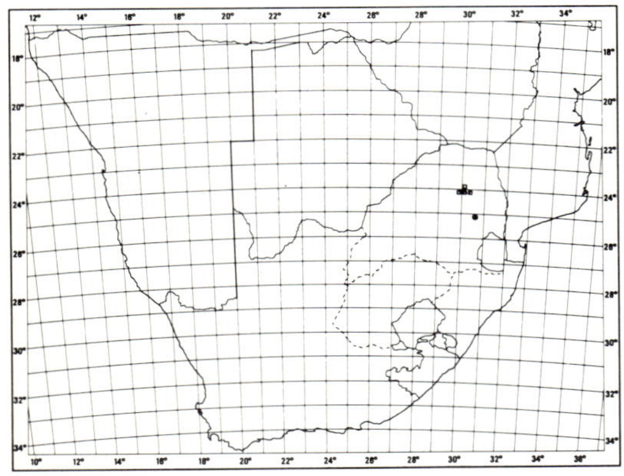

MAP 209.— ● **Helichrysum summo-montanum**
    □ **Helichrysum junodii**

Voucher: *Galpin* s.n. (PRE).

215. **Helichrysum junodii** *Moeser* in Bot. Jb. 44: 334 (1910). Type: Eastern Transvaal, Shilouvane, *Junod* 934 (G!).

Stoutly branched low-growing shrub, ultimate branchlets thinly white-woolly, closely leafy throughout. *Leaves* spreading, later reflexed, 12—22 × 1—3 mm, linear or linear-lanceolate, acute, apiculate, base slightly narrowed, half-clasping, margins more or less revolute, upper surface glandular-puberulous, thinly greyish-white cobwebby as well, lower thickly white-woolly. *Heads* heterogamous, campanulate, c. 15 mm long, double that across the radiating bracts, solitary at branchlet tips. *Involucral bracts* in c. 10 series, graded, imbricate, inner much exceeding flowers,

outer series rich glossy rose pink, outermost with a brown overcast, inner white, acuminate. *Receptacle* shortly honeycombed. *Flowers* c. 350—460, 30—50 ♀, 325—425 ☿. *Achenes* not seen, ovaries glabrous. *Pappus* bristles many, equalling corolla, barbellate above, bases not cohering. Fig. 57: 4.

Recorded only from the mountains between the Wolkberg and The Downs, NE. Transvaal, among rock outcrops in grassland; flowering between April and July. A most handsome plant with silvery leaves and rose pink and white heads. Has been confused with *H. reflexum* (no. 218), but easily distinguished by its bigger softer leaves and bigger heads. Map 209.

Vouchers: *Müller & Scheepers* 141 (PRE); *Rogers* 20107 (J; PRE; Z).

216. **Helichrysum wilmsii** *Moeser* in Bot. Jb. 44: 333 (1910); Compton, Fl. Swaziland 636 (1976). Lectotype: Eastern Transvaal, between Spitzkop & Komati River, July 1884, *Wilms* 707 (E!; G!).

Soft-wooded subshrub, branches long, loose, straggling or sprawling, glandular-pubescent, white-cobwebby as well, particularly on young parts, leafy. *Leaves* mostly 12—30 × 3—9 mm, smaller and more distant upwards, narrowly panduriform to oblong becoming lanceolate-acuminate upwards, apex obtuse to acute, apiculate, base broad, more or less cordate-clasping, margins scarcely revolute, upper surface glandular-pubescent, sometimes greyish-white cobwebby as well, lower white-woolly or cobwebby. *Heads* heterogamous, campanulate, c. 13—20 mm long, nearly double that across the radiating bracts, solitary at the tips of long laxly leafy then bracteate branchlets. *Involucral bracts* in c. 10 series, graded, imbricate, inner much exceeding flowers, acute, glossy, outer pale or rich chestnut brown sometimes with a rosy overcast, inner white sometimes tinged rose. *Receptacle* shortly honeycombed. *Flowers* 219—483, 18—39 ♀, 188—445 ☿. *Achenes* 0,75 mm long, glabrous. *Pappus* bristles many, equalling corolla, barbellate above, bases not cohering.

Ranges from Magoebaskloof and Woodbush in the NE. Transvaal south along the mountains to the Barberton Mountains and the mountains near Mbabane in Swaziland. Forms straggling tangled clumps in rough rocky grassland and forest margin herbage. Flowering recorded between November and July, at its peak in April and May. Map 210.

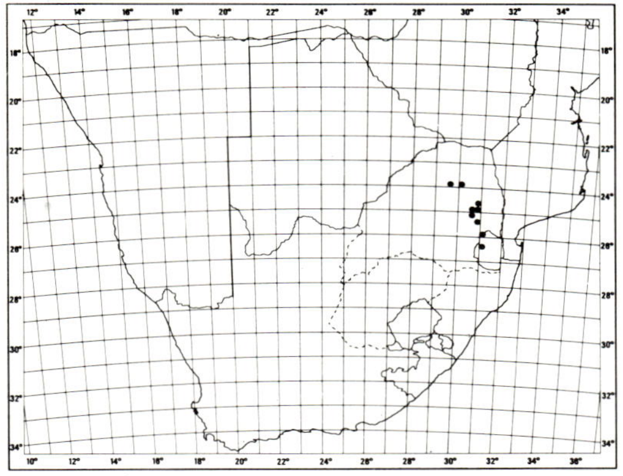

MAP 210.— **Helichrysum wilmsii**

Vouchers: *Codd 7854* (PRE); *Compton 25680* (NBG; PRE); *Galpin 14613* (PRE); *Hilliard & Burtt 14277* (E; K; M; NU; PRE; S).

Compton 31491 (NBG; PRE) from Emlembe Mountain, Barberton Mts, is possibly *H. wilmsii* × *H. cooperi* (no. 240) or a species allied to *H. cooperi*. *Hilliard & Burtt 14319* (E; K; NU; PRE; S), from the summit of the Wolkberg, is a hybrid between *H. wilmsii* and *H. confertifolium* (no. 220), both of which also grow on the summit of the Wolkberg. The hybrid (one plant only seen) has the habit of *H. wilmsii* and was in full flower when *H. wilmsii*, growing nearby, was only budding. The foliage is intermediate between that of *H. wilmsii* and *H. confertifolium* and the heads are borne either singly or in loose open corymbs, as in *H. confertifolium*.

217. **Helichrysum argyrolepis** *MacOwan* in Bot. J. Linn. Soc., 25: 387 (1890); Moeser in Bot. Jb. 44: 333 (1910); Compton, Fl. Swaziland 628 (1976); Hilliard, Compositae in Natal 227 (1977). Lectotype: Transkei, Mt Malowe, c. 1 830 m, *Tyson 2788* (K!; PRE!).

Soft-wooded subshrub with tangled masses of branching stems from a woody stock, stems greyish-white felted, closely leafy throughout. *Leaves* linear or linear-lanceolate, c. 6−25 × 1−2 (−3) mm, smaller and more distant upwards, appressed or spreading, eventually reflexed, tips acute or subacute, apiculate, base broad, half-clasping, margins more or less revolute, upper surface glandular-pubescent and thickly greyish-white cobwebby or woolly, lower more closely felted. *Heads* heterogamous, rarely homogamous, campanulate,

13−15 mm long, c. 25 mm across the fully radiating bracts, solitary at branch tips. *Involucral bracts* in c. 10 series, graded, loosely imbricate, inner much exceeding flowers, acute, glossy white, mostly palest to dark chestnut brown above, rarely rose-pink. *Receptacle* shortly honeycombed. *Flowers* 111−232, (0−) 5−25 ♀, 98−227 ☿. *Achenes* 1−1,5 mm long, glabrous. *Pappus* bristles many, equalling corolla, barbellate above, bases not cohering.

Ranges from Mount Sheba in the Transvaal Drakensberg to the high country in the SE. Transvaal and neighbouring Swaziland and through the Midlands of Natal to Umzimkulu district in the Transkei and the Suurberg near Weza in southernmost Natal. Grows in large tangled clumps in grassland, particularly among outcropping rocks; flowering between November and April. Map 211.

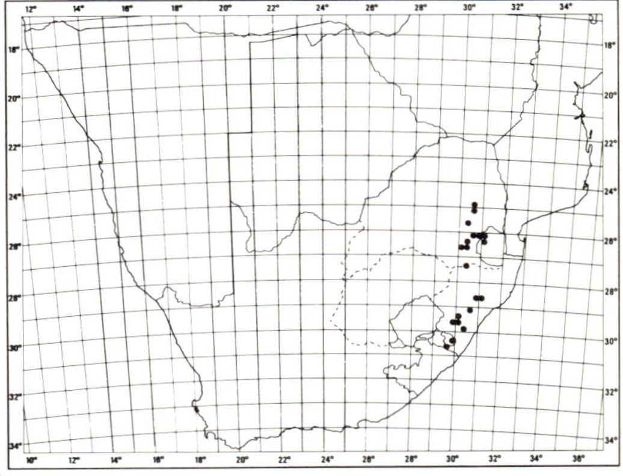

MAP 211.— **Helichrysum argyrolepis**

Specimens in the southern part of the range have distinctly glandular as well as woolly leaves and heads mostly c. 15 mm long; in the Transvaal and Swaziland, leaves are generally smaller and less markedly glandular and heads are c. 13 mm long.

Vouchers: *Acocks 12744* (PRE); *Compton 26713* (NBG; PRE); *Hilliard 5491* (E; K; MO; NU; S); *Junod 4392* (PRE).

218. **Helichrysum reflexum** *N.E. Br.* in Kew Bull. 1894: 356 (1894); Compton, Fl. Swaziland 634 (1976). Type: Transvaal, Barberton, Saddleback Mt., 5 000 ft, 11 v 1890, *Galpin 947* (K, holo.!; BOL; PRE; Z, iso.!).

Twiggy shrublet 300–600 mm tall, young stems thinly white-woolly, glandular-pubescent, closely leafy throughout. *Leaves* spreading, later reflexed, 5–11 × 0,75–1,5 mm, linear or linear-lanceolate, very acute, apiculate, bases broad, half-clasping, margins strongly revolute, glandular-pubescent above, sometimes thinly greyish-white woolly as well, densely white-woolly below. *Heads* heterogamous or rarely homogamous, campanulate, c. 10–12 mm long, double that across the fully radiating bracts, solitary at branchlet tips. *Involucral bracts* in c. 10 series, graded, imbricate, inner much exceeding flowers, acute, glossy white, outer tinged light brown often with a rosy cast. *Receptacle* shortly honeycombed. *Flowers* 124–222, (0–) 3–31 ♀, 97–196 ☿. *Achenes* 1 mm long, glabrous. *Pappus* bristles many, equalling corolla, barbellate above, bases not cohering.

Recorded from the mountains of the SE. Transvaal and Swaziland, from Hebronberg above the Blyde River Canyon south to the Steenkampsberge and the Barberton Mountains. Locally common among rock outcrops; flowering mainly between February and April. Easily recognized by its stiff strongly reflexed leaves with markedly revolute margins. Map 212.

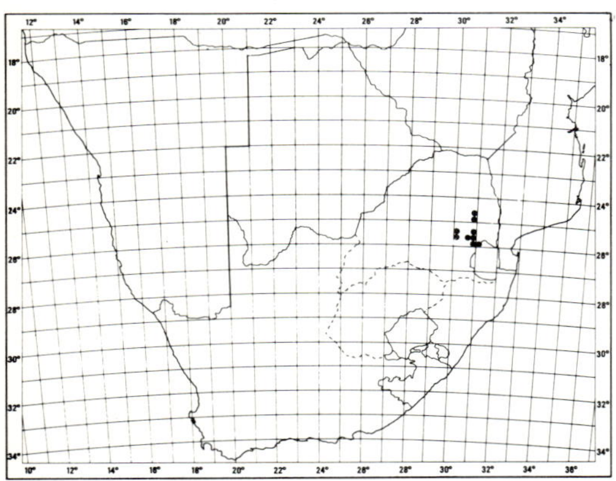

MAP 212.— **Helichrysum reflexum**

Vouchers: *Codd* 6430 (PRE); *Compton* 27684 (NBG; PRE); *Galpin* 14399 (BM); *Hilliard & Burtt* 14166 (E; K; M; MO; NU; PRE; S); *Rogers* 19940 (PRE).

219. **Helichrysum mariepscopicum** *Hilliard* in Notes R. bot. Gdn Edinb. 40: 258 (1982). Type: Transvaal, Pilgrim's Rest distr., Mariepskop, c. 2 300 m, 23 i 1968, *Hilliard* 4726 (NU holo.!; E; K; MO; PRE; S, iso.!).

Tufted perennial herb or shrublet c. 150–300 mm tall, stems glandular pubescent, closely leafy throughout. *Leaves* mostly 10–30 × 2–5 mm, becoming smaller and more distant upwards and passing into scale-tipped bracts, oblong or lanceolate-oblong, apex obtuse to acute, apiculate, base broad, half-clasping, both surfaces glandular-pubescent, occasionally also cob-webby below, margins and main vein greyish-white woolly or wool wanting. *Heads* heterogamous, campanulate, c. 12 mm long, double that across the radiating bracts, solitary at branchlet tips. *Involucral bracts* in c. 10 series, graded, imbricate, inner much exceeding flowers, acute, glossy white, all but innermost tipped brown, sometimes with a rosy cast. *Receptacle* shortly honeycombed. *Flowers* 164–277, 10–31 ♀, 134–252 ☿. *Achenes* 1 mm long, glabrous or hairy, not myxogenic. *Pappus* bristles barbellate above, bases not cohering. Fig. 57: 3.

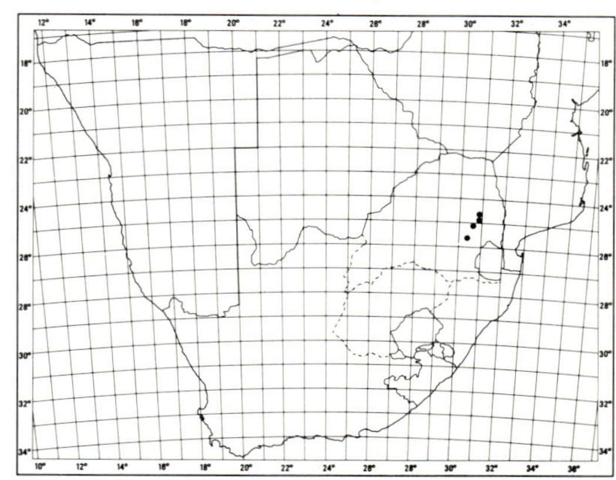

MAP 213.— **Helichrysum mariepscopicum**

FIG. 57.—1, **Helichrysum milfordiae,** part of plant, × 1; 1a, hermaphrodite flower, × 8; 1b, female flower, × 8; 1c, pappus bristle, × 8 (*Hilliard* 5295). 2, **H. summomontanum,** part of plant, × 1 (*Galpin* 13748). 3, **H. mariepscopicum,** part of plant, × 1 (*Hilliard* 4726). 4, **H. junodii,** part of plant, × 1; 4a, leaf, × 2,3 (*Rogers* 17739).

1
1c
1b
1a
2
3
4a
4
Wonda-du Toit

Recorded from a small area of the Transvaal Drakensberg from Mariepskop to Mount Anderson and Elandshoogte near Machadodorp. Grows in damp mountain grassland particularly among rock outcrops. Flowers between September and April, at its peak between October and January. Map 213.

Vouchers: *Codd & De Winter* 3336 (K; NU; PRE); *Galpin* 14532 (NU; PRE); *Hilliard & Burtt* 6011 (E; K; M; MO; NU; PRE; S); *Merxmüller* 570 (K).

220. **Helichrysum confertifolium** *Klatt*

in Bull. Herb. Boissier 4: 835 (1896); Moeser in Bot. Jb. 44: 333 (1910); Hilliard, Compositae in Natal 228 (1977). Lectotype: Transkei, Insizwa Mtn, 2 072 m, *Schlechter* 6476 (G, holo.!; BOL; K; PRE; Z, iso.!).

Mat-forming perennial herb with prostrate rooting branches radiating from a stout woody stock and giving rise to innumerable crowded leaf rosettes, flowering stems terminal, 50—400 mm tall, leafy. *Radical leaves* 10—40 (−70) × 2—3 (−5) mm, linear-lanceolate, acute, apiculate, base broad, clasping, margins weakly revolute, both surfaces silvery white-felted, lower paler than upper which is sometimes merely cobwebby, prominently 3-5-nerved; *cauline leaves* similar but often shorter, closely imbricate. *Heads* heterogamous, campanulate, 12—15 mm long, c. 25 mm across the fully radiating bracts, mostly solitary, occasionally 2 or 3 corymbosely arranged. *Involucral bracts* in c. 8 series, graded, loosely imbricate, inner much exceeding flowers, glossy white, very pale to dark chestnut brown or reddish brown outside in upper half. *Receptacle* shortly honeycombed. *Flowers* 153—443, 15—28 ♀, 138—424 ☿. *Achenes* 1,5 mm long, glabrous. *Pappus* bristles many, equalling corolla, barbellate above, bases not cohering.

Ranges from the Duiwelskloof-Woodbush area, NE. Transvaal, south along the mountains to the highlands of the SE. Transvaal, the low Drakensberg along the Transvaal-Orange Free State-Natal borders, the highlands of northern Natal and Zululand, the Natal Midlands and Drakensberg as far south as Mount Insizwa in Transkei, descending to 600 m in Natal. Favours short rocky grassland, rock sheets and other bare areas; flowering between December and July, chiefly in February and March. Map 214.

Plants growing over Table Mountain Sandstone in Natal tend to produce tall flowering stems with pale heads, but similar plants turn up sporadically elsewhere (e.g. Platberg near Harrismith, and Billy's Vlei near Carolina), just as richly coloured heads are also sometimes found over T. M. S. (e.g. at Dumisa, southern Natal).

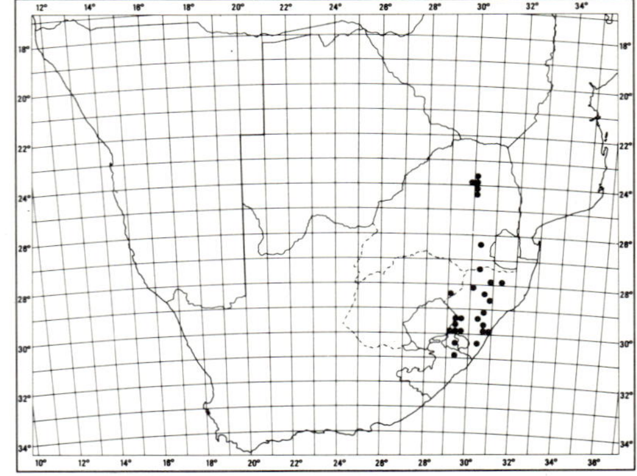

MAP 214.— **Helichrysum confertifolium**

Vouchers: *Codd* 9582 (PRE); *Hilliard* 2666 (E; K; M; NU; PRE; S); *Hilliard & Burtt* 6544 (E; K; MO; NU; PRE; S); *Tyson* 1256 (PRE).

221. **Helichrysum milfordiae** *Killick*

in Bothalia 7: 414 (1960); Hilliard, Compositae in Natal 229 (1977). Type: Natal, Bergville distr., Cathedral Peak Forest Station, top of Organ Pipes Pass, 9 700 ft, *Killick* 2322 (PRE, holo.!; BM; K; M, iso.!).

Prostrate mat-forming subshrub, old stems woody, gnarled, up to 8 mm in diam., young stems very slender (c. 1 mm diam.), stoloniferous, leafy, giving rise to innumerable congested leaf rosettes, flowering stems terminal, erect, up to 120 mm tall, closely leafy. *Rosette leaves* obovate to subspathulate, up to 14 × 9 mm, densely appressed silvery silky-woolly; *cauline leaves* similar but lanceolate to subspathulate, bracteate below the heads and tipped with a coloured scarious scale. *Heads* heterogamous, campanulate, c. 15 mm long, double that across the fully radiating bracts, solitary at the branch tips. *Involucral bracts* in c. 10 series, graded, loosely imbricate, much exceeding flowers, glossy white, mostly tipped crimson or dark brown. *Receptacle* very shortly honeycombed. *Flowers* 138−233, 18−29 ♀, 120−208 ☿. *Achenes* 1,5 mm, glabrous. *Pappus* bristles many, equalling corolla, barbellate above, bases not cohering. Fig. 57: 1.

Recorded only from the high Drakensberg on the Natal-Lesotho border from Mont aux Sources and the Oxbow area to Sani Pass and Hodgson's Peaks, between 2 985 and 3 500 m, forming mats on rocks, or growing from cracks in cliff faces. Flowers from late December to February. Map 215.

Vouchers: *Hilliard* 5295 (E; K; M; NU; S); *Nordenstam* 2130 (NU); *Wright* 438 (E; K; NH; NU).

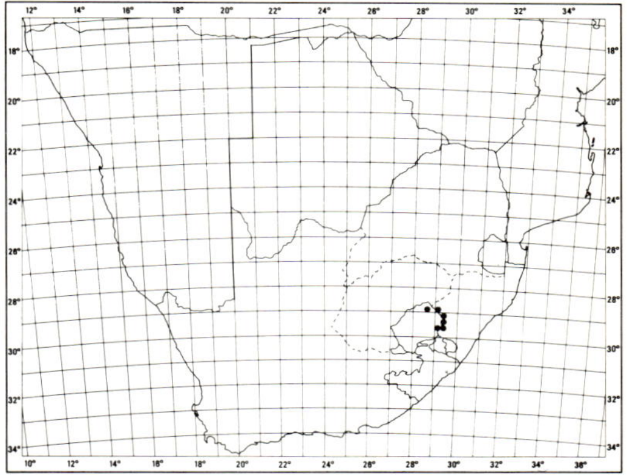

MAP 215.— **Helichrysum milfordiae**

Group 30

Coarse biennials, herbaceous perennials or shrubs; *leaves* medium-sized to large; *heads* heterogamous, c. 8−20 mm long, mostly depressed-globose, sometimes broadly campanulate, usually few to many in corymbose panicles, sometimes solitary; *involucral bracts* radiating, mostly yellow, often brown outside, sometimes tawny, light brown, white, or (one species only) red (pink) and white; *receptacle* very shortly honeycombed; *flowers* c. 300−2350, 15−350 ♀, ♀ flowers sometimes outnumbering ☿; corolla of ☿ flowers cylindric, of ♀ flowers narrowly cylindric, limb conspicuous; *achenes* glabrous; *pappus* bristles scabrid, tips barbellate to subplumose, bases either smooth or with patent cilia, cohering or not.

Species 222 to 242, widespread, from the Cape Peninsula to Ethiopia, the Yemen and West Africa, though very few species cross the Limpopo and Zambezi intervals. The taxonomy is difficult and the older names have been grossly misapplied. Grow mostly in grassland or rough herbage on forest margins, occasionally in marshy places. At least two species readily become weeds.

Many species are similar in facies, but are distinguished by differences in habit and leaf shape in particular; depauperate or fragmentary specimens will be difficult or impossible to key out. Hybrids are thought to occur, and these are not catered for.

1a Involucral bracts white, or red (pink) and white, creamy, straw-coloured or brown:

  2a Involucral bracts white or red (pink) and white:

    3a Herbaceous perennial with tufts of radical leaves on the stock, heads usually solitary, involucral bracts white ....222. *H. aureum* var. *candidum*

    3b Tall biennial (?) herb, heads many in a large corymbose panicle, involucral bracts white or red (pink) and white ....237. *H. elegantissimum*

  2b Involucral bracts creamy, straw-coloured or brown:

    4a Involucral bracts creamy, sometimes with a brownish overcast outside:

      5a Bushy perennial herb or subshrub, leaves mostly elliptic, generally white-woolly on both surfaces, heads c. 30−35 mm across the radiating bracts, few (up to c. 7) in a compact corymbose cluster ....222. *H. tenax* var. *pallidum*

      5b Biennial herb, stem simple or sparingly branched from the base, leaves lanceolate, acute to acuminate, lower surface thinly white-tomentose, heads c. 15−25 mm across the radiating bracts, many in a large leafy corymbose panicle ....236. *H. foetidum*

    4b Involucral bracts brown or straw-coloured:

6a Heads c. 15−20 mm across the radiating bracts; rhizomatous perennial herb, radical leaves thinly greyish-white woolly contrasting with the green, glandular stem leaves ....225. *H. fulvum*

6b Heads c. 32−38 mm across the radiating bracts; plants shrubby:

    7a Leaves oblong or oblong-lanceolate, obtuse to acute, greyish-white woolly on one or both surfaces on sterile twigs, merely glandular on flowering twigs, heads often solitary, sometimes few, at the tips of the twigs. Flowers in October and November .... 228. *H. milleri*

    7b Leaves lanceolate, very acute to acuminate, glandular, wool, if present, only on margins and midline, heads several to many in a large corymbose panicle. Flowers in March and April .... 229. *H. transmontanum*

1b Involucral bracts yellow (lemon to golden) at least inside; sometimes brown outside:

    8a Bases of pappus bristles without patent cilia; grey-leaved shrub with lemon-yellow heads. Southern Natal .... 227. *H. citricephalum*

    8b Bases of pappus bristles with patent cilia, lightly cohering or not:

        9a Perennial herbs or subshrubs with a stout woody caudex crowned with one or more leaf rosettes:

            10a Cauline leaves oblong or lanceolate ....222.*H. aureum* (see under species for vars.)

            10b Cauline leaves ovate becoming broadly lanceolate only near the heads ....223. *H. lesliei*

        9b Biennial herb with radical leaves rosetted in the first year of growth; or perennial herbs, subshrubs or shrubs, sometimes with vegetative buds or small leaf tufts at the base of the stem, but lacking a stout caudex:

            11a Subshrub with leaf rosettes terminating the branches, flowering stems lateral to them ....224. *H. tenax*

            11b Biennial or perennial herbs, subshrubs, or shrubs; leaf rosettes, if present, radical:

                12a Leaf bases not decurrent in stem wings (character not always visible in upper leaves):

                    13a Heads mostly 30−40 mm across the radiating bracts; biennial herb with greyish-white woolly radical leaves and green cauline leaves. Natal Drakensberg .... 238. *H. heterolasium*

                13b Heads 15−30 mm across the radiating bracts:

                    14a At least some of the cauline leaves grey- or white-woolly on one or both surfaces:

                        15a Leaves lanceolate (ignore uppermost reduced leaves):

                            16a Leaves generally silvery on both surfaces, or occasionally wool confined to lower surface; bushy perennial herb. Transvaal .... 226. *H. albilanatum*

                            16b Leaves green above, white-woolly below; biennial. Cape and Transkei .... 236. *H. foetidum*

                        15b Leaves oblong, oblong-lanceolate or ovate (ignore uppermost reduced leaves):

                            17a Leaves mostly 18 to 40 mm broad; heads 28−30 mm across the radiating bracts .... 230. *H. edwardsii*

                            17b Leaves mostly 4−13 (−24) mm broad; heads 18−28 mm across the radiating bracts .... 233. *H. mutabile*

                    14b Leaves glandular-setose, wool, if present, confined to margins and midline:

                      18a Involucral bracts wholly yellow:

                        19a Leaves mostly oblong-ovate becoming elliptic-lanceolate upwards; inflorescence branches glandular-setose, occasionally thinly white-woolly near the heads 231.*H. setosum*

                        19b Leaves oblong or linear-oblong; stem and inflorescence branches thinly cobwebby or white-woolly ....232. *H. tongense*

                    18b Involucral bracts yellow inside, brown outside:

                        20a Leaves on the inflorescence branches large enough to overlap and so produce a very leafy compound inflorescence .... 231. *H. setosum*

                        20b Leaves on the inflorescence branches reduced and distant towards the heads:

                            21a Leaves mostly 25−75 × 4−13 mm ....233. *H. mutabile*

                            21b Leaves mostly 50−90 × 15−38 mm .... 234. *H. aureolum*

            12b Leaf bases decurrent on the stem in short or long wings (character not always visible in upper leaves):

22a Biennials, producing a rosette of leaves in the first season, growing out into a flowering stem in the second season:

23a All the cauline leaves thinly cobwebby above, thinly greyish-white woolly below .... 235. *H. decorum*

23b Cauline leaves at most woolly on the margins and midline, only some of the old radical leaves sometimes woolly on one or both surfaces:

24a Heads mostly 30—40 mm across the radiating bracts, radical leaves greyish-white woolly, clinging, old and withered, at the base of the flowering stem, involucral bracts usually brownish outside .... 238. *H. heterolasium*

24b Heads mostly 15—32 mm across the radiating bracts; bracts usually yellow inside and out, rarely brownish outside and then heads 15—25 mm across:

25a Inflorescence branches persistently white-woolly or cobwebby to well below the first forking; flowering mainly from September to November .... 239. *H. ruderale*

25b Inflorescence branches either without wool or woolly close to the heads only; plants flowering between December and April .... 240. *H. cooperi*

22b Perennial herbs with one or several stems from the base (vegetative buds at the base, or old stems, indicate perennation):

26a Leaves mostly 50—70 (−80) × 8—20 (−26) mm .... 241. *H. difficile*
26b Leaves mostly 65—100 × 18—32 mm .... 242. *H. molestum*

222. **Helichrysum aureum** *(Houtt.) Merrill* in J. Arnold Arb. 19: 372 (1938); Hilliard, Compositae in Natal 249 (1977). Type: Cape, *Thunberg* (specimen given by Thunberg to Houttuyn now in herb. Burmann, G!).

Gnaphalium aureum Houtt., Nat. Hist. 2, fasc. 10: 590, t. 67, fig. 3 (1779).

Xeranthemum fulgidum L. f., Suppl. 365 (1781); Curtis's Bot. Mag. t. 414 (1798). *Helichrysum fulgidum* (L. f.) Willd., Sp. Pl. 3: 1904 (1804); Thunberg, Fl. Cap. 664 (1823); Less., Syn. Comp. 285 (1832); DC., Prodr. 6: 187 (1838); Harv. in F.C. 3: 232 (1865); Moeser in Bot. Jb. 44: 338 (1910). *Gnaphalium fulgidum* (L. f.) Zuccagni in Roem., Collect. 153 (1809). Type: Cape of Good Hope, *Thunberg* (LINN 990.23!).

Perennial herb, very variable in stature, in leaf-size and indumentum, and in head size, stock becoming very stout and woody, crown up to c. 25 mm diam. producing one or several leaf rosettes, flowering stems several, lateral to the rosettes, c. 100—400 (−800) mm high, simple or sparingly branched towards the top, branches simple, glandular-setose, often cobwebby or loosely white-woolly as well, leafy throughout. *Radical leaves* c. 30—100 (−270) × 10—25 (−60) mm, narrowly or broadly elliptic-spathulate, apex obtuse to subacute, base broad, clasping, both surfaces glandular-setose, lower or both surfaces sometimes woolly as well, sometimes wool confined to margins and midline; *cauline leaves* up to 70 × 20 mm, decreasing in size upwards, oblong, oblong-lanceolate to lanceolate upwards, correspondingly obtuse, acute to acuminate, apiculate, base more or less cordate-clasping, very shortly decurrent in the larger leaves, glandular-setose, sometimes cobwebby or loosely woolly as well, wool sometimes confined to margins or margins and main veins. *Heads* heterogamous, depressed-globose, 17—45 mm across the radiating bracts, solitary or up to c. 6 subracemosely or corymbosely arranged at the stem tip on long or short leafy peduncles. *Involucral bracts* in c. 7—9 series, graded, imbricate, much exceeding flowers, radiating, glossy, bright yellow, outer sometimes overlaid palest brown or buff, occasionally snow-white. *Receptacle* scarcely honeycombed. *Flowers* 311—1856, 34—284 ♀, 251—1671 ☿, yellow. *Achenes* 0,75— 1 mm long, barrel-shaped, glabrous. *Pappus* bristles many, equalling corolla, tips barbellate, bases with patent cilia not or scarcely cohering.

Six varieties are recognized:

1a Involucral bracts yellow, outer sometimes pale brown or buff:

2a Heads mostly 15—20 mm long measured from base of outer involucral bracts to tips of longest inner ones, c. 35—45 mm across the fully radiating involucral bracts; if only 30 mm across, then on cliffs in the Drakensberg:

3a Flowering stems loosely woolly or cobwebby mostly throughout; flowering mainly between August and December:

4a Plants growing in grassland or scrub:

   5a Cauline leaves only very lightly woolly, or wool confined to margins and midline .... (a) var. *aureum*

   5b Cauline leaves thickly and persistently silver-grey woolly . (f) var. *argenteum*

4b Plants growing on cliff faces or on rocky terrain immediately below cliffs in the Drakensberg................(b) var. *scopulosum*

3b Flowering stems woolly or cobwebby only in the uppermost part, flowering mainly between January and April .... (c) var. *serotinum*

2b Heads mostly 8−15 mm long, up to 32 mm across the radiating bracts, growing in grassland .... (d) var. *monocephalum*

1b Involucral bracts snow-white...... (e) var. *candidum*

(a) var. **aureum.**

Helichrysum fulgidum var. *angustifolium* DC., Prodr. 6: 187 (1838). Type: Cape, Albany, *Drège* 3739 (G-DC, holo.!).

H. fulgidum var. *heterotrichum* DC., Prodr. 6: 187 (1838). Type: Natal, between Umkomaas and Umzimkulu Rivers, *Drège* 5021 (G-DC, holo.!; BM; E; K; TCD, iso.!).

Flowering stems c. 150−800 mm high, simple or sparingly forked above, loosely white-woolly or cottony. *Radical leaves* c. 70−100 (−270) × 10−25 (−60) mm, very thinly to densely greyish-white woolly above and below, wool sometimes eventually confined to margins and main veins; *cauline leaves* up to 70 (−100) × 15 (−20) mm, glandular-setose, often cobwebby or lightly woolly at first, wool later confined to margins and main veins. *Heads* 16−20 mm long, c. 35−45 mm across the radiating bracts, solitary or up to 6 racemosely or subcorymbosely arranged. *Involucral bracts* yellow, outer becoming brownish with age. *Flowers* 1488−1642, 199−284 ♀, 1204−1443 ☿.

Ranges from about Sezela on the Natal coast south of Durban to about Humansdorp in the E. Cape, and inland to Tabankulu Mountain, Transkei, and the Stutterheim area, the Amatola Mountains, Boschberg and the Koudeveld Mountains in the E. and central Cape. Grows in grassland or short scrub; flowering mainly between July and November. Map 216.

The distinguishing features of var. *aureum* are the large heads and green, woolly-margined cauline leaves.

Much of the E. Cape coastal material in herbaria lacks radical leaves, but these, when present, may be green with woolly margins or thickly to thinly grey-woolly. The original material of *H. aureum* probably came from the Humansdorp-Port Elizabeth area, the furthest east that Thunberg collected, and radical leaves are wanting on his specimens. Plants from coastal Natal, whence came Drège's specimen of *H. fulgidum* var. *heterotrichum* (in syn., above), have grey-woolly radical leaves and green cauline leaves. It seems that variation in the indumentum of the radical leaves must be accepted as characteristic of var. *aureum*.

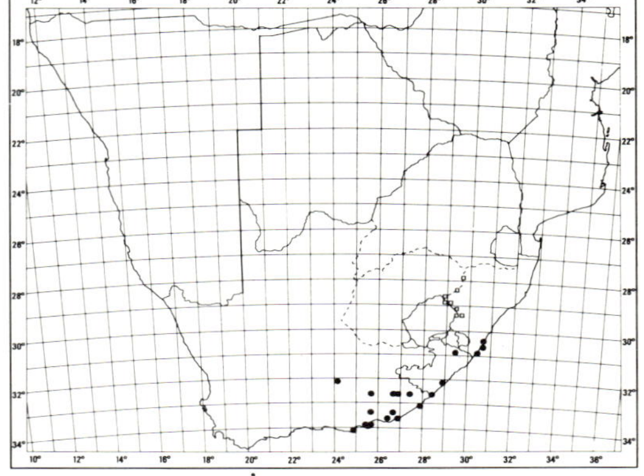

MAP 216.— ● **Helichrysum aureum** var. **aureum**
   □ **Helichrysum aureum** var. **scopulosum**

Vouchers: Acocks 9142 (PRE); *Bayliss* 2346 (NBG); *Flanagan* 869 (PRE); *Hilliard & Burtt* 10954 (E; K; MO; NU; S); *Pegler* 676 (PRE).

(b) var. **scopulosum** (*M.D. Henderson*) Hilliard

in Notes R. bot. Gdn Edinb. 34: 80 (1975), Compositae in Natal 252 (1977). Type: Natal, Bergville distr., Cathedral Peak Forest Research Station, 7 400 ft, *Killick* 1554 (PRE, holo.!; K, iso.!).

H. scopulosum M.D. Henderson in Bothalia 6: 423 (1954).

Flowering stems c. 50−300 mm high, simple or subsimple, loosely white-woolly or cottony. *Radical leaves* c. 30−80 (−110) × 20−30 (−40) mm, thickly greyish-white woolly, or wool sometimes sparse; *cauline leaves* up to 40 (−70) × 10 (−20) mm,

---

FIG. 58.−1, **Helichrysum aureum** var. **argenteum**, whole plant, × 0,2; 1a, tip of flowering stem, × 1 (*Devenish* 1951). 2, **H. lesliei**, whole plant, × 0,26; 2a, part of flowering branch, × 1; 2b, hermaphrodite flower, × 10; 2c, female flower, × 10; 2d, pappus bristle, × 10 (*Hilliard & Burtt* 14171).

2c
2b
2c
1a
1
2a
2
H. Weida-du Toit

loosely woolly at first, wool eventually confined to margins and main veins. *Heads* (15−) 20 mm long, (30−) 35−45 mm across the radiating bracts, usually solitary, occasionally 2−3 subcorymbosely arranged on long leafy peduncles. *Involucral bracts* yellow, outer usually overlaid buff or pale brown. *Flowers* 516−1253, 36−199 ♀, 360−1054 ☿.

Recorded only from the Natal Drakensberg between Muller's Pass west of Newcastle and the Giant's Castle area, Estcourt district, between 1 645 and 2 750 m above sea level. Grows mainly on rock faces, often in inaccessible places, sometimes on the boulders fallen from cliffs or occasionally on the steep rubble and grass slopes below cliffs, when it may then have less woolly leaves than cliff-face plants. Flowers mainly between July and December, with one record in April. Map 216.

Vouchers: *Esterhuysen* 15552 (BOL; PRE); *Wright* 293 (E; K; NU; S).

(c) var. **serotinum** *Hilliard* in Notes R. bot. Gdn Edinb. 34: 80 (1975), Compositae in Natal 253 (1977). Type: Natal, Impendhle distr., 'Storm Heights', 7 100 ft, 9 i 1968, *Wright* 375 (NU, holo.!; E, iso.!).

Flowering stems c. 100−400 mm high, glandular-setose, cobwebby or loosely woolly only in the upper part. *Radical leaves* (30−) 60−150 × 5−15 mm, glandular-setose, margins and sometimes main veins woolly, or wool wanting; *cauline leaves* up to 70 × 10 mm, glandular-setose, margins not or scarcely woolly. *Heads* 15−20 mm

long, c. 35−40 mm across the radiating bracts, solitary or up to 4 subcorymbosely arranged on long leafy stalks. *Involucral bracts* yellow, outer sometimes tinged brown. *Flowers* 1079−1856, 76−185 ♀, 1 003−1 671 ☿.

Along the Drakensberg from about Cathedral Peak Forest Reserve in Natal to Barkly Pass, E. Cape, the high Lesotho mountains, and Mount Gilboa, the highest point of the Karkloof Range, a spur of the Drakensberg, c. 1 800 to 2 600 m above sea level. Grows in open grassland on mountain slopes; flowering between December and May. This late flowering sets it apart from all the other, spring-flowering, components of *H. aureum,* some of which may grow with it. Map 217.

Vouchers: *Hilliard & Burtt* 8999 (E; NU; S); *Wright* 409 (E; K; NH; NU).

(d) var. **monocephalum** *(DC.) Hilliard* in Notes R. bot. Gdn Edinb. 34: 80 (1975), Compositae in Natal 253 (1977). Type: Cape, Witteberg, *Drège* 866 (G-DC, holo.!).

H. fulgidum var. *monocephalum* DC., Prodr. 6: 188 (1838); Harv. in F.C.3: 232 (1865).

H. fulgidum var. *nanum* DC., Prodr. 6: 188 (1838); Harv. in F. C. 3: 232 (1865). Type: Cape, Witteberg, *Drège* 3738 (G-DC, holo.!).

Flowering stems c. 50−300 mm high, simple or rarely subsimple, glandular-setose, often thinly woolly or cobwebby as well, particularly above. *Radical leaves* c. 30−120 × 10−20 mm, greyish-white woolly or wool confined to margins; *cauline leaves* up to 30 (−50) × 5 (−10) mm, glandular-setose, margins woolly or not. *Heads* 8−15 mm long, 17−32 mm across the radiating bracts, usually solitary, very rarely 2 or 3 corymbosely arranged on long leafy peduncles. *Involucral bracts* yellow. *Flowers* 290−1341, 32−149 ♀, 251−1192 ☿.

Widespread, from the E. highlands and Highveld of Zimbabwe to the E. highlands and Highveld of the Transvaal, western Swaziland, Natal (where it is widespread from near sea level to c. 2 300 m), the high parts of Lesotho and the neighbouring O.F.S., Transkei, and E. Cape as far west as the Witteberg and the mountains about Queenstown and Stutterheim. Also in the highlands of Angola. Grows in open grassland, flowering between July and November, often conspicuous in spring after grass-burning. Map 218.

Var. *nanum* was distinguished by its woolly stem and leaves and larger heads (c. 14 mm long, 30 mm across the radiating involucral bracts), var. *monocephalum* by its glandular-setose rather than woolly

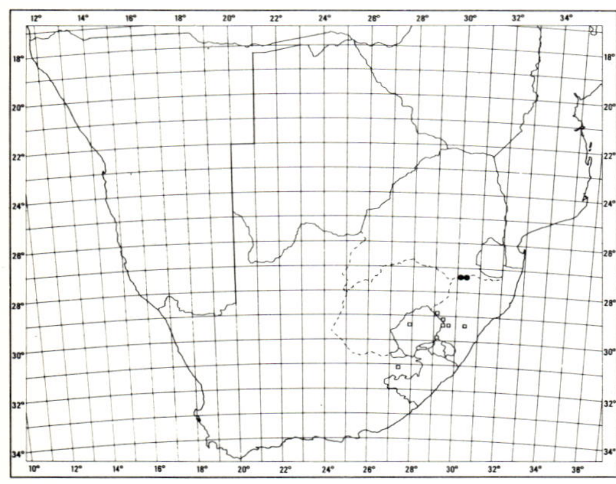

MAP 217.— □ **Helichrysum aureum** var. **serotinum**
● **Helichrysum aureum** var. **argenteum**

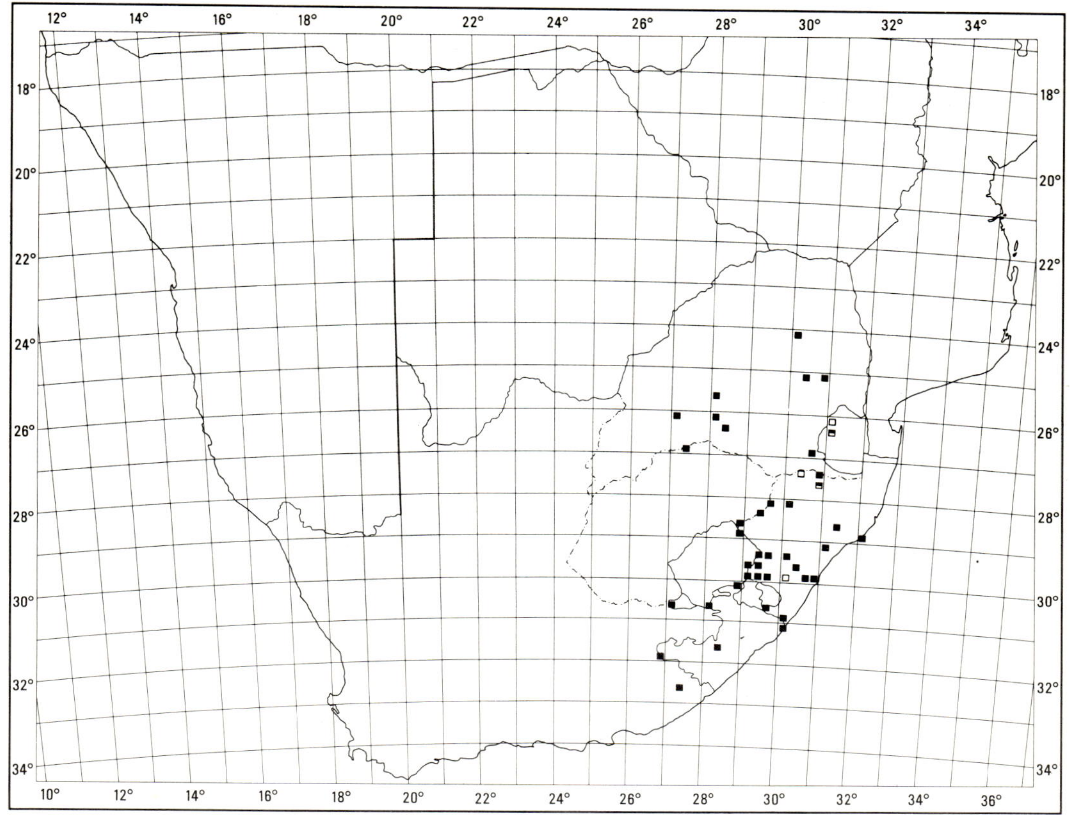

MAP 218.— ■ *Helichrysum aureum* var. **monocephalum**
  □ *Helichrysum aureum* var. **candidum**
  ▰ sympatric

stems and leaves and smaller heads (c. 8—10 mm long, c. 20 mm across the radiating bracts). However, with the wealth of material now available, it proves difficult to distinguish them on these or any other characteristics, and intermediates are commonplace.

Vouchers: *Compton* 29279 (PRE); *Galpin* 1614 (PRE); *Hilliard & Burtt* 8794 (E; K; MO; NU; PRE; S).

(e) var. **candidum** *Hilliard* in Notes R. bot. Gdn Edinb. 34: 80 (1975), Compositae in Natal 254 (1977). Type: Swaziland, near Mbabane, Mhlambanyati, 1968, *Cook* s.n. (NU, holo.!; mixed with var. *monocephalum*).

Recorded from Pigg's Peak and Mbabane districts in Swaziland, Wakkerstroom district in southern Transvaal and the Utrecht, Paulpietersburg and Richmond districts in Natal. Differs from var.

monocephalum in its snow-white bracts, and the two varieties may grow together. Sometimes mistaken for *H. elegantissimum* (no. 237). Map 218.

Vouchers: *Acocks* 11745 (PRE); *Codd* 1728 (NU; PRE); *Devenish* 1563 (E; K; NU).

Compton 26990 and 28009 (NBG; PRE), from grassland near Forbes Reef, Swaziland, appear to be a variant with grey-woolly cauline leaves.

(f) var. **argenteum** *Hilliard* in Notes R. bot. Gdn Edinb. 40: 252 (1982). Type: Transvaal, Wakkerstroom distr., Wakkerstroom Commonage, c. 1 800 m, 20 xii 1979, *Devenish* 1951 (NU, holo.!; E, iso.!).

Flowering stems 200—400 mm high, simple or forking above into the compound inflorescence, loosely silvery-grey woolly. *Radical leaves* up to 75 × 15 mm, oblanceolate, acute, both surfaces loosely silvery-grey woolly; *cauline leaves* up to c. 50 × 10 mm, lanceolate, acute to shortly acuminate, margins often crisped-undulate, both surfaces loosely silvery-grey woolly. *Heads* c. 15−18 mm long, c. 30−35 mm across the radiating bracts, solitary or up to 6 corymbosely arranged. *Involucral bracts* bright golden-yellow. Fig. 58: 1.

Recorded only from Wakkerstroom in the SE. Transvaal and a neighbouring farm, Oshoek, growing in montane grassland and flowering between December and February, c. 1 800−2 000 m above sea level. Map 217.

A robust plant, rendered distinctive by its silvery woolly stems and leaves and large golden-yellow heads often in small corymbs.

Vouchers: *Beeton* 199 (PRE; SAM; STE); *Devenish* 1217 (PRE).

223. **Helichrysum lesliei** *Hilliard* in Notes R. bot. Gdn Edinb. 40: 257 (1982). Type: Transvaal, Lydenburg distr., 16 miles S. of Lydenburg, Kemp's Heights, 6 200 ft, 21 i 1954, *Codd* 8313 (PRE, holo.!; K, iso.!).

Soft-wooded subshrub, stock stout, up to 20 mm diam., woody, crowned with one or several leaf rosettes, flowering stems several, lateral to the rosettes, erect, up to 600 mm, simple below the compound inflorescence, thickly and loosely white woolly, closely leafy. *Radical leaves* up to 150 × 40 mm, oblong-lanceolate, apex subacute, margins somewhat undulate; *cauline leaves* mostly 35−50 × (15−) 20−40 mm, smaller on the inflorescence branches, ovate or broadly lanceolate, apex very obtuse to subacute, base broad, cordate-clasping, all leaves with both surfaces loosely greyish-white woolly. *Heads* heterogamous, 10−14 mm long, 20−30 mm across the radiating bracts, depressed-globose, 2 to several in a leafy open corymbose panicle. *Involucral bracts* in c. 9 series, graded, imbricate, inner much exceeding the flowers, glossy, bright yellow, outer overlaid palest brown outside. *Receptacle* shortly honeycombed. *Flowers* c. 700−1 000, 140 ♀, 560−830 ☿. *Achenes* c. 1 mm long, barrel-shaped, glabrous. *Pappus* bristles

many, equalling corolla, tips barbellate, shaft scabrid, bases with minute patent cilia, cohering or not. Fig. 58: 2.

Recorded from Lydenburg, Belfast and Barberton districts, and on hills (Swaziland System) 17 miles N. of Johannesburg. Possibly always on quartzites. Flowers between January and April. Map 219.

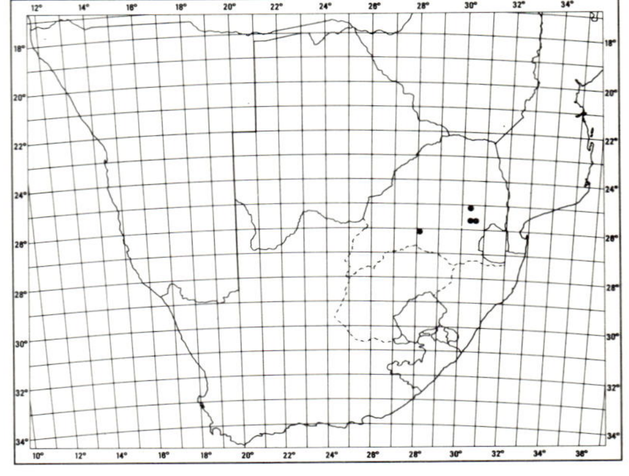

MAP 219.— **Helichrysum lesliei**

The broad, cordate-clasping loosely woolly grey cauline leaves and the very open and leafy inflorescence are characteristic.

Vouchers: *Hilliard & Burtt* 14171 (E; K; M; MO; NU; PRE; S); *Jacobsen* 4831 (PRE); *Richardson* 12 (E; K; NU).

224. **Helichrysum tenax** *M.D. Henderson* in Bothalia 6: 423 (1954); Hilliard, Compositae in Natal 248 (1977). Type: Natal, Bergville distr., Cathedral Peak Forest Research Station, 5 800 ft, *Killick* 1632 (PRE, holo.!; K; NH; NU, iso.!).

H. fulgidum (L.f.) Willd. var. *subnudatum* DC., Prodr. 6: 187 (1838). Type: Cape, Zwart Kei, in the Mountains, *Drège* 5733 (G-DC, holo.!).

Bushy subshrub up to c. 1,8 m high, viscid and aromatic particularly when fresh, stem base thick and woody, branching (and sometimes rooting) there or higher, branches spreading, rough with leaf bases, glandular and with traces of white wool, each terminating in a large leaf rosette subtended by dry withered leaves, flowering stems (peduncles) one or several from the axils of the spent leaves below the fresh terminal growth, c. 300 (−450) mm long, glandular, thinly to thickly appressed white-woolly as well, closely leafy. *Rosette leaves* up to c. 180 (−250) × 80 mm, narrowly to broadly elliptic or obovate, narrowed to a broad clasping base, apex acute to obtuse, apiculate, margins of young leaves often undulate, both surfaces very glandular, in var. *pallidum* greyish woolly as well, at least below; *peduncle leaves* erect, imbricate, up to 80 × 20 (−30) mm, diminishing in size upwards, oblong-lanceolate to lanceolate, apex subacute to acute to acuminate, apiculate, base broad, cordate-clasping, sometimes very shortly decurrent, margins crisped-undulate in upper half, both surfaces glandular-pilose with traces of white wool on the margins and over the main veins to densely greyish-white woolly in var. *pallidum. Heads* heterogamous, depressed-globose, c. 25−35 mm across the radiating involucral bracts, solitary or up to c. 12 in a compact corymbose cluster at the peduncle tip, loosely white-woolly at the base. *Involucral bracts* in 10 series, somewhat graded, imbricate, much exceeding the flowers, radiating, acute, glossy, bright or pale yellow, or creamy white overlaid palest buff in var. *pallidum. Receptacle* shortly honeycombed. *Flowers* 701−1148, 80−156 ♀, 592−1048 ☿, yellow. *Achenes* 0,75−1 mm long, barrel-shaped, glabrous. *Pappus* bristles many, equalling corolla, tips barbellate, bases cohering lightly by patent cilia.

Two varieties are recognized:

1a Involucral bracts yellow; leaves without white wool except sometimes on the margins and over the veins ....(a) var. *tenax*

1b Involucral bracts creamy-white, outer overlaid palest buff; leaves greyish-white at least on lower surface ....(b) var. *pallidum*

(a) var. **tenax.**

Ranges from the mountainous northern part of Lesotho along the face of the Natal Drakensberg and its outliers (including the Karkloof range) to the mountains about Kokstad and the face of the Drakensberg in the Transkei to the mountainous parts of the E. Cape as far west as the Wildschutsberg near Molteno, Katberg, and the Suurberg N. of Port Elizabeth.

Grows on grassy mountain slopes particularly near rocks, or in the boulder beds of mountain streams, and readily becomes a weed colonizing bare or disturbed areas. Flowers between October and December. At Katberg and Hogsback, the rosette leaves are much

narrower than in specimens from further north, and the heads are often solitary. However, these rosettes of very sticky leaves are diagnostic. But even detached flowering stems are easily recognized by the loose white wool on the stem and enveloping the base of each head, and the crisped-undulate leaf margins. Map 220.

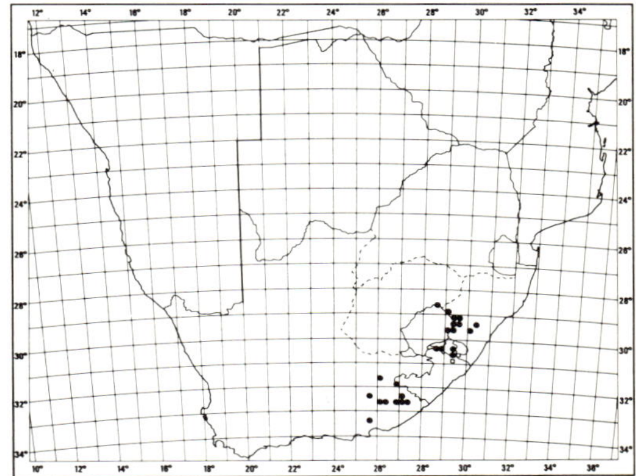

MAP 220.— ● **Helichrysum tenax** var. **tenax**
◻ **Helichrysum tenax** var. **pallidum**

Vouchers: *Acocks* 9316 (PRE); *Barker* 1161 (NBG); *Hilliard & Burtt* 7369 (E; K; MO; NU; S); *Hilliard & Burtt* 11031 (E; K; NU; S); *Wright* 356 (E; NH; NU).

(b) var. **pallidum** *Hilliard & Burtt* in Notes R. bot. Gdn Edinb. 34: 84 (1975); Hilliard, Compositae in Natal 248 (1977). Type: Transkei, Tabankulu distr., Mt Tabankulu, c. 1 525 m, *Hilliard & Burtt* 7336 (NU, holo.!; E; K; MO; NBG; PRE; S, iso.!).

Recorded only from Tabankulu and Insizwa Mountains in the Transkei, and Ngeli Mtn and the Suurberg near Weza in Natal. Grows on steep rocky terrain, often rooted in rock crevices, and near Weza it has colonized the steep rock banks of road cuttings. Flowers in October and November. Map 220.

Vouchers: *Hilliard & Burtt* 7308 (E; K; NU); *Hilliard & Burtt* 7238 (E; K; MO; NU).

225. **Helichrysum fulvum** *N.E. Br.* in Kew Bull. 1895: 146 (1895); Moeser in Bot. Jb. 44: 338 (1910); Hilliard, Compositae in Natal 242 (1977). Lectotype: Natal, Drakensberg, edge of brook near Van Reenen's Pass, 5 000−6 000 ft, *Wood* 4533 (K!; NH; Z, isolecto.!).

Rhizomatous, strongly aromatic perennial herb, rhizome slender, woody, flowering stem up to 1 m tall, simple below the inflorescence branches, glandular-pubescent, leafy throughout. *Radical leaves* rosetted, withered at flowering, up to 350 × 55 mm, oblanceolate, apex subacute or obtuse, narrowed to a broad, flat, clasping petiole-like base, both surfaces thinly grey-woolly; *cauline leaves* up to c. 120 × 30 mm, decreasing in size upwards, elliptic-lanceolate, apex acute to acuminate, base broad, cordate-clasping, very shortly decurrent, both surfaces minutely glandular-pubescent. *Heads heterogamous*, depressed-globose, c. 7−8 mm long, c. 15−18 mm across the radiating bracts, many in a large spreading leafy corymbose panicle. *Involucral bracts* in 6 series, graded, imbricate, exceeding flowers, radiating, glossy light brown, very acute. *Receptacle* honeycombed. *Flowers* 423−731, 26−97 ♀, 378−634 ☿, orange-yellow. *Achenes* 0,75 mm long, barrel-shaped, glabrous. *Pappus* bristles many, equalling corolla, scabridulous, bases with patent cilia, lightly cohering or not.

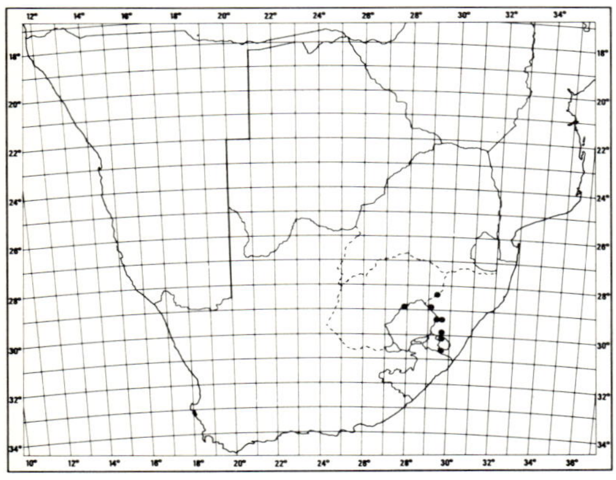

MAP 221.— **Helichrysum fulvum**

Ranges along the Drakensberg and its outliers from Van Reenen's Pass in Natal to Insikeni Mountain in Umzimkulu district, Transkei, and Ngeli Mountain on the Natal-Transkei border. Also recorded from near Ficksburg in the NE. Orange Free State. Grows along streams, between c. 1 400 and 2 100 m above sea level, often in the shallows, rooted in mud; flowering between October and January, but mainly in November and December. The tuft of grey-woolly radical leaves is produced in the first season, the terminal flowering stem with glandular but not woolly leaves the following year, together with several woody runners each terminating in a new leaf rosette. The small heads with brown pointed bracts allied to the markedly discolorous radical and cauline leaves make this a very distinctive species. Map 221.

Vouchers: *Hilliard* 2491 (NU); *Hilliard & Burtt* 3429 (E; NU); *Trauseld* 863 (NU; PRE); *Wright* 321 (E; K; M; NBG; NU; PRE).

226. **Helichrysum albilanatum** *Hilliard* in Notes R. bot. Gdn Edinb. 40: 249 (1982). Type: Transvaal, 2530 BA, 8 km from Lydenburg on Sabie road, rocky grassland, 5 iii 1981, *Hilliard & Burtt* 14193 (NU, holo.!; E; K; M; PRE; S, iso.!).

Subshrub up to 1 m tall, stock with vegetative buds, stems many from the base forming rounded clumps, woody, brittle, virgate, forking above into the compound inflorescence, thinly silvery-grey appressed woolly, long red patent glandular hairs as well, closely leafy. *Leaves* mostly 35−60 × 11−20 mm, 20−40 × 6−15 mm on the inflorescence branches, smaller near the heads, lanceolate, acute to shortly acuminate, base more or less cordate-clasping, lower surface thickly silvery-grey appressed woolly, long red patent glandular hairs as well, upper surface generally appressed woolly, but wool sometimes wanting and then surface merely glandular-setose. *Heads* heterogamous, (10−) 12−13 mm long, (20−) 25−28 mm across the radiating bracts, depressed-globose, several in a very open terminal leafy corymbose panicle, or subsolitary on side branches. *Involucral bracts* in c. 9 series, graded, imbricate, inner much exceeding the flowers, glossy, bright yellow, sometimes overlaid pale brown outside. *Receptacle* shortly honeycombed. *Flowers* c. 750−950, 90−125 ♀, 640−850 ☿. *Achenes* 1 mm long, glabrous. *Pappus* bristles many, equalling the corolla, tips subplumose, shaft scabrid, bases cohering by patent cilia.

Recorded only from the highlands of the E. Transvaal, from about Mariepskop in the north to Belfast district in the south, in rocky montane grassland and marginal to forest patches. Flowers between January and June. Map 222.

H. albilanatum is frequently confused with *H. decorum* (no. 235) but is easily distinguished by its differently shaped and more silvery leaves, and its smaller heads. Furthermore, *H. decorum* is confined to the coastal areas of the Transkei, Natal and Mozambique.

H. albilanatum can be confused with grey-leaved plants of *H. mutabile* (no. 233), but that species has oblong rather than lanceolate leaves, mostly 4−13 mm broad, not 11−20 mm.

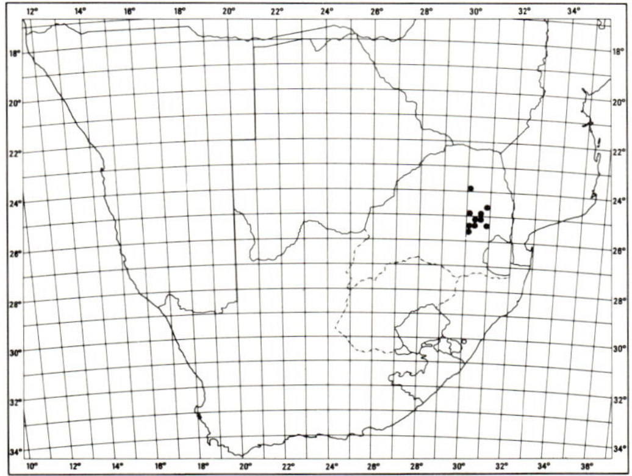

MAP 222.— ● **Helichrysum albilanatum**
      ○ **Helichrysum citricephalum**

Vouchers: *Hilliard & Burtt* 14202 (E; K; NU; PRE); *Hutchinson* 2745 (BOL; K; PRE); *Smuts* 19 (K; PRE); *Wilms* 712 (BM; E; G).

Hilliard & Burtt 14370 (E; NU; PRE) from Long Tom Pass, Mount Anderson, appears to be a hybrid between *H. albilanatum* and *H. molestum* (no. 242), while *Hilliard & Burtt* 14371 (E; K; NU; PRE) appears to be a hybrid between *H. albilanatum* and *H. aureolum* (no. 234); all the putative parents were growing in close proximity. *Hilliard & Burtt* 14396 (E; K; NU; PRE) and *Hilliard & Burtt* 14204 (E; NU) represent a number of plants that are probably *H. albilanatum* × *H. difficile* (no. 241); they grew on the grass slope between a dolerite outcrop (*H. albilanatum*) and a marsh (*H. difficile*). All these supposed hybrids were intermediate in character between the probable parents.

227. **Helichrysum citricephalum** *Hilliard & Burtt* in Notes R. bot. Gdn Edinb. 34: 259 (1976); Hilliard, Compositae in Natal 249 (1977). Type: Natal, Ixopo distr., Ixopo to Umzimkulu road, straggling shrub in forest margin type vegetation on steep bank at roadside, 26 ii 75, *Hilliard & Burtt* 8045 (NU, holo.!; E; K; MO; PRE; S, iso.!).

Shrub up to 1 m tall and as much across, main branches bare, rough with persistent leaf bases, flowering shoots grey-felted, glandular, very closely leafy. *Leaves* mostly 30−50 × 7−18 mm, oblong-lanceolate to elliptic-lanceolate, uppermost lanceolate, apex acute to acuminate, apiculate, base narrowed, rounded, sessile, both surfaces silvery-grey felted, glandular as well. *Heads* heterogamous, campanulate, c. 15 mm long, 25 mm across the radiating bracts, 4−20 solitary at the tips of long (50−120 mm) leafy peduncles corymbosely arranged. *Involucral bracts* in 6−12 series, graded, lanceolate, acute to very acute, much exceeding flowers, glossy, pale lemon-yellow. *Receptacle* honeycombed. *Flowers* 465−484, 16−19 ♀, 446−468 ☿, yellow. *Ovary* 0,75 mm long, glabrous. *Achenes* not seen. *Pappus* bristles many, about equalling the corolla, tips barbellate, shaft smooth, bases not cohering.

Known only from the type collection and one made by Krook in the same area nearly a hundred years ago. Grows in coarse herbage, probably on forest margins; flowering in February. Easily recognized by its shrubby habit, grey leaves and lemon-yellow heads, more like the tropical species *H. petersii* Oliv. & Hiern than any S. African species. Map 222.

The only known locality at Ixopo has been destroyed by road-widening.

Voucher: *Krook* in herb. Penther 995 (BM; M).

228. **Helichrysum milleri** *Hilliard* in Notes R. bot. Gdn Edinb. 40: 259 (1982). Type: Transvaal, Barberton distr., Emlembe Mountain, Devil's Bridge, 5 200 ft. 26 x 1963, *Miller* 8509 (PRE, holo.!).

Shrub up to 1 m tall, branches brittle, glandular-setose, thinly white-woolly as well when young, wool often persisting under the heads, leafy. *Leaves* mostly (15−) 20−25 × 6−12 mm, oblong or oblong-lanceolate, obtuse to acute, apiculate, base broad, half-clasping, both surfaces greyish-white woolly, wool persistent on sterile twigs, shed on flowering twigs to reveal the glandular-setose surface. *Heads* heterogamous, c. 15 mm long, c. 32 mm across the radiating bracts, depressed-globose, solitary or few at the tips of the lateral flowering twigs. *Involucral bracts* in c. 9 series, graded, imbricate, much exceeding the flowers, glossy, light brown. *Receptacle* shortly honeycombed. *Flowers* c. 600−700, 30−60 ♀, 570−650 ☿. *Achenes* 1,25 mm long, glabrous. *Pappus* bristles equalling corolla, tip barbellate, shaft scabrid, base smooth.

Recorded only from the Barberton Mountains, at Havelock and Emlembe Mountain on the Swaziland-Transvaal border, and from the Lydenburg area. Probably always in coarse herbage on the margins of forest patches; flowering in October and November. Map 223.

H. milleri is unusual in this group in its spring flowering. It is easily recognized by its leaves that remain persistently grey-woolly on the sterile twigs but on the flowering twigs gradually shed the wool to reveal the green and glandular-setose surfaces, brown heads solitary or few at the tips of the lateral flowering twigs, and pappus bristles without patent cilia at the base.

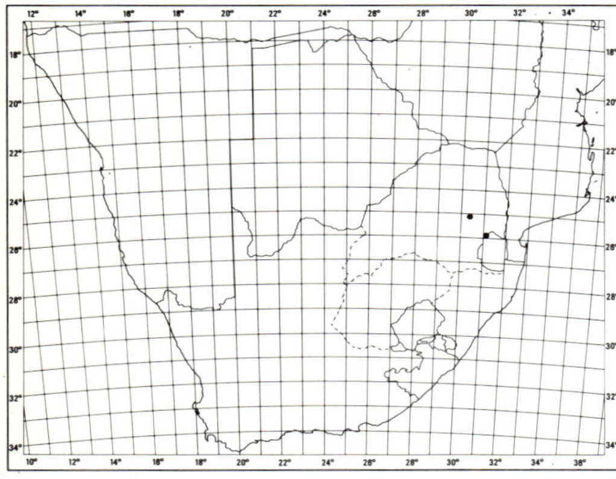

MAP 223.— **Helichrysum milleri**

Vouchers: *Anderson* A80 (PRE); *Miller* 3035 (PRE).

229. **Helichrysum transmontanum** *Hilliard* in Notes R. bot. Gdn Edinb. 40: 268 (1982). Type: Transvaal, 2330 CC, Letaba distr., near Haenertsburg, opposite Magoebaskloof Hotel, 13 iii 1981, *Hilliard & Burtt* 14332 (NU, holo.!; E; K; M; PRE; S, iso.!).

Subshrub, stems many from the base, stout, erect to c. 1,5 m, mostly simple below the inflorescence branches, glandular-setose, young parts thinly cobwebby as well, closely leafy. *Leaves* mostly 50−80 × 20−28 mm, smaller on the inflorescence branches, lanceolate, acute to acuminate, base broad, half-clasping, both surfaces and margins glandular-setose, midline and margins sometimes thinly cobwebby as well. *Heads* heterogamous or rarely homogamous, 17−20 mm long, about double that across the radiating bracts, broadly campanulate, several in a very open, leafy, terminal corymbose-panicle. *Involucral bracts* in c. 9 series, graded, imbricate, much exceeding the flowers, glossy, straw-coloured to light brown. *Receptacle* shortly honeycombed. *Flowers* c. 600−950, (0−) 5−75 ♀, 550−920 ☿. *Achenes* 0,75−1 mm long, glabrous. *Pappus* bristles many, equalling the corolla, tip barbellate, shaft barbellate above, scabrid below, base smooth. Fig. 59: 1.

Recorded only from the highlands of the E. Transvaal and W. Swaziland from Duiwelskloof and Woodbush in the north south to Forbes Reef and Mbabane districts in Swaziland. Grows in montane grassland; flowering in March and April. Map 224.

Easily recognized by its very acute to acuminate leaves, large brownish heads and pappus bristles without patent cilia at the base.

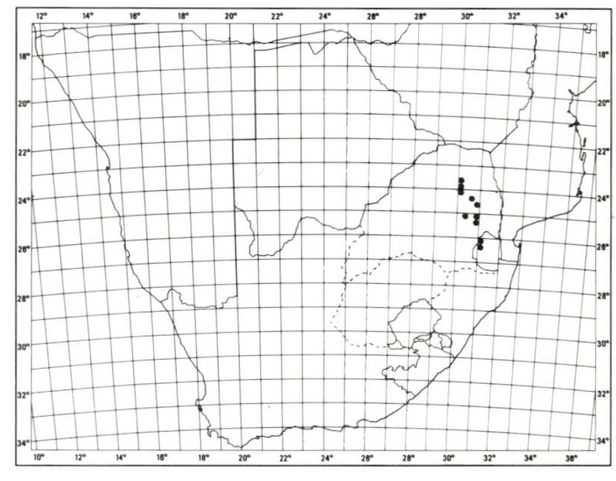

MAP 224.— **Helichrysum transmontanum**

Vouchers: *Codd* 7891 (K; PRE); *Compton* 28553 (K; NBG; PRE); *Meeuse* 10127 (K; M; PRE); *Scheepers* 623 (BM; K; M; PRE).

FIG. 59.−1, **Helichrysum transmontanum**, part of whole plant, × 0,06; 1a, branch of flowering panicle, × 1 (*Hilliard & Burtt* 14332). 2, **H. setosum**, part of compound inflorescence, × 1 (*Hutchinson* 15).

1a
1
2
H. Weitz-du Toit

230. **Helichrysum edwardsii** *Wild* in Notes R. bot. Gdn Edinb. 40: 254 (1982). Type: Rhodesia [Zimbabwe], Inyanga, World's View, 6 500 ft, 9 iv 1966, *Edwards 950* (SRGH, holo.!).

Subshrub up to 1 m tall, stock with woolly vegetative buds or leaf tufts, stems several to many from the base, stout, woody, simple or sparingly branched below the inflorescence branches, glandular-setose, often thinly white woolly-cobwebby as well, densely leafy. *Radical leaves* (seen in one specimen only) rosulate, oblong, acute, c. 50−55 × 8−15 mm, thinly greyish-white woolly on both surfaces; *cauline leaves* mostly 40−75 (−100) × 18−28 (−40) mm, smaller on the inflorescence branches, oblong or ovate, acute, mucronate, base broad, cordate-clasping, both surfaces glandular-setose, lower leaves at least initially greyish-white woolly on both surfaces as well, upper surface usually glabrescent, but upper leaves, particularly those on the inflorescence branches, often without wool or with wool confined to the margins and midline. *Heads* heterogamous, depressed-globose, c. 14−17 mm long, 28−36 mm across the radiating bracts, several corymbosely arranged, the inflorescence branches simple or subsimple, closely leafy up to the heads. *Involucral bracts* in c. 9 series, graded, imbricate, inner much exceeding the flowers, glossy, pale yellow inside, light golden-brown outside. *Receptacle* shortly honeycombed. *Flowers* c. 830−2 350, 100−165 ♀, 700−2 200 ☿. *Achenes* c. 1 mm long, barrel-shaped, glabrous. *Pappus* bristles many, about equalling corolla, tips barbellate, shaft scabrid, bases with patent cilia, cohering lightly or not.

Recorded from Inyanga in Zimbabwe, the eastern part of the Soutpansberg and the mountains in the NE. Transvaal, about Sibasa, Haenertsburg, Potgietersrust, the Wolkberg and The Downs. Grows in shrub communities among rock outcrops or on mountain slopes, often near the margins of forest or forest relicts; flowering mainly in March and April. Map 225.

H. edwardsii can be recognized by its large yellow and brown heads and green leaves on the inflorescence branches and upper parts of the stems contrasting with the wholly or partially woolly lower leaves. In this it resembles *H. milleri* (no. 228), which differs in its pappus and flowering time.

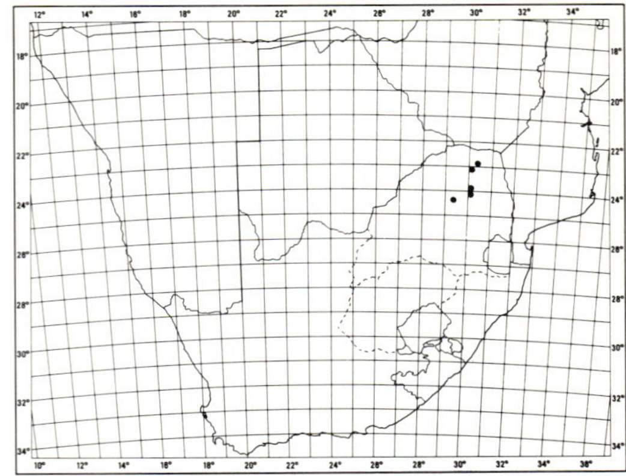

MAP 225.— **Helichrysum edwardsii**

In the herbarium it is almost impossible to distinguish *H. edwardsii* from *H. setosum* (below) when only the inflorescence and upper part of the stem is present and field notes are wanting. The following specimens from the NE. Transvaal are difficult to place: *Codd 2899*, PRE (Hanglip Mountain Range), *Moffett 1778*, PRE (Wolkberg-Strydpoort Mountains) *Galpin 9003*, PRE (near Potgietersrust). The problem will yield only to field studies.

Vouchers: *Crundall in PRE 36894* (PRE); *Hemm 584* (PRE); *Hilliard & Burtt 14318* (E; K; NU; PRE); *Rogers 20118* (NH).

Hilliard & Burtt 14328 (E; K; NU; PRE; S), from Magoebaskloof, has the foliage of *H. edwardsii*, but the involucral bracts are pale brown and the pappus bristles lack cilia at the base, both features of *H. transmontanum* (no. 229), which was growing nearby. It is not unlikely that some crossing has taken place.

231. **Helichrysum setosum** *Harv.* in F. C. 3: 231 (1865); Moeser in Bot. Jb. 44: 337 (1910); Brenan in Mem. N. Y. bot. Gdn 8,5: 474 (1954); Hilliard, Compositae in Natal 246 (1977). Lectotype: Transvaal, on the Vaal River, May, *Burke 98* (TCD!; BM; K; Z; fragment PRE, isolecto.!).

Gnaphalium setosum (Harv.) O. Kuntze, Rev. Gen. Pl. 3,2: 154 (1898).

G. setosum var. *hoffmannii* O. Kuntze, Rev. Gen. Pl. 3,2: 154 (1898). Type: Natal, Glencoe, Highlands Station, *Kuntze* (K!).

Herbaceous perennial or subshrub, branched from the base, stock with vegetative buds or small leaf tufts, stems up to 1 m tall, but often only c. 500 mm, woody, brittle, glandular-setose, uppermost parts sometimes white-cobwebby as well, closely leafy. *Leaves* mostly 20−50 × 10−20 mm, slightly smaller on the inflorescence branches, mostly oblong-ovate often becoming elliptic-lanceolate to lanceolate on the inflorescence branches, obtuse to acute, mucronate, base broad, more or less cordate-clasping, both surfaces and margins glandular-setose, margins and sometimes the midline above sometimes woolly as well, young leaves sometimes cobwebby all over. *Heads* heterogamous, 10−12 (−14) mm long, about double that across the radiating bracts, depressed-globose, solitary at the tips of long leafy branchlets corymbosely arranged. *Involucral bracts* in c. 9 series, graded, imbricate, inner much exceeding the flowers, glossy, yellow often overlaid pale brown outside. *Receptacle* shortly honeycombed or nearly smooth. *Flowers* c. 480−1 100, 65−160 ♀, 400−950 ☿. *Achenes* 1 mm long, glabrous. *Pappus* bristles many, equalling the corolla, tips barbellate, shaft scabrid, base with patent cilia cohering or not. Fig. 59: 2.

Widespread in the Transvaal from the Soutpansberg south to Potgietersrust in the east and the Waterberg in the west, then widespread on the Highveld, and reaching Parys and Vredefort in the Orange Free State and Newcastle, Dundee, Ngotshe and Klip River districts in Natal. Also in Zimbabwe and Malawi. Map 226.

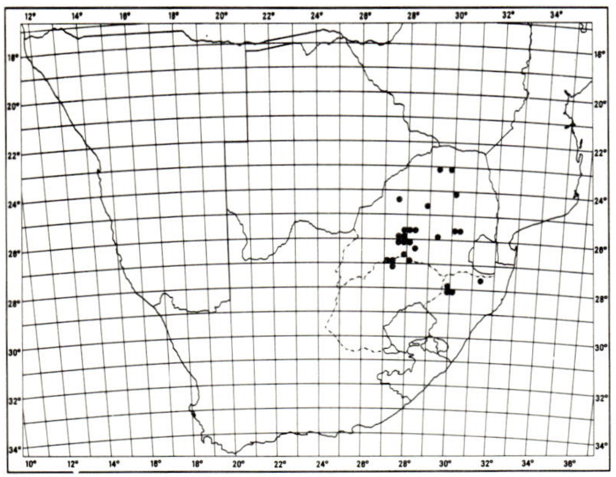

MAP 226.— **Helichrysum setosum**

The name has been much misused. The species is characterized by its bushy habit, glandular-setose

leaves mostly ovate or oblong-ovate, obtuse or acute, and heads solitary at the tips of long leafy branches arranged in a corymbose panicle; these branches are glandular-setose like the leaves; any white woolly hairs are confined to the area immediately below the heads and to the leaf margins and main vein. Although the leaves under the heads are a little smaller than those further back down the branches, they are nevertheless relatively large and crowded, often equalling or exceeding the involucre, and contribute to the very leafy appearance of the compound inflorescence. The arrangement of the heads alone will immediately distinguish *H. setosum* from *H. cooperi* (no. 240), with which it is frequently confused: *H. cooperi* has a well-branched compound inflorescence, and furthermore the main stem leaves are usually at least shortly decurrent. *H. tongense* (below) and specimens of *H. mutabile* (no. 233) lacking wool on the leaves have also been confused with *H. setosum,* but the leaves on the inflorescence branches are reduced and distant near the heads, not close-set and relatively large as in *H. setosum.*

O. Kuntze remarked of his varietal name 'a form with somewhat woolly leaf margins as in *H. fulgidum*' (that is, *H. aureum;* no. 222); it is not worth upholding.

Vouchers: *Acocks* 21029 (PRE); *Hilliard & Burtt* 14423 (E; K; M; MO; NU; PRE; S); *Pegler* 1772 (PRE); *Strey* 3756 (BM; K; M; PRE); *Vahrmeijer* 1548 (PRE).

232. **Helichrysum tongense** *Hiliiard* in Notes R. bot. Gdn Edinb. 40: 267 (1982), in Compositae in Natal 247 (1977) sub *Helichrysum* sp. Type: Natal, Ingwavuma distr., Ndumu, Usutu floodplain, 20 ix 1968, *Pooley* 76 (NU, holo.!; E, iso.!).

Herbaceous, probably biennial, up to 1,2 m tall, aromatic, stem forking, often low down, into long virgate branches, glandular-pubescent, loosely and thinly white-woolly at first, persisting on the upper parts, leafy. *Leaves* mostly 30−80 × 7−20 mm, smaller on the inflorescence branches, reduced near the heads, linear-oblong or oblong, apex acute, base broad, half-clasping, both surfaces glandular-pubescent, margins and sometimes the midline white-woolly. *Heads* heterogamous, depressed-globose, c. 12−14 mm long, (20−) 25−30 mm across the radiating bracts, solitary at the tips of long (mostly 50−300 mm) leafy branches corymbose-paniculately arranged. *Involucral bracts* in c. 9 series, graded, imbricate, inner much exceeding the flowers, glossy, bright yellow. *Receptacle* shortly honeycombed. *Flowers* c. 1 000−1 400, 100−340 ♀, 700−1 200 ☿. *Achenes* 1 mm long, barrel-shaped, glabrous. *Pappus* bristles many, equalling corolla, tips barbellate, bases with patent cilia, cohering or not.

Nearly confined to the Tongaland plain, from about the Limpopo in Mozambique south to about Empangeni and Richards Bay in Natal, with a few records from the Swaziland Lowveld. Grows in sandy grassland or open woodland, and becomes a weed in old fields. Flowers mainly between January and June. Map 227.

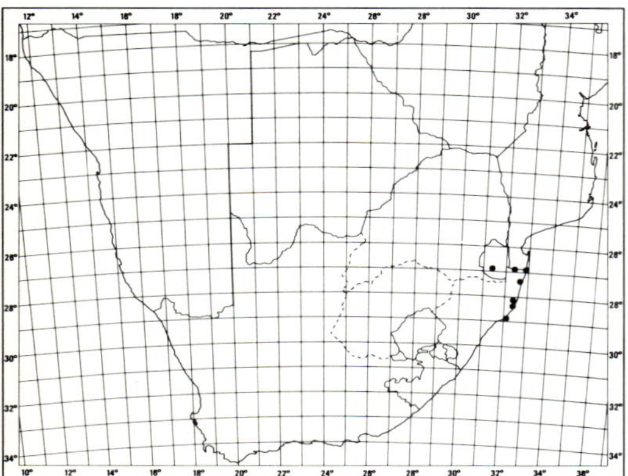

MAP 227.— **Helichrysum tongense**

Much confused with *H. setosum* (above) but distinguished by its differently shaped leaves, persistently white-woolly ultimate branches and leaves much reduced near the heads.

Vouchers: *Bourquin* 418 (NU); *Gerstner* 2724 (PRE); *Lawson* 503 (NH); *Vahrmeijer* 677 (PRE).

233. **Helichrysum mutabile** *Hilliard, sp. nov.* H. tongensi *Hilliard affinis, sed habitu caulibus inferne simplicibus superne ramosis (nec inferne furcatis ramis longis virgatis), saltem foliis paucis pagina una vel utrinque griseo-albidis lanatis (nec lana ad margines et interdum ad costam restricta) involucri bracteis externe laete brunneis (nec omnino vivide luteis).*

Type: Transvaal, 2530 CB, Waterval Boven, old road near tunnel, 4 iii 1981, *Hilliard & Burtt* 14189 (NU, holo.!; E; K; M; MO; PRE; S, iso.!).

Perennial herb up to c. 1 m tall, stems one or several from the stock, simple below, branching above into the compound inflorescence, glandular-setose, often thinly and loosely white-woolly as well, leafy. *Leaves* mostly 25−75 × 4−12 mm, sometimes larger (up to 90 × 24 mm) on the lower part

of the stem, and often smaller on the inflorescence branches and further reduced near the heads, oblong or the upper ones oblong-lanceolate, apex acute, base broad, more or less cordate-clasping, both surfaces glandular-pubescent, often thinly to thickly greyish-white woolly as well either on both surfaces, or the lower surface, or the lower leaves woolly on one or both surfaces, the upper leaves merely glandular. *Heads* heterogamous, depressed-globose, c. (10−) 12−14 mm long, (18−) 20−28 mm across the radiating bracts, solitary at the tips of the long (mostly 45−155 mm) leafy branches of a spreading corymbose panicle. *Involucral bracts* in c. 9 series, graded, imbricate, inner much exceeding the flowers, bright yellow, outer light brown outside. *Receptacle* shortly honeycombed. *Flowers* c. 850−1 200, 100−130 ♀, 760−1 100 ☿. *Achenes* 1 mm long, barrel-shaped, glabrous. *Pappus* bristles many, equalling the corolla, tips barbellate, shaft scabrid, bases with patient cilia, cohering or not.

Recorded mainly from the Lowveld of the Transvaal and Swaziland, from the environs of Tzaneen in the north to Barberton, Waterval Boven and the Swaziland Lowveld in the south, with one record from Premier Mine east of Pretoria. Favours rocky sites in woodland, or in grassland near forest patches, or rarely marshy places near rocks. Flowers mainly between March and May. Map 228.

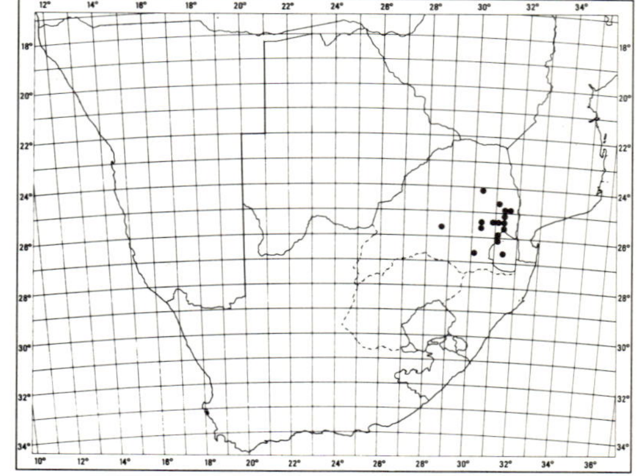

MAP 228.— **Helichrysum mutabile**

Allied to *H. tongense* (above) from which it is distinguished by its habit, at least some leaves generally greyish-white woolly on one or both surfaces, and brown outer involucral bracts. The leaf indumentum is very variable: frequently the lower leaves are woolly on one or both surfaces but the upper leaves, on the flowering twigs, are merely green and glandular; sometimes all the leaves are woolly below, glandular above; sometimes all the leaves lack wool; occasionally all the leaves are persistently woolly on both surfaces. At Witklip near Nelspruit, plants with woolly leaves, others with green leaves, still others with lower leaves woolly, upper green, grow together; elsewhere, all the plants in a colony may be wholly green, or wholly woolly, or woolly below, green above.

See also *H. aureolum* (below).

Vouchers: *Buitendag* 503 (K; NBG; PRE); *Compton* 26844 (K; NBG; PRE); *Van der Merwe* 1254 (K; PRE); *Van der Schijff* 2655 (K; PRE).

234. **Helichrysum aureolum** *Hilliard* in Notes R. bot. Gdn Edinb. 40: 251 (1982). Type: Transvaal, 2530 BA, Sabie to Lydenburg, Long Tom Pass, Whisky Spruit, c. 1 980 m, 15 iii 1981, *Hilliard & Burtt* 14365 (NU, holo.!; E; K; PRE, iso.!).

Subshrub, stock stout, woody, producing vegetative buds, flowering stems several to many from the base, simple below, branching above into the compound inflorescence, up to c. 1 m tall, glandular-setose, hairs red or purple, young parts thinly white-woolly as well, closely leafy, aromatic. *Leaves* mostly 50−90 × 15−38 mm, slightly smaller on the inflorescence branches, lanceolate or oblong-lanceolate, apex acute or obtuse, apiculate, base broad, cordate-clasping, both surfaces glandular-setose, margins and midline often white-woolly as well. *Heads* heterogamous, depressed-globose, c. 12−13 mm long, (20−) 24−30 mm across the radiating bracts, several in an open spreading corymbose panicle, the branches often simple, or branching once or twice, leafy. *Involucral bracts* in c. 9 series, graded, imbricate, inner much exceeding the flowers, glossy, bright golden-yellow inside, golden-brown outside. *Receptacle* shortly honeycombed. *Flowers* c. 900−1 200, 125−200 ♀, 775−975 ☿. *Achenes* c. 1 mm long, barrel-shaped, glabrous. *Pappus* bristles many, about equalling corolla, tips barbellate, shafts scabrid, bases with patent cilia, cohering or not.

Recorded from the highlands of the E. Transvaal, from Magoebaskloof and New Agatha Forest Reserve near Tzaneen south to the Palwane Hills, Mbabane district, Swaziland, in grassland and scrub, often on damp streamsides or other marshy or well-watered places; flowering between February and July. Map 229.

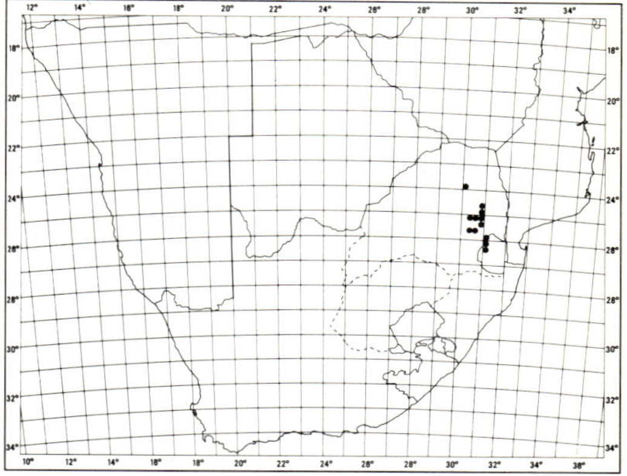

MAP 229.— **Helichrysum aureolum**

Characterized by its leaves, not decurrent on the stem, and large heads with involucral bracts bright yellow inside, golden-brown outside. These characters distinguish it from *H. cooperi* (no. 240), with which it is frequently confused. It has also been confused with *H. setosum* (no. 231), from which it may be distinguished by its inflorescence branches with red-purple glandular hairs and reduced, distant leaves. It can be distinguished from green-leaved plants of *H. mutabile* (above) most readily by its generally longer and broader leaves; it is an altogether more robust plant than *H. mutabile*.

Vouchers: *Compton* 25878 (K; NBG; PRE); *Hilliard & Burtt* 14206 (E; K; NU; PRE); *Kluge* 61 (NBG; PRE); *Muller & Scheepers* 59 (PRE); *Rogers* 18641 (PRE).

See under *H. albilanatum* (no. 226) for possible hybrid.

235. **Helichrysum decorum** *DC.*, Prodr. 6: 188 (1838); Harv. in F.C. 3: 232 (1865); Moeser in Bot. Jb. 44: 336 (1910); Hilliard, Compositae in Natal 245 (1977). Type: Natal, Port Natal [Durban], *Drège* 5022 (G-DC, holo.!).

Gnaphalium decorum (DC.) Sch. Bip. in Bot. Ztg 3: 171 (1845).

Biennial or short-lived perennial herb up to 1,3 m tall, stem stout (base up to 10 mm diam.), usually simple below the inflorescence branches, sometimes forking once or twice near the base, thinly greyish-white woolly, leafy. *Radical leaves* rosetted in the first year of growth, wanting at flowering, up to 150 × 30 mm, elliptic, narrowed to a broad clasping base, apex obtuse or subacute, apiculate, both surfaces thinly greyish-white woolly; *cauline leaves* up to 80 × 30 mm, diminishing in size upwards, oblong-lanceolate or elliptic-lanceolate, apex usually acute, base clasping, all but the uppermost at least shortly decurrent, upper surface glandular-setose, thinly cobwebby, lower thinly greyish-white woolly. *Heads* heterogamous, depressed-globose, c. 25—32 mm across the radiating bracts, many in a large corymbose panicle. *Involucral bracts* in c. 9 series, graded, imbricate, much exceeding the flowers, acute, glossy, bright yellow, radiating. *Receptacle* shortly honeycombed. *Flowers* c. 900—1 260, 35—110 ♀, 850—1 150 ☿, yellow. *Achenes* 1 mm long, barrel-shaped, obscurely ribbed, glabrous. *Pappus* bristles many, equalling corolla, tips barbellate, bases sometimes cohering lightly by patent cilia. Fig. 60: 2.

Ranges from about Inhambane in Mozambique to the mouth of the Bashee River in the Transkei, also on the Lebombo Mountains in Natal and Swaziland. Grows in sandy grassland or open woodland from near sea level to c. 900 m; flowering mainly between November and April. Map 230.

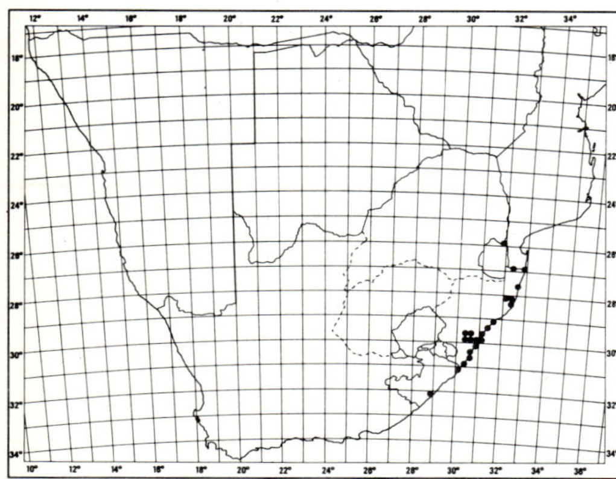

MAP 230.— **Helichrysum decorum**

The chief distinguishing mark of *H. decorum* is its oblong or elliptic-oblong leaves, generally broadest in the upper half, cobwebby above, tomentose below, most shortly decurrent on the stem. It lacks the persistent foetid odour of *H. foetidum* (below) with which it is often confused, and the heads are larger.

Vouchers: *Bayliss* 2198 (NBG); *Hilliard* 4896 (E; NU); *Pooley* 234 (E; K; NH; NU; S).

236. **Helichrysum foetidum** *(L.)* *Moench*, Meth. Pl. 575 (1794); Cass. in Dict. Sci. nat. 25: 469 (1822); Less., Syn. Comp. 284 (1832); DC., Prodr. 6: 187 (1838); Harv. in F.C. 3: 232 (1865); Moeser in Bot. Jb. 44: 336 (1910); Levyns in Adamson & Salter, Fl. Cape Penins. 783 (1950); Clapham in Tutin et al., Fl. Europ. 4: 130 (1976). Type described from Africa, *Gnaphalium* no. 13 in herb. Hort. Cliff. (BM!).

Gnaphalium foetidum L., Sp. Pl. 851 (1753); Sims in Curtis's Bot. Mag. t. 1987 (1818). *Anaxeton foetidum* (L.) Gaertn., Fruct. 2: 406, t. 166 (1791); Lam., Illus. t. 692 fig. 1 (1823).

Gnaphalium argenteum Mill., Dict. Gard. edn 8 no. 14 (1768). Type: no specimen found in BM, but contemporary specimens from Chelsea Physic Garden are a pale headed form of *H. foetidum*.

Xeranthemum corymbosum Lam., Encycl. 3: 242 (1789). Type: Cape of Good Hope (P-LAM!).

Helichrysum foetidum var. *citreum* Less., Syn. Comp. 285 (1832); DC., Prodr. 6: 187 (1838). Type: no specimen found.

H. foetidum var. *pallidum* Less., Syn. Comp. 285 (1832); DC., Prodr. 6: 187 (1838). Type: no specimen found.

Stout biennial herb, foetid, stem simple or sparingly branched from the base, up to c. 1 m tall, glandular-pubescent, young parts thinly white-tomentose as well, leafy. *Radical leaves* more or less rosulate in first year of growth, withered at flowering, up to c. 120 × 50 mm, elliptic, apex obtuse to subacute, mucronate, base auriculate-clasping, upper surface sparsely pubescent, lower thinly white-tomentose; *cauline leaves* mostly 40—90 × 10—25 mm, smaller on the inflorescence branches, lanceolate, apex very acute to acuminate, base broad, cordate-clasping, upper surface rather harshly glandular-pubescent, lower thinly white-tomentose. *Heads* heterogamous, depressed-globose, c. 8—12 mm long, c. 15—25 mm across the radiating bracts, many in a large, leafy, spreading corymbose panicle. *Involucral bracts* in 8—9 series, graded, imbricate, inner much exceeding the flowers, tips acute, glossy, deep to pale lemon-yellow or creamy. *Receptacle* shortly honeycombed. *Flowers* 396—835, 54—262 ♀, 201—694 ☿, ♀ sometimes outnumbering ☿. *Achenes* 0,75—1 mm long, glabrous. *Pappus* bristles many, equalling corolla, scabrid, bases cohering lightly by patent cilia.

Recorded from the environs of the Umtata River, Transkei, to the hills around Komgha, King William's Town and Stutterheim, the Amatola Mountains and Katberg, thence along the coastal mountain ranges to the Cape Peninsula and northwards through Worcester and Tulbagh to the Cedarberg. Often a constituent of shrubby growth on hill and mountain slopes, particularly damp places along streams or on forest margins; can become a weed. Flowers between October and May. Map 231.

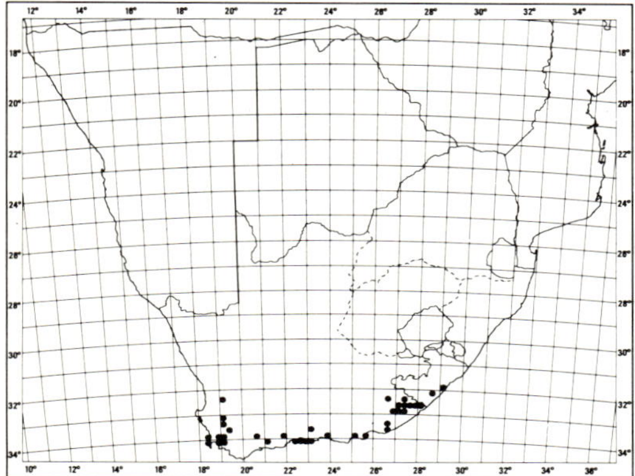

MAP 231.— **Helichrysum foetidum**

Naturalized in W. Europe, Madeira and Hawaii. Records from tropical Africa are either introductions or misdeterminations: *H. foetidum* var. *microcephalum* A. Rich is a distinct species (probably *H. glutinosum* A. Braun fide Wild, pers. comm.).

H. foetidum is closely allied to *H. cooperi* (no. 240), which gradually replaces it from the Amatola Mountains and Transkei northeastwards to the summer rainfall area. *H. cooperi* can be distinguished by its leaves, often decurrent on the stem and lacking white wool below.

Vouchers: *Acocks* 21717 (PRE); *Compton* 10698 (NBG); *Flanagan* 600 (PRE; SAM); *Hilliard & Burtt* 10814 (E; K; NU); *Pillans* 8494 (BOL).

237. Helichrysum elegantissimum *DC.*, Prodr. 6: 179 (1838); Harv. in F.C. 3: 229 (1865); Moeser in Bot. Jb. 44: 336 (1910); Hilliard, Compositae in Natal 241 (1977). Type: Cape, Witteberg, *Drège* 3736 (G-DC, holo.!; BM; E; K; S; SAM; TCD; fragment PRE, iso.!).

Gnaphalium elegantissimum (DC.) Sch. Bip. in Bot. Ztg 3: 170 (1854).

Stout biennial herb, stem up to c. 1 m high, simple below the inflorescence bran-ches, glandular pubescent, sometimes cobwebby as well, leafy throughout. *Radical leaves* up to 150 × 40 mm, oval-oblong, slightly narrowed to the base, glandular above, woolly or cobwebby below; *cauline leaves* up to c. 90 × 30 mm, diminishing in size upwards, oblong-lanceolate to lanceolate, apex acute to acuminate, base broad, cordate-clasping, shortly decurrent in lower leaves, both surfaces glandular-pubescent. *Heads* heterogamous, depressed-globose, c. 20−25 mm across the radiating involucral bracts, many in a large, spreading leafy corymbose panicle. *Involucral bracts* in c. 10 series, slightly graded, imbricate, much exceeding flowers, acute, glossy, white sometimes tipped crimson or flushed pink, radiating. *Receptacle* nearly smooth. *Flowers* c. 362−745, 26−110 ♀, 336−635 ♀. *Achenes* 0,75−1 mm long, barrel-shaped, glabrous. *Pappus* bristles many, equalling corolla, tips subplumose, bases with patent cilia, not cohering.

Recorded only from the Witteberg near Lady Grey in the Cape, the nearby heights above the Kraai River, the southern corner of Lesotho in Mohaleshoek and Quthing districts, and Bulwer in Natal, between c. 1 370 and 2 470 m, in scrubby growth on mountain slopes and along river banks. Flowers between January and April. The white or pink and white heads make *H. elegantissimum* distinctive among its yellow-headed allies, but it is rarely collected. Map 232.

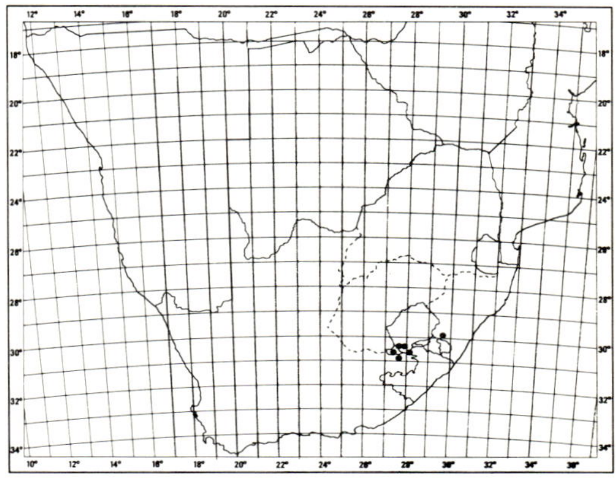

MAP 232.— **Helichrysum elegantissimum**

Vouchers: *Dieterlen* 1210 (PRE; SAM); *Galpin* 6679 (BOL; PRE); *Haygarth* s.n. (NH); *Hilliard & Burtt* 12254 (E; NU); *Nordenstam* 2066 (LD; NU);

238. Helichrysum heterolasium *Hilliard* in Notes R. bot. Gdn Edinb. 34: 81 (1975), Compositae in Natal 242 (1977). Type: Natal, Giant's Castle Game Reserve, Langalibalele Pass, c. 2 680 m, *Wright* 1247 (NU, holo.!; E; K; PRE; S, iso.!).

Biennial herb, aromatic when fresh, flowering stems up to c. 800 mm tall, simple below the inflorescence branches or sparingly forked at the base, glandular pubescent, thinly white-woolly as well mainly towards the tips of the inflorescence branches. *Radical leaves* rosetted in the first year of growth, persisting, dried and matted together, at the stem base in the second season, up to 100 × 350 mm, elliptic, narrowed in the lower third to a broad clasping base, apex obtuse to subacute, mucronate, both surfaces with long and short gland-tipped hairs, thinly greyish-white woolly as well; *cauline leaves* up to 90 × 25 mm, diminishing in size upwards, elliptic to lanceolate upwards, apex obtuse to acute to acuminate, base cordate-clasping, margins glandular-setose and white-woolly, both surfaces glandular-setose, only the lowermost derived from the basal rosette woolly as well. *Heads* heterogamous, depressed-globose, mostly 30—40 mm across the radiating bracts, 2—12 corymbosely arranged on leafy peduncles up to c. 120 mm long. *Involucral bracts* in c. 9 series, graded, imbricate, much exceeding flowers, radiating, glossy, inner bright yellow, outer overlaid pale golden-brown, rarely with a pink overcast. *Receptacle* nearly smooth. *Flowers* 786—1909, 71—261 ♀, 646—1648 ☿, yellow. *Achenes* 1 mm long, barrel-shaped, obscurely ribbed, glabrous. *Pappus* bristles many, equalling corolla, tips subplumose, bases with patent cilia, cohering lightly.

Recorded only from the high Drakensberg in Natal, from Royal Natal National Park to Garden Castle Forest Reserve. Grows socially in tall rough grassland on steep slopes and in boulder-strewn gullies and valleys between c. 2 280 and 2 850 m above sea level. Flowers between January and April. Map 233.

Easily recognized by its grey-woolly basal leaves contrasting with green, glandular stem leaves, and large heads with golden involucral bracts often washed light brown outside.

Vouchers: *Hilliard* 8142 (E; K; NU); *Hilliard* 5333 (E; K; NU; PRE; S); *Wright* 1146 (E; MO; NU).

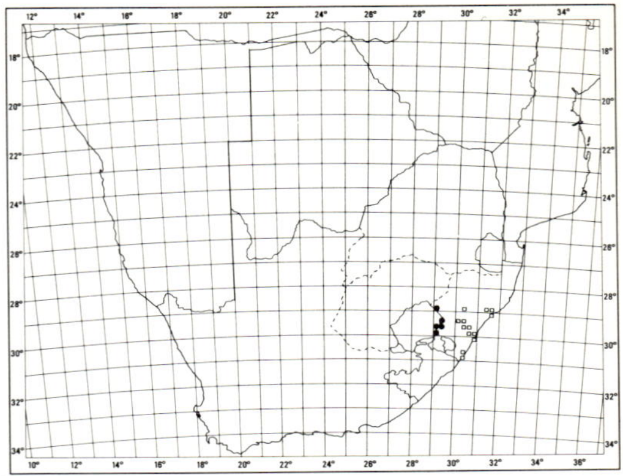

MAP 233.— ● **Helichrysum heterolasium**
        □ **Helichrysum ruderale**

239. Helichrysum ruderale *Hilliard & Burtt* in Notes R. bot. Gdn Edinb. 34: 82 (1975); Hilliard, Compositae in Natal 244 (1977). Type: Natal, Pinetown distr., Everton, Eskotene, c. 600 m, *Hilliard & Burtt* 7216 (NU, holo.!; E; K; PRE, iso.!).

Biennial herb, aromatic when fresh, up to c. 1 m tall, stem stout, base up to 10 mm diam., simple below the inflorescence, glandular-setose, the uppermost part and the inflorescence branches thinly white-woolly as well, leafy throughout. *Radical leaves* wanting at flowering; *cauline leaves* up to 180 × 65 mm, decreasing in size upwards, elliptic or lanceolate-elliptic, uppermost lanceolate, apex acute to acuminate, apiculate, base broad, cordate-clasping, all but the uppermost decurrent in long or short wings, both surfaces glandular-setose, margins often white-woolly as well. *Heads* heterogamous, depressed-globose, c. (22—) 25—32 mm across the radiating bracts, many in a large corymbose panicle, the branches obliquely ascending. *Involucral bracts* in c. 9 series, graded, imbricate, much exceeding flowers, radiating, glossy, bright yellow. *Receptacle* honeycombed. *Flowers* 811—1053, 121—226 ♀, 681—754 ☿, yellow. *Achenes* c. 1 mm long, barrel-shaped, glabrous. *Pappus* bristles many, equalling corolla, tips barbellate, bases with patent cilia, cohering or not.

Recorded only from Natal, from Eshowe, Mtunzini and Umvoti districts in the north to Port Shepstone district in the south, from near sea level to c. 1 050 m. A weed, often forming dense stands along roadsides and in old fields; flowering in September and October, fading in November. Map 233.

Closely allied to *H. cooperi* (below) and not easy to distinguish in the herbarium. The persistent white wool on the inflorescence branches to well below the first forking is a useful character to distinguish *H. ruderale*; white wool is usually wanting in *H. cooperi* except sometimes immediately below the heads. *H. ruderale* flowers in spring, *H. cooperi* in summer. In the field, the obliquely-ascending inflorescence branches of *H. ruderale* immediately distinguish it from *H. cooperi* with inflorescence branches wide-spreading to nearly horizontal.

A specimen in PRE (*Kioch* s.n., Boshoff's Road Siding, Pietermaritzburg) is annotated 'Fibre produced from stems' and the fibre sample mounted on the sheet seems to be excellent.

Vouchers: *Rogers* 24413 (PRE); *Strey* 6934 (NU; PRE); *Wood* 12491 (NH; NU).

240. **Helichrysum cooperi** *Harv.* in F.C. 3: 231 (1865); Moeser in Bot. Jb. 44: 338 (1910); Hilliard, Compositae in Natal 243 (1977). Type: Orange Free State, near Drakensberg, *Cooper* 1117 (TCD, holo.!; BM; E; K; PRE; W; Z, iso.!).

Biennial herb up to 1,5 m tall, or rarely perennating, aromatic, stem stout, base up to 10 mm diam., simple below the inflorescence, glandular-setose, inflorescence branches sometimes thinly white-cobwebby as well, leafy. *Radical leaves* usually wanting at flowering, but rosetted in first year of growth, up to c. 150 (−250) × 60 mm, elliptic, tapering to a broad clasping base, glandular or very thinly woolly; *cauline leaves* mostly 65−130 × (15−) 20−45 mm, smaller upwards, oblong-lanceolate to lanceolate, apex acute or sometimes acuminate on uppermost leaves, apiculate, base cordate-clasping, the main stem leaves in particular usually decurrent in long or short wings, both surfaces glandular-setose, margins often white-woolly as well, very rarely some leaves thinly woolly below. *Heads* heterogamous, depressed-globose, c. (8−) 10−12 mm long, c. 15−25 mm across the radiating bracts, many in a large spreading leafy corymbose panicle. *Involucral bracts* in c. 9 series, slightly graded, imbricate, much exceeding the flowers, glossy bright yellow, backs of outer rarely overlaid palest brown. *Receptacle* shortly honeycombed. *Flowers* c.

650−1 200, 50−160 ♀, 600−1 100 ☿. *Achenes* 1 mm long, barrel-shaped, glabrous. *Pappus* bristles many, equalling corolla, tips barbellate, bases with patent cilia, lightly cohering or not. Fig. 60: 1.

Recorded with certainty from the low Drakensberg on the Transvaal-Natal border thence south through the mountainous parts of the Orange Free State, Lesotho, and the Midlands and Uplands of Natal to the Transkei and the eastern Cape, between Maclear and Naude's Nek. Map 234.

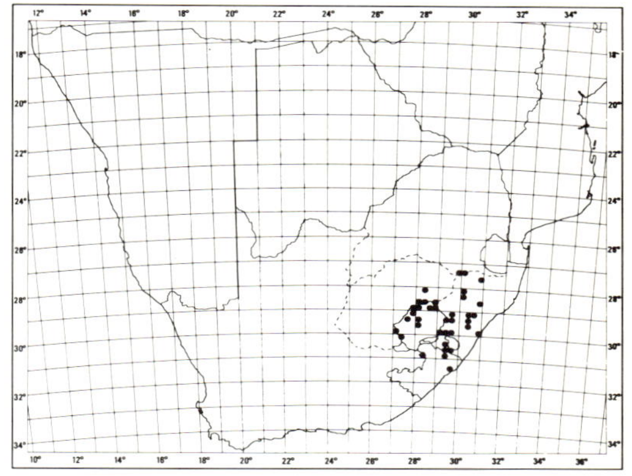

MAP 234.— **Helichrysum cooperi**

Grows in grassland, often in the coarse herbage on or near forest margins, or, in the Drakensberg in particular, along drainage lines or in scrub among boulders and on scree. Flowers between December and April.

The name *H. cooperi* is frequently misapplied to plants in the Transvaal and tropical Africa. It is not clear, however, if *H. cooperi* occurs further north than the low Drakensberg on the Transvaal-Natal border. It is characterized by its biennial habit, green leaves shortly to strongly decurrent on the stem, heads in a spreading corymbose panicle, and involucral bracts usually bright yellow inside and out.

Vouchers: *Devenish* 571 (PRE); *Hilliard & Burtt* 10093 (E; K; MO; NU; S); *Tyson* 1640 (SAM); *Wright* 481 (E; K; MO; NU; S).

241. **Helichrysum difficile** *Hilliard* in Notes R. bot. Gdn Edinb. 40: 253 (1982). Type: Transvaal, 2430 DD, Graskop, 2 km from town on road to Blyde River Canyon, marsh, 14 iii 1981, *Hilliard & Burtt* 14335 (NU, holo.!; E; K; M; MO; PRE; S, iso.!).

2a
1
2
1a
1b
H. Wanda. du Toit

Perennial herb, stock often with vegetative buds, stems solitary or 2−6 from the base, usually simple below the inflorescence branches, mostly c. 750 − 1 500 mm high, precocious plants shorter, glandular setose, upper parts sometimes thinly cobwebby as well, leafy throughout. *Leaves* mostly 50−70 (−80) × 8−20 (−26) mm, smaller on the inflorescence branches, oblong or oblong-lanceolate, uppermost lanceolate, acute to acuminate, mucronate, base cordate-clasping, shortly decurrent, both surfaces glandular-setose, margins and midvein below often white-woolly as well. *Heads* heterogamous, depressed-globose, c. 10−15 mm long, c. (20−) 23−28 mm across the radiating bracts, few to many in a leafy spreading corymbose panicle. *Involucral bracts* in c. 9 series, graded, imbricate, much exceeding the flowers, glossy, bright yellow. *Receptacle* shortly honeycombed. *Flowers* c. 720−1 130, 55−100 ♀, 800−1 025 ☿. *Achenes* 1 mm long, barrel-shaped, glabrous. *Pappus* bristles many, equalling the corolla, tips barbellate, shaft scabrid, bases cohering by patent cilia.

Recorded from the Transvaal Highveld and E. Highlands, from Rustenburg in the W. to Graskop in the E., in the Lowveld around Plaston and Nelspruit, and W. Swaziland near Oshoek, Mbabane and Forbes Reef. Grows in marshes and along wet streambanks; flowering principally in March and April. Map 235.

Much confused with *H. cooperi* (above), but distinguished by its perennial habit, generally shorter and narrower cauline leaves, and generally larger heads. See also *H. molestum* (below).

Vouchers: *Hilliard & Burtt* 14409 (E; K; NU; PRE; S); *Jacobsen* 956 (PRE); *Lambrechts* 306 (PRE); *Mogg* 17305 (NU; PRE).

The following specimens, which have leaves thinly white-woolly below, are possibly no more than variants of *H. difficile*, but only field studies can determine that: *Bryant* D58, PRE (Johannesburg, Turffontein); *Codd* 2887, K; NU; PRE (10 km SW. of Nylstroom); *Jenkins* TM 12721, PRE (Olifants River); *Walker* 47, J (W. Rand, Witkoppen). Similar plants occur in the E. Highlands of Zimbabwe and in Mozambique.

See under *H. albilanatum* (no. 226) for possible hybrids.

242. **Helichrysum molestum** *Hilliard* in Notes R. bot. Gdn Edinb. 40: 260 (1982). Type: Transvaal, 2530 BA, Sabie to Lydenburg road, Long Tom Pass, Whisky Spruit, marshy ground by stream, 15 iii 1981, *Hilliard & Burtt* 14364 (NU, holo.!; E; PRE, iso.!).

Perennial herb up to c. 1 m tall, stems several from the base, simple below, branching above into the compound inflorescence, stout, glandular-setose, uppermost parts sometimes cobwebby as well, closely leafy. *Leaves* mostly 65−100 × 18−32 mm, smaller on the inflorescence branches, oblong or oblong-ovate, uppermost ovate-lanceolate, apex acute, mucronate, base cordate-clasping, shortly decurrent, both surfaces glandular setose, margins and midline sometimes thinly white-woolly (sometimes one or both surfaces thinly woolly: see notes below). *Heads* heterogamous, depressed-globose, c. 11−14 mm long, 22−30 mm across the radiating bracts, many in a large leafy spreading corymbose panicle. *Involucral bracts* in c. 9 series, graded, imbricate, much exceeding the flowers, glossy, bright yellow. *Receptacle* shortly honeycombed. *Flowers* c. 1 000−1 300, 65−135 ♀, 1 035−1 235 ☿. *Pappus* bristles many, equalling corolla, tips barbellate, shaft scabrid, bases cohering by patent cilia.

Recorded with certainty from the highlands of the eastern Transvaal, from Mount Anderson south to the

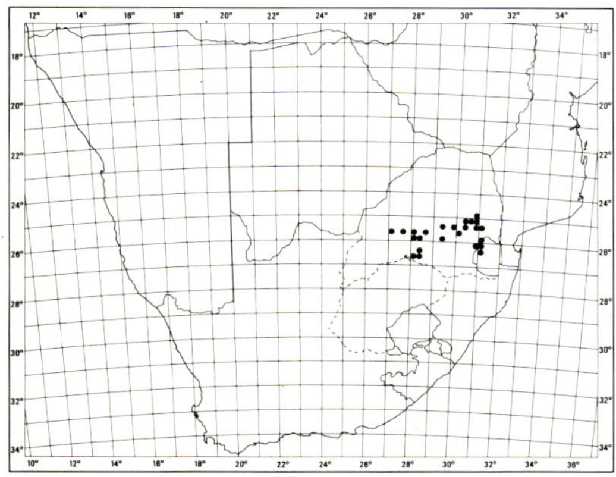

MAP 235.— **Helichrysum difficile**

FIG. 60.−1, **Helichrysum cooperi,** compound inflorescence, × 0,3; 1a, detail of heads, × 1; 1b, leaf, broadest at base, decurrent, × 1 (*Hilliard & Burtt* 14041). 2, **H. decorum,** compound inflorescence, × 0,3; 2a, leaf, broadest in upper part, decurrent, × 1 (*Hilliard* 4896).

Barberton mountains, but possibly ranges as far north as the Soutpansberg, west along the Magaliesberg, and south to northern Natal (see notes below). Favours moist ground in rough grassland or at the margins of scrub or forest patches; flowering mainly in March and April. Map 236.

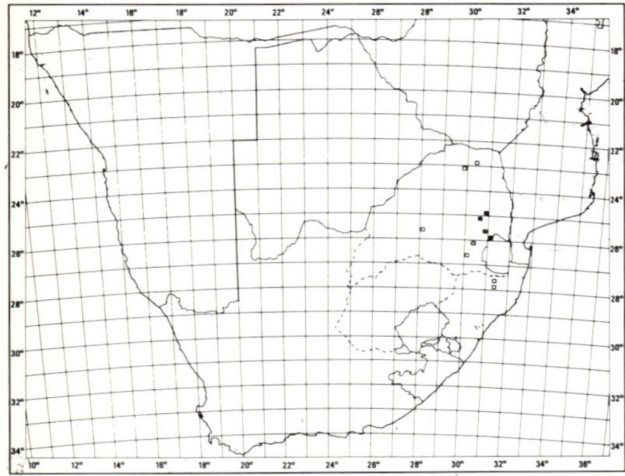

MAP 236.— ■ **Helichrysum molestum**
        □ possibly **Helichrysum molestum**

H. molestum can easily be confused with *H. difficile* (no. 241), but is usually more robust, with broader leaves; it is distinguished from *H. cooperi* (no. 240) by its perennial habit and usually larger heads.

The leaves of *H. molestum* are usually green below. However, *Hilliard & Burtt* 14207 (E; NU; PRE) is probably no more than a variant with leaves thinly white-woolly below; it grew in the vicinity of normal green-leaved plants. *Hilliard & Burtt* 14162 (E; K; NU; PRE; S) from Elandshoogte near Machadodorp has leaves thinly woolly on both surfaces. Hybrids are also known: see under *H. albilanatum* (no. 226).

Vouchers: *Hilliard & Burtt* 14209 (E; K; NU; PRE); *Kluge* sub *Hilliard & Burtt* 14240 (E; K; NU; PRE); *Taylor* 1902 (PRE).

In the absence of basal parts or field notes it is impossible to place many specimens. For example, there are a number of sheets of a plant from the forests north of Louis Trichardt, on the Soutpansberg: the leaves are scarcely decurrent and the heads may be only c. 20 mm across. Pending good collections, the plant is included here, with reservation (*Breyer* s.n. sub TM 22113, *Junod* 4929, *Koker* 16, *Letty* 250, *Rogers* 12433, all PRE; *Blenkiron* sub *Moss* 14483, J). However, several collections from Tate Vondo Forest Reserve, Sibasa district, Venda, have larger heads (c. 22−25 mm across) and may well represent true *H. molestum*, but both basal parts and notes on habit are wanting (*Van Wyk* 4159, PRE; *Hemm* 86 and 161, both J and PRE).

Then there are several sheets from the Barberton Mountains and the Highveld around Ermelo representing only the tops of plants and lacking notes on habit (e.g. *Leendertz* s.n. TM 7863, *Potter* 1791, both PRE,

and *Galpin* 1313, BOL; PRE). They are probably *H. molestum* rather than *H. cooperi* (no. 240) because the heads measure c. 25 mm across (the upper limit for *H. cooperi*), but adequate collecting to establish the habit of plants in these areas is particularly desirable because this is where the geographical ranges of *H. cooperi* and *H. molestum* are likely to meet.

Two collections from Pretoria district (*Acocks* 11326, K and PRE, Rietvallei 221, and *Mogg & Dyer* s.n. in PRE 44088, Wonderboom Poort, PRE) can be included with more certainty because Acocks recorded 'shrub'.

Plants from further south, in northern Natal, at Itala Nature Reserve near Louwsburg, and nearby Ngome Forest Reserve, differ from *H. molestum* in having a single very short main stem branching near the base, not several stems from the base; leaves may be wholly green, or lightly white-woolly below, or lightly woolly on both surfaces (*Hilliard & Burtt* 9934, E; NU; *Hilliard & Burtt* 10000, E; K; NU; S). Similar plants occur on Gorongoza Mountain in Mozambique, and two specimens from Swaziland, regrettably only parts of inflorescences, probably also belong here (*Burtt Davy* 2154, PRE, and *Stewart* in TM 9000, PRE). Clearly much field work is needed to sort out the species in this whole difficult group.

Excluded species

(only those species not dealt with elsewhere in this volume are listed; *Helipterum* is under revision by Prof. B. Nordenstam, Stockholm).

H. argenteum (Thunb.) Thunb. = 'Helipterum'
H. canescens (L.) Willd. = Helipterum canescens *(L.) DC.*
H. citrinum Less. = Helipterum citrinum *(Less.) Harv. & Sond.*
H. dykei H. Bol. = 'Helipterum'
H. ericoides (Lam.) Pers. = Dolichothrix ericoides *(Lam.) Hilliard & Burtt*
H. eximium (L.) Less. = Helipterum eximium *(L.) DC.*
H. ferrugineum (Lam.) Pers. = Helipterum ferrugineum *Sond. & Harv.*, fide F.C. 3: 258 (1865).
H. ferrugineum Less. = Helipterum citrinum *(Less.) Harv. & Sond.*, fide F.C. 3: 258 (1865).
H. gemmiferum H. Bol. = Atrichantha gemmifera *(H. Bol.) Hilliard & Burtt*
H. lepidopodium H. Bol. = 'Helipterum'
H. paniculatum (L.) Willd. = 'Helipterum'
H. phlomoides (Lam.) Spreng. = Helipterum phlomoides *(Lam.) DC.*
H. recurvatum (L.f.) Thunb. = 'Helipterum'
H. scleranthoides S. Moore = Ifloga molluginoides *(DC.) Hilliard*
H. sordescens DC. = 'Helipterum'
H. speciosissimum (L.) Willd. = Helipterum speciosissimum *(L.) DC.*
H. staehelinoides Less. = Helipterum gnaphaloides *(L.) DC.*
H. striatum (Thunb.) Thunb. = 'Helipterum'
H. variegatum Thunb. = Helipterum variegatum *(Thunb.) DC.*
H. vestitum (L.) Willd. = 'Helipterum'
H. virgatum Willd. = Helipterum virgatum *(Willd.) DC.*

9006a   **Genus nov.?**

Apparently a bushy perennial herb, only upper part present, main stem woody, c. 3 mm diam., glandular-setose, thinly white-woolly, glabrescent, branches virgate, young parts closely leafy. *Leaves* up to 30 × 4 mm, lanceolate, apex very acute, apiculate, base broad, half-clasping, upper surface shortly glandular-setose, cobwebby at first, lower white-felted. *Heads* heterogamous, campanulate, c. 7 × 9 mm, solitary on long leafy twigs, 2 or 3 corymbosely arranged at the branch tips. *Involucral bracts* in c. 6 series, graded, closely imbricate, lanceolate to oblong-lanceolate inwards, very acute, horny, backs white-woolly, stereome fenestrate, tips of 2 innermost series glabrous, exceeding the flowers, glossy, straw-coloured. *Receptacle* honeycombed. *Flowers* c. 120, c. 15 ♀, the rest ⚥. *Corolla* of ⚥ flowers 4 mm long, cylindric, scarcely widened upwards, 5-lobed, large glands on backs of lobes; ♀ flowers similar but tube narrower. *Anthers* with a lanceolate apical appendage, base tailed. *Style* branches truncate and penicillate. *Ovaries* 0,75 mm long, cylindric, glabrous. *Achenes* not seen. *Pappus* bristles many, equalling corolla, scabrid, tips barbellate, bases not cohering. Fig. 61: 1.

Known only from a single collection: Zululand, Nkandla, 26 iii 1903, *Wylie* in herb. Wood 9012, K; NH. Allied to *Helichrysum*, but differing in its thick-textured involucral bracts without a clearly defined stereome. The plant should be sought again in northern Natal and Zululand.

FIG. 61.—1, **Genus nov.** ?, habit of plant, × 0,3; 1a, flowering twig, × 1; 1b, head, × 3,3; 1c, pappus bristle, × 10; 1d, female flower, × 10; 1e, hermaphrodite flower, × 10 (*Wylie* in herb. Wood 9012).

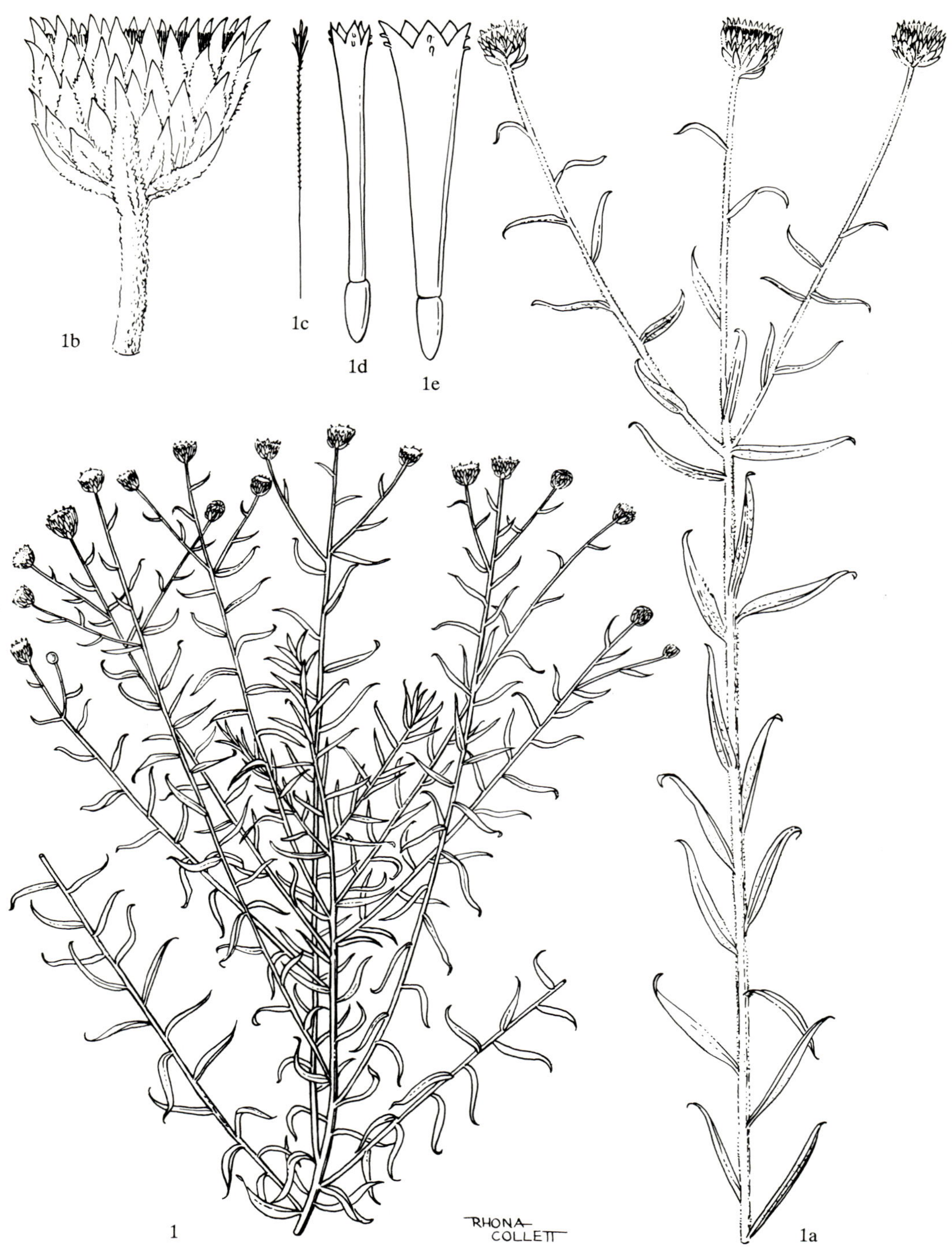
1b
1c
1d
1e
1
1a
RHONA
COLLETT

9006b   **EDMONDIA**

Edmondia *Cass.* in Bull. Soc. philom. 1818: 75 (1818); Hilliard & Burtt in Bot. J. Linn. Soc. 82: 216 (1981). Lectotype species: *E. splendens* Cass. = *E. sesamoides* (L.) Hilliard.

Aphelexis D. Don in Mem. Wern. nat. Hist. Soc. 5: 546 (1826). Type species: *A. sesamoides* (L.) D. Don.

Helipterum section *Edmondia* (Cass.) DC., Prodr. 6: 214 (1838).

Helichrysum section *Edmondia* (Cass.) Harv. in F.C. 3: 219 et 255 (1865).

Suffrutices, stems tufted from the base, 100−750 mm long, simple or sparingly and loosely branched, sprawling, ascending or erect, stout, closely leafy throughout. *Leaves* rigid, ascending, erect or appressed, imbricate, subulate to linear-lanceolate, often rounded abaxially, very variable in size, sometimes uniformly short, often long on vegetative shoots, short on flowering shoots, apex very acute to more or less obtuse, often bract-tipped below the involucre, base broad, clasping, margins strongly involute and hiding the upper surface in long leaves, flat in short leaves, white-tomentose above, glabrous and shining below. *Heads* homogamous or heterogamous, broadly campanulate, large, often solitary at the branch tips, or rarely 2 to several. *Involucral bracts* in many series, graded, loosely imbricate, radiating, sterome either clearly fenestrated or with small thin patches adjoining the side nerves, lamina lanceolate, acute to acuminate, inner much exceeding the flowers, white, pink or yellow, outer sometimes pale dark brown. *Receptacle* fimbrilliferous. *Flowers* c. 85−315, 0−19 ♀, corolla of ♀ flowers narrowly tubular, of ♀ flowers campanulate above, 5-lobed, all flowers glandular-hairy on backs of lobes. *Anthers* with a lanceolate apical appendage, tails about equalling the filament collar; staminodes often present in ♀ flowers. *Style branches* truncate and penicillate. *Achenes* either terete with myxogenic duplex hairs, or *(E. pinifolia)* flattened, winged, glabrous. *Pappus* bristles shortly plumose above, barbellate below, fused at the base in a smooth ring, also cohering by small patent cilia.

Endemic, 3 species confined to the W. and SW. Cape, all similar in facies and easily confused.

1a Achenes terete, with conspicuous myxogenic hairs, fimbrils much exceeding the ovaries; heads c. 25−27 mm long:

  2a Peduncles conspicuously scaly only immediately below the heads, involucral bracts white, pink or pale yellow (outermost often brownish)..1. *E. sesamoides*

  ·2b Peduncles conspicuously scaly for 20−60 mm below the heads, involucral bracts bright golden-yellow (outermost often brown) ....2. *E. fasciculata*

1b Achenes flattened, winged, glabrous, fimbrils about half as long as ovaries; heads c. 30 mm long
.... 3 *E. pinifolia*

    1. Edmondia sesamoides *(L.) Hilliard* in Bot. J. Linn. Soc. 82: 217 (1981). Lectotype: specimen forming basis of Burman, Rar. Afr. Pl. 181, t. 67, fig. 2. (G!).

Xeranthemum sesamoides L., Sp. Pl. 859 (1753). *Helichrysum sesamoides* (L.) Willd., Sp. Pl. 3: 1908 (1804); Less., Syn. Comp. 322 (1832); Harv. in F.C. 3: 255 (1865); Moeser in Bot. Jb. 44: 340 (1910); Levyns in Adamson & Salter, Fl. Cape Penins. 785 (1950). *Aphelexis sesamoides* (L.) D. Don in Mem. Wern. nat. Hist. Soc. 5: 546 (1826); *Helipterum sesamoides* (L.) DC., Prodr. 6: 214 (1838). *Helichrysum sesamoides* var. *willdenowii* Harv. in F.C. 3: 255 (1865). *Gnaphalium sesamoides* (L.) O. Kuntze, Rev. Gen. Pl. 3,2: 154 (1898).

Xeranthemum heterophyllum Lam., Encycl. 3: 239 (1789), nom. illegit., quoad spec. excl. syn. *Helipterum heterophyllum* DC., Prodr. 6: 214 (1838). *Helichrysum sesamoides* var. *heterophyllum* Harv. in F.C. 3: 255 (1865).

Edmondia splendens Cass. in Dict. Sci. nat. 14: 253 (1819). Type: herb. Jussieu (P).

FIG. 62.−1, **Edmondia sesamoides**, flowering branchlets, × 1; 1a, leaf to show involute margins, × 4,6 (*Hilliard & Burtt* 13036). 2, **E. pinifolia**, achene, compressed and winged, × 20 (*Rodin* 3178). 3, **E. fasciculata**, head on scaly peduncle, × 1; 3a, hermaphrodite flower, × 10; 3b, part of ring of pappus bristles, × 10 (*Stokoe* SAM 56575).

RHONA
COLLETT
1a
1
2
3
3b
3a

Edmondia bicolor Cass. in Dict. Sci. nat. 14: 254 (1819). Type: herb. Jussieu (P).

Helichrysum pseudofasciculatum Schrank in Denkschr. K. Akad. Wiss. Münch. 8: 167 (1824). Type: Cape Town, *Brehm* (S, iso.!).

Aphelexis filiformis D. Don in Mem. Wern. nat. Hist. Soc. 5: 547 (1826). *Helichrysum filiforme* (D. Don) Less., Syn. Comp. 323 (1832). *Helipterum filiforme* (D. Don) DC., Prodr. 6: 215 (1838). *Helichrysum sesamoides* var. *filiforme* (D. Don) Harv. in F.C. 3: 256 (1865). Types: Cape of Good Hope, *Niven* and *Roxburgh* in herb. Lambert.

The salient characters of *E. sesamoides* are the vegetative *stems* clad in long (mostly 20—50 mm) ascending *leaves* and sharply demarcated from the flowering shoots clad in short (c. 4—7 mm) appressed leaves, these leaves normally bract-tipped only immediately below the heads; *heads* c. 25—27 mm long; *involucral bracts* pure white, cream, primrose yellow or pink, the outer bracts often partly or wholly pale to dark brown or deep pink, sometimes changing colour with age (e.g. pink when young, pale yellow with age); *receptacle* with fimbrils much exceeding the ovaries; *achenes* terete, with conspicuous myxogenic duplex hairs. Fig. 62: 1.

Recorded from the Kamiesberg, then a disjunction to the Koue Bokkeveld, thence south, on the mountains as well as on the flats, to the Cape Peninsula and east to Robinson's Pass and Mossel Bay. Grows in rocky places or in sand, in shrub communities; flowering mainly between September and December. Common. Map 237.

Colour variation in the involucral bracts is remarkable and several different colour forms may grow together. Although there is no clear-cut geographical patterning to colour variation, only plants with white, or white faintly tinged pink or brown, bracts have been recorded at the northern and eastern extremes of the range.

See also *E. fasciculata* (below).

Vouchers: *Esterhuysen* 26956 (BOL); *Hilliard & Burtt* 13036 (E; K; MO; NU; PRE; S); *Pillans* 9332 (BOL); *Schlechter* 1352 (PRE); *Schlieben* 12411 (PRE).

2. **Edmondia fasciculata** *(Andr.) Hilliard,* comb. nov. Type: a plant introduced from the Cape by Niven, probably no specimen kept.

Xeranthemum fasciculatum Andr., Bot. Rep. t. 242 (1802). *Aphelexis fasciculata* (Andr.) D. Don in Mem. Wern. nat. Hist. Soc. 5: 546 (1826). *Helichrysum fasciculatum* (Andr.) Willd., Sp. Pl. 3: 1909 (1804). *Helipterum fasciculatum* (Andr.) DC., Prodr. 6: 214 (1838). *Helichrysum sesamoides* var. *fasciculatum* (Andr.) Harv. in F.C. 3: 255 (1865).

Closely resembles *E. sesamoides* (above) in foliage, fimbrils and achenes, but differs in its more robust habit, peduncles scaly for 20—60 mm below the heads, and bright golden-yellow involucral bracts (the outermost bracts may be brown). Fig. 62:3.

Recorded from the Kamiesberg, then a disjunction to the mountains about Worcester and Ceres, Jonkershoek, and Riviersonderend Mountains. Grows in shrub communities; at the peak of its flowering in December and January. Map 238.

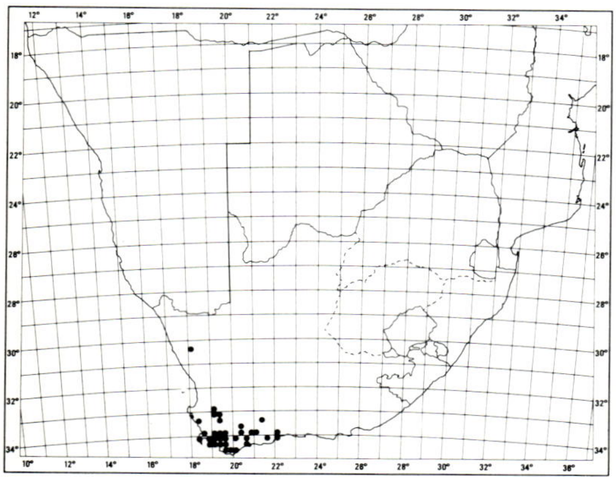

MAP 237.— **Edmondia sesamoides**

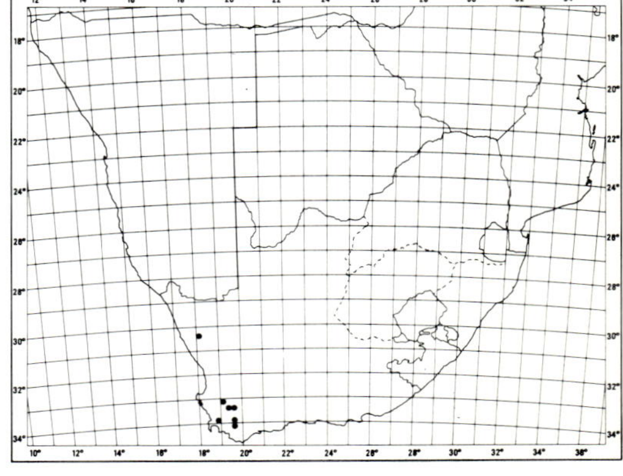

MAP 238.— **Edmondia fasciculata**

The late flowering of *E. fasciculata* further distinguishes it from *E. sesamoides,* which flowers in spring. The two species are frequently sympatric, and we have seen them on Jona's Kop, in the Riviersonderend Mountains, in September, when *E. fasciculata* was in tight bud, *E. sesamoides* in full flower.

Vouchers: *Esterhuysen* 20895 (BOL; PRE); *Hilliard & Burtt* 13035 (E; K; NU; S); *Stokoe* SAM 56575 (PRE; SAM); *Strauss* 72 (NBG).

3. **Edmondia pinifolia** *(Lam.) Hilliard* in Bot. J. Linn. Soc. 82: 217 (1981). Type: Cape of Good Hope (P-LAM!).

Xeranthemum pinifolium Lam., Encycl. 3: 240 (1789). *Helipterum humile* var. *pinifolium* (Lam.) DC., Prodr. 6: 214 (1838). *Helichrysum pinifolium* (Lam.) Schrank in Denkschr. K. Akad. Wiss. Münch. 8: 168 (1824); Levyns in Adamson & Salter, Fl. Cape Penins. 785 (1950); Hilliard & Burtt in Notes R. bot. Gdn Edinb. 32: 358 (1973).

X. squamosum Jacq., Collectanea 3:279, t. 20, fig. 2 ('1789', but late 1791 fide Stafleu); *Helichrysum squamosum* (Jacq.) Thunb., Fl. Cap. 661 (1823) quoad syn.; *Gnaphalium squamosum* (Jacq.) Sch. Bip. in Bot. Ztg 3: 171 (1845). Type: Cape of Good Hope (no specimen cited).

X. humile Andr., Bot. Rep. 10, t. 652 (1812). *Aphelexis humilis* (Andr.) D. Don in Mem. Wern. nat. Hist. Soc. 5: 547 (1826). *Helichrysum humile* (Andr.) Less., Syn. Comp. 322 (1832); Harv. in F.C. 3: 255 (1865); Moeser in Bot. Jb. 44: 340 (1910). *Helipterum humile* (Andr.) DC., Prodr. 6: 214 (1838). *Gnaphalium andrewsii* O. Kuntze, Rev. Gen. Pl. 3: 150 (1898), non *G. humile* Thunb.

Type: cult. spec. introd. from Cape of Good Hope by Niven (lost?).

Helichrysum spectabile Lodd., Bot. Cab. 1, t. 59 (1817). Type: the plate cited.

Edmondia bracteata Cass. in Dict. Sci. nat. 14: 254 (1819). Type: herb. Jussieu (P).

Helichrysum longebracteatum Schrank in Denkschr. K. Akad. Wiss. Münch. 8: 168 (1824). Type: Cape of Good Hope, *Brehm* s.n. (M, holo.!).

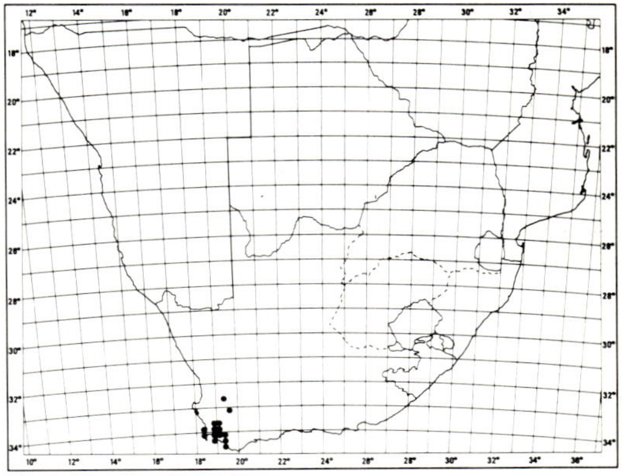

MAP 239.— **Edmondia pinifolia**

The distinguishing features of *E. pinifolia* are: *Heads* large, c. 30 mm long; *involucral bracts* white or pale to deep pink, the outermost pale to medium brown, and always descending on the peduncle and passing into scale-tipped leaves for 20—60 mm below the head; *receptacle* with fimbrils about half as long as the ovaries; *achenes* compressed, with broad semipellucid marginal wings and an incipient wing up each face, glabrous. Fig. 62: 2.

Recorded only from the mountains of the SW. Cape, from the south Cedarberg to the Peninsula and Kogelberg, Caledon division. Grows in rocky places, often in the crevices of cliffs; flowering between September and December. Map 239.

The peculiar achenes make recognition easy.

Vouchers: *Fair* in herb. Bolus 7945 (BOL; PRE); *Boucher* 895 (PRE); *Esterhuysen* 33423 (BOL; PRE); *Stokoe* SAM 53951 (PRE; SAM).

INDEX